The Growth Story of the 21st Century

The Economics and Opportunity of Climate Action

Nicholas Stern

LSE Press

LSE Global School of Sustainability

Published by
LSE Press
10 Portugal Street
London
WC2A 2HD
press.lse.ac.uk

First published 2025

Cover design by Diana Jarvis

Print and digital versions typeset by PDQ Media

ISBN (Paperback): 978-1-911712-47-3
ISBN (PDF): 978-1-911712-48-0
ISBN (EPUB): 978-1-911712-49-7
ISBN (Mobi): 978-1-911712-50-3

DOI: https://doi.org/10.31389/lsepress.tgs

This book has been peer-reviewed to ensure high academic standards. For
our full publishing ethics policies, see https://press.lse.ac.uk

This book was published with financial support from the Global School of
Sustainability at the London School of Economics and Political Science.

Suggested citation:
Stern, Nicholas (2025) *The Growth Story of the 21st Century: The Economics
and Opportunity of Climate Action*, London: LSE Press.
https://doi.org/10.31389/lsepress.tgs. Licence: CC BY-NC-ND 4.0

To read the free, open access version of this book online, visit https://doi.
org/10.31389/lsepress.tgs or scan this QR code with your mobile device:

For Orlando, Casper, Joe, Rosa, and Theo

Contents

PART I. FOUNDATIONS: A WORLD RE-DRAWN AND AN URGENT AGENDA FOR ACTION

PART II. THE NEW GROWTH STORY: INVESTMENT, INNOVATION, AND FUNDAMENTAL STRUCTURAL CHANGE

PART IV. GALVANISING ACTION

List of figures and tables

Figures

Tables

Foreword

Twenty years ago, Nick, now Lord Nicholas Stern, reshaped how we think about climate change and the economy. The presumption used to be that lowering greenhouse gas emissions would mean slower economic growth. *The Economics of Climate Change: The Stern Review* upended this notion, drawing on cutting-edge scientific research to demonstrate that unmitigated climate change would be so economically catastrophic that investing in decarbonising our economies was cheap in comparison. The message was quietly revolutionary: there is no trade-off between climate action and economic development. Quite the opposite, in fact: growth can only be sustained if it is environmentally sustainable.

Since then, the world has taken steps forward on carbon pricing, green innovation and investment, and international cooperation, but not nearly at the speed and scale necessary. Global emissions have continued to rise, and the climate system is nearing crucial tipping points. The climate crisis threatens to become unmanageable in many parts of the world, with the poorest people least equipped to cope. There is an urgent need for action and few excuses for not doing so. In fact, the window for decisive action to reduce carbon emissions and restore biodiversity is shrinking. But even as further foot-dragging becomes ever more dangerous, elevated economic anxiety and cost-of-living concerns worldwide make policy change harder – and that was true even before the recent attacks on environmental science.

That is why this book is so timely and so exciting. It convincingly shows that a better growth and development model is ours for the taking – one in which prosperity and the planet go hand in hand. In this new growth story, investment and innovation to combat the climate and biodiversity crises will drive and sustain economic output, development, and the creation of quality jobs. Growth will be more resource efficient and more productive, with fewer negative externalities for the health of people and ecosystems alike.

Technological changes in the past decade have made this 21st-century growth story possible. As Lord Stern puts it in his introduction, for many technologies, 'the clean is now cheaper than the dirty'. Just as much of Africa and Asia skipped fixed-line telephony and went straight to mobile phones, developing countries can now leapfrog to renewable electricity without sinking additional capital into more expensive and polluting fossil-fuel-based energy. The breakthroughs go beyond renewable energy. Artificial intelligence promises to help small-scale farmers increase yields while optimising input use.

New techniques can make cement and concrete cheaper, more durable, and less carbon intensive. What is exciting is that poor countries can benefit from this new growth story if, collectively, we get our policy and investment priorities right.

But we are a long way from where we need to be. Africa has more than half the Earth's best solar energy resources, yet only around 2% of its solar investment. Excessive risk aversion and high costs of capital are holding back investments that would otherwise be good for the economy and the environment.

This illustrates the challenge – and the opportunity – before us: capitalising on the new growth story involves trillions of dollars of private investment in all countries, and especially in emerging markets and developing economies. Incentivising that investment demands smart public policy, paired with catalytic public investment by development banks and governments. At a time of fiscal constraints it makes sense to repurpose to this end some of the roughly US$2 trillion in current direct annual public spending on environmentally damaging fossil fuel subsidies, trade-distorting agriculture subsidies, and harmful and wasteful water and fisheries subsidies.

As Director-General of the World Trade Organization (but speaking on a personal basis), my agreement with the book's emphasis on the importance of international collaboration and coherence will surprise no one. Ensuring that trade and investment can flow freely across borders is vital for the growth agenda set out in the book. Diffusing low-carbon technologies around the world and driving down their cost through increased competition and scale is only part of the story.

Open trade and investment is necessary to harness green comparative advantages, in which energy-intensive activities shift to places with abundant and cheap renewable energy. If Africa can use solar and wind energy and green hydrogen to add more value to critical minerals mined on the continent, it would be good for global supply resilience of these in-demand products, good for African growth and development, and good for the planet.

At a moment when too many companies and countries are retreating from environmental commitments in the name of economic growth, this book offers a welcome corrective: climate action is where the growth opportunities are. The alternative is environmental and economic catastrophe. Delay is a luxury we cannot afford.

Dr Ngozi Okonjo-Iweala, Director-General of the WTO
June 2025

About the author

Nicholas Stern is IG Patel Professor of Economics and Government, Chair of the Grantham Research Institute on Climate Change and the Environment, and Chair of the Global School of Sustainability at the London School of Economics and Political Science (LSE). He has held posts at other UK and overseas universities and as Chief Economist at both the European Bank for Reconstruction and Development (EBRD) and the World Bank. He was at the UK Treasury, 2003–2007, where he was Second Permanent Secretary and Head of the UK Government Economic Service, led the writing of the report of the Commission for Africa (2005), and produced the Stern Review on the economics of climate change in 2006. He has served as President of the European Economic Association (2009), of the Royal Economic Society (2018–2019), and of the British Academy (2013–2017). He was elected Fellow of the Royal Society (June 2014). He was knighted for services to economics (2004), made a life peer (2007), and appointed Companion of Honour for services to economics, international relations, and tackling climate change in 2017. He has published more than 25 books and over 200 articles.

Preface

It is almost two decades since the Stern Review, *The Economics of Climate Change*, was published. Its key conclusion from the examination of the science and economics of climate change was that the costs of inaction far outweigh the cost of action. Two years after the Stern Review, I published *A Blueprint for a Safer Planet* distilling the Stern Review messages ahead of COP15 in Copenhagen in 2009, setting out a plan of action, and arguing that such action could create a new era of progress and prosperity. Ten years after the Stern Review, I published *Why Are We Waiting?* which argued not only that that the key conclusion was still stronger, but also that the investment in the new low-carbon economy could be a major driver of growth. I emphasised still more strongly the dangers of delay, hence the title. At the end of the year that it was published, 2015, the world created the landmark Paris Agreement[1] at COP21 with its temperature target of well below 2 °C and efforts to hold to 1.5 °C.[2] The argument that tackling the climate crisis embodied a tremendous investment and growth opportunity began to gain traction in the years before Paris; indeed it was an important foundation in delivering the Paris Agreement.

We are now a decade on from the Paris Agreement and *Why are We Waiting?*. Emissions have gone on rising, the climate crisis is deepening, and it is ever more clear that it is intertwined with the crisis of biodiversity loss. Nevertheless, those 10 years have seen remarkable progress, with around three out of four countries having net zero targets, covering around 80% of world emissions. Many cities, provinces, private companies, universities, and other institutions have such targets. The costs of renewable power have dropped dramatically. So too those of batteries and electric vehicles. The AI revolution is with us and together with the drive to the low-carbon economy will shape and deliver *the growth story of the 21st century*.

This book anticipates and traces that future and how it can be realised. Beyond the previous books, it highlights not only the opportunities in this unique moment in history but also the dynamics, investment, innovation, and structural change which will be necessary for the delivery of the new growth story. And I go much more deeply than before into the policies, action, and mechanics that can make it happen. There is a strong focus on policies and finance both in theory and in practice. When I wrote the earlier books the processes of change were just beginning. Now we are grappling with the detail. This book can be seen as the Stern Review 2.0. Like the original it tries to look across the whole landscape and focus analytically on both understanding and on action.

It will not be easy. It will need a great deal of investment and there will be significant dislocation. But it is a very attractive future. And it is much more realistic in my view than inaction and dithering, which would lead to catastrophic outcomes for large fractions of the world's population. It makes little sense, in my view, to associate pragmatism or realism with meandering down a road to disaster.

To generate this new story of growth the world will have to overcome many obstacles. Given the remarkable advances in technology, we can now see that the majority of those obstacles lie in economy, society, and politics. It will be economics and the other social sciences that will help us tackle the difficulties and chart the way forward. It is that understanding which has led the London School of Economics (LSE) to create its Global School of Sustainability (GSoS). This was co-founded by Lei Zhang and myself and began work in January 2025. Lei, a much-valued alumnus, has been a major donor to the GSoS.

This book builds on the Lionel Robbins Lectures which I gave at the LSE in March 2024. I also gave the Lionel Robbins Lectures in 2012, and they were the foundation of *Why Are We Waiting?*. In the preface to that book, I suggested that with his acute theoretical mind, his empiricism, and his policy focus, Lionel Robbins would have been grappling with these issues had he been with us. Robbins was not only an outstanding economist, and central to the development of the LSE's Department of Economics, he knew the importance of entrepreneurship and investment for growth and change and recognised the crucial role of finance for global public goods. His views on education embodied a deep belief in the right to personal development for all. The idea of our common humanity runs through the themes of this book together with the right to sustainable development. The crises and opportunities we now face demand the attention of the outstanding economists of this generation.

The 2024 lectures and this book draw on my own work over the last 20 years on the economics of climate change. Throughout that time, I have seen the building of the low-carbon economy as essentially part of the story of sustainable development. Climate, development, and growth are not different subjects. They are intertwined and action on climate will drive growth and development. And without climate action, growth and development will be undermined. That is the key message of the book.

In my work of the last two decades, and in attending all the United Nations climate COPs since 2006, I have come across many perspectives and learned many lessons that I carry with me and are reflected in the story told in this book. From 2013 Felipe Calderón and I chaired the Global Commission on the Economy and Climate, collaborating with researchers from the World Resources Institute (WRI), the Brookings Institution, and beyond. Andrew Steer, President of the WRI, played a key role. We helped build the argument ahead of COP21 in Paris that climate action and development come together. I have co-chaired the Independent High-Level Expert Group on Climate

Finance with Amar Bhattacharya and Vera Songwe, since COP26 in 2021, working with the UNFCCC and with the COP presidencies. I worked with the G20, including on the Eminent Persons Group on Global Financial Governance established under the German G20 presidency of 2017 and chaired by Tharman Shanmugaratnam, and the Independent Expert Group on Strengthening the Multilateral Development Banks (MDBs), established by the Indian G20 presidency in 2023 and co-chaired by N.K. Singh and Larry Summers. I worked with President Macron on his initiatives around the Pact for Prosperity, People, and the Planet and also with the UK and German G7 presidencies of 2021 and 2022. I have worked closely with Christine Lagarde and Kristalina Georgieva as Managing Directors of the IMF, and Ajay Banga, President of the World Bank and some of his predecessors. Throughout the period I have collaborated with the Executive Secretaries of the United Nations Framework Convention on Climate Change (UNFCCC), including Christiana Figueres, Patricia Espinosa, and Simon Stiell. The UN Secretaries-General Antonio Guterres and Ban Ki-moon have championed these issues and I have interacted and worked with them. So too the UN Deputy Secretary-General Amina Mohammed. There are and have been many strong leaders and initiatives working towards a better future. But moving from persuasive arguments to real action is a tough and detailed task. This book is an attempt to help this process along the way.

This is a transformation within which the private sector, and its investments and innovations, are of great importance. My interactions with the private sector, for example through COP26 in Glasgow and the EBRD (as Chief Economist), have been of great importance to my own understanding. As part of that I have learned much from my experience as Chair of Systemiq over the last seven years.

I emphasise these international collaborations because this is a subject that is quintessentially international. Nations must make their own choices and increasingly they recognise that the clean route to growth carries great advantages for their own nation. But collaboration is crucial across nations. Through the G20, the international financial institutions (IFIs), the UN and in many other ways, it must go beyond the climate institutions and tackle world economic and finance issues alongside and together with climate and biodiversity.

The next decades of global development are critical, as choices made now will either lock us into a high-carbon, catastrophic future or set us on a new low-carbon path of growth and development. In particular, this fork in the road is of special importance to emerging markets and developing countries (EMDCs) where most of world growth will occur, and most of the new infrastructure will be built. Embracing low-carbon growth could see EMDCs leapfrog straight to cleaner and more efficient technologies and structures, to both great benefit for their own development and to the benefit of the planet. The argument is strengthened still further by the abundance of renewable potential in EMDCs.

This is also a moment of special importance as a result of the turmoil associated with the Trump presidency in the USA which began in January 2025. There is no doubt that the hostility of his administration to climate action is a setback. But when a great power steps back, here on all of trade, aid, finance, health, security, and climate, other powers can and will step forward. That process is beginning. Further, much of the US public and private sectors will continue to go forward as they did under the previous Trump presidency. Many investors put their money into solar and wind power generation because it is cheaper. Change aligned with climate action in the USA will continue, although it will no doubt be somewhat slowed by the actions of the presidency.

The purposes of the book are to describe the actions that will be necessary to realise the growth story, to set out a coherent agenda for action and policy, and to show how that agenda can be delivered. In so doing we will also establish a research agenda, particularly in economics. However we must be clear that this is an area where action and research have to go together. We cannot wait for the research to be completed prior to taking action. But at the same time action will need serious analytical guidance.

The audience for the book follows in large measure from its purposes. I hope it will be useful to all those interested in the challenges of climate change, biodiversity, and economic development and poverty reduction. And how policies and action to take on these challenges can be researched, articulated, and implemented. That audience would include civil servants and politicians, and more generally, all those interested in the making of policy. Further, I hope it will be of interest to the private sector, NGOs and the public at large. This is a subject in which all should be interested and where effective action requires the involvement of all.

It would also, I hope, include university students across the world. I have tried to create a book which can be used in courses on the economics of climate change. Also, in courses on climate change and the social sciences more generally. I hope it would be of value to those studying climate in science and engineering as well: these students will need to have some understanding of economy, politics, and society if their own ideas are to have traction. Further, and of particular importance, I hope it will be of value to courses on economic development and economic policy. Indeed, it is in large measure a text on sustainable development.

The first key message of the book is the growth story. There is no horse race between climate action on the one hand and economic development and poverty reduction on the other. They go hand-in-hand. For clarification at the outset let me emphasise two points. First, climate action and biodiversity action are intertwined. But they are not the same. Nevertheless, action on these two fronts should be integrated. When I say climate action, it should be understood to also involve careful attention to biodiversity action. Second, by growth, I mean growth of a very different kind to the dirty, destructive models of the past. By growth, I usually mean growth and development across all dimensions of well-being. And I mean growth over the next few decades, not over some vague notion of the indefinite future.

The second message is that delay is deeply dangerous. This point was central to the Stern Review and at the core of *Why Are We Waiting?*. It remains central now. Severe impacts are being felt already, and the dangers will only escalate if climate is unmanaged or managed ineffectively or managed with delay. They would likely lead towards mass movement of people. And potentially widespread conflict.

The third message is that we can now describe the investments and innovations for the transformation of the economy that will be necessary. At least we can do that sufficiently clearly to set off in a purposeful way down the road we must travel. We will, of course, learn along the way and must design policies and actions so that we do learn.

The fourth is that all this will involve not only great investments and innovations but also structural change and dislocation. Careful design of policies and actions will be critical for a just transition and to avoid or reduce potential opposition. This should be a story in which all can participate.

The fifth is that much of what countries can and should do to tackle climate change is unambiguously in their own interests, even narrowly interpreted. However, international collaboration will be of the essence. That is why our international institutions, including the G20, the UN, the development banks, the IMF, and the WTO will be so important.

This volume (taking the Stern Review and *The Blueprint* together as the first) is essentially the third in 20 years, with 10 years between the first and the second (*Why Are We Waiting?*), and with 10 years between the second the third. Thus, I am led to wonder whether I will need to write a fourth, 10 years from now. One measure of the success of this book would be that the fourth turns out to be unnecessary. I am indeed confident and optimistic about what we can do. I worry very much about what we will do. This book is in large measure part of the process which tries to move from possibility to delivery.

Finally, I should note that this is one of the earliest volumes of LSE Press. I am delighted that I have the opportunity to contribute at an early stage in the work of the Press. I am very grateful to Sarah Worthington for all her encouragement and her leadership as Chair of the LSE Press. And I see this also as one of the first major publications for the Global School of Sustainability. For these reasons and because I wanted the book to be freely available to students, it is of great importance that this book is available as open access.

Notes

[1] This was the twenty-first session of the 'Conference of the Parties' of the United Nations Framework Convention on Climate Change.

[2] Temperature here means the increase in global mean surface temperature relative to the second half of the 19th century.

Acknowledgments

This book builds on the Lionel Robbins Lectures given in March 2024 which can be found on the Grantham Research Institute website (Stern, 2024). I am grateful to the Lionel Robbins family, the lecture committee, and its Chair, Richard Layard, for the honour of asking me to give those lectures for a second time. The first time was in 2012 which led to the publication of *Why Are We Waiting?* (Stern, 2015). The acknowledgements section there recognises the many individuals across the world who have helped and guided me in my work on development and growth over many decades and in more recent decades on climate. I would like to thank them again and refer to those acknowledgments. I would like also to thank again all who helped and guided the Stern Review (Stern, 2006) and refer to those acknowledgments. The last 10 years have sadly seen the passing of a number of people I thanked in those earlier volumes with whom I have collaborated closely and from whom I learned so much. I think particularly of my dear friends and collaborators Tony Atkinson, Pete Betts, Kemal Dervis, Claude Henry, Marty Weitzman, and my mentors Ken Arrow, Jim Mirrlees, Manmohan Singh, Bob Solow, and Jim Wolfensohn.

At the Grantham Research Institute (GRI), the support, guidance, and creativity of Delfina Godfrid, Roberta Pierfederici, Maria João Pimenta, and Éléonore Soubeyran, my collaborators in the preparation of the lectures, the book, and beyond has been outstanding. I am immensely grateful to them and have been extremely fortunate to have had the opportunity to work with them. Bob Ward has been a guide and companion at GRI over nearly two decades and has provided much wise advice on this book and so much more. I am very grateful to him. I have benefited greatly too from collaborations at the GRI on related work with Kamya Choudhary, Rob Macquarie, Charlotte Taylor, and Chunping Xie. My thanks go also to the Directors of the GRI, including Sam Fankhauser and Elizabeth Robinson over the past decade.

The GRI, STICERD, and the Department of Economics at the LSE have been wonderful homes for me in the more than 40 years that I have been with the LSE. I am very fortunate to have been with the LSE for so long and thank all the Directors (now Presidents) of the LSE I have worked with, most recently, Larry Kramer and Minouche Shafik. I am very grateful to them and to the directors, chairs, and convenors of those units for their collegiality and research support. Since January 2025 we have had the Global School of Sustainability (GSoS) at the LSE, which I chair, and I am already benefitting greatly from its support. I am very grateful to the co-founder, with me, of the GSoS, Lei Zhang, an LSE alumnus and a key donor for GSoS. This book is the first in a series flowing from GSoS. For the last seven years, I have chaired the board of Systemiq, an analytical and advisory firm, and have learned so much; I am grateful to them, particularly Jeremy Oppenheim, the founding partner.

In the Preface, I described my many collaborations over the last two decades in international bodies, groups, and commissions. There are too many individuals from whom I have learned in the process to name them all, but I should mention some of them with whom the collaboration has been particularly close. Amar Bhattacharya has been at the heart of so much of our work on international climate finance and I am hugely grateful to him. I have been particularly close to Masood Ahmed, Homi Kharas, Hans Peter Lankes, Mattia Romani, Josué Tanaka throughout all this work; special friends and colleagues. Ngozi Okonjo-Iweala, Mari Pangestu, Tharman Shanmugaratnam, N.K. Singh, Vera Songwe, Andrew Steer, Larry Summers, have all been chairs or co-chairs on commissions or groups on which I have served. They are good friends, fine colleagues, and real leaders, and I am very grateful to them. I am very grateful to Ngozi for kindly providing a foreword.

At international institutions, particularly the World Bank, IMF, and UN, I have, amongst others, been privileged to work on the issues of the book with Ajay Banga, Stéphane Hallegatte, Martin Raiser, Juergen Voegele (World Bank), Christine Lagarde, Kristalina Georgieva, Ceyla Pazarbasioglu (IMF), and at the UN, Christiana Figueres, Patricia Espinosa, Simon Stiell (UNFCCC) and Amina Mohammed as UN Deputy Secretary-General. And with the Chief Economists of the World Bank and IMF over the years. Jacques de Larosière has been a mentor and friend over three decades and I am very grateful to him.

My collaborations and interactions have been particularly intense in Europe; India, China, and the USA. Amongst the many from whom I have learned I want to thank Emmanuel Guérin, Ruth Kattumuri, Caio Koch-Weser, Amélie de Montchalin, Luiz Pereira da Silva, and Laurence Tubiana (Europe); Ranjit Barthakur, Amitabh Kant, Ajay Mathur, S. Ramadorai, Bittu Sahgal, Montek Singh Ahluwalia, N.K. Singh, and Jayant Sinha (India); Liu He, Zou Ji, Zou Jiayi, and Zhu Min (China); and Al Gore, John Kerry, Jonathan Pershing, and Todd Stern (USA). N.K. Singh (India) and Zhu Min (China) have for long been my guides and collaborators on the two most important countries in the world and beyond.

There are too many people in the UK with whom I have collaborated and from whom I have learnt for me to name them all. This is a moment, however, to celebrate and thank Pete Betts, climate diplomat extraordinary, who sadly died in October 2023. I have enjoyed working with environment and energy ministers from all major parties including Ed Davey, David Miliband, Ed Miliband, and Amber Rudd. Also with Hilary Benn and the late John Prescott. I thank them and all the splendid civil servants who worked with them. I would also like to thank the chairs, chief executives, and staff of the exemplary Climate Change Committee in the UK. And Tony Blair and Gordon Brown who asked me to work on the Stern Review in 2005.

In preparing this book in its later stages, Halsey Rogers provided outstanding guidance on an early version of the text, and I am immensely grateful to him. Stéphane Hallegatte and Cameron Hepburn provided very helpful reviews, and I am very grateful to them for these and for many constructive interactions over the years. The whole LSE Press and Communications teams have been tremendous colleagues throughout the process of turning an initial typescript into a book and I would like to thank Simone Chiara van de Merwe, Justin Clark, Philippa Grand, Alice Park, Ellie Potts, and Sarah Worthington.

On the science front, I refer again with gratitude to my continuing interactions with Brian Hoskins at our sister institution, the Grantham Institute at Imperial College, and to Myles Allen, Nicola Ranger, Johan Rockström, John Schellnhuber, and Keith Shine. My economics colleagues at LSE are a constant source of ideas and guidance, and I should mention particularly, within STICERD, Oriana Bandiera, Tim Besley, and Robin Burgess. I have learned greatly from my interactions on economics and climate over the years with Ehtisham Ahmad, Mark Carney, Partha Dasgupta, Simon Dietz, Peter Diamond, Peter Hammond, Geoff Heal, David Newbery, Joe Stigltiz, Adair Turner, Anna Valero, Martin Wolf, and Dimitri Zenghelis. And on both economics and philosophy I owe a huge debt to Amartya Sen.

On communications, I have learned much from Nik Gowing and Bob Ward. On some particular issues I would like to thank Agnieszka Smoleńska and Joseph Feyertag (on prudential regulation), Kate Laffan (behavioural science), and Nicola Ranger (nature and resilience).

Funding is gratefully acknowledged from the Grantham Foundation for the Protection of the Environment, UK Economic and Social Research Council, Children's Investment Fund Foundation, Rockefeller Foundation, European Climate Foundation, ClimateWorks Foundation, Quadrature Climate Foundation, Energy Foundation China, Vincent Cheng, Michelle Liem, Cherie Nursali, and Lei Zhang.

Finally, very special thanks go to Eva Lee Britton and Kerrie Quirk. Their skills, support, judgement, patience, wisdom, and guidance are hugely appreciated.

Acronym list

AFOLU	agriculture, forestry, and other land use
AI	artificial intelligence
AMOC	Atlantic Meridional Overturning Circulation
ASCM	the WTO's Agreement on Subsidies and Countervailing Measures
AUC	African Union Commission
BAU	business-as-usual
BRICS	Brazil, Russia, India, China, and South Africa
CBAM	the EU's Carbon Border Adjustment Mechanism
CBDR	common but differentiated responsibilities
CBEs	consumption-based emissions
CCDRs	the World Bank's Country Climate and Development Reviews
CCS	carbon capture and storage
CCUS	carbon capture, usage, and storage
CDR	carbon dioxide removal
CERN	European Organization for Nuclear Research
CFMCA	Coalition of Finance Ministers for Climate Action
CO$_2$	carbon dioxide
CO$_2$e	carbon dioxide equivalent
COP	Conference of the Parties
CPI	Climate Policy Initiative
CREA	Centre for Research on Energy and Clean Air
DC	direct current
EBRD	European Bank for Reconstruction and Development

EMDCs	emerging markets and developing countries
ETC	Energy Transitions Commission
ETS	the EU's Emissions Trading Scheme
EV	electric vehicle
G20	Group of 20
G7	Group of 7
G77	Group of 77
GBF	The Kunming–Montreal Global Biodiversity Framework
GCEC	Global Commission on the Economy and the Climate (but usually termed the New Climate Economy, NCE)
GDP	gross domestic product
GFANZ	Glasgow Financial Alliance for Net Zero
GHG	greenhouse gas
GNP	gross national product
HDI	Human Development Index
IAM	integrated assessment model
IEA	International Energy Agency
IEG	G20 Independent Expert Group
IFIs	international finance institutions
IHLEG	Independent High-Level Expert Group on Climate Finance
IIGCC	Institutional Investors Group on Climate Change
IMF	International Monetary Fund
IPCC	Intergovernmental Panel on Climate Change
IRA	Inflation Reduction Act
IRENA	International Renewable Energy Agency
JETPs	Just Energy Transition Partnerships
LCOE	levelised cost of electricity
LDCs	Least Developed Countries
LULUCF	land use, land-use change, and forestry
MAT	mean annual temperature
MDBs	multilateral development banks

MDGs	Millenium Development Goals
NbS	nature-based solutions
NCE	New Climate Economy
NCQG	New Collective Quantified Goal on Climate Finance (adopted at COP29)
NDC	nationally determined contribution
NGFS	Network for Greening the Financial System
ODA	Official Development Assistance
OECD	Organisation for Economic Co-operation and Development
PBEs	production-based emissions
PPAs	power purchase agreements
PPP	purchasing power parity
PV	photovoltaic
R&D	research and development
R,D&D	research, development, and deployment
REN21	Renewable Energy Policy Network for the 21st Century
RMI	Rocky Mountain Institute
SBTi	Science-Based Targets Initiative
SDGs	Sustainable Development Goals
SDRs	Special Drawing Rights
SPG	subpolar gyre
SRM	solar radiation management
TEDA	Tianjin Economic-Technological Development Area
UNCCD	United Nations Convention to Combat Desertification
UNCTAD	United Nations Trade and Development
UNDESA	United Nations Department of Economic and Social Affairs
UNDP	United Nations Development Programme
UNEP	United Nations Environment Programme
UNFCCC	United Nations Framework Convention on Climate Change
UNIDO	United Nations Industrial Development Organization
V20	The Vulnerable 20 Group

VCMs	voluntary carbon markets
WEF	World Economic Forum
WHO	World Health Organization
WMO	World Meteorological Organization
WRI	World Resources Institute
WTO	World Trade Organization
WWA	World Weather Attribution
WWF	World Wide Fund for Nature

Introduction:
if not us, who? If not now, when?

We are at a critical moment in history. The next decade is decisive. It is this generation that must act, and it must act urgently and at scale to reduce greenhouse gas (GHG) emissions and protect and restore biodiversity. Delay is dangerous. The consequences could be catastrophic. For many, perhaps hundreds of millions or billions, they could be existential. The science has been clear for decades on the potential scale of the risks. As emissions go on rising and the evidence gets ever stronger, the alarm signals are unmistakeable. The science sets a fierce timetable for action.

The rewards from urgent action, however, go way beyond the avoidance of immense risk, fundamental though that is. Whilst that argument was made in the *The Economics of Climate Change: The Stern Review* (hereafter the Stern Review) in 2006, it has become ever more clear, particularly since 2015: the investment, innovation, and structural change that can deliver the radical change in economic activities necessary to reduce emissions and protect biodiversity can also deliver a much more attractive story of growth and development than the dirty, destructive growth models of the past. We have in our hands a new growth story for the 21st century. The urgency demands that we create this new path in the second quarter of this century.

That growth story is more resource efficient, creates lower-cost energy, kills and maims far fewer people from pollution, yields robust and fruitful ecosystems, makes our land more productive, focuses investment on the most innovative areas, and raises investment itself. Clarity on the necessary scale of action and the risks of delay is essential to our argument. But the core of this book, and I trust its new contribution, is the focus on the growth story embodied in the action necessary, on its key elements and drivers, and on the policies and actions required to create it. It is this story that has become ever more clear to me in the two decades since the publication of the Stern Review in 2006.[1] It is not simply a story of destruction avoided; it is a new story of growth. That new story will be built on remarkable technological change, particularly since 2015; the clean is now cheaper than the dirty across a wide range of economic sectors. The new story of growth, what it looks like, and how it can be achieved is the core of this book. In many ways this book is the Stern Review 2.0.

Because this new growth is driven by investment and innovation, it will in large measure be created by the private sector. But public action will be critical in creating a favourable environment for the investment, in fostering creativity and innovation, in promoting research and development, and in carrying through vital public sector investment. That will involve a role for the state very different from that associated with the market fundamentalism so popular at the end of the last century.

The first and primary purpose of this book is to set out a new view of the economics of growth and development which has sustainability at its core as both a key requirement and a key driver. Second, in so doing I offer a research agenda for and some challenges to my own discipline, economics. Third, I offer a reflection on the Stern Review after two decades. Fourth, I hope that it can serve as a course book – open access – for courses on sustainability and development economics.

It is the concentration of GHGs in the atmosphere that causes global warming (see Chapter 2). Carbon dioxide is the most important of these gases. Since a kilogram of carbon dioxide emitted from Johannesburg has the same climate effect as one from London, or anywhere else, the response must take place across the world. This is a challenge for international collaboration and coherence. That challenge is not easy – particularly in today's turbulent and often fractious world – but the difficulties are moderated by the fact that so much of the necessary action is in the best interests, even if narrowly interpreted, of each country.

Setting out this growth story and showing how it can be realised is the primary purpose of the book. It is sketched briefly in this introduction, together with a summary of the scale of the risks faced by the world in terms of the climate and biodiversity crises and thus the necessity for action with urgency and scale. It is followed by a sketch of the book's structure and logic.

This book carries a message of hope. But the path that turns hope into reality is full of difficulties and obstacles. I believe that they can be overcome and this book will try to show how. One thing is clear: it will take substantial investment, innovation, and systemic and structural change if that hope is indeed to become a reality. And, to stress again, delay is dangerous.

A new growth story: drivers and potential

The key argument of this book is that the investment and innovation necessary for a strong and effective response to the climate and biodiversity crises will drive a new growth story, much more attractive than the dirty, destructive models of the past. That investment is therefore both an imperative and an opportunity. We begin here with a summary of the opportunity: the growth story. And then of the imperative, given the scale and nature of the risks. To be crystal clear on two points: this is growth which is very different from the past; and we are talking about the next two or three decades, not some vague notion of indefinite growth (see Chapter 5).

There are six key drivers of the new growth story, and investment is at the core of that story:

1. **Lower costs, learning by doing, induced innovation:** Investment will be focused where innovation is vibrant and costs are low and falling.

2. **Increasing returns to scale in new technologies:** Much of the investment will take place in areas where there are powerful increasing returns to scale as new initiatives are started and expanded, for example solar panels or LED lightbulbs.

3. **Improved resource efficiency:** Investments in resource efficiency have great potential, including the circular economy. Energy efficiency is of special importance. Efficiency means productivity means growth.

4. **Stronger system productivity:** The investment, innovation, and structural change necessary to strengthen the productivity of big systems – energy, transport, cities, land, and water – will increase productivity across the board. For example, interconnected transport systems, with smooth links between public and private and greater scope for cyclists and pedestrians, create cities that are much more efficient and productive than those associated with traffic jams and frustrated travellers and commuters.

5. **Improved health:** Improving health can drastically reduce the deaths and disability associated with air and other pollution in both urban and rural areas. The health benefits from reduced pollution will surely be of great value to well-being in their own right, but they also make people more productive. Enhancing human capital enhances productivity and growth.

6. **Increased share of investment:** The new investment itself adds to productive capacity and drives growth. In this context, infrastructure is of special importance.

Artificial intelligence (AI) will play a powerful role across all these drivers, particularly in innovation, discovery, efficiency, and management of systems. We are fortunate that the need for fundamental change and AI have arrived at the same time.

All of these effects simultaneously reduce emissions and damage to our ecosystems and create much more productive economies and societies. They do that by lowering costs, fostering innovation, improving efficiency, creating better functioning of the key systems, improving health, and raising productive investment itself.

As I shall argue in this book, most macro modelling and analyses of output growth largely ignore the six crucial drivers, with the exception of the last. And generally, they also ignore the losses from failing to adapt. It is a new growth story both in theory and in practice. Further, adaptation and the building of resilience to a climate that is changing will increase output and productivity, relative to strategies which act as if climate change is not happening or is minimal. Avoiding destruction from flooding increases productivity relative to failing to do so. Clearly adapting farming techniques to changing seasons and rainfall patterns is more productive than carrying on as if nothing is happening.

Most of the investment and innovation in this new form of growth will be from the private sector. Indeed, it has been described as the biggest investment opportunity since the Industrial Revolution.[2] But strong public policy and leadership are central to creating favourable conditions for this investment and innovation. At the regional level, the EU Green Deal is doing this, through a strong package of cohesive long-term policies.[3] At a more local level, it is also, for example, what the Basque Government is doing through the Net-Zero Basque Industrial Super Cluster, a green hub where the public sector works as a facilitator, aligning business needs and technological supply to clean priorities and economic growth (World Economic Forum [WEF], 2025). Making it easier to use or build the new and clean can increase consumer demand. For example, the German government's elimination of permitting rules allowed residents to install solar panels simply by hanging them and plugging them into a socket. Balcony solar panels increased rapidly – a convenient way to reduce electricity bills. Around 1.5 million German balconies now have solar panels (Burgen, 2024).

Strategic thinking and acting on sustainable investments can be found across the world. In Latin America, for example, Chile is seeking to position itself among the largest green hydrogen exporters by 2040 (Ministry of Energy of Chile, 2022). On the African continent, countries such as Namibia, Morocco, and South Africa are building on the continent's clean energy potential in an exciting way by developing plans to position themselves as green hydrogen production hubs, potentially exporting to European countries or using it themselves (Radford and Field, 2023). Green skills were mainstreamed into the school curricula of Rwanda, as part of its Green Growth and Climate Resilience Strategy (REN21, 2025). And Kenya's Energy Transition and Investment Plan (2023) outlines a pathway to meet the country's energy needs and socioeconomic development objectives simultaneously, aiming to reach net zero by 2050 (SEforALL, 2025). In Asia, for example, India is crafting a roadmap to green their steel sector (Ministry of Steel, Government of India, 2024). And the Lao People's Democratic Republic (PDR)–Thailand–Malaysia–Singapore Power Integration Project is setting the example in energy multilateralism: this cross-border electricity trading initiative facilitates transferring renewable energy electricity across borders, particularly from the Lao PDR, which is highly abundant in hydroelectricity, to Singapore, through Thailand and Malaysia. This helps participating nations to both drive development forward and meet their sustainability goals, while also creating an important precedent for future multilateral energy trading projects (REN21, 2024).

In some sectors jobs will be created. In others they will be lost. And across the economy many jobs will change. There will be great opportunities but also dislocation for workers and consumers. Public action must recognise that the rewards of success can be immense but also make sure that the difficulties and dislocations are managed to protect those who might suffer. To fail to do so would not only be unjust but also undermine support that is necessary for change. As the economies of the world are transformed, more jobs are expected to be created than lost. But workers who face transitions will need support, particularly in relation to skills. Investments will be needed in people and places where dislocation occurs. Consumers will see some prices rise alongside others that will fall and some poorer consumers particularly may need protection, including through transfers (Stern and Stiglitz, 2023). The participation of civil society will be of great importance in managing change.

The new growth story unlocks new employment opportunities for employees, the self-employed, and entrepreneurs. This is of particular importance for the development of low- and middle-income countries. There are real opportunities in the new approach to growth to create quality employment opportunities, of special importance in a context where many promising high-skilled youths may otherwise decide to emigrate (African Union Commission [AUC] and Organization for Economic Co-operation and Development [OECD], 2024; International Labour Organization [ILO], 2024). Examples of practical measures to take such opportunities include the 'Boosting Green Employment and Enterprise Opportunities' programme in Ghana, whose integrated approach both provides useful technical skills and connects the unemployed with opportunities through employment as well as self-employment (Ambasz et al., 2024; GrEEn, 2020).

The developing world is at centre stage in the new growth story. It is where the majority of world growth will occur, and infrastructure and other investments will be made in the coming two decades. For example, most of the infrastructure of developing nations is yet to be built; even in a middle-income country like India nearly 70% of the urban infrastructure likely to exist in 2047 will be built between now and then (Kouamé, 2024). If these investments look anything like the old, the risks to our future will be immense. Delaying action brings the risk of locking-in emissions from carbon-intensive, long-lasting dirty capital.

The next few decades will see strong urbanisation in the developing world and it will be crucial for sustainable growth to avoid the great congestion and pollution locked into many existing city structures. It is indeed possible to find other ways to do things. Medellín, Colombia, introduced cable cars as part of their mass transport system, reducing emissions from mobility while integrating marginalised areas on hillsides and improving the links between residents and jobs (Dávila and Daste, 2013). The lion's share of solar and wind energy opportunities in the world reside in developing nations, hitherto largely untapped. But examples are emerging and among the 12 nations that, as of 2023, were already generating virtually all their electricity from renewables (>98%), eight are from the Global South. Among these clean powerhouses are Albania, Bhutan, Nepal, Paraguay, Iceland, Ethiopia, and the Democratic Republic of Congo (Jacobson, 2025).

For most developing nations there is a huge opportunity to leapfrog the destructive and inefficient models of the past, avoiding the outdated and dirty, in much the same way that Africa and Asia skipped traditional fixed-line telephony to move directly to modern mobile phones. Kenya has increased access to electricity dramatically, reaching around 80% of its population in 2023, by relying extensively on clean sources instead of fossil fuels, with nearly 90% of its electricity being renewable based (International Energy Agency [IEA], 2025). Around 20%

of this electricity is being generated with grid-scale wind and solar technology (Thurber, 2024). In the words of William Ruto (2022), Kenya's president: 'Rather than trudging in the fossil-fuel footsteps of those who went before, we can leapfrog this dirty energy and embrace the benefits of clean power.'

The scope for innovation across all sectors is immense. For example, in agriculture we can now see a future where AI helps determine planting dates, irrigation techniques, fertiliser use, and market opportunities – and via accessible devices for small-scale farmers, such as smartphones. This future is already beginning with startups in developing nations, combining the latest technology with local expertise in agricultural systems. The infrastructure of the future could be built using lower-carbon materials backed by favourable economics. For example, by substituting clinker with calcined clays, it appears possible to reduce the ratio of clinker (responsible for over 80% of the overall emissions of cement production) in cement or concrete mixes, thus decreasing process- and energy-related production emissions. In comparison to clinker, producing calcined clay requires lower temperatures (approximately 800 °C vs 1400 °C), thereby reducing production costs by around 25%. These costs could decrease even further as calciners for calcined clay, which are still at the early stages of their development, improve over time (Systemiq, 2024).

So too we are learning from the past: scientists are learning from ancient structures that have passed the test of time. It seems Romans used concrete-manufacturing strategies with self-healing functionalities, qualities that enabled buildings such as the Pantheon, the largest unreinforced concrete dome in the world, to survive millennia – while many modern concrete structures endure only a few decades (Chandler, 2023; Seymour et al., 2023). Imagine the widespread use of concrete that can self-heal its cracks, just like bones do, extending the life of structures and cutting down concrete use and repair costs.[4]

Some of the new approaches to, and forms of, development will involve changing lifestyles, including changing behaviour to reduce waste and pollution, and altering diets. During 2022, I had the privilege to be part of the Indian Government's launch of the Lifestyle for Environment (LiFE) movement, an initiative that fosters behavioural and consumption changes of individuals through simple acts, for example around waste, while also leveraging social networks to influence collective norms in favour of sustainability. Ultimately, this initiative is about generating a mass movement for 'mindful and deliberate utilization, instead of mindless and destructive consumption', said Narendra Modi, Prime Minister of India, at COP26 (Modi, 2021).

Achieving a global transformation of this magnitude will be challenging. The pace and scale of the investment increases and transformations that need to take place are historically unprecedented outside war time. In addition to the profound changes necessary in institutional structures and behaviours, the macroeconomic and public finance challenges will be substantial. All this is feasible, in my view, but we must not underestimate the difficulties. We must think ahead, prepare for them, tackle them, and be ready to learn and adjust along the way. There will be surprises, failures, disappointments, and false starts. But the pace must be maintained. The prize of success is immense and the consequences of hesitation or failure potentially devastating.

Rising emissions, rising risks

Under current policies, temperatures are headed to potentially catastrophic warming of close to 3 °C by 2100 (Climate Action Tracker, 2023).[5] Our planet has not not experienced average temperatures that are 3 °C higher than pre-industrial levels for around three million years. Sea levels were then 5–25 metres higher than today (Miller et al., 2012; IPCC, 2023). This was long before our current civilisations established themselves in the period after the last ice age. Since then, humanity has lived – and, on the whole, flourished – in a fairly stable climate and temperature. It was this benign period, known as the Holocene Epoch, which covers the last 11,000 or 12,000 years, that allowed settled agriculture.[6] Grasses became cereal crops and there were settlements to enable the nurturing and protecting of crops whilst waiting for harvests. There were stocks of grain, which meant hunting and gathering was no longer necessary on a daily or weekly basis. And there were surpluses, allowing much greater non-agricultural activities. During the Holocene Epoch, temperatures are estimated to have been within the range of plus or minus 1 °C relative to the end of the 19th century – until the last couple of decades. *Homo sapiens*, present for a quarter of a million years, has not experienced anything close to temperature increases of 3 °C.

Rising emissions have already, according to the World Meteorological Organisation (WMO), pushed the world to a long-term temperature increase in 2024 of 1.34–1.41 °C compared with 1850–1900 (WMO, 2025a). The rate of warming between 1991 and 2024 was more than 0.2 °C per decade (WMO, 2025b). At that rate of increase, warming of 1.5 °C will be reached within the next decade, and 2 °C by the 2060s. Future risks will escalate rapidly with every fraction of a degree of warming, particularly since 'tipping points' could be passed, such as destabilisation of major ice sheets and collapse of the Amazon forest system. These could create powerful dynamic feedbacks, including further temperature increases. We stabilise temperatures by stabilising concentrations of GHGs, since that determines the magnitude of the warming effect. We stabilise concentrations or stocks by balancing sources and sinks for the emissions or flows (see United Nations Framework Convention on Climate Change [UNFCCC], 2015, Article 4). This balance has come to be known as 'net zero' (see Section 1.4 in Chapter 1).

These risks could lead to large-scale movements of people, potentially hundreds of millions or billions and, as a consequence, severe and extended conflict (Brown, 2008; Clement et al., 2021; Institute for Economics and Peace, 2020). In the words of leading climate scientist Michael Mann (in Carrington, 2023, paragraphs 10–11):

> If we can keep warming below 1.5 °C then we can preserve this fragile moment [i.e. the more benign Holocene conditions]. But if we go beyond 3 °C, it's likely we can't. In between is where we're rolling the dice … 1.5 °C is already really bad but 3 °C is potentially civilisation-ending bad.

And remember that those who cannot move could suffer even more. Already at 1.5 °C, for example, 950 million people across the world's drylands will experience increased risks of water stress, heat stress, and desertification (Mirzabaev et al., 2019). The share of the global population exposed to flooding will continue to rise sharply. Heatwaves that, on average, arose once every 10 years in a climate with little human influence, will likely occur around four times more frequently with 1.5 °C of warming and close to six times with 2 °C (Intergovernmental Panel on Climate Change [IPCC], 2021).

The actions we take today will shape not only the climate realities of future generations, but also those of many people currently alive. The observed changes

Figure I.1: The potential generational impacts of climate change

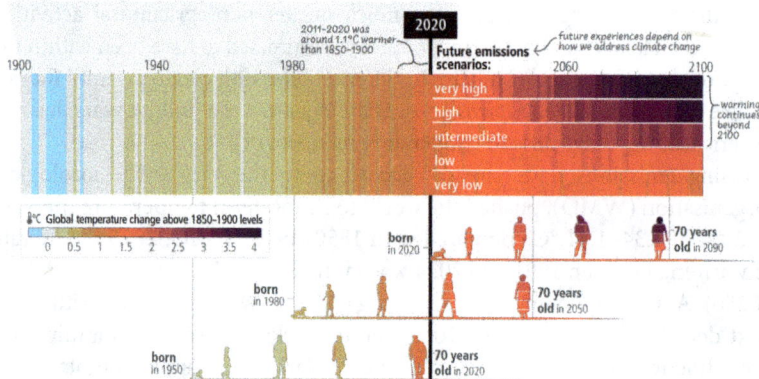

Note: Observed (1900–2020) and projected (2021–2100) changes in global surface temperature (relative to 1850–1900), which are linked to changes in climate conditions and impacts, illustrate how the climate has already changed and will change along the lifespan of three representative generations (born in 1950, 1980, and 2020). Future projections (2021–2100) of changes in global surface temperature are shown for very low (SSP11.9), low (SSP1–2.6), intermediate (SSP2–4.5), high (SSP3–7.0), and very high (SSP5–8.5) GHG emissions scenarios. Changes in annual global surface temperatures are presented as 'climate stripes', with future projections showing the human-caused long-term trends and continuing modulation by natural variability (represented here using observed levels of past natural variability). Colours on the generational icons correspond to the global surface temperature stripes for each year, with segments on future icons differentiating possible future experiences.
Source: Figure SPM.1 Panel (c) in IPCC (2023, p. 7). Copyright 2023 by IPCC. Reproduced with permission. Use of IPCC figure(s) is at the User's sole risk. Under no circumstances shall the IPCC, WMO or UNEP be liable for any loss, damage, liability or expense incurred or suffered that is claimed to have resulted from the use of any IPCC figure(s), without limitation, any fault, error, omission, interruption or delay with respect thereto. Nothing herein shall constitute or be considered to be a limitation upon or a waiver of the privileges and immunities of WMO or UNEP, which are specifically reserved.

in global surface temperatures since the 1850–1900 period, along with projections through 2100, portray a deeply concerning trend of escalating climate impacts over a single lifetime. Figure I.1 illustrates this for individuals born in 1950, 1980, and 2020, showing how different emissions scenarios directly influence warming levels, which are depicted in distinct temperature stripes. Those born today are at major risk of being plunged into catastrophic climate change of 3 or 4 °C unless our generation acts swiftly and strongly.

Climate impacts are not only a future problem; they are happening right now and are increasingly part of our lived experience. They include rising risks of wildfires, floods reaching coastal properties or storm surges inundating whole regions, extreme heat in densely populated areas such as the Indo-Gangetic plain, the southerly advance of the Sahara, and increases in the frequency and intensity of extreme weather events everywhere. Across the world climate change and biodiversity loss are damaging our lives and livelihoods.

The world is already heading for temperature increases above 1.5 °C. Every extra centigrade decimal of temperature involves still more danger. Temperatures will not stabilise unless the concentrations or stocks of GHGs stabilise. That means net zero emissions or flows of GHGs. The earlier we achieve net zero, the lower the temperature at which we stabilise.[7] Given that we are already into dangerous territory where tipping points may be close, the agreed targets of the Paris Agreement – well below 2 °C and efforts towards 1.5 °C – require net zero by 2050. Breaching of those targets involves very serious risks of outcomes which would be catastrophic for billions of people (see Chapter 2).

Sketch of book structure and its logic

The book, in large measure, follows the logic of the arguments set out above and in Chapter 1. The four chapters of Part I together provide the foundations, principles, and scaffolding – from the science to the politics – for the building of the story of the book. Chapter 1 traces our route to the present and its crises, together with how ideas have changed and how international action has started to emerge. And it begins the discussion of new opportunities for growth and development. It is 'how we got here, and where we need to go'.

Chapters 2 and 3 briefly examine some fundamentals crucial to the theses of this book. The science and the scale and nature of necessary responses are examined in Chapter 2. Key issues and strands in politics, economics, and ethics are the subject of Chapter 3. It notes in particular how the contribution of economic analysis in this area has often been misguided, and also how it must change, whilst recognising that the change has begun. It is in the economy, politics, and society that the main obstacles will lie. Social scientists have a crucial responsibility to help chart ways forward.

These chapters on the foundations are especially important considering the widespread misinformation and deliberate disinformation that obscure or dismiss the immense risks of inaction and, similarly, exaggerate the costs and downplay the many benefits of action. We examine and tackle common fallacies and confusions in Part IV. These distortions of analysis and data are, all too often, designed to undermine action and to protect the status quo and associated vested interests. They are dangerous in terms of the massive climate risks we face. And they hinder the taking of the great opportunities in the new growth story. We are confronting complex challenges in a fast-paced environment and in that context there is fertile ground for simplistic narratives. These attempts to muddle and divert make thoughtful analysis and action more important than ever. While disagreement is central to healthy debate, it is essential that discussions are grounded in a robust understanding of the fundamentals and a respect for data and analysis.

Chapter 4 focuses on opportunities, particularly in relation to how the world, its challenges, and the understanding of those challenges have evolved since the mid-2000s. This encompasses the journey from the Stern Review in 2006 to the UNFCCC COP21 in Paris in 2015, and then from Paris to the present. Before Paris, extrapolations based on current policies indicated the possibility of temperature increases of around 4 °C. After Paris, and a new sense of direction, these increases dropped to around 3 °C and they are now between 2.5 and 3 °C. We are still in a very dangerous place. The acceleration of action is now critically urgent. Chapter 4 sets the action agenda.

Part II sets out the central thesis that the response to climate change embodies the growth story of the 21st century; the opportunity. It is essential to respond quickly and at scale to avoid the most severe risks of climate change and to adapt to changes that we cannot prevent; the imperative. Chapter 5 shows how we can act on the risks while creating a new path of growth and development. The response begins with understanding the nature and the specifics of the challenges and requirements for action. There will be real difficulties, competing interests, different perspectives, mistakes, and learning along the way, but collectively, we now know enough to move quickly and purposefully. At the heart of progress will be an understanding that the actions necessary can foster the growth story of the 21st century, much more attractive than the environmentally damaging approaches to growth of the past. That argument is at the core of this book, and successful action depends on understanding it. Developing countries have a special opportunity to leapfrog the dirty, destructive paths followed in the past by developed countries. The new approach to growth can both achieve sustainable development across its many dimensions and respond forcefully to the climate and biodiversity crises. Identifying the necessary investment in the new technologies, fostering that investment, and financing it will be crucial. Governments will set strategies, policies, and frameworks, but the private sector will be in the vanguard of investment.

Chapter 6 offers an analysis of key elements of that growth story, focusing on investment, innovation, and systemic and structural change, and particularly on policies and actions to foster the necessary change. It examines actions, policies, institutions, and behaviours that are necessary to deliver transformation at the country level. Investment in natural and social and cultural capital, as well as physical and human capital, will all be critical. The systemic changes necessary involve recasting the key systems of energy, transport, cities, land, and water, all of which are themselves intricately connected.

Chapter 7 recognises and explores the consequences of the action agenda for the role of the state in fostering and enabling the fundamental sectoral, systemic, and technological changes embodied in the necessary new path of development and growth. It also examines how political economy and vested interests present obstacles to change. Closely associated with these obstacles are the ways in which preferences and behaviours evolve, or fail to evolve, with changing circumstances, with experiences, and with public discussion of new ideas. It is in part a synthesis, from a policy and institutional perspective, of the arguments of the second part of the book.

Climate action requires international commitment across the world. Such action will inevitably transform international relations and the world's economic geography. Part III of the book concerns these international issues and dynamics, exploring how differences across and relationships between nations and regions will reshape the world economy, taking into account the decline of fossil fuels. These differences, interdependencies, and the new economic geography are the focus of Chapter 8. It highlights too how the developing world is moving to centre stage in shaping the issues and setting the agenda.

Creating this new growth story will require both international and national action. Richer countries have an obligation to support poorer countries in their pursuit of a new path of growth and development; the richer countries prospered in large measure through environmentally harmful growth, but it is the poorer countries that suffer the most from climate change despite having contributed the least to the causes of it. Mobilising support from the rich world, often in the face of reluctance or opposition within their politics, will be critical to a successful transition. Overcoming that reluctance requires an understanding of the risks of inaction, of mutual interdependence, and of the opportunities in the new growth story. Moral imperative, responsibility, and self-interest point in the same direction. The necessity of collaboration across richer and poorer nations, especially around finance, is examined in Chapter 9.

Finance, both private and public, will be at the heart of international collaboration. Multilateral development banks (MDBs) play a central role in creating the conditions for investment and mobilising finance. They are powerful instruments of public policy if used to their full potential and offer the best 'value for money' in providing international support. Reforming these institutions is a core issue, one in which, with close collaborators, I have been actively engaged through public discussion and written proposals.[8] That involvement has been based on my experience as Chief Economist of both the World Bank and the European Bank for Reconstruction and Development (a regional development bank) and senior economist in the UK Treasury (Second Permanent Secretary and Head of the Government Economic Service), from where I led the Stern Review. Increasing the scale of and the collaboration amongst MDBs will be crucial. Improving their impact and efficiency and their collaboration with the private sector will also be key to success.

The final part of the book, Part IV, draws arguments together and asks how action can be galvanised. We begin in Chapter 10 by emphasising the fallacies and confusions that are encountered in public discussion of climate change and how they can be confronted. It also recognises and highlights the major obstacles which must be overcome. It is, in an important sense, a summary of many of the arguments of this book.

The final chapter, Chapter 11, reflects on the evolution of ideas, including those since the Stern Review, to show both how far we have come and how far we must go. It summarises the research agenda embodied in the arguments and conclusions set out in the book. It concludes with a call to action, stressing the need for immediate and large-scale efforts. A much better path for growth and development is in our hands. We should be confident about what we can achieve but recognise that our abilities to manage the politics of change and to collaborate will be tested. This will be a challenging transition. Translating what we *can* do, in principle, into what we *will* do, is fraught with difficulty. Optimism about the former is very different from optimism about the latter. The risk of failure is real, with potentially devastating consequences. Success means creating dynamic, sustainable, resilient, and equitable development. A world of shared prosperity.

The spirit of the book is well summarised by Brutus in Shakespeare's *Julius Caesar* (Act 4, Scene 3):

> There is a tide in the affairs of men which, taken at the flood, leads on to fortune. Omitted, all the voyage of their life is bound in shallows and in miseries. On such a full sea are we now afloat. And we must take the current when it serves. Or lose our ventures.

Brutus was not necessarily on the side of virtue, nor was he always successful, but the metaphor of the tide does capture well the idea that we are at a moment of crucial decision. Delay is dangerous.

Notes

[1] See Stern (2006, 2007). Published online in October 2006 and by Cambridge University Press in January 2007.

[2] Mark Carney (2020, p. 2) has said: 'Achieving net zero will require a whole economy transition – every company, every bank, every insurer and investor will have to adjust their business models. This could turn an existential risk into the greatest commercial opportunity of our time.' And Philipp Hildebrand and Edwin Conway of BlackRock: 'The transition to a low-carbon economy presents historic investment opportunities and challenges for clients – on par with the rise of emerging markets and digitization in recent decades' (in Segal, 2022, paragraph 6).

[3] For more details on the delivery of the EU's Green Deal see European Commission (2023).

[4] For an overview on self-healing concrete technologies see Zhang et al. (2024).

[5] These are average global surface temperatures measured as increases relative to the second half of the 19th century, assumed to correspond to temperatures preceding the industrial revolution.

[6] Some scientists argue that we left the Holocene around 1950 and have entered a period that some term the Anthropocene (Ellis, 2024).

[7] We should, of course, also recognise that the shape of the path to net zero matters since it is the whole path, particularly the summation of flows over time, that determines concentrations.

[8] See Songwe et al. (2022) and Summers and Singh (2023).

Part I.

Foundations: a world re-drawn and an urgent agenda for action

1. How we got here, and where to now

Our starting point is a world facing climate and biodiversity crises and severe poverty in many countries. The challenge is to deal with those crises, overcome poverty, and build a more sustainable, resilient, and equitable world. An effective response requires careful analyses of the crises and origins as foundations for coherent and effective plans for action. And that, in turn, requires an understanding of the nature, origins, and determinants of the crises, together with clarity on the objectives that will be used to assess actions and outcomes. In so doing, we will also recognise how international action to tackle the crises has emerged, how objectives have changed in response to growing understanding of the crises, and how approaches to understanding the nature of growth have been changing in economics and elsewhere.

I begin that analysis of where we are now and where we need to go, in a somewhat personal way, by describing briefly, in Section 1.1, the changes in my ideas since the Stern Review was published in 2006. As noted this book is in many ways the Stern Review 2.0.

Objectives have changed over the last decades, particularly around well-being and sustainability. This is the topic of Section 1.2. All this has been in the context of remarkable advances in incomes per capita and life expectancy, and thus population, since the Second World War, which are explored in Section 1.3. Section 1.4 narrates the international response to the crises since their beginnings, and Section 1.5 reviews past growth theories, how we must go beyond them, and structural and technological changes that are key in the new growth story.

1.1 Lessons from the two decades following the Stern Review

The world as a whole has, for too long, underestimated the risks of climate change. Many still do. However, this has changed since we began work on the Stern Review in 2005. In the face of overwhelming evidence, for most people and countries, there is much less science denial in public discussion.[1] Yet a lingering reluctance to act at the pace required persists among much of the public and many decision-makers.

There is often an implicit science denial of a new type, sometimes called 'lukewarmism', which started becoming prominent in about 2010. This argument suggests delay in action because 'it might not get as bad as you suggest'. This is anti-science or sloppy thinking. The science is clear that unmanaged climate change could be catastrophic and existential for many. The magnitude of potential consequences points to strong action even though the probability of catastrophe (however defined) may be less than 100%. As emphasised in the Introduction and in Chapter 10, clarity on the fundamentals is crucial to confronting the misinformation and distortion that is all too common on these issues. We will return to the evidence on climate risks and on acceptance of the science in Chapters 2 and 3. We should note, however, that outright denial of the science and the assertion that it is all a 'huge hoax' is still a hallmark of some on the political right.

This section reflects on the Stern Review and how my understanding of climate risks and climate action has changed since then.[2] Note that it was a review of the economics of climate change and took account of economic analysis until that point (2006) and the then-current state of climate science. At the time of the Stern Review there were three Assessment Reports (ARs) available, which had been conducted by the Intergovernmental Panel on Climate Change (IPCC) (AR 1, 2, 3), the UN body that monitors and assesses climate change science (see IPCC, 1990, 1996, 2001). The fourth was a work in progress. By 2025, there were six, plus the 2018 special report on global warming of 1.5 °C and 2 °C. Whilst IPCC reports have always been clear that there is a major issue around the risks and dangers of climate change, they have, on the whole, been cautious about drawing strong conclusions. As an intergovernmental institution looking to find broad cross-nation agreement, many of their statements have been 'lowest common denominator' in terms of negotiation and often heavily qualified.

As time has gone by, the ARs have been ever stronger in their statements, as concentrations of greenhouse gases (GHGs) have continued to rise, as the evidence of the potential magnitude of the risks has become ever stronger, and as the scientific analyses have become ever broader and deeper. For example, in contrast to 1990, when the IPCC could not confirm that climate change was human-induced, the 2021 AR6 report states that there is 'overwhelming' evidence of climate change and identifies human activities as its primary cause. This shift is due to improved data and analysis (Chen et al., 2021). Nevertheless, a reading of the evidence then, as we started to work on the Stern Review, made it clear that the climate risks were likely very large. In retrospect, whilst I was accused of alarmism by many commentators, climate sceptics, and some economists when it was published, the Stern Review in fact *understated* the risks of climate change.

Further, in 1990 technological alternatives to fossil fuels looked costly. Indeed, it was generally assumed that low-carbon sources of energy and activity were *more costly* than fossil fuel sources. The idea that the clean was in all cases clearly more costly than the dirty was 'baked in' to the United Nations Framework Convention on Climate Change (UNFCCC) in 1992.[3] Thus, for the UNFCCC, the question of who pays for the extra cost of the clean was central. For many, that assumption persists today, notwithstanding powerful evidence that the clean is already cheaper than the dirty across a wide, and increasing, range of

economic sectors given the strong technological advances of the last two decades, particularly the last. Those advances continue at a rapid rate.[4]

The Paris Agreement came a decade after we began work on the Stern Review.
One reflection of the increase in the assessment of risks was that the Paris Agreement looks to a balance of sources and sinks – that is, net zero (also see Section
1.4 on the Paris Agreement). The Stern Review had taken the weaker criteria of
cutting emissions by 80% between 1990 and 2050, whereas net zero is a 100%
reduction. An important feature of a net zero objective overall is that, for its realisation, it becomes necessary for all to set and achieve a net zero objective, since the
amount of net negative emissions is likely to be small. With an 80% target, all too
many could see themselves in the remaining 20% – see Section 2.2 in Chapter 2.

Notwithstanding the understating (in retrospect) of climate risks in the climate science, and the under-anticipation, as it turned out, of the future technological advances, the three central conclusions of the Stern Review were very
clear: first and foremost, 'the benefits of strong and early action far outweigh
the economic costs of not acting' (Stern, 2007, p. xv). Second, the 'world does
not need to choose between averting climate change and promoting growth and
development' and 'tackling climate change is the pro-growth strategy for the
longer term' (Stern, 2007, p. xvii). Third, 'climate change is the greatest market
failure the world has ever seen' and 'a range of options exists to cut emissions;
strong, deliberate policy action is required to motivate their take-up' (Stern,
2007, pp. xviii, xvii). There was much more, of course, including on the dangers
of delay, international collaboration, the analysis of the necessary policy instruments, and so on.

Whilst there are nearly 700 printed pages in the Stern Review, with
27 chapters and a technical annex, these are the three broad conclusions
that we highlighted then and which I would highlight now. They have stood
the test of time; indeed they have been strengthened still further, as the science has further emphasised and demonstrated the immense risks of climate
change and as technology has advanced so rapidly.

In relation to my perspective on the climate challenge, the most significant
change since the mid-2000s is the strengthening of my recognition that *climate
action drives growth*. It is not simply something that can be combined with
growth; it is a core driver. Further, it is a much better form of growth. And it can
last. It is sustainable. That argument is gathering momentum and recognition,
but it has not yet been won. Hence my enthusiasm and commitment to foster
the idea and set it out in this book. The discussion should move on to the practicalities and difficulties, which are real and substantial, of delivering the emissions reductions and the new growth story. The time to suggest delay is long
gone, even though there are many who persist with that proposition.

A key perspective in driving forward is to see action as an investment rather
than a cost. The investment increases must be large and rapid. But these are, in
large measure, investments with high returns beyond the advantages associated with emissions reduction, which are of course of fundamental importance.
There is thus an investment opportunity as well as the immediate imperative for

rapid action to stop climate change. I would now emphasise still more strongly investment for growth and change rather than 'cost of action'.

Another major change is that, whilst the remit of the Stern Review was climate change, in retrospect I would give stronger emphasis to natural capital as a whole, particularly biodiversity, which is addressed in depth in Chapters 2 and 3. Indeed, now, as this book demonstrates, I would speak of sustainability rather than narrowly of climate. Sustainability is about offering to future generations opportunities at least as good as ours (assuming future generations behave similarly to those who follow them) and that is clearly broader than climate (see Section 1.2).

The growth and investment argument has to be set out clearly and strongly because, even in 2025, the difficulties in investment, in change, and in pursuing the new growth story, which are indeed substantial, are still seen as reasons for moving slowly. This sometimes masquerades as 'realism'. However, choosing inaction or delay is to choose to strongly increase the likelihood of climate catastrophe. Promoting catastrophe should, in my view, be regarded as a profoundly unrealistic position.

Good economic analysis should recognise the essential substance of the problem: the huge climate change risks associated with inaction *and* the vast clean investment opportunities that can come with strategic action. We must transform not only our economies, but also economics, as will be emphasised at a number of points in the book. Increasingly, the change in economics is happening, but it is time to accelerate both the change in the subject of economics and the imperative for action in the world. Indulging in hesitation and delay is the opposite of promoting development. It risks the reversal of development.

The consequences of delay are underlined by the observation that we now have to do over 20 years what we could have done over 40 years. This implies that the increase in the investment rates, starting now, to stabilise climate at an acceptable level is substantially higher than it would have been then. One upside is, however, that technological advance has meant current investments are even more conducive to growth than those of 20 years ago would have been. That is not, however, an argument for further delay because we are now so close to 2 °C and dangerous climate change, and to tipping points. Note further that starting 20 years ago would have brought the technological advances to an earlier date. And would have made stabilisation at lower temperatures more likely, reducing risk substantially.

At the same time, economic analyses should examine closely the challenges of implementation. The new story of growth takes us to a much better path of development, but the transition will not be easy. Dislocation will be substantial, and the political economy of opposing forces will pose serious problems. Good policies and constructive and reliable institutions can help create opportunities and benefits for all across the short and medium term as well as the long. But policy detail will matter. So does the narrative, which shapes commitment and strategies.

1.2 The new objective: from growth to sustainable development

The new growth story is not about economic growth in its narrow and crude sense, but about sustainable *development*. In the last two or three decades, we have seen a deepening and broadening of ideas around the meaning and goals of development. The goals are now seen as way beyond income and output, as measured for example by GDP, and now embrace health, education, and environment. They also embrace distributional issues, including in relation to gender, rights, and income and wealth. This section presents this history, explaining how development objectives have been examined and discussed over the last 80 or 90 years. Clarity on goals is at the core of the design of strategy and policy. And it is central to understanding that the new path of growth is much more attractive than the dirty, destructive models of the past.

From GDP to development

We begin in the 1930s when, with John Maynard Keynes and Simon Kuznets, the measurement of aggregate economic activity got underway with a real sense of purpose. A timeline is set out in Figure 1.1.

Figure 1.1: Timeline of key events, discussions, and concepts around growth and development

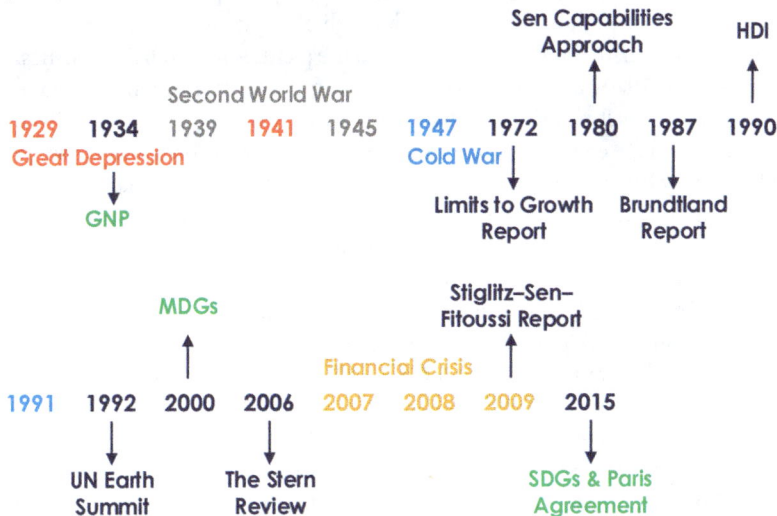

Source: Author's elaboration.
Note: HDI is the Human Development Index, MDGs the Millennium Development Goals, and SDGs the Sustainable Development Goals (see text).

The Great Depression led to a concern to understand the causes of aggregate economic activity, which had slumped so badly, and thus how to understand, define, and measure it became a crucial issue. A key aspect of the desire to measure economic activity was the direct concern with levels of employment, and with the desperate circumstances associated with the mass unemployment of the 1930s, particularly in the USA and Europe. Thus, the focus in measuring output was not necessarily to measure well-being as such, but on activity and the demand for labour. It was here that the idea of GDP or gross national product (GNP) was born; output and activity was the focus.[5]

Keynes led the theorising about the causes of the level of activity and many contributed. Kuznets, Richard Stone, James Meade, and others measured economic activity in the form of GDP in the USA, the UK, and beyond. Though the inventors of GDP emphasised the dangers of misunderstanding when complex realities are simplified into a single number, GDP or output increasingly became the focus for thinking about governmental success in terms of the ability of the economy to generate advances in living standards.[6]

Politicians – from Harold Macmillan in the 1960s in the UK, to Ronald Reagan and Bill Clinton in the 1980s and 1990s in the USA, and Donald Trump in this century – have asked versions of the question, 'Are you better off now, as a result of my administration and leadership, than when I started in office?' Or, when attacking an incumbent, they would ask whether people have become worse off during their tenure. Often changes measured in real income per capita of relevant groups were and are used in these discussions, although perceptions were and are sometimes more powerful than and different from the pictures from official statistics.

During the 1970s and 1980s, concern was growing in relation to broader dimensions of well-being, or standard of living, including on inequality of income and on the environment. And discussion of the distribution of outcomes extended way beyond that of income to all the dimensions of standard of living or well-being. The *Limits to Growth* report, 1972, focused on the possibility of exhausting natural resources and the increasing pressure on the environment (Meadows et al., 1972). The report of the Brundtland Commission on the Environment in 1987, titled *Our Common Future*, had a similar focus. That report contained a definition of sustainability as follows '[meeting] the needs of the present without compromising the ability of future generations to meet their own needs' (Brundtland Commission, 1987, p. 15). The Rio Earth Summit followed in 1992, which launched the UNFCCC.[7]

The concept of sustainability (introduced briefly in Section 1.1) that I shall suggest and use is similar in spirit and language to Brundtland but with important differences. I define sustainability as offering future generations *opportunities* at least as good as the opportunities available to the current generation, assuming that future generations behave similarly toward those that follow. This definition and the one of the Brundtland Commission are similar, but I think the notion of 'opportunity' captures the issues more clearly than 'needs'. The idea of sustainability runs through the book, and we come back to it in many places.

The definition proposed here is in the spirit of the work of Amartya Sen. During the late 1970s and 1980s Sen was developing his 'capabilities approach', which suggested that the purpose of development was not narrowly about outcomes or output, but the opportunities people had to shape their own lives. It was also about the freedom to act in pursuit of lives that people had reason to value; see, for example, his seminal work *Development as Freedom* (Sen, 1999). These capabilities are shaped, in large measure, by health and education as well as material assets and output, and also by the constraints associated with social conventions; in this context, of particular importance are the opportunities for women and social groups facing constraints and discrimination. Running through this approach, from the perspectives of freedom and capabilities, was the concern to understand the meaning of well-being and of poverty – and the distribution of well-being more generally.

In a time when many, understandably, highlight the importance of freedom, it is necessary to remember the imperative of achieving a sustainable development path that can create, and not destroy, the freedoms of generations to come. Following a dirty fossil-fuel path now means reducing the opportunities and freedoms for future generations to develop and thrive.

From the Millenium Development Goals to the Sustainable Development Goals

Through these processes of reflection and enquiry and public discussion, perspectives on the goals of development widened and deepened. The statistical measures used to calibrate development progress moved on accordingly. The UN Human Development Report, led by Mahbub ul Haq, and benefiting from the guidance of Amartya Sen, produced, in 1990, the Human Development Index (HDI), an aggregate of life expectancy, an education measure, and GDP per capita (see United Nations Development Programme [UNDP], 1990). In my view the aggregation into one number, the HDI, creates difficulties of interpretation, but the key point is the introduction of the dimensions of health and education both as crucial development outcomes and as routes to the expansion of income and other dimensions.

The Millenium Development Goals (MDGs) were agreed at the UN in 2000 and had eight broad dimensions, based, similarly to the Human Development Report, on income or consumption, health and education, but also with the addition of environment (Goal 7) and the development of a universal partnership for development (Goal 8). However, their focus was on family income, health, and education. This broadening of the agenda and measurement did make a difference to the behaviour of international development institutions. For example, when I was Chief Economist of the World Bank (2000–2003), I made sure that progress on the MDGs appeared right at the front of the data presentation, for example in the flagship report of the World Bank, the World Development Report. That did, I think, help raise the awareness of health and education as objectives and sharpen the focus of the World Bank and other development organisations on these goals and issues.

The focus on the environment had been developing from the 1970s and 1980s with Brundtland in 1987 and Rio in 1992. The Stern Review was published in 2006. Environment finally made its way strongly and prominently into international development goals with the Sustainable Development Goals (SDGs) in 2015. By 2015 the lived experience of climate change and environmental degradation had become more intense. And the understanding of the climate and biodiversity crises, particularly the former, was deepening.

Adopted in 2015, the SDGs aim to be universally applicable; they apply to all countries, whereas the MDGs applied only to developing countries (see United Nations [UN] General Assembly, 2015). That universality is a crucial point and implicit within the idea of sustainability where the future environment is influenced by the actions of all. They incorporated new dimensions such as equity, empowerment, peace, and inclusion. This extensive set of 17 goals and 169 targets was developed through a participatory process, reflecting a shared global undertaking on the need for an approach to measurement that recognises the importance and interdependence of a whole range of factors shaping well-being and development, both as goals and instruments or means. Of the 17 goals, 11 refer to the environment, sustainability, or climate explicitly. One way to understand the 17 goals, shown in Figure 1.2, is to see the first 10 as building on the MDGs but with a much stronger focus on distributional issues, the next five as focused on environment, and the last two focused on how we work together as countries and peoples.

Figure 1.2: The SDGs

Source: UN (2023a). Copyright 2023 UN. The content of this publication has not been approved by the United Nations and does not reflect the views of the United Nations or its officials or Member States.

The shift from GDP to the MDGs and then to the SDGs reflects an evolution in our collective understanding of what constitutes development. It underscores a global recognition that development must not only yield increases with respect to some narrow economic indicators but also improve human well-being, broadly understood, and protect the environment for future generations. As with health and education, environment is a goal in itself as well as an instrument that enhances health and material output. At the heart of this agenda is the strong argument that countries in the Global South have the right to develop. In other words, they have the right to the capacity to enhance the well-being of their citizens, now and in the future.

It is the right to development that underpins the SDGs. And their preamble acknowledges the relationship between development and the environment.[8] This right is also recognised in the Paris Agreement, which states that nations should 'respect, promote and consider' the right to development when taking climate action (UNFCCC, 2015, p. 1). This right was reflected in the principle of common but differentiated responsibilities under the UNFCCC, which led to different obligations for developed and for developing countries (Kaltenborn et al., 2020).

A critical point here, as already emphasised, is that as soon as we think about sustainability, we have to include all countries. The bigger the output, the bigger the potential consequences for the environment. Thus, combining growth and development with sustainability requires finding ways of increasing well-being whilst acting sustainably. That this is possible is a central argument of this book. Indeed, the argument is stronger than that. The quest and drive for sustainability, particularly around climate and biodiversity, will drive development and yield much more attractive paths than those of the past. As we will see in the next section, the growth paths of the past, while yielding great benefits, have failed to deal with a great many challenges and have created intense pressures, indeed crises, for our future sustainability.

1.3 Extraordinary advances and deep challenges

The last seven decades have seen extraordinary advances in life expectancy, education, income growth, and to some extent, until the last few years, democracy and human rights. This period has witnessed rapid and large falls in global poverty and falls in global inequality in health and education. But a 13-fold increase in total economic output (narrowly measured), and dirty and destructive methods of production, particularly around energy and fossil fuels, has put extreme pressure on our climate, biodiversity, and the environment. That 13-fold increase has been made up of, roughly, a quadrupling of output per head and a tripling of population. The latter has, in large measure, been a consequence of extraordinary increases in life expectancy, by around 26 years, or 50%, since 1950. I take it that most people would regard these advances as impressive and welcome in and of themselves. Some statistics on change in key outcomes globally are summarised in Figure 1.3.

Figure 1.3: Global advances since the Second World War

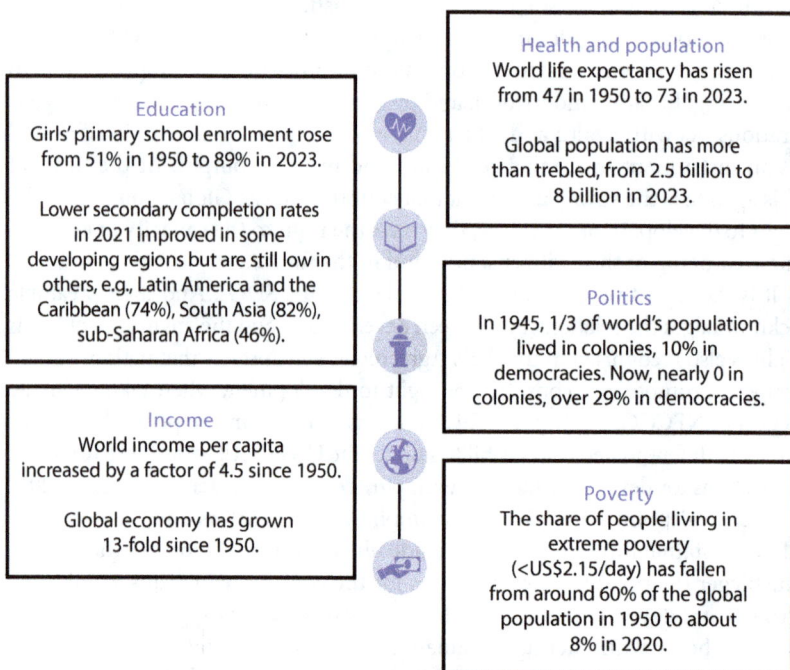

Education
Girls' primary school enrolment rose from 51% in 1950 to 89% in 2023.

Lower secondary completion rates in 2021 improved in some developing regions but are still low in others, e.g., Latin America and the Caribbean (74%), South Asia (82%), sub-Saharan Africa (46%).

Income
World income per capita increased by a factor of 4.5 since 1950.

Global economy has grown 13-fold since 1950.

Health and population
World life expectancy has risen from 47 in 1950 to 73 in 2023.

Global population has more than trebled, from 2.5 billion to 8 billion in 2023.

Politics
In 1945, 1/3 of world's population lived in colonies, 10% in democracies. Now, nearly 0 in colonies, over 29% in democracies.

Poverty
The share of people living in extreme poverty (<US$2.15/day) has fallen from around 60% of the global population in 1950 to about 8% in 2020.

Source: Author's elaboration using the following sources on education: Our World in Data (2023); World Bank (2024a, 2024b). Health and population: United Nations [UN] (2024a). Politics: UNCCD (2024); Herre (2022). Income: Roser (2019). Poverty: Yonzani et al. (2022).

While advances in overall outcomes for this period are striking, there remains fragility and great variation across regions. Advances in democracy have stalled or been set back in the last two decades (Freedom House, 2023). Many countries have seen the rise of autocratic figures and populism. There is much that is deeply troubling, including democratic backsliding, declines in press freedom, and the continued marginalisation, demonisation, or persecution of religious, racial, social, and national groups in many countries.

Economic progress has been less strong in Africa (e.g., on life expectancy and infant mortality). For example, the lower secondary completion rate in 2021 was much weaker in sub-Saharan Africa (46%) compared to other developing regions such as Latin America and the Caribbean (74%) and South Asia (82%) (World Bank, 2024a). After a strong decline between 1910 and 1980, within-country inequality on income dimensions has, in most cases, risen since 1980, particularly when measured in terms of the shares of income and wealth going to the top 1% (Chancel et al., 2022).

One should recognise, however, that overall world inequality in income has decreased since the 2000s – although it remains remarkably high. This results from within-country inequality stabilising between 2010 and 2020, while the decline in between-country inequality accelerated – partly due to rapid growth in many, including large, developing countries, and relatively poor economic growth in developed nations after 2008 (Chancel et al., 2022).

Many people who have been lifted out of poverty remain highly vulnerable to falling back into it. For example, the Covid-19 pandemic pushed more than 70 million more people into extreme poverty in 2020 (World Bank, 2022). At present, almost 700 million people live in extreme poverty,[9] including around 400 million in sub-Saharan Africa and 150 million in South Asia (World Bank Poverty and Inequality Platform, 2023). They are very vulnerable to both extreme weather events and longer-term processes such as desertification.

The combined impacts of the pandemic, geopolitical conflicts, and the crises of climate and environment have severely damaged prospects for delivery of the SDGs. Indeed, the turbulence and crises have threatened to reverse decades of advancement in various areas, pushing more populations into extreme poverty and exacerbating global inequalities (World Bank, 2024b). Food insecurity has worsened, with the number of people facing hunger rising in large measure as a result of the pandemic, war, and climate change (UN, 2023b). The pandemic has led to significant setbacks in education, with millions of children missing critical learning periods. The uncertainties at the time of writing, Spring 2025, about three months into the second presidency of Donald Trump, particularly about the Trump administration's approach to international economic relations, are creating still more turbulence.

Gender equality has also suffered; there was a surge in gender-based violence during Covid-19 lockdowns (UN Women, 2021). Progress towards gender equality is worryingly slow, with extrapolation of current trends indicating it would take centuries to achieve parity in several key areas such as closing gaps in legal protection and removing discriminatory laws (UN, 2023c). For example, nearly 50% of married women in the world do not have decision-making power over their reproductive health and rights, severely impairing their social and economic autonomy (UN, 2023c). Making progress on gender equality is not only of fundamental importance in its own right but also can create communities that are more resilient and can make better use of resources for development (Erman et al., 2021; UN Women, 2021).

Whilst we focus on the difficulties and crises we now face, we should not lose sight of the great advances in the decades since the Second World War. The growth of incomes has lifted billions out of poverty. People around the world tend to live longer and many deadly illnesses have been overcome or beaten back. The challenge now is to go still further in combating poverty and creating more healthy ways of living across the world, particularly where deprivation is most intense. That will require, and must go hand in hand with, tackling the climate and biodiversity crises and building growth and development that is sustainable, resilient, and equitable.

These are global crises, and the new paths must be created by the actions of nations across the world. We now turn to international interactions and agreements which can help foster and shape national and international action.

1.4 International agreements: the significance of Paris, COP21

The 2015 Paris Agreement, with its commitment to limiting temperature increases to well below 2 °C and to pursue efforts towards 1.5 °C, was a landmark. Along with the SDGs, it is a guiding light of what must be done to achieve sustainability. It brings a commitment towards net zero. It stands as a testament to what can be achieved when countries recognise a shared threat and new opportunities and are ready to adapt their domestic agendas towards a common global goal. See Figures 1.4 and 1.5 for the celebration of what at times seemed an unlikely outcome.[10]

To grasp why the adoption of the Paris Agreement represented such a huge and celebrated international advance, some insights on the difficulty of developing this kind of international agreement are offered in this subsection. How it happened helps us understand some basic principles of this and other agreements. It must be seen in the context of the history of international negotiations around these issues.

In negotiating these agreements there are three key dimensions.[11]

1. The legal strength of agreements: Stronger legal agreements mean countries are more likely to take their commitments more seriously and stick with them, even when governments change. However, trying to make agreements 'legally binding' can make some countries wary of losing control over their own policies, which can discourage them from joining or committing fully. In my view, it is a mistake to lay excessive stress on 'legally binding', if it is expressed in terms of invoking international sanctions on those who break agreements. It is implausible that such sanctions could ever be really strong and effective. Who would make the judicial assessments and who would do the enforcing? But, as I will describe, national courts and legal structures, based on domestic interpretations of the obligations arising from international agreements, such as the Paris Agreement, can be powerful.

2. How agreements are structured (architecture): Agreements could, in principle, be structured in a way that spells out exactly what each country must do (top-down), while others could let countries set their own goals (bottom-up). While bottom-up can encourage more countries to join by giving them more flexibility, it may result in less ambitious aggregate goals. Conversely, top-down can lead to higher goals but might discourage participation. The way top-down and bottom-up are combined matters greatly. The success of COP21 was founded on the way the overall temperature agreement, essentially a top-down but aggregate goal, was coupled with bottom-up country targets for emissions reductions set by countries themselves. And these were complemented by mechanisms for ratcheting-up targets over time, given that the sum total of emissions reductions from countries (from the bottom-up) was insufficient to deliver the aggregate (top-down) temperature target.

Figure 1.4: Moment of approval of the Paris Agreement

Note: (from left to right, front row) Laurence Tubiana, Christiana Figueres, Ban Ki-moon, Laurent Fabius, and François Hollande.
Source: UN Climate Change (2015a). Copyright 2015 UN Climate Change. Reproduced with permission.

Figure 1.5: Standing ovation for adoption of the Paris Agreement at COP21, 2015

Note: (from left to right, front row) The author, Al Gore and Ségolène Royal.
Source: UN Climate Change (2015b). Copyright 2015 UN Climate Change. Reproduced with permission.

3. Fair distribution of responsibilities: The UNFCCC introduced the principle of 'common but differentiated responsibilities' (CBDR) in the original agreement of 1992, recognising that not all countries are historically equally responsible for climate change or equally capable of action to tackle climate change. There is broad agreement on this principle, but its interpretation has changed through time, a matter addressed in Chapter 3. And there are differences in interpretation across countries. Further, as we will examine in Chapter 9, the mutual responsibilities include, at centre stage, international finance to support action in poorer countries.

Different countries and groups have different approaches to agreements and different emphases on what they regard as important or crucial. For example, in some contexts, the EU might push for greater formality of obligations, while the USA might prefer more flexibility and market-driven approaches.[12] The G77 (the main negotiation group of developing countries) and China emphasise their own interpretation of fairness (in relation to CBDR), stressing obligations arising from historical emissions and thus for wealthier countries to provide strong financial support.

Given these difficulties and the complexities and tensions of international politics, it is indeed remarkable how far the world has come: in creating the UNFCCC in 1992; in getting the Paris Agreement of 2015; in advances in subsequent COPs, particularly COP26, where ambition was strengthened and the private sector took the stage in a strong way; and in integrating climate into sustainable development, for example in multilateral development banks (MDBs). Collaboration and action are indeed possible. However, we must constantly emphasise that we have been going much too slowly. And there have been and will be setbacks. Indeed, as I write in Spring 2025, international collaboration has weakened, in particular as a result of Trump's second presidency.

From Rio to Paris: a brief history

The journey of formal agreements on climate action began with the UNFCCC, created in 1992. That in turn drew on the creation of the IPCC in 1988 and the Rio Earth Summit of 1992. Early on during the UNFCCC negotiations, there was a tension between the EU, which pushed for specific emission limits for countries, and the USA, which argued for a focus on national strategies and measures. This resulted in a hybrid system where all countries reported their climate policies (a bottom-up approach) while only developed nations aimed to reduce emissions to 1990 levels by 2000 (a top-down target) (Bodansky, 2011).[13]

As climate science advanced, coupled with direct experience of climate change and extreme events, the need for strong and urgent action became ever more apparent. The persistent and effective advocacy of small island states vulnerable to climate change helped keep a focus on the importance of continuing to make progress in the climate discussions. The Kyoto Protocol in 1997 took a more stringent stance with 'legally binding' targets for developed countries. But it ultimately struggled with participation. It also, over time, showed the fragility of the idea of 'internationally legally binding' in this context, where the term is understood as the attempted enforcement of the obligation to meet targets for a particular period. In principle, under Kyoto, a country that failed to meet its targets in a given period would have to add the extent to which it had missed its target in the given period to its target for the next period. But it had little strength. For example, when Canada undershot its target – and thus, in principle, faced stronger targets in the next period – it simply left the Kyoto Protocol.

The 2009 Copenhagen Conference (COP15) was expected to be a landmark event, but international tensions were severe and it ended with only a broad political statement, which was not formally adopted by the COP; it is known as the Copenhagen Accord. Despite its limitations, the Copenhagen Accord was an important advance in some key respects. It set a 'below 2 °C' global warming limit and called for financial support for climate action in developing nations. For the first time, China and other developing countries committed explicitly to emission cuts, alongside the world's developed economies, a move that broadened participation. However, only a limited group of countries drafted the Accord.

The Copenhagen Accord of COP15 in 2009 was taken forward into COP16 in Cancún in 2010 and was in large measure formally adopted there. Interestingly, the shared shock of the quarrelsome and chaotic atmosphere in Copenhagen created a desire for calmer discussion and finding agreement in 2010. Some also remarked, only partially light-heartedly, that the warmth of Cancún was more conducive to agreement than the cold of Copenhagen. Cancún also benefited from the strong leadership of Mexican President Felipe Calderón and Foreign Minister Patricia Espinosa.

The success of the Paris Agreement at COP21 in 2015 was founded on a flexible yet ambitious architecture. Nations would set their own climate action commitments, known as nationally determined contributions (NDCs), which they would enhance over time. This approach was a substantial innovation, as it combined national flexibility with a clear and strong collective goal and regular global reviews to assess progress and increase ambition. It married self-determined commitments with global objectives (Bodansky et al., 2017). It created a durable framework adaptable to the changing capacities and circumstances of nations, institutionalising an iterative process for shaping medium- and long-term climate action.

The transparency and accountability mechanisms within the Agreement build pressure for countries to intensify their efforts over time. It was a political success that was driven very skilfully by the Presidency of France and the leadership of the Secretariat of the UNFCCC. The top tier of French negotiating leadership was Laurent Fabius, President of the COP and Foreign Minister of France, and Laurence Tubiana, as France's Climate Change Ambassador and Special Representative to the conference, who led the French team. Christiana Figueres was Executive Secretary of the UNFCCC. All three showed great wisdom and strong leadership. I had the privilege of working closely with them and saw at first hand the importance of personal relationships, trust, and the right kind of coalitions. For example, Pete Betts,[14] who was lead negotiator for Europe (and the UK), had built strong relationships with other negotiators over many years and was a key figure in bringing people together. Very sadly, Pete died in 2023. The energies and persuasiveness of the group of small island states were very impressive and they played a key role. The account of Paris set out by Todd Stern, the USA's lead negotiator, is a valuable personal and analytical history (see Stern, 2024).[15]

Another key foundation for success at Paris was the increasing understanding that the transition could be a growth story. The Global Commission on the Economy and Climate (New Climate Economy), which I co-chaired with Felipe Calderón at that time, helped build the case.[16] Felipe and I had worked together on the successful COP16 in Cancún, 2010. The growth story had not really gained traction there in any strong way, but it had begun. It was indeed in the Stern Review in 2006. We both felt that in the drive to the next level of commitment it would have to be central. Andrew Steer (then head of the World Resources Institute [WRI]), Felipe, and I worked, with many friends and colleagues, to establish the Commission in 2013.

There were two other important factors contributing to the success of the Paris Agreement. One was the work of Lima (Peru) COP20 (impressively led by Manuel Pulgar-Vidal), which was in close harmony with those leading COP21. It was a strong two-year process. Another was the leadership of Presidents Xi Jinping and Barack Obama, who announced their NDCs jointly, in November 2014, one year ahead of Paris COP21.

Table 1.1: Characteristics of the Paris Agreement, 2015 (COP21)

Characteristics	Description
Global scope	Applies to all countries (195 members) while recognising their different capabilities
Progressive action	Expects stronger action over time, adaptable to national circumstances
Iterative process	Establishes a regular cycle for updating and enhancing NDCs
Transparency and accountability	Implements a framework to track progress and hold countries accountable
Legally binding	Creates legal obligations for countries to participate in a process to continually improve their climate action

Source: Author's elaboration drawing on Bodansky (2016, pp. 290–291) and Stern (2024).

That Obama–Xi leadership was in turn facilitated by the longstanding and close collaboration of John Kerry, Todd Stern, and Xie Zhenhua, the lead envoys and negotiators for the USA and China. Personal relationships and trust do matter and they are built over time and around shared goals.

Table 1.1 reflects the characteristics of the Paris Agreement. Paris was a landmark. A key element in reaching agreement was the recognition that whilst rich countries had been the main drivers of past emissions, emerging markets and developing countries (EMDCs) would be major drivers of future emissions. It was time to come together.

Contributions of the COPs

I have attended all the COPs since 2006, from COP12 in Nairobi to COP29 in Azerbaijan. Since the Paris Agreement of 2015, COPs have seen nations solidify their commitments into law, while the Agreement itself has catalysed rapid technological changes, especially in renewable energy. Countries have been ratcheting up their ambitions, with the shared goal of aligning national actions with global targets. Broadly speaking, some momentum has been maintained. There are two key factors at work driving this momentum: ever-growing strength of climate impacts and ever-stronger recognition of the potential of the growth story. However, when pointing to achievement, we must emphasise that progress has been far too slow. And we know that there will be setbacks along the way. Those setbacks include the withdrawal of the USA from the Paris Agreement by the first and second Trump administrations. Yet, the resilience of international commitment has been impressive, with the big majority of countries maintaining or increasing commitments since Paris, and many states and cities of the USA declaring that 'we are still in' during the first Trump presidency.

Finally, in the Paris Agreement, we have the introduction of the idea of 'net zero' emissions. The language used in the Paris Agreement was the balance between sources and sinks of GHGs (Article 4.1). The idea of net zero subsequently became very influential. Prior to Paris, many targets were expressed in terms of, for example, cutting emissions by 80% from 1990 to 2050 (see, e.g., Stern, 2007). As noted above, it is remarkable how many emitters thought they were in the remaining 20% and could get away with minimal action. If there are only a few places which are negative, then most will have to be around net zero emissions. Concepts which can be broadly understood – that net zero flows are necessary to stabilise emissions – and which can apply generally are key foundations of agreement.

Through time, the COPs have become much more than negotiating platforms for countries. City representatives, the private sector, and civil society come together too. COP26 in Glasgow in 2021 was particularly important in advancing the participation and commitment of the private sector and tightening overall temperature targets (on the back of the work of the IPCC in comparing 1.5 °C and 2 °C, see IPCC, 2018). It was led strongly by

Alok Sharma (now Lord Sharma) as president, and well supported by the UK government. That participation of the private sector and their emphasis on the investment opportunities was an indication of the increasing traction of the growth story.

COP26 was two years in the making because the Covid-19 pandemic prevented a COP in 2020. It was in Glasgow that Amar Bhattacharya and I first showed that delivering climate finance on the scale necessary would involve tripling finance from MDBs. That led to the establishment of the Independent High-Level Expert Group on Climate Finance (IHLEG), which produced reports for subsequent COPs 27, 28, 29, and 30. Mark Carney played an outstanding role in forming the Glasgow Financial Alliance for Net Zero (GFANZ), within which many private financial institutions committed to net zero.

It is in many ways remarkable how far the climate COPs have come. Since the successful Cancún COP16 in 2010, there have been real successes at Paris COP21 in 2015 and Glasgow COP26 in 2021. Interestingly, they were all two-year COPs in some shape or form and roughly, took place every five years. There are good signs that Brazil's COP30, due to be held in Belém in November 2025, and on which Brazil has been working for some time, will also carry some successes, including around finance and forests. There have been important steps forward in other COPs too, for example around a fund for loss and damage in Sharm el-Sheikh COP27, and the clarity of the Global Stocktake (showing how far action was behind what was necessary) in Dubai COP28, together with its explicit recognition – for the first time – of movement away from fossil fuels and commitments to improve energy efficiency and triple renewables. The COPs can make progress.

At the same time, the COPs have become large and lumbering affairs. They do bring all nations together, but with nearly 200 countries and a principle of unanimity, progress is difficult. In the future, in my view, the focus should be more on delivery. That will involve more participation of non-state actors such as MDBs. Particular tasks and responsibilities could be associated with particular groups, for example MDBs on finance. The Global Stocktake could be the responsibility of a group drawn from the IPCC. Private finance institutions, building on GFANZ and the creation of 'country platforms' (see Chapter 9), could work on how to bring investment possibilities into real investment-grade propositions. There are new ways forward building on, not rejecting, the COP system, which could help provide the pick-up in pace and contribute to the greater international solidarity needed.

1.5 Growth: received theories, change, and the new vision

To rise to the challenges of the climate change and biodiversity crises, to tackle poverty and to move forward on the SDGs, we must keep growing and developing. Progress on these dimensions requires resources and investment in capital in all its forms. At the same time, the core argument of this book is that investment for a sustainable future will itself drive growth, and that growth will be in a different form from the dirty, destructive models of the past. We shall also argue that this requires an analytical approach to growth which goes beyond standard theories and focuses on different drivers.

The new analytical approach is set out in Chapter 5. And we dispose there of two related mistaken assertions. First, that there is an inevitable trade-off between growth and environmental responsibility and second, that achieving sustainability requires stopping growth. I should underline that the new kind of growth emphasised and analysed in this book is not the one depicted in the long-run infinite-horizon models of traditional growth theory, with exogenous technical progress,[17] indefinite growth, and little structural change. Our story here is one of clean growth and structural transformation that must take place at speed over the next few years. We are talking about transformation and growth over the next two or three decades. This does not mean indefinite, eternal, infinite or 'forever' growth in the sense of the traditional growth model. In those models there is a portrayal of growth that goes on, at a given rate, say 2% per annum, forever. That means that with the constant growth rate the economy becomes larger indefinitely. Thus, in these models, given long enough, output would surpass any given level you may wish to specify. That really is a fairy tale, but it is how traditional growth models portray the long run.

To understand the difference from past growth theories, and how we must move beyond them, it is important to have a brief description of those theories in front of us. That is the first purpose of this section. And in doing so we emphasise that there is real substance and insight in received growth theories, notwithstanding the peculiarities of the infinite horizon. The second purpose of this section is to highlight the pivotal changes in structures and technologies, which are at work in the transformation and which must be recognised, built on, and integrated into the new growth story.

Received theories of growth

Understanding the conceptual foundations of the older approach is an essential part of the new growth story. Some of this modelling can get rather mathematical and technical and I promise that there will be very little formal economic or mathematical analysis in this book. There will, however, be quite a lot of economics.

Although the history of enquiring into growth is long, including in the classics of Smith, Ricardo, and Marx, we begin with theories emerging around the Second World War, associated particularly with Roy Harrod and Robert Solow. The conceptual approaches of Harrod and Solow were and remain very influential in shaping economic thinking. But in more recent years macro and growth modelling have been evolving to integrate industrial dynamics and environmental factors as essential components of future growth and development.

The Harrod model (1939) explains a country's growth rate with reference to (a) its level of saving, which in turn shapes the level of investment, and (b) the amount of investment necessary for producing an extra unit of output. In other words, the more you invest, the faster you grow, and the more productive the extra capital, the faster you grow. Whereas Harrod had a shorter-run focus, Solow (1956) was interested in the path of growth rates over time and the determinants of long-run steady growth. In simple terms, an economy grows if capital, labour,[18] or technology augment, or improve. However, in the models, capital accumulation does not by itself sustain growth indefinitely; it has diminishing returns. What leads to sustained long-term growth is technological progress, which allows for continuous and indefinite improvements in efficiency. Essentially, the Solow-type models in this sense envisage growth forever. However, both the Harrod and Solow models are silent on the causes of technical progress – this is why they are called exogenous growth models. It is important to note that the infinite horizon here is not meant to be taken literally. The idea of the long run here is a state of affairs that persists over a long period of time with a settled structure.

The understanding of technical progress and its relationship with past paths of investment and growth was powerfully influenced by Kenneth Arrow's (1962) theory of 'learning by doing'. In the 1980s and 1990s, Arrow's insights were used to link micro learning and macro growth, under the 'endogenous growth' strand of research.[19] Paul Romer (1986) put forth a model where technological advance is a by-product of capital accumulation. More endogenous growth models followed. Of particular importance is the work of Philippe Aghion and Peter Howitt (1992), who formalised Joseph Schumpeter's ideas of creative destruction.

Schumpeter's (1942) influential story is that dominant firms in markets will at some point have their market power overturned by firms employing new technologies and ideas, hence 'creative destruction'. Theories and models in the tradition of both Arrow and Schumpeter have real power in making technological advances endogenous or explained within the model itself. Some of these approaches to understanding and fostering technical progress will be further explored in Part II of this book, and are helpful in shaping ideas around the driving forces of the new growth story.

Whilst the approach to modelling the drivers of growth embodied in the Harrod and Solow models has long shaped economic analysis, perspectives on the processes of economic development, structural change, and associated policies have fluctuated and changed. Development economics moved from a focus on sectoral shifts and planning in the 1950s and 1960s to a focus, in the 1980s and 1990s, on market-driven policies and processes. There has been, more recently, what I regard as a much needed and welcome re-emphasis on structural and technological change in the context of increasing emphasis on sustainability.

In the early days of development economics, in the sense of the economics of developing countries, there was a strong focus on how changing sectoral structures were at the heart of growth and development, including through the growth of manufacturing or sectors where capital was important. In the 1950s, Arthur Lewis (1954, 1955) put the question of how savings and investment in poor countries could increase to drive growth, which was generally recognised to be at the core of the subject of economic development. He highlighted the movement out of low-productivity subsistence sectors as a driving force in increasing savings, investment, and overall productivity. In the 1940s and 1950s P.C. Mahalanobis, who was very influential in Indian planning, emphasised the importance of 'machines to make machines' in achieving rapid growth; this approach embodied a focus on sectoral priorities, particularly in relation to capital goods. Ragnar Nurkse (1953) argued for a 'big push' across all sectors given the importance of complementarity of activities and a potential shortage of effective demand. Paul Rosenstein-Rodan also advocated a 'big push' theory of development across all sectors together both to overcome fixed costs, thereby exploiting economies of scale, and to guarantee the availability of complementary inputs for each sector. There was also a strong Keynesian element in relation to aggregate demand generating aggregate supply. Albert Hirschman (1958) favoured creating 'unbalanced growth' to induce entrepreneurship and investment. The mechanism he advocated for this inducement was the creation of demand for key inputs in an economy for which domestic supply was not as yet on a substantial scale in order to create clear signals to foster entrepreneurship and investment. All of these approaches, although very different, embodied the idea of an active state and structural change in driving economic growth.

The necessary climate action will require and will generate powerful structural change in all the world's economies and will recast large parts of the world's economic geography (see Chapter 8). In some cases, it will create opportunities for leapfrogging, across the dirty stages of development, especially in EMDCs. To leverage and accelerate such change requires understanding both the processes of economic change and the conditions which enable them to happen. The foundations of Harrod/Solow, of the development economics of the early post-war period, and of the new approaches to Schumpeterian ideas, are all of real value in creating and understanding the new growth story. All these insights will contribute. But we have to go beyond – for more on this, see Chapter 5.

Changing structures

The developing world has been, and is, rising as a share of global economic activity. China's rapid industrial growth, which started in the late 1970s, has fundamentally reshaped global production and trade patterns, challenging traditional economic alliances and redefining the structure of world trade. India, Brazil, Indonesia, and many other countries, large and small, have seen real progress. We do indeed now have a multi-polar world.

Structural change has involved much more than the structure of the world's output: aside from a doubling of output, the next two or three decades will likely see a doubling of global infrastructure and a doubling of the urban population.

The majority of these increases will occur in developing countries, and they are all central to the new growth story. For example, the growth in the share of the population at working ages in the next two or three decades, which will occur in many developing countries, can become an opportunity for accelerated clean economic growth. The ranking of the most populous nations will likely change, as shown in Table 1.2, with the USA being surpassed by Pakistan, Nigeria, and the Democratic Republic of Congo in the second half of this century (United Nations Department of Social and Economic Affairs [UNDESA], 2024).

Table 1.2: Projected rankings of most populous countries in 2024, 2054, and 2100

	2024	2054	2100
1	India	India	India
2	China	China	China
3	USA	**Pakistan**	Pakistan
4	Indonesia	*USA*	**Nigeria**
5	Pakistan	**Nigeria**	**Dem. Rep. of Congo**
6	Nigeria	*Indonesia*	*USA*
7	Brazil	**Ethiopia**	Ethiopia
8	Bangladesh	**Dem. Rep. of Congo**	*Indonesia*
9	Russian Federation	*Bangladesh*	**United Rep. of Tanzania**
10	Ethiopia	*Brazil*	*Bangladesh*

Note: Country names set in bold are those that have increased in rank; country names set in italic are those that have fallen in rank. The others have retained the same ranking for two years.
Source: Author's elaboration adapted from UNDESA (2024).

In the face of rapidly changing birth rates, in different ways in different places, estimates of future populations have some uncertainty, but a peak of around 10 billion or a little more in the two or three decades following 2050 is seen as likely (UNDESA, 2024). However, the consequences of unmanaged climate change could potentially bring those numbers down in a dramatic and tragic way. It is striking that population forecasts rarely take account of the potential effects of climate change.

In the case of Africa, the population aged between 15 and 64 years old is expected to nearly double by 2050 (African Union Commission [AUC] and Organisation for Economic Co-operation and Development [OECD], 2024). Indeed, while most countries in the world grapple with an ageing population, it is likely that youth (aged 15–24) in Africa will represent 40% of the global total of young people by 2030 (UNDESA, 2015; World Economic Forum [WEF], 2024).[20] Absorbing this young population into the African continent's workforce is a huge development challenge. Failure to generate opportunities might create great social tensions. If the availability of attractive quality jobs does not grow, many highly skilled workers might decide to emigrate (AUC and OECD, 2024). In Asia, where an increasing number of young people are getting higher education, the types of jobs available do not match their skills, leading to rising graduate unemployment (International Labour Organization [ILO], 2024). It is in the developing world that the workforce of the future will grow. It must – and can – be sustainable growth that will provide them with the opportunities.

These increases in populations and changing structures of workforces in key developing countries will be accompanied by rapid urbanisation, with powerful demands for infrastructure (see Figure 1.6). Choices made on infrastructure and capital now will either lock us into high emissions or set us on a low-carbon growth path which can be sustainable, resilient, and inclusive. If the growth in capital, jobs, and infrastructure in these rapidly growing regions looks anything like the past, we cannot meet the goals of the Paris Agreement and will be headed for temperature increases closer to 3 °C than 2 °C, with devastating consequences. The new growth story embodies an investment imperative as well as an investment opportunity.

Figure 1.6: The critical need for decreased emissions over the next three decades

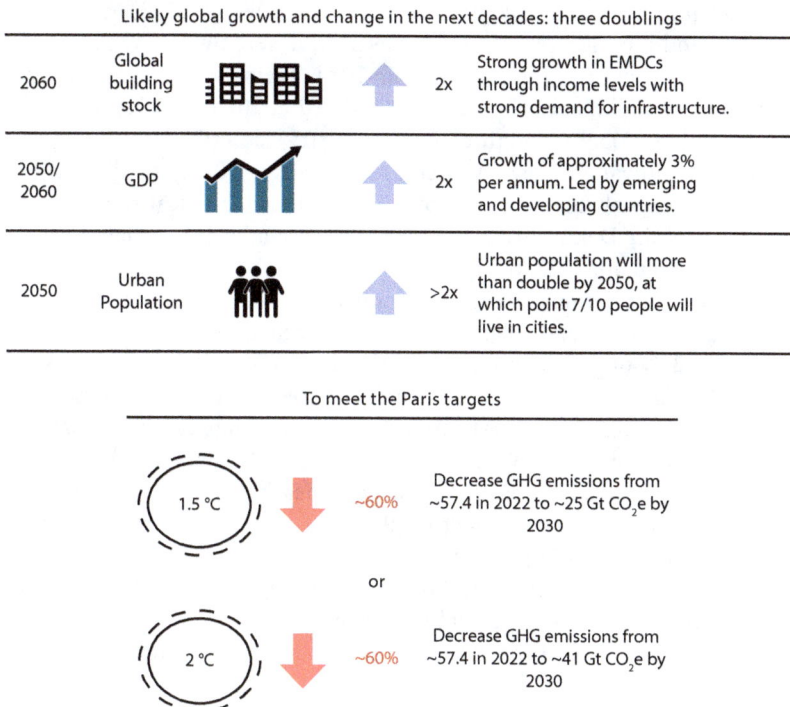

Likely global growth and change in the next decades: three doublings

2060	Global building stock		2x	Strong growth in EMDCs through income levels with strong demand for infrastructure.
2050/ 2060	GDP		2x	Growth of approximately 3% per annum. Led by emerging and developing countries.
2050	Urban Population		>2x	Urban population will more than double by 2050, at which point 7/10 people will live in cities.

To meet the Paris targets

1.5 °C	~60%	Decrease GHG emissions from ~57.4 in 2022 to ~25 Gt CO_2e by 2030
or		
2 °C	~60%	Decrease GHG emissions from ~57.4 in 2022 to ~41 Gt CO_2e by 2030

Note: GHG reduction needs estimates from UNEP (2023). Source: Author's elaboration.

Changing technologies

As mentioned throughout this chapter, and it is a core theme of the book as a whole, the world has seen extraordinary advances in green technologies over the last one or two decades, particularly around electricity generation and road transport. The clean is now cheaper than the dirty in sectors accounting for one-third of emissions, and that proportion is likely to grow rapidly in the next decade (Systemiq, 2020). These issues are central to the new growth story and are discussed in greater detail in the next part of this book, particularly in Chapter 5.

Fortunately, whilst the difficulties in creating the new growth story are substantial and might at times seem daunting, the timing for seeking transformational change is opportune in one important sense: there are changes in technologies that are affecting the global economy, especially around digitisation and artificial intelligence (AI). The development of AI is likely to accelerate powerfully into the 2030s, and it will surely play a crucial part in this new growth story. For example, there is huge potential for AI in helping reshape the crucial systems in cities, energy, transport, land, and water, which will be at the heart of the transition to the new economy (Stern et al., 2025; Stern and Romani, 2023, 2025; Zenghelis et al., 2024).

Further, AI can be critically important for economic development by driving innovation itself: AI technologies can automate activities to accelerate scientific and technological development (OECD, 2023). Sir Demis Hassabis and John Jumper of Google DeepMind were two of the three winners of the Nobel Prize for Chemistry in 2024. Their award was granted for creating AlphaFold2, an AI model that fulfilled in 2020 what Heiner Linke (Chair of the Nobel Committee for Chemistry) described as a '50-year-old-dream' (Royal Swedish Academy of Sciences, 2024): predicting the three-dimensional structure of proteins from their chemical sequence. In the words of Sir Demis, 'AI can deliver scientific advance at digital speed' (Stern and Romani, 2023, p. 5).[21]

In developing nations, access to foundational technologies such as the internet, computers, and smartphones is still limited though expanding. In 2022, only 25% of people in low-income countries used the internet (World Bank, 2024c). Such access is crucial to unlocking economic opportunities. For example, the availability of fast internet access can increase the probability that an individual is employed by up to 13.2%, raise total employment per firm by up to 22%, and enhance firm exports nearly four-fold (World Bank, 2024c). In agriculture, where self-employment is pervasive, small farmers with smartphones can improve productivity and profitability through guidance on weather, agriculture, techniques, and marketing. Digitisation, the internet, and AI will be of special importance in the green transformation given the importance of interconnectors, systems management, discovery, and innovation. And digitisation requires electricity.

1.6 Concluding remarks: towards sustainable development

The crises we now face have their origins in huge growth in both income per capita and life expectancy in the eight decades since the end of the Second World War. Those advances have been of fundamental importance to well-being across the world and billions have been lifted out of poverty. But that growth has been driven by energy from fossil fuels and by running down our natural capital, for example through deforestation, land degradation, and pollution of rivers and oceans. We now see that we must follow a very different approach to growth in the future. The last two or three decades have seen a deepening understanding of the processes at work that have led to the crises; to a deeper understanding that national and international objectives should reflect our concerns with sustainability, resilience, and equity; and, on these foundations, how we should act to deliver the new and much more attractive approach to growth and development. That is the purpose of this book. The book also reflects how my own ideas on growth and development, and how growth and development can be shaped and fostered, have deepened over the two decades since the Stern Review.

As the world pursues climate action and moves towards sustainability, the transformation of the global economy will be shaped by outcomes which follow from the decisions and circumstances of each nation. Many of the poorer nations have a tremendous opportunity to embark on a very different and better path than the destructive route taken by rich countries. The potential to leapfrog is of huge significance. But the cost and availability of capital will be crucial in the ability to take these opportunities. Understanding these differential challenges and opportunities (see Chapter 8) is vital for the analysis of international action and policy which will be a key focus of the book (see Chapter 9).

We can build a new approach to development that is sustainable, resilient, and equitable – for all countries and communities. The investments, innovations, and structural and systemic change necessary to tackle climate change and create the new growth story involve dynamic and strategic choices. The argument goes way beyond narrow, static cost–benefit analyses of pros and cons. The climate science sets a severe time constraint if we are to avoid devastating damage. Thus, we must invest, innovate, change, and collaborate with urgency and with scale.

Notes

[1] Vlasceanu et al. (2024) find, in their study with more than 59,400 participants across 63 countries, that the average belief in climate change is high (over 85%), with only modest variation among nations. This is a shift from the public confusion, uncertainty, and denial of the 2000s (Capstick et al., 2015).

² In Chapter 3, I shall comment in more detail on a number of issues of relevance to reflections on the Stern Review, including the failure of much of the literature to recognise the severity of key market failures. And it has, on the whole, given inadequate attention to the central importance of the dynamics of innovation, investment, and structural and systemic change. These are in addition to the problems noted here of much of the economics literature failing to take account of the potential magnitude of the risks.

³ It is embedded throughout the document (see UNFCCC, 1992). For example, Article 3 emphasises the need to balance the priorities of economic development and environmental goals, implying there is a trade-off in resource use.

⁴ The clean is already cheaper than the dirty across sectors responsible for a third of emissions, a proportion that will likely rise quickly (Systemiq, 2020).

⁵ GNP refers to the income of a nation whereas GDP refers to income generated within its borders.

⁶ In the words of Kuznets (1934, pp. 5–6): 'The valuable capacity of the human mind to simplify a complex situation in a compact characterization becomes dangerous when not controlled in terms of definitely stated criteria. With quantitative measurements especially, the definiteness of the result suggests, often misleadingly, a precision and simplicity in the outlines of the object measured. Measurements of national income are subject to this type of illusion and resulting abuse, especially since they deal with matters that are the centre of conflict of opposing social groups where the effectiveness of an argument is often contingent upon oversimplification.'

⁷ And two other UN Conventions: on biodiversity and combating desertification.

⁸ The preamble states: 'We are determined to end poverty and hunger in all their forms and dimensions, and to ensure that all human beings can fulfil their potential in dignity, equality, and a healthy environment' (UNGA, 2015, p. 2).

⁹ Living on less than US$2.15 per day (in 2017 purchasing power parity values) (the extreme poverty line).

¹⁰ The Agreement set the goal of 'holding the increase in the global average temperature to well below 2 °C above pre-industrial levels and pursuing efforts to limit the temperature increase to 1.5 °C above pre-industrial levels'. These limits on the rise of global temperature were set after intense negotiations with the aim of reducing climate change risks and impacts.

[11] Discussion based on Bodansky (2011, 2016).

[12] And the USA withdrew from the Paris Agreement during the first and second Trump presidencies.

[13] In the original UNFCCC agreement of 1992 there are broadly two categories of country: Annex 1, mainly the developed, and non-Annex 1, the rest.

[14] See his book *The Climate Diplomat: A Personal History of the COP Conferences*. Profile Books. 2025.

[15] Todd Stern is not a relation, but he is a friend.

[16] See, e.g., the reports of the New Climate Economy (2014, 2015, 2016, 2018, 2020) generated by the Global Commission on the Economy and Climate.

[17] Exogenous technical progress in this context means technical progress determined outside the model not by forces within the model. It is sometimes referred to as 'manna from heaven' in the sense that, as far as the model is concerned, it is progress that just appears from nowhere.

[18] Labour tends to be treated in these models as a constant, or growing at a fixed rate, whereas capital can be augmented by investing.

[19] Endogenous technical progress here means that technological advance is explained within the model – in this case via past experience as expressed through cumulated investment.

[20] Africa will be the major source of future population growth. While in 1960 Africa comprised around 10% for the world's population, in 2050 it is expected to be the home for around 30% (United Nations Economic Commission for Africa [UNECA], 2024). For example, the populations of nine nations, including Angola, Central African Republic, Democratic Republic of Congo, Niger, and Somalia, are expected to at least double in size between 2024 and 2054.

[21] This work is part of a much bigger story of extraordinary advances in biochemistry, with great potential to shape change. Indeed, the revolution that AI technology can bring is just beginning, with AlphaFold2 code having been made freely available. It has already been applied by millions of scientists to conduct their own work, from research seeking to develop resilient crops to new vaccines (O'Donnell, 2024).

2. Some fundamentals: science and nature

Tackling the immense risks of the climate and biodiversity crises by investing, on scale and with urgency, delivers a new path for growth and development. This is much more attractive than the dirty, destructive models of the past. That is the story of this book. It was sketched in the introduction and Chapter 1, which provided an outline of our argument. Chapters 2, 3, and 4 complete Part I of the book with some fundamentals underlying the analysis (Chapters 2 and 3) and, in the context of the crises and a changing world, an agenda for action with urgency and scale (Chapter 4). The response to the challenges will then be discussed in the rest of the book. Parts II and III set out the kind of investment and structural and technological change that is necessary, show how this can lead to the new growth story, examine how that investment can be fostered and financed, and analyse the challenges in international collaboration. All this must be built on an understanding of the nature of the challenges and of the analytical principles that are required for a serious response. It is these fundamental underpinnings that constitute the subject of this chapter and the following, with Chapter 2 examining science and nature, and Chapter 3, economics, politics, and society; the natural sciences followed by the social sciences.

In this chapter, we discuss the challenges arising from the science, which describes the nature of the risks, dictates the sense of urgency, determines the scale of response, and sets the timelines. These are not political deadlines arising from discussion and negotiation, but real timelines established in terms of consequences and dangers of delay. Timetables for cutting emissions are in this sense crucially different from other development goals. If a health goal is achieved later than we would wish, that is troubling. But if emissions goals are achieved late, then all other goals are potentially undermined. This observation does not imply some intrinsic ranking, in terms of importance or values, climate above health, but follows from the practical realities of the problem.

Climate change drives biodiversity loss and biodiversity loss drives climate change. Our ecosystems underpin industries such as agriculture, fisheries, and forestry, and provide services that support life itself, like clean water and air. In many cases, this natural capital cannot be substituted, and once destroyed is very difficult or impossible to restore. These issues go beyond climate change. But the climate and biodiversity crises are interwoven. Thus, whilst they are not the same, we have to tackle them in an integrated way.

The forces at work in climate change identified by the science, starting with Fourier's basic insight 200 years ago on the trapping of heat, through to our current understanding of the immense risks that climate change involves, are set out in Section 2.1. The potential increase of those risks in the next few decades, together with the likelihood of overshooting the Paris targets and the dangers of tipping points, are examined in Section 2.2. The necessity of large-scale adaptation given the presence and intensification of the effects of climate change are analysed in Section 2.3. The interweaving of the twin crises of climate and biodiversity is discussed in Section 2.4.

Clarity on the foundations of the science and biodiversity (this chapter) and foundations in politics, history, economics, and ethics (next chapter), is essential for the analysis and delivery of effective action. The story embodied in these summaries of relevant natural science and social science is in large measure one of obstacles, difficulties, and great risk. Some of the more positive forces at work are set out in Chapter 4.

2.1 The forces and the dangers

The science of climate change is founded in basic physics: greenhouse gases (GHGs) trap heat. It is, in its essence, clear, simple, and long-standing. The predictions of the theory have matched the temperature increases subsequently observed. And the consequences of the warming have become ever more troubling. The climate science and its history are summarised briefly here.

Two hundred years of science

The science of climate change has grown more sophisticated in recent decades, but the basic scientific story of the existence of a warming effect is two centuries old. The cause, in the form of GHGs, was identified in the middle of the 19th century and the first substantive prediction of the quantitative effects of rising concentrations of GHGs was at the end of that century, around 130 years ago. As the science has advanced and concentrations and temperatures have risen, the data have validated the theory in a comprehensive and striking way. The theory came very early in the story, and the data, subsequently arising from measurement and experience, have been consistent with that theory. In the 1980s scientific consensus on anthropogenic global warming stated to emerge. By 2021, the Intergovernmental Panel on Climate Change (IPCC) noted, in its Sixth Assessment Report (AR6):

> The first IPCC report, released in 1990, concluded that human caused climate change would soon become evident, but could not yet confirm that it was already happening. Today, evidence is overwhelming that the climate has indeed changed since the pre-industrial era and that human activities are the principal cause of that change. (IPCC, 2021, p. 244)

Figure 2.1 shows in simple form the basic history of the key contributions to the atmospheric physics of climate change.

Figure 2.1: Climate change science timeline

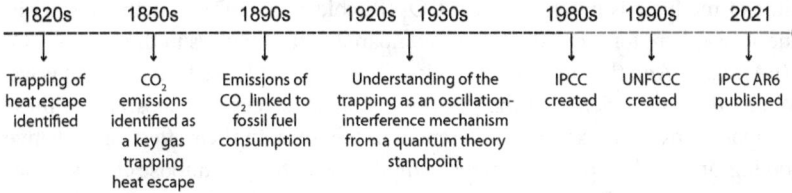

1820s	1850s	1890s	1920s 1930s	1980s	1990s	2021
Trapping of heat escape identified	CO_2 emissions identified as a key gas trapping heat escape	Emissions of CO_2 linked to fossil fuel consumption	Understanding of the trapping as an oscillation-interference mechanism from a quantum theory standpoint	IPCC created	UNFCCC created	IPCC AR6 published

Note: The red line denotes the period from which scientific consensus on anthropogenic global warming emerged.
Source: Author's elaboration.

The research conducted in 1824 by French mathematician Joseph Fourier was the start of the science of climate change. He concluded, by analysing inflows and outflows of energy, and the surface temperature of the Earth, that there must be something trapping the escape of heat from the atmosphere. Without that, his calculations showed the Earth would be much cooler than it is. In the 1850s, Eunice Newton Foote in the USA, and a few years later, physicist John Tyndall in Ireland, independently showed that the key gas trapping the escape of heat was carbon dioxide.[1] In 1896 Svante Arrhenius, a Swedish scientist, linked the emissions of CO_2 to fossil fuel consumption, and did the first calculation of the likely effect of doubling the concentration of CO_2 in the atmosphere on temperature.[2] The effect on temperature of a doubling is known as 'climate sensitivity'. It is generally taken to be in the range of 2 °C to 5 °C (IPCC, 2021). The AR6 best estimate is 3 °C. It is usually estimated from modelling or from historical data. So far, since the second half of the 19th century, the concentration of CO_2 has gone up by around 50% with the big majority of the increase in the last half-century (see the discussion on carbon budgets below).

During the 1920s and 1930s, an explanation of the mechanism of the trapping of heat – in terms of certain gases, the 'greenhouse gases', interfering with the escape of infra-red energy from the atmosphere – was developed using quantum theory (two of the pioneers were Niels Bjerrum and David Dennison). By the early 1930s a fairly complete quantum picture of CO_2 infra-red spectroscopy was in place. More recently, in 2024, the physicist Keith Shine and collaborators diagnosed more precisely the impact of 'Fermi resonance' on CO_2 radiative forces – further advancing our understanding of carbon dioxide's heat-trapping effect at the molecular level.[3]

Different GHGs make different contributions to the greenhouse effect in terms of their trapping of heat; they have different effects on 'radiative forcing', meaning the amount of energy from the sun that enters our atmosphere, relative to the amount leaving (MIT, 2024a). To measure the total effect of different GHGs, and given that CO_2 is the most important, overall flows and stocks are often converted to CO_2e or CO_2 equivalent, to give an aggregate flow or stock of GHGs, where the conversion is done via their relative radiative forcing. One kilogram of methane, for example, traps about 84 times more heat than one kilogram of CO_2 averaged over 20 years – but 28 times as much over 100 years since the atmosphere lifetime of methane is much less than CO_2 (Ritchie et al., 2024).[4] Methane stays in the atmosphere for around 12 years, compared with centuries in the case of CO_2 (IEA, 2022). And they are absorbed by oceans, land, and forests in different ways. Whilst the basic science of the greenhouse effect is very simple, some of these complications – for example, differences in lifetimes, in their effect on radiative forcing, and in absorption – can make modelling of the greenhouse effects somewhat complicated. But the scientific essentials of the underlying story are straightforward, in terms of the trapping of heat, how the trapping works, and which gases are responsible. This is simple, robust, and basic science.

The warnings from scientists of the potential magnitude of global warming started to build around the 1980s. In June 1988, US physicist James Hansen testified before the US Congress about global warming and the responsibility of humans. The IPCC was created in the same year – a UN body that brings scientists together from across the world to assess the science associated with climate change (see Chapter 1). These assessments, intended to be used by policymakers, synthesise the state of the scientific knowledge on climate change for that purpose. Indeed, the IPCC does not produce new research; rather, it collects and evaluates existing conclusions, identifying which findings are widely supported by the scientific community. The reports undergo lengthy reviewing processes that include several reviewing stages, aiming to build objectivity in assessments. As such, they are institutionally cautious.

Though the IPCC has always been careful to make statements in a way which takes account of uncertainty, its warnings on the strength of the effects and the confidence in the conclusions have grown ever stronger. In 2021 it declared, 'it is unequivocal that human influence has warmed the atmosphere, ocean and land' (IPCC, 2021, p. 4).

Given that the science is so old, and well founded, it is natural to ask why we as a world have been so slow to accept it, let alone move on it. Part of the answer lies in inertia and resistance to change. But part also lies in the work of vested interests to undermine the science. These have been the 'merchants of doubt' (see Oreskes and Conway, 2010). According to Banerjee (2015), as far back as the mid- to late-1970s, fossil fuel giants such as Exxon and Shell were conducting internal research that corroborated the effects of carbon emissions on climate change. Yet major oil companies spent decades working to create confusion and doubt around the issue.

Ever more worrying impacts: the lived experience

The history of IPCC reports has been one of estimates of the potential severity of impacts being revised upwards. Indeed, some of the effects of climate change are coming through at greater speed, scale, and intensity than anticipated in each earlier report. The lived experience and the detail of the accumulating scientific evidence point to ever-increasing risks at a scale that should be of deep concern. However, the economics, as we will see in Section 3.2, has struggled to keep up with these conclusions. As noted in Chapter 1, when the Stern Review was published in 2006, the prospects were already very worrying; they look far worse now. And, as noted, the IPCC's processes make it institutionally cautious.

The levels of CO_2 concentrations have been rising strongly since the second half of the 19th century – the usual benchmark in terms of the beginning of the Industrial Revolution. The flows of emissions have risen very strongly since the Second World War (see Figure 2.2).

Figure 2.2: Annual CO_2 territorial emissions from fossil fuels and industry since 1850

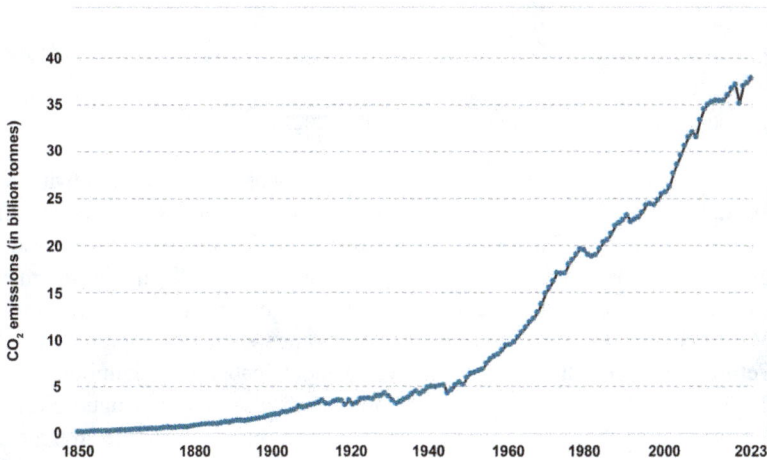

Note: Measured in tonnes. Land-use change is not included.
Source: Author's elaboration using data from Our World in Data (2024).

Concentrations are now at levels the world has not seen for at least 800,000 years, and likely three million years (see Figure 2.3). Whilst flows of emissions are net positive, concentrations will go on rising. The usual baseline for calculating the rise in average global surface temperature is the period 1850–1900, when the concentration of CO_2 was between 280 and 290 parts per million (ppm). In 2025, it was about 425 ppm. Given that the concentration was around 310 ppm in 1950, it is clear that the powerful increases have been driven by the rapid and carbon-intensive growth of the last seven or eight decades.[5] At present, global annual GHG emissions are still rising, with a 1.3% increase to 57.1 billion tonnes of carbon-dioxide-equivalent recorded between 2022 and 2023 (UNEP, 2024b).

Rising concentrations mean rising temperatures and climate impacts. The world has witnessed increasing sea levels and desertification (in some places). Such effects are termed 'slow onset'. And we have seen an increased frequency and intensity of some extreme weather events in many parts of the world – heavy rainfall, heatwaves, and droughts.

Figure 2.3: The rise in CO$_2$ concentrations in the last 800,000 years

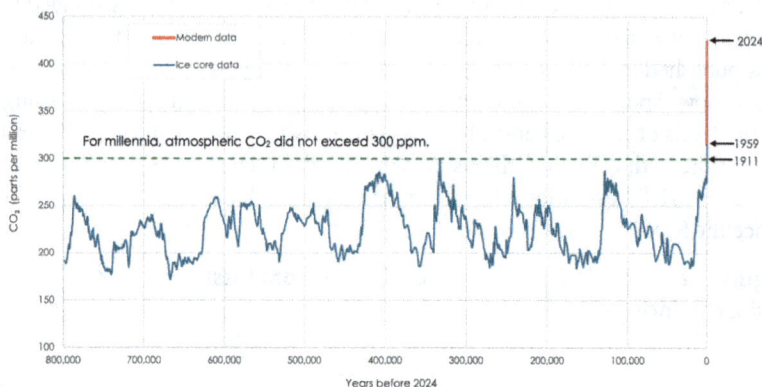

Source: Author's elaboration, using ice core data from Etheridge et al. (2001) and Luthi et al. (2008), and modern data from Lan and Keeling (2025)..

The physical impacts of climate change directly affect economic activity, for example in agriculture, and destroy property and livelihoods, for example, in floods and wildfires. Figure 2.4 shows some examples, including the devastating impacts that occurred in Los Angeles, California, in 2024 due to the massive Palisades and Eaton wildfires, which engulfed more than 16,000 structures. Many lives, homes, and businesses were lost. Some were insured, some were not. Future insurance will be very expensive or unobtainable (Madakumbura et al., 2025; WWA, 2025a). Figure 2.4 also shows the effects of fires in August 2024 in São Paulo state, Brazil, which occurred following a period of extreme drought, as temperatures in the country rose to between 5–10 °C higher than the 1991–2020 average. This contributed to the rapid spread of thousands of fires in the state of São Paulo, destroying many thousands of hectares of crops and causing severe health problems from smoke (Fowle, 2024). In the town of Cobargo, Australia, on New Year's Eve 2019, a devastating fire destroyed much of the town and took many lives. A month earlier, John Mullins, the former chief of the New South Wales Fire and Rescue, had told *The Guardian* (in Zhou, 2019):

> Just a 1 °C temperature rise has meant the extremes are far more extreme, and it is placing lives at risk, including firefighters. Climate change has supercharged the bushfire problem.

Figure 2.5 shows further examples, all of which occurred in 2024: during October, in Valencia in Spain, nearly a year's worth of rain fell in just eight hours, taking with it the lives of more than 200 people. Around that time Hurricane Milton hit Florida, USA, after rapidly intensifying into one of the strongest Atlantic hurricanes on record, and causing long-lasting impacts on both communities and their economies.[6] Nearer the end of 2024, Super Typhoon Man-yi hit the Philippines, causing a terrifying storm surge several metres high

Figure 2.4: Extreme event examples: massive fires

Note: top: A firefighter watches the flames from the Palisades Fire burning homes on the Pacific Coast Highway amid a powerful windstorm on 8 January 2025 in Los Angeles, California, bottom: Destroyed crops at a sugarcane farm following wildfires in Ribeirão Preto, São Paulo state, Brazil, on Tuesday 27 August 2024. An unprecedented outbreak of fires that hit sugar-cane fields in the world's top exporter Brazil over the weekend is set to impact global supply of sweetener and elevate prices.
Source: top: Apu Gomes/Getty Images via Getty Images, bottom: Victor Moriyama/Bloomberg via Getty Images.

Figure 2.5: Climate-related flooding examples from 2024

Note: top: Scenes of devastation in the streets of Benetusser after the passing of the flood; army, firefighters, police and volunteers help to normalise the situation, Valencia, Spain on 11 November 2024, bottom: An aerial view shows submerged homes at a village in Ilagan, Isabela province on 18 November, 2024, due to continuous heavy rains from Super Typhoon Man-yi. Filipinos cleared fallen trees and repaired damaged houses on 18 November after the sixth major storm to batter the Philippines in a month smashed flimsy buildings, knocked out power and claimed at least one life.
Source: top: David Carbajo/NurPhoto via Getty Images, bottom: VILLAMOR VISAYA/AFP via Getty Images.

Figure 2.6: The complex interconnections of climate change impacts

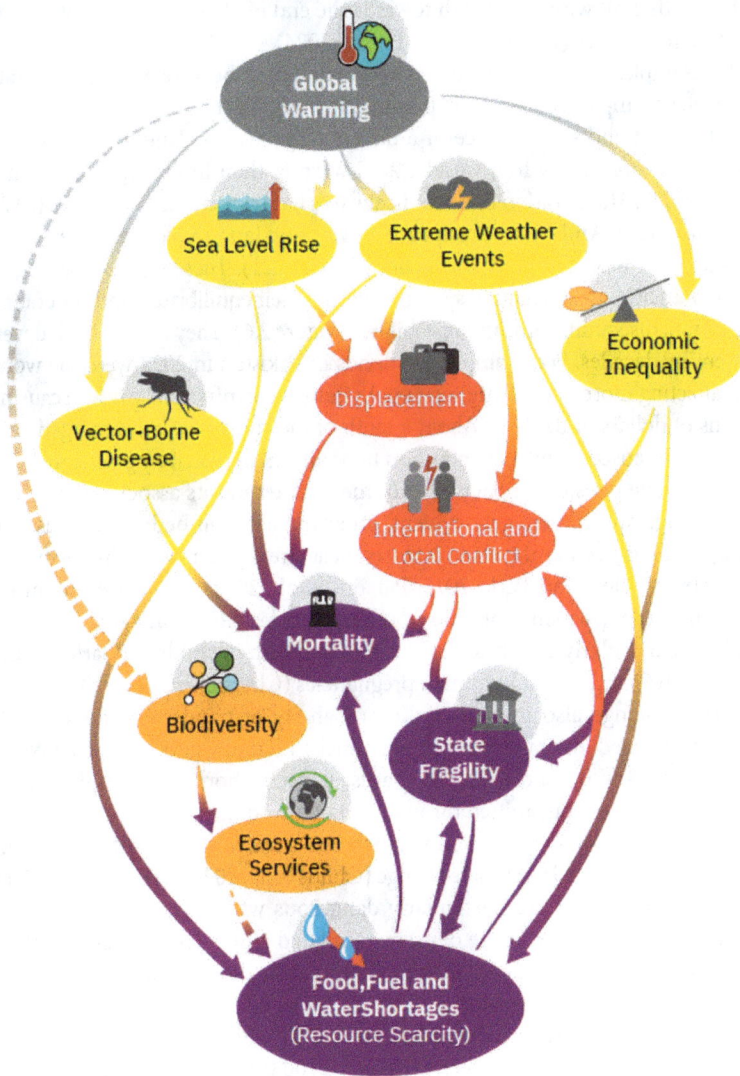

Note: This is a causal loop diagram, in which a complete line represents a positive polarity (e.g., amplifying feedback; not necessarily positive in a normative sense) and a dotted line denotes a negative polarity (meaning a dampening feedback).
Source: Figure 3 in Kemp et al. (2022, p.7), p. 7. Copyright 2022 The Authors, CC BY 4.0.

on the shores of Catanduanes, requiring 650,000 people to flee (see Figure 2.5 for an image of a submerged village in Ilagan, in the Isabela province). The six typhoons that hit within a month towards the end of 2024 in that country took more than 160 lives (Agence France-Presse, 2024).

The examples chosen here are very recent and have been selected to illustrate the immense impacts on lives and livelihoods. Such effects are becoming increasingly frequent and severe. Once-in-a-decade heatwaves on land are now nearly three times more likely in a world 1.2 °C warmer than in pre-industrial times (IPCC, 2021). The duration of ocean heatwaves has tripled since the 1940s (Marcos et al., 2025). And heavy precipitation events on land that occurred every 10 years are now more frequent and wetter (IPCC, 2021). These impacts propagate throughout human and natural systems, altering their equilibrium and generating disruption, disequilibrium, and instabilities (Figure 2.6). They can set back development by decades. For example, the floods in Pakistan in 2022 were the worst ever, affecting more than 30 million people, destroying infrastructure and causing billions of dollars of damage. About a tenth of the country was submerged. The floods were caused by intense monsoon rains and exacerbated by glacial melt.

While some physical impacts of extreme weather events associated with climate change are immediately evident, other impacts can be less obvious. For example, it has been found that increased heat stress lowers work intensity and productivity (Heal and Park, 2015; Parsons et al., 2022),[7] decreases cognitive performance (Seppänen et al., 2006), and impairs learning (Park et al., 2021). It can have particularly significant effects on the health of children (Carleton and Hsiang, 2016) and so too in human pregnancies (Chersich et al., 2020).

Climate change also influences the probability of pandemics, for example through changed interactions amongst birds, animals, and humans (World Bank, 2024). And it is already causing a rise in vector-borne diseases, for example dengue, a virus transmitted through certain mosquito species. At present, in parts of Asia and South, Central, and North America, nearly one-fifth of dengue cases can be attributed to climate change (Childs et al., 2025) (see also Table 2.1).

High temperatures are particularly dangerous when combined with high humidity. For example, in the densely populated North China plain, vital for food production, projections indicate that, mainly due to climate change but also the effect of irrigation for agriculture, wet bulb temperatures (a combination of temperature and humidity) of 35 °C, which can be fatal after six hours, could be reached between 2070 and 2100 under a very high emissions ratio (Kang and Eltahir, 2018). That is a projection which is particularly worrying for the health of farmers, who work outside without air conditioning. Such effects are also likely in the Indo-Gangetic plain, where occurrences of extreme heat are increasing. Population densities in the North China and Indo-Gangetic plain are high. These are the homes of hundreds of millions of people. Their environments could become uninhabitable in the second half of the century. Population movements could be immense.

Those working inside are not always protected from the effects of heatwaves. Brownouts in electricity systems due to increased demand for cooling during hot weather generate serious costs, with cascading impacts across urban systems (Aivalioti, 2015; Zuo et al., 2015). Many energy systems across the world are not prepared for the high energy demands arising from temperature peaks. For example, in March 2025, massive blackouts occurred during an extreme heatwave in Buenos Aires, Argentina,[8] that affected hundreds of thousands of users, causing disruptions across the city (including due to non-functioning traffic lights) (WWA, 2025b; Xinhua, 2025).

Over 80% of international trade volume is carried by sea (UNCTAD, 2021), and climate change threatens to disrupt the functioning of waterways, canals, and seaports. For example, in 2023, the Panama Canal authorities had to impose restrictions on ships' depth – meaning they could transport fewer goods – due to drought conditions lowering water levels (Reuters, 2023). Similar examples can be found around the world. Small island states and landlocked countries with a limited number of trade routes are particularly vulnerable to such disruptions from sea level rise, droughts, and floods.

The interconnectivity of our world means that climate change impacts can ripple across economies, disrupting the lives of many far and wide. An illustrative example is the 2011 floods in Thailand, which left key manufacturing hubs for the automotive and electronic industries submerged in water, causing economic losses of US$47 billion, equivalent to 12% of Thailand's GDP (World Bank, 2018). This led to a global shortage of hard drives and delayed car production for months, which increased prices and affected businesses internationally.

These examples are offered to illustrate the nature and intensity of the impacts of climate change which are already occurring. At the aggregate level, the economic losses associated with extreme weather, climate, and water events are estimated to have increased eight-fold from the 1970s to the 2010s (WMO, 2023). Between 1991 and 2024, global temperature has risen by around 0.2 °C each decade (WMO, 2025). There is no doubt that these losses will increase sharply as temperatures continue to rise.

The challenges that arise from the climate science dictate both the urgency of and scale at which action must occur – to avoid the immense risks of a changing climate. The argument for acting in this way is greatly strengthened when we recognise that climate action will bring the multiple benefits of driving economic growth, protecting and creating resources, and fostering a healthier environment. Climate action must be at the heart of and a priority in each nation's sustainable development story – both to avoid catastrophic outcomes and reap the opportunities that come from moving to a more sustainable, resilient, and equitable world.

2.2 Risks, urgency, overshooting, tipping points, and carbon budgets

Under current policies, temperatures are headed towards an increase of close to 3 °C by 2100 (Climate Action Tracker, 2023) and even higher next century. Whole areas of the world might become uninhabitable as a result of inunda-

Figure 2.7: Projected temperature increases up to 2100

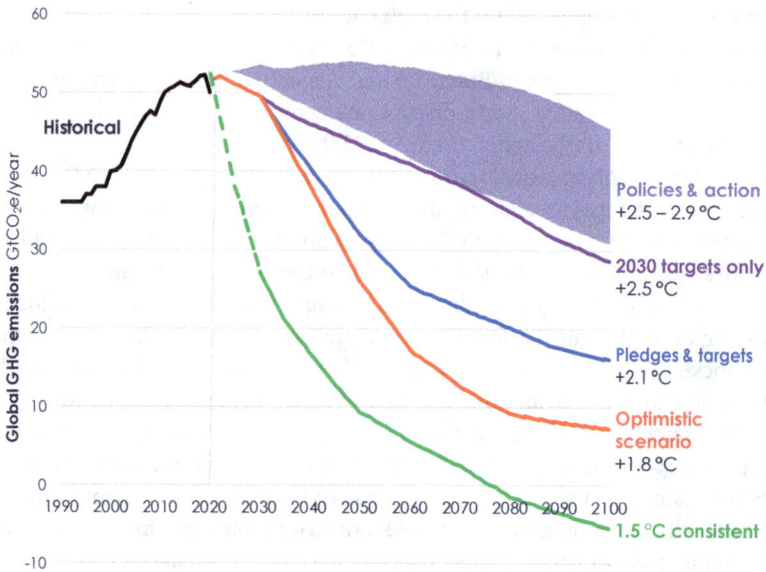

Note: GHG emissions are measured in gigatonnes (Gt). The policies and action scenario is based on current policies. The 2030 targets-only scenario is based on 2030 NDC targets. The pledges-and-targets scenario is based on 2030 NDC targets submitted and binding long-term targets. The optimistic scenario is the best-case scenario assuming the full implementation of announced targets, including net zero targets, long-term strategies, and NDCs. Source: Author's elaboration using Climate Action Tracker (November 2023) data.

tion, desertification, extreme events, or extreme heat. Growth and development for many would be set in reverse. Such temperatures could lead to large-scale movements of people, likely hundreds of millions or billions, potentially leading to major and persistent conflicts (Brown, 2008; Clement et al., 2021; Institute for Economics and Peace, 2020). These risks, our proximity to tipping points, and the exhaustion of the 'carbon budgets' (see below) demonstrate the acute urgency of action.

It is now clear that we will pass 1.5 °C in the next few years. Thus, if we maintain that as a target, there will be 'overshooting' in the sense that we will pass it and, over time, will have to return. Figure 2.7 illustrates some possible paths, the lowest of which would allow for some overshooting of 1.5 °C. Whether we follow a path that takes us closer to 1.5 °C or closer to 3 °C depends on actions on emissions, by countries and communities across the world, delivery of country commitments under the Paris Agreement, expressed, as noted in Chapter 1, as nationally determined contributions (NDCs).

Prior to the Paris Agreement in 2015, the then current policies looked as if they would take us closer to 4 °C by 2100, so there has been progress. But far too little to get us to the Paris target of well below 2 °C. Nevertheless, if all the net zero pledges and targets were to be delivered, the outcome would be close to 2 °C. That would still be dangerous, but much less than the immensely dangerous 4 °C that was likely before Paris and the highly dangerous 3 °C to which we are headed on current policies.

To retain the 1.5 °C target, there will have to be a period of negative emissions. And given the increasing dangers of tipping points beyond 1.5 °C (see the next subsection), it does indeed make sense to plan for this. If emissions become net negative, for example as a result of carbon dioxide removal, then concentrations and temperatures will fall.[9]

Exceeding 1.5 °C would generate powerful impacts across the world. At 1.5 °C, the annual longest heatwave duration in the world's largest cities would average around 16 days, and at 3 °C nearly a month (24 days).[10] At 3 °C it is likely that more than one-sixth of the biggest cities in the world, where more than 300 million people live, would be exposed to at least one heatwave with a one-month duration (or longer) annually (Wong et al., 2024). Europe, a region relatively more prepared for climate change than many developing nations, already saw almost 50,000 excess deaths attributed to heat in 2023, at global warming of around 1.2 °C (Gallo et al., 2024).

As shown in Table 2.1, hotter days may mean more days with optimal temperatures for disease-carrying mosquitos: arboviruses are expected to become more widespread.[11] The incidence of dengue is already rising strongly (WHO, 2024). Cities in low-income nations would be disproportionately affected, being expected to experience over 70 peak arbovirus days each year at 1.5 °C, and nearly 90 peak arbovirus days each year at 3 °C (Wong et al., 2024).

In the last 6,000 years, human beings have farmed, worked, lived, and thrived mostly in areas within a narrow band of mean annual temperature (MAT) of around 11 °C to 15 °C, with a relatively smaller proportion in hotter areas corresponding to around 20 °C to 25 °C (Xu et al., 2020). However, research indicates that, if our emissions are kept high, by 2070:

> 3.5 billion people will be exposed to MAT ≥29.0°C, a situation found in the present climate only in 0.8% of the global land surface, mostly concentrated in the Sahara, but in 2070 projected to cover 19% of the global land. (Xu et al., 2020, paragraph 6)

This exposure can result in increased morbidity and mortality, with many of the expected 3.5 billion people living in those areas possibly having to move, a number which represents 30% of the projected global population (Figure 2.8).

Predicting numbers of people who will be forced to move due to climate change, both internally and across borders, is not straightforward.[12] Nevertheless, the IPCC, back in the early 1990s, already had enough evidence to warn the world that the 'gravest effects of climate change may be those on human migration as millions are displaced by shoreline erosion, coastal flooding and severe drought' (IPCC, 1992, p. 103). The IPCC states in its latest report (AR6) that it has high confidence that climate hazards are a 'growing driver of

Table 2.1: Examples of climate change impacts in the world's largest cities at 1.5 °C vs 3 °C

Change from 1.5 °C to 3 °C	Annual longest heatwave duration in the % of days from 1.5 °C to 3 °C	Annual number of heatwaves in the % of heatwave count from 1.5 °C to 3 °C	Annual arboviruses peak transmission days in the number of peak transmission days
East Asia and Pacific	▲ 66%	▲ 30%	▲ 6.9 days
South Asia	▲ 58%	▲ 30%	▼ –3.8 days
Sub-Saharan Africa	▲ 58%	▲ 56%	▲ 25.2 days
Middle East and North Africa	▲ 60%	▲ 13%	▼ –0.7 days
Latin America and the Caribbean	▲ 25%	▲ 48%	▲ 12.8 days
North America	▲ 34%	▲ 23%	▲ 5.9 days
Europe and Central Asia	▲ 36%	▲ 8%	▲ 3 days

Note: Heatwaves are defined as three or more consecutive days on which the maximum temperature is equal to or exceeds the local 90th percentile of daily maximum temperature over a 40-year period (1980–2019). The development, survival, reproduction, and biting rates, and the rate at which mosquitoes acquire and transmit viruses are affected by temperature (Mordecai et al., 2022). Annual arboviruses peak transmission days are those that reach the average temperature optimal for the vector mosquito species *Aedes aegypti* and Ae. *albopictus*.
Source: Author's elaboration drawing on data from Wong et al. (2024).

Figure 2.8: Expansion of extremely hot regions in a high-emissions scenario

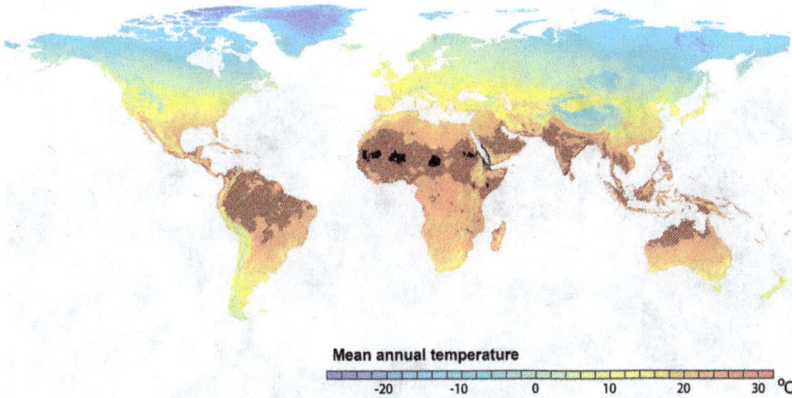

Mean annual temperature

Note: In the current climate, MATs >29 °C are restricted to the small dark areas in the Sahara region. In 2070, such conditions are projected to occur throughout the shaded area following the RCP8.5 scenario. Absent migration, that area would be home to 3.5 billion people in 2070 following the SSP3 scenario of demographic development. Background colours represent the current MATs.
Source: Figure 3 in Xu et al. (2020, p. 11,352). Copyright 2020 The Authors, CC BY-NC-ND 4.0.

involuntary migration and displacement' (Cissé et al., 2022, p. 1044). It is the poorest households who are particularly at risk when areas start to become uninhabitable because many will lack the resources to move (Birkmann et al., 2022). The IPCC also notes it has high confidence in climate hazards being a 'contributing factor to violent conflict' (Cissé et al., 2022, p. 1044).[13]

With global warming of around 1.5 °C, West Africa and the Sahel could see an estimated 40% reduction of the area suitable for maize production (Birkmann et al., 2022). Globally, it has been estimated that there could be a loss of 7–10% of rangeland stock at an increase of 2 °C. Food security risks in many regions will increase strongly as global warming advances (Hoegh-Guldberg et al., 2019).

The overall risks from passing 1.5 °C are illustrated by the following estimates concerned with heat and floods. Lenton et al. (2023) conclude that decreasing global warming from 2.7 °C to 1.5 °C could reduce the number of people exposed to MAT above 29 °C by around five times. And on floods, the IPCC highlights that with 'every additional one degree Celsius of warming, the global risks of involuntary displacement due to flood events are projected to rise by approximately 50%' (Cissé et al., 2022, p. 1046).

The effects of global warming on biodiversity could be catastrophic, as illustrated by the experience of Pantanal, in Brazil, the largest tropical wetland in the world, and immensely rich in biodiversity. In 2024, it experienced dry and windy conditions that drove devastating wildfires which were found by scientists to be '40% more intense due to climate change' (WWA, 2024). If warming increased to 2 °C, the weather conditions that lead to these huge wildfires would be twice as likely to occur than at present (see Figure 2.9).

Figure 2.9: Recent fires and expected recurrence of associated conditions in the Brazilian Pantanal

How often should we expect similar June fire weather conditions in the Brazilian Pantanal?

Before climate change – 1850	Today with 1.2°C warming – 2024	Future with 2°C warming – Around 2060 under current policies
1 in 161-year event	1 in 35-year event	1 in 18-year event

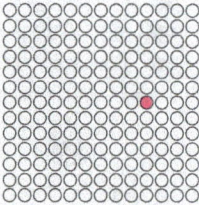

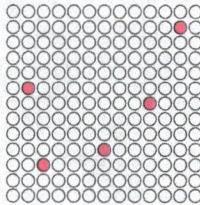

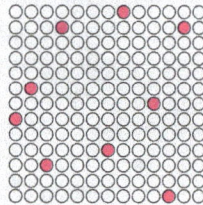

World Weather Attribution

Note: top: A National Security Force vehicle is travelling along a road bordered by areas scorched by forest fires that have been devastating the southern Pantanal in Mato Grosso do Sul for a month in Corumba, Brazil, on 2 July, 2024.
Source: top: Gustavo Basso/NurPhoto via Getty Images, bottom: World Weather Attribution (2024). Copyright 2024 World Weather Attribution. Reproduced with permission.

Tipping points: dynamic, irreversible, and destructive processes

We may be dangerously near tipping points, which could unleash unstable feedback loops and rapid and destructive change. A tipping point is where a small alteration in a system can cause abrupt changes because of amplifying feedback processes. Some of these changes may be very difficult, or impossible, to reverse. Further, triggering one tipping point could trigger another, potentially setting off further dangerous dynamic instabilities. Many of the consequences are difficult to predict but potentially catastrophic.

Figure 2.10: Climate tipping elements and their sensitivity to global warming

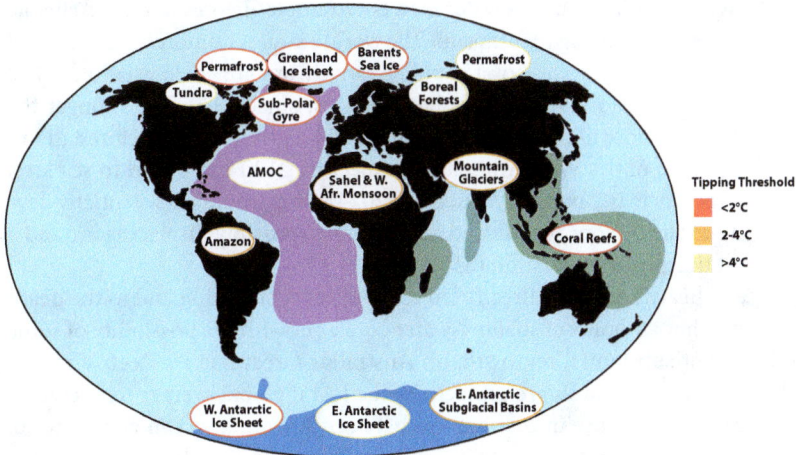

Note: Tipping elements are categorised as cryosphere (blue), biosphere (green), or circulation (purple). Colours of labels denote temperature thresholds categorised into three levels of global warming above pre-industrial (key on the right), with darker red indicating lower temperature thresholds (greater urgency). Permafrost appears twice as some parts are prone to abrupt thaw (at lower temperatures) and some (organic-rich Yedoma) to self-propelling collapse (at higher temperatures).
Source: Adapted from Figure 1 in Lenton et al. (2024, p. 5). Copyright 2024 The Authors, CC 4.0.

The IPCC warned of tipping points in its third Assessment Report (AR) more than two decades ago (IPCC, 2001). During that period, these 'large-scale discontinuities' in the climate system were considered probable only if global warming surpassed 5 °C compared to pre-industrial levels. Over successive IPCC reports, estimated trigger temperatures have repeatedly been revised downwards. It is alarming that we may already be near critical tipping points at current levels of global warming (Lenton et al., 2019). The most recent IPCC Special Reports (IPCC, 2018, 2019) suggest that some tipping points are likely to be triggered between 1.5 and 2 °C of warming. This is hugely worrying given that 1.5 °C could be reached within the next decade. We may be close to abrupt and irreversible change: the urgency for action is intense.

This subsection elaborates briefly on five potential tipping points and their possible implications. Figure 2.10 indicates that a number of major systems are at risk of being tipped at temperatures below 2 °C. Recent studies suggest that five major global tipping points could be triggered, possibly within a decade (e.g., Lenton et al., 2023b). The five major tipping points we may be near to crossing, and that we explain briefly below are: (1) the demise of the West Antarctic ice sheet; (2) the die-off of tropical coral reefs; (3) the irreversible thaw of the Arctic permafrost; (4) the slowing of an ocean current known as the North Atlantic subpolar gyre (SPG); and (5) the loss of the Greenland ice sheet.

There are other tipping points which could occur as a combination of biodiversity loss and climate change. They are potentially of immense importance and might occur in the next few decades. For example, deforestation and climate change are destabilizing the Amazon, the world's largest rainforest. Estimates of where an Amazon tipping point could lie range from 40% deforestation to just 20% forest-cover loss. About 17% has been lost since 1970. The stronger the warming, the more likely the system collapse for a given level of deforestation.

The demise of the West Antarctic ice sheet by 2100 could lead to sea level rises of as much as 2 metres (Moseman, 2024). This would be potentially devastating to the billions who live in coastal areas, possibly displacing around a tenth of the global population (SMC, 2024).

Ocean heatwaves have already led to mass coral bleaching and to the death of significant amounts of shallow-water coral, but with the possibility of some subsequent substantial regrowth, on Australia's Great Barrier Reef. Virtually all (99%) tropical corals are projected to be lost if global average temperature rises by 2 °C, because of interactions between warming, ocean acidification, and pollution. Warm water coral reefs are particularly vulnerable to global warming and are crucial elements in biodiversity. They are home to one-quarter of marine biodiversity and provide critical services to approximately 330 million people, who depend directly on reefs for protection from storm surges, and as sources of food and livelihood (Cooley et al., 2023; WWF, 2024).

Permafrost, which is ground that remains frozen, is primarily found in the Arctic, a region that is warming three times faster compared to the global average. The warming is causing the permafrost to begin thawing, a process likely to be irreversible in the foreseeable future. As a result, future generations will face the ongoing challenge of offsetting these emissions. When permafrost thaws, it releases CO_2 and methane, with methane being a GHG that is around 30 times more potent per kg than CO_2 over a 100-year period. Current models suggest that permafrost is already contributing to global warming, and at current emission levels, the Arctic is heading towards contributing similar levels of CO_2 and methane as a large industrialised country (International Cryosphere Climate Initiative, 2021; MIT, 2024b). By 2100, some estimates suggest that additional warming caused by thawing of the permafrost could be as high as 0.7 °C.

The SPG is a system of circulating currents located south of Greenland that is linked to the main Atlantic Meridional Overturning Circulation (AMOC) – a central current that works as the main heat transport engine in the Atlantic and one of the biggest on Earth (Swingedouw et al., 2021). The current known as the Gulf Stream in the UK is part of the AMOC. The slowing of the SPG could cause regional cooling, for example, affecting agriculture in Europe and North America, and changing potentially dangerously, extreme weather event patterns, including storms. The threshold for the SPG collapse appears to be around 1.8 °C (1.1 to 3.8 °C). This could happen within a decade or so.

The SPG collapse could also accelerate the weakening of the AMOC and, though this would take more time, the consequences would be catastrophic and unpredictable (International Cryosphere Climate Initiative, 2021; Rahmstorf,

2024). Such an event would result in substantial cooling of the northern hemisphere and warming of the southern. It would disrupt the Amazon rainforest ecosystem, possibly leading to cascading tipping points, which were described above in this section. Sea levels would rise quickly in areas such as the East Coast of the USA, and the severity of storms would intensify (MIT, 2024c; van Westen et al., 2024).

The Greenland ice sheet is melting increasingly rapidly. If it were to melt completely, it could raise sea levels by 7 metres over thousands of years. Some models suggest that at 1.5 °C of global warming, which will likely occur within the next decade, the Greenland ice sheet may already be doomed. If it passes such a threshold, 'as the elevation of the ice sheet lowers, it melts further, exposing the surface to ever-warmer air' (Lenton et al., 2019, p. 592). Whilst the full effect might take place over thousands of years, substantial impacts could appear this century (Aschwanden et al., 2019; Gautier, 2023).

The science is not yet definitive on which systems have tipping points, how close we are to crossing them, and what the impacts would be if we do cross them. But it can, and has, identified great risks and it is possible that impacts could be still more catastrophic than currently understood. The risks around these, and other, tipping points send a powerful message that delaying action is extremely dangerous; potentially existential for many.

Carbon budgets, urgency, overshooting, and negative emissions

A carbon budget describes the total amount of future emissions that could be consistent with staying below a given temperature limit (e.g., 1.5 °C or 2 °C) with a specified probability. It is a useful way to express the scale and urgency of action, and it embodies the idea that it is the path and accumulation of emissions over time that determines concentrations of GHGs and thus warming. It is often related to a historical budget starting at a much earlier date, to illustrate how much we have already consumed, and how much (or little) remains, as shown in Figure 2.11. If the 'total carbon budget' is exceeded and no additional actions to remove atmospheric CO_2 are pursued, the result is higher global warming than the specified limit or target.[14]

As we saw in Chapter 1, and earlier in this chapter, our past polluting model of growth powered by fossil fuels has already raised concentrations of CO_2 to 50% above levels of the second half of the 19th century. Budgets for 1.5 °C are close to or already exhausted and there are very limited budgets for 2 °C. It took 107 years to use up the first quarter of the carbon budget for 1.5 °C that existed in 1850. However, only 33 years were needed to exhaust the second quarter, and just 21 years for the third quarter. The remaining carbon budget in 2024 to stay under 1.5 °C is 150 $GtCO_2$ and from 2025 for a 50% probability to stay under 1.5 °C it is 130 $GtCO_2$ (Forster et al., 2025). This will be depleted in the next few years, given that the current annual rate of CO_2 emissions is around 40 Gt. Further, the remaining carbon budget from 2025 for a 50% probability to remain under 2 °C is 1050 $GtCO_2$, which will be exhausted within around three decades at the current emissions rate (Forster et al., 2025).[15]

Figure 2.11: What are carbon budgets?

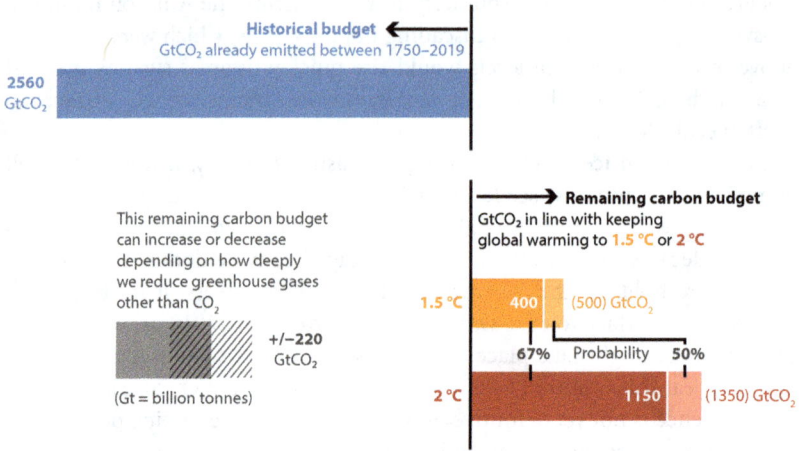

Note: Historical cumulative carbon dioxide (CO_2) emissions determine to a large degree how much the world has warmed to date, while the remaining carbon budget indicates how much CO_2 could still be emitted while keeping warming below specific temperature thresholds. Several factors limit the precision with which the remaining carbon budget can be estimated. Therefore, estimates need to specify the probability with which they aim at limiting warming to the intended target level (e.g., limiting warming to 1.5 °C with a 67% probability). Diagram refers to budgets as calculated in 2020. A carbon budget, from 2025 for a 50% probability of staying below 1.5 °C would be around 130 $GtCO_2$, and for 2 °C would be around 1050 $GtCO_2$ (Forster et al., 2025).
Source: FAQ 5.4, Figure 1 in Canadell et al. (2021), p. 778. Copyright 2023 IPCC. Reproduced with permission. Font and colours adjusted.

The chance to limit global warming to 1.5 °C without overshooting has gone, and limiting to well below 2 °C without overshoot, as in the Paris Agreement, will be very difficult. Risks escalate with the degree and duration of the overshoot, increasing the likelihood of hitting tipping points that could push systems into states of no return. It is important therefore to keep the degree of overshooting as small as possible and the period as short as possible.

The first priority is to bring down emissions rapidly so that the problem of overshooting does not become unmanageable. Nevertheless, it is now clear that strong action to remove GHGs will be necessary. The role of negative emissions is twofold: they can offset stubborn residual emissions, such as from cattle, and they can enable net negative emissions, crucial for reversing temperature overshoots. In 2018, the IPCC Special Report on global warming of 1.5 °C above pre-industrial levels showed that carbon dioxide removal (CDR) is essential in all pathways that meet the Paris Agreement's temperature goals (IPCC, 2018). Methods and costs for CDR[16] are examined in Chapter 9, together with geoengineering and associated issues and challenges.

Figure 2.12: CO_2 emissions from 1850 to 2022

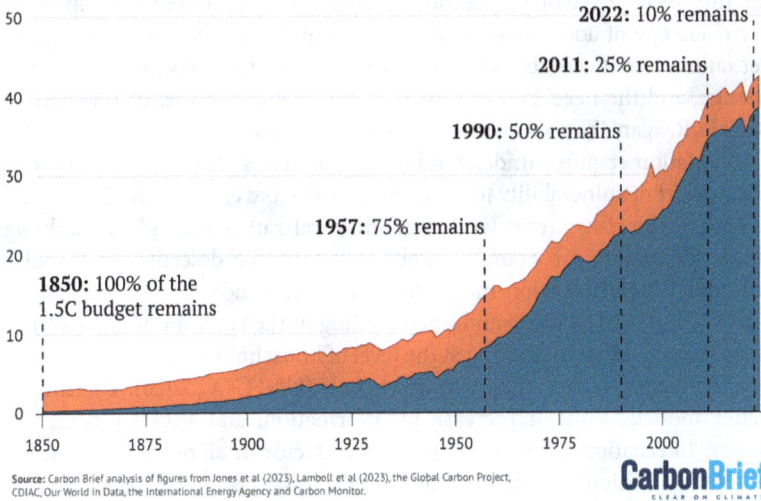

Early CO_2 emissions were driven mainly by deforestation, not fossil fuels

Annual global CO_2 emissions from land and fossil fuels, billion tonnes

Source: Carbon Brief analysis of figures from Jones et al (2023), Lamboll et al (2023), the Global Carbon Project, CDIAC, Our World in Data, the International Energy Agency and Carbon Monitor.

CarbonBrief
CLEAR ON CLIMATE

Note: Annual global CO_2 emissions from fossil fuels and cement (dark blue) as well as from land use, land-use change, and forestry (red), 1850–2023, billions of tonnes. The figure uses a different baseline of 1850 for the start of the carbon budget compared with a baseline of 1750 for Figure 2.11.
Source: Evans and Viisainen (2023). Copyright 2023 Carbon Brief. Reproduced with permission.

Geoengineering, particularly solar radiation management (SRM),[17] presents a set of supplementary measures that can temporarily, if maintained, offset some of the warming effects of GHGs. However, these methods introduce a whole slew of new risks and uncertainties. They do not address the root cause of climate change – the accumulation of GHGs – and their effects last only as long as their deployment. Further, they do not reduce CO_2 in the atmosphere which itself can have direct damaging effects, including acidification of the oceans. And governance could present real problems as single nations, institutions, or even individuals could pursue such methods unilaterally, notwithstanding any collateral damage. Lastly, it is possible that the adoption of these methods could diminish the motivation for reducing emissions.

As we brace for a future with overshoot, negative emissions technologies such as CDR and carbon capture, usage, and storage (CCUS) emerge as indispensable tools. Their judicious deployment, coupled with an unwavering commitment to emissions reduction, should, in my view, form a key element of our path to a stable climate. Geoengineering would likely be hugely problematic, in my view, carrying great risks. But given that overshooting is likely to be a major problem, these risks from geoengineering require deeper understanding.

2.3: Adaptation, hazards, vulnerability, and development

Mitigation is about avoiding the unmanageable; adaptation is about managing the unavoidable. The faster and stronger action on emissions is, the less the future challenge of adaptation. However, given past hesitation on action, the challenge of adaptation is already daunting. As GHG emissions are not yet falling to net zero, but are actually still rising, the risks posed by climate change, and the need to adapt, will get much more severe. In the words of Ronald Reagan: 'You ain't seen nothin' yet.'[18]

Adaptation requires understanding climate risks. These risks come from hazards, from vulnerability to the hazards, and from exposure to them (IPCC, 2022b).[19] The risks are in large measure local rather than global, although some adaptation, for example, around flooding or desertification, will be regional. This localisation means that the science needs to be as precise as possible if it is to facilitate an understanding of the risks and how to adapt to them. Sound decision-making at the level of household, factory or farm needs much greater precision than is available now (UNFCCC, 2020). This involves better models, a major increase in information, and stronger computing power. International collaboration is a key factor in all of these. Further, as adaptation is highly context specific, support for good adaptation interventions requires understanding the specific difficulties, needs and opportunities of communities. That understanding will in large measure come from the communities themselves.

Effective adaptation involves not only dealing with existing threats but also building resilience to future threats. That requires forward-looking planning and long-term capacity building. In essence, resilience means developing the capacity to prepare, respond, and recover from climate impacts. For example, in the case of risks of flooding, this could involve improving drainage, early warning, and repairing or re-building. That requires a strategic approach to strengthening the social, human, natural, physical and financial capacities, and infrastructure of communities (Mehryar, 2022).

Whilst adaptation is mainly designed to reduce climate risks it can generate multiple development benefits beyond the reduction of those risks even if climate change impacts are less severe than expected. For example, strengthening energy and irrigation systems plays a critical role in adaptation and has powerful development benefits in terms of productivity beyond those associated with reducing climate risks (Birkmann et al., 2022; Hammill and McGray, 2018).

Overall, adaptation and the building of resilience are basic economics and common sense. Indeed, it makes no sense to build houses and infrastructure or to follow cropping patterns as if the climate is unchanged when we know it has changed and is changing. It is in this way that adaptation and development are intrinsically linked: building effective adaptive capacity is good development, and good development must incorporate adaptation and resilience.

There are many adaptation measures that can be profitable and attract private investment. Improving guttering and water resistance in a building in the face of more heavy rainfall would be an example. However, there are also many adaptation investments that are not financially attractive to private sector investors, particularly where there are difficulties in capturing returns in one place. Adaptation investment in flood control across a city or region, for example, can deliver great public benefits in avoided losses in that city or region, but not necessarily direct financial returns to a private investor.[20]

Benefits from adaptation measures do not come only from avoiding losses. For example, mangrove forests provide more than US$80 billion annually in avoided losses from coastal flooding (Global Commission on Adaptation, 2019) (see Figures 2.13 and 2.14), but also generate additional economic, social and environmental benefits, from fisheries and tourism (e.g., watching Bengal tigers in the Sundarbans in India and Bangladesh). And they are very effective at capturing carbon. In a completely different context, urban development in East London was enabled by the flood protection provided by the Thames Barrier (Global Commission on Adaptation, 2019). In the case of LoGIC in Bangladesh (see Box 2.1), help with changing cropping patterns in the face of salinity, together with a resilience fund, generated profits which financed expansion in agricultural activity.

Box 2.1: The economics of adaptation: a farming example

The Local Government Initiative on Climate Change (LoGIC) is a multi-donor initiative working in Bangladesh to support the most vulnerable 500,000 households in hard-to-reach areas. One of their success stories took place in Tildanga, a rural village in Khulna. Due to their land being exposed to increased salinity because of high tides and cyclones exacerbated by climate change, crop production was being lost. The LoGIC project, under the leadership of local authorities and cooperatives, intervened by providing a Community Resilience Fund (CRF). The CRF, which is aimed at supporting vulnerable women, introduces new technologies and provides capacity-building to apply them; as well as financial risk support for investments. For example, CRF's support enabled local farmers to invest in sunflower crops, which are more saline resistant than other crops, substituting them for previous, more vulnerable, crops. This generated enough profits to expand cultivation and achieve a stable income source. The provision of knowledge, technology, financial capacity, and risk financing are all key actions to enable local communities to try and test novel adaptation alternatives. Information, investment, technology, and finance are key.

Source: UNDP (2021).

Adaptation and development

Planting mangroves (Box 2.2) and restoring degraded land are important examples of development, mitigation, and adaptation coming together in the same investments. So too public transport and decentralised solar power. There are many more. This will be discussed in Chapters 5 and 6. It is a conceptual and practical mistake to insist that a climate-related investment is either mitigation or adaptation. It is so often both, and usually also development. Like mitigation, adaptation involves investment in physical, human, social, and natural capital. Climate action through investment – and the powerful returns from that investment – is a constant message running through this book.

Box 2.2: The economics of adaptation: a mangrove example

Mangroves are tropical and sub-tropical shrubs or trees which play a central role in local livelihoods by supporting fisheries and providing raw materials, coastal protection, and carbon dioxide storage services. Since 2021, the government of Indonesia has recognised the important role that mangroves play by paying local people to grow them. This policy not only supports local communities through monetary return for their work but has led to increased fish and crab farm outputs, thereby boosting local profits. In Indonesia, the coastal protection services provided by mangroves are especially vital in areas like Java or Lombok, where many properties are at risk of coastal flooding, erosion, and storm damage. In certain high-value areas of Indonesia, annual mangrove coastal protection benefits surpass US$10,000 per hectare annually – see Figure 2.13 for an example of how mangroves provide protection. These substantial benefits are not unique to Indonesia; they can be found worldwide but are particularly critical in areas where socioeconomic activities are concentrated along the coast, such as in tropical islands like Jamaica. There, a significant portion of the economy is based in coastal areas, making mangroves an essential aspect of the nation's protection of built capital: they protect 50% of residential capital stock, 14% of industrial facilities, and 35% of service capital stock (e.g., health, education, and public service buildings). Total avoided damages to infrastructure due to mangroves are significant (Figure 2.14).

Source: Castano Isaza et al., 2019; World Bank, 2021; World Bank, 2023.

Figure 2.13: Behind mangroves surge is decreased, alleviating floods

Mangroves give protection against storm surges

With mangroves

Without mangroves

Source: Author's elaboration drawing on The Nature Conservancy (2024).

Figure 2.14: Current flood risk and annual expected benefits from mangroves for flood risk reduction in Jamaica: avoided property damages

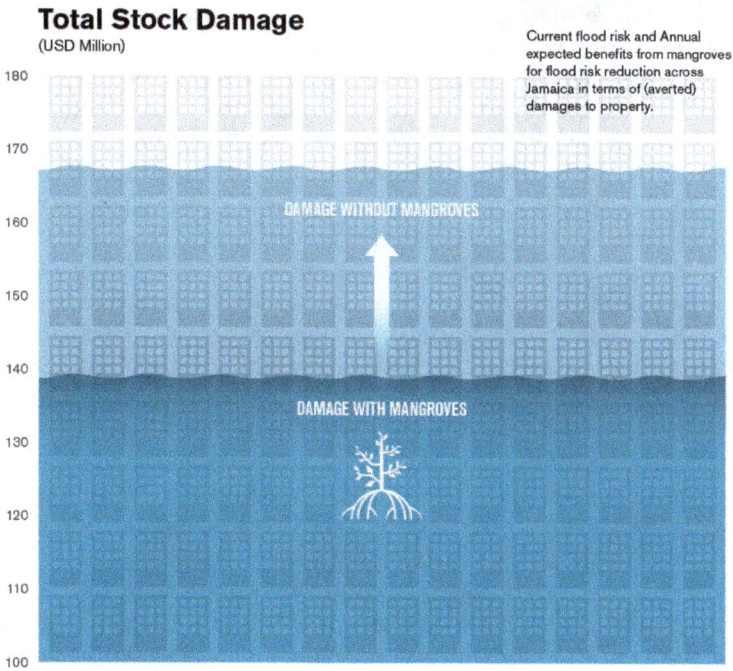

Total Stock Damage
(USD Million)

Current flood risk and Annual expected benefits from mangroves for flood risk reduction across Jamaica in terms of (averted) damages to property.

DAMAGE WITHOUT MANGROVES

DAMAGE WITH MANGROVES

Source: Figure 15 in Castano Isaza et al. (2019, p.56). Copyright 2019 World Bank. Reproduced with permission.

Climate hazards tend to hit hardest those nations and countries that are the poorest. Between 1970 and 2021, least developed countries (LDCs) have faced weather-, climate-, and water-related hazards with several disasters causing economic damages up to nearly 30% of a country's GDP (WMO, 2023). For small island developing states, the financial impact of disasters has at times surpassed 100% of a country's GDP. In contrast, developed nations have not experienced any such disasters where losses exceed 3.5% of a country's GDP, despite these countries accounting for over 60% of the reported financial losses from disasters globally. In addition, 90% of reported deaths from such disasters in the last 50 years took place in developing countries (WMO, 2023).

Developing nations suffer disproportionally from the effects of climate change, relative to developed nations. This is not only due to their geographical and climatic location and conditions but also their limited capacity to invest to adapt and build resilience to changing environmental realities, and their reliance on climate-dependent sectors. And in some cases, their ability to respond is weakened by a history of conflict and weak governance. Climate change can trigger social tensions and conflicts in areas receiving migration flows because of increased competition over land, jobs, and/or public service access (Hendrix et al., 2023; Sharifi et al., 2021). The effects of climate change act as aggravators of existing sociopolitical challenges. Climate change is a dangerous threat multiplier which deepens pre-existing problems, as observed in many drought-affected regions in the African continent, where climate change, migration, and conflict intertwine. Such is the case of Darfur, Sudan, where extended droughts and desertification mixed with land control disputes and low institutional capacity have contributed to violent long-term disputes (Henrico and Doboš, 2024; Oxfam, 2014).

In 2019, eight of the 10 countries most affected by extreme weather were low- or lower-middle-income countries, with half of these being LDCs (Eckstein et al., 2021). Between 2010 and 2020, human mortality from floods, droughts, and storms was 15 times higher in highly vulnerable regions than in low-vulnerability ones (IPCC, 2022a). We must recognise that these numbers and examples represent past losses over a period when global temperature increases were around 1 °C. Far worse is coming if we head towards 2 °C or beyond.

This pattern of vulnerability is present across various developing nations in Africa, Asia, and Latin America and the Caribbean. A study by Zhang et al. (2024) found that compound drought–heatwave events (when both extreme events occur simultaneously) between 1981 and 2020 have significantly increased in close to 60% of global land areas, home to nearly four billion people. The majority of those affected reside in Africa (40%), East Asia (33%), Latin America (8%), and the Middle East (4%). One study found that, from 2000 to 2015, the proportion of the global population exposed to floods had already increased by 20–24% and that 2030 climate change projections suggest that this proportion will keep increasing (Tellman et al., 2021). Nearly 90% of the flood-exposed people in the world live in low- and middle-income countries. Also, nearly half of the 170 million individuals in extreme poverty (living on less than US$1.90 per day) and threatened by high flood risk live in sub-Saharan Africa (Rentschler et al., 2022). Asia and the Pacific are particularly at risk from rising sea levels, being home to 70% of the global population susceptible to this threat (UNDP, 2024). These events can wipe out infrastructure, destroy buildings, lives and livelihoods, and set development back decades (WEF, 2024). We have already remarked on the devastating floods in Pakistan in 2022 (and see Climate Overshoot Commission, 2023).

Climate change intensifies existing inequalities across many dimensions, including income, gender, social class, race, ethnicity, disability, age, and migrant status. It places some at higher risk from climate impacts than others, especially when these identities intersect. For example, women and girls often face heightened risks due to their roles and responsibilities in many societies. They often have responsibilities, particularly in rural areas, for collecting water and fuel, and climate change can make journeys longer and more dangerous, including vulnerabilities to personal attack. Further, they are often more vulnerable to extreme events. A striking example is that of the 2004 Indian Ocean Tsunami: of the 230,000 people killed, 70% were women (Rahiem et al., 2021). They are often the last to leave an affected area when a catastrophic event approaches.

Similarly, indigenous communities frequently inhabit areas that are more susceptible to environmental hazards, such as low-lying islands vulnerable to sea-level rise, flood plains, or arid regions prone to droughts. A specific example is the Aymara Indigenous Peoples in Central and South America who have experienced decreasing rain and snow, leading to degraded and dry peatland pastures, contributing to out-migration, over-grazing, and the loss of ancestral practices (Caretta et al., 2022). Many minority communities face not only physical risks but also barriers to aid and recovery efforts, due to systemic discrimination or cultural misunderstandings. For example, after natural disasters the Dalit community in India, a historically marginalised group outside the Indian caste system, face difficulties in accessing relief camps, and tend to receive poorer or fewer relief goods due to caste discrimination (IDSN, 2013).

Even the most resilient structures can be overwhelmed. Aside from risk resilience, a third key concept related to climate change impacts, alongside mitigation and adaptation, is loss and damage (see Figure 2.15). The scope for adaptation and resilience is limited by the capacity of both the natural environment and human societies, especially as global warming intensifies (see, e.g., IPCC, 2023, p. 2).[21] The term 'loss and damage' is the focus of Article 8 of the Paris Agreement, which draws attention to the vulnerability of developing countries through the Warsaw International Mechanism for Loss and Damage associated with Climate Change Impacts. The Agreement states:

> Parties recognize the importance of averting, minimizing and addressing loss and damage associated with the adverse effects of climate change, including extreme weather events and slow onset events, and the role of sustainable development in reducing the risk of loss and damage (UNFCCC, 2015, p.12).

Figure 2.15: Averting, minimising, and tackling losses and damages

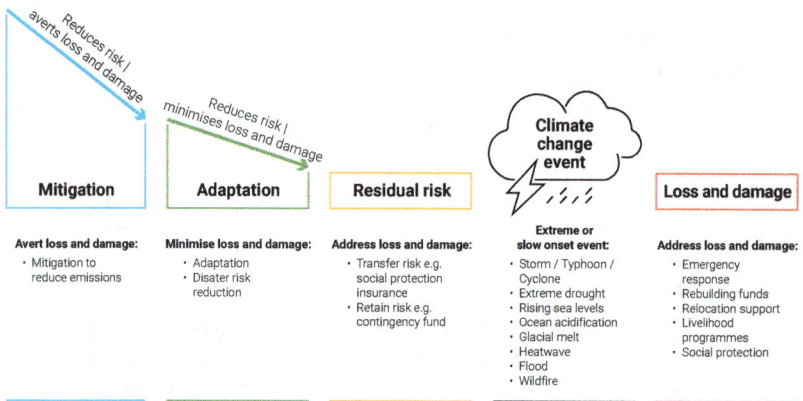

Source: Figure ES.6 in UNEP (2023b, p. xviii). Copyright 2023 UNEP. Reproduced with permission.

Though in this section we have given particular attention to developing nations, which, on the whole, are hit earliest and hardest, developed regions too need to accelerate action for adaptation: risks to them are also increasing faster than their response (see, e.g., EEA, 2024, on Europe). We saw the examples of the deadly Los Angeles fires and Valencia floods in 2024. Only around half of European cities have dedicated adaptation plans (EEA, 2024), with the most disadvantaged groups within the countries being always, in both developed and developing nations, the most vulnerable to climate change impacts (Birkmann et al., 2022). Indeed, though adaptation effort trends have been positive, with planning and implementation increasing (e.g., only 26% of European cities had dedicated adaptation plans in 2018, compared to 51% in 2024), the ambition, scope, and pace of adaptation efforts in both developed and developing countries remain inadequate in relation to the increasing climate risks (C3S and WMO, 2024; IPCC, 2022a; IPCC, 2023).

The IPCC (2022a) found that observed adaptation efforts often operate at small scales, responding only to short-term risks. There is a need for more transformative actions aimed at adapting to long-term risks. Such systemic adaptation may include integrating adaptation into urban planning or developing large-scale coastal protection infrastructure, including 'nature-based solutions' (UNEP, 2023b), such as mangroves.

2.4 Nature: biodiversity and climate

The biodiversity and climate crises are closely intertwined. It makes little sense to try to tackle them as if they were separate. They have shared causes, particularly the burning of fossil fuels. Indeed, biodiversity loss and climate change are accelerating together, feeding into each other in destructive cycles. Deforestation weakens carbon sinks; rising temperatures stress ecosystems;

and land degradation undermines food production. They are both closely related to natural capital and the environment. These crises demand urgent action at local, national, and international levels. They are not, however, the same thing. And not all causes are the same. The differences matter both for understanding and for policy.

The biodiversity crisis

Biodiversity refers to the variety of life at all levels of the biosphere, from genes to species, and up to the ecosystems and biomes[22] that they form. It underpins ecosystem services that sustain human life by supporting natural sustainability and resilience (Dasgupta, 2021, Section 2.1),[23] with an estimated 55% of global GDP (around US$58 trillion) being moderately or highly dependent on nature and its services (Evison et al., 2023). Nevertheless, the rate of global change in nature experienced since the 1970s appears to be unprecedented (IPBES, 2019).

When an ecosystem is degraded, it becomes more vulnerable to tipping points, and often irreversible change. Ecological degradation combined with climate change can turn forests into grasslands, and grasslands into deserts. Species population decline is in many of these transitions an early warning signal of reduced ecosystem resilience (WWF, 2024). And the vast majority of indicators that monitor the status of nature at the global level show a decline. In the period 1970–2020, the average size of monitored wildlife populations in the Living Planet Index (LPI) decreased by 73% – the LPI includes nearly 35,000 population trends and 5,495 species (WWF, 2024). Among the origins of this degradation are the unsustainable extraction of natural resources, including deforestation, overfishing, and the impacts of the exploitation and use of minerals and fossil fuels.

As ecosystem services decline or disappear, their degradation can cascade into economic sectors, generating either gradual or sudden disruptions, which can sometimes be permanent. There are certain economic sectors, such as agricultural, food, beverages, and construction, which are particularly sensitive to biodiversity breakdown (WEF, 2020).[24] As the Dasgupta Review (2021) emphasises, nature is not an external factor but a core foundation and envelope for our economic systems. Our lives, livelihoods, and economic growth are inextricably embedded within it. Yet, the way we measure economic success consistently undervalues natural capital, leading to over-exploitation and degradation. Our pattern of growth results in what Dasgupta calls the 'impact inequality' – where human demand for nature's services exceeds nature's ability to provide sustainable supply.

We are already experiencing substantial land degradation driven by the combination of climate change; destructive, contaminated, and unsustainable agricultural practices; pollution from industry; urban sprawl; and some infrastructure development. This not only leads to the loss of agricultural land but also intensifies food insecurity and poverty. And it can translate into the use of further energy-intensive products. For example, a study found that in maize production in the USA, fertiliser would need to increase by around 30% to compensate for long-term fertility loss due to soil degradation or erosion (Jang et al., 2020).

Environmental degradation extends beyond climate change and biodiversity loss, encompassing a spectrum of issues that can damage the Earth's ecosystems and human health. Pollution remains one of the most pervasive problems in environmental degradation. Air, water, and soil pollution affect millions of people globally. This pollution results from many sources, including burning fossil fuels, industrial discharges, agricultural runoff and crop burning, and inadequate waste disposal. An example is the overuse and pollution of water, which can result in scarcity that affects billions of people worldwide (Gleick and Cooley, 2021; UNEP, 2024a). That scarcity is exacerbated by climate change, which alters hydrological cycles, affecting everything from local agriculture to global food supplies.

It is, in general, very difficult to substitute other forms of capital for ecosystem services. If we push an ecosystem to collapse or beyond tipping points, we cannot simply replace the essential services it provides with other forms of capital. It cannot be understated how important biodiversity is in supporting, not only natural life, but economic life as we know it. There is a crucial sense in which our economy is a wholly owned subsidiary of our environment and its biodiversity.

The interweaving of climate and biodiversity

There are intricate and dynamic relationships between climate and biodiversity. They are interwoven. Climate change and biodiversity loss are driven by similar human activities that have dramatically altered the natural world. The burning of fossil fuels both causes climate change and damages plants, animals, and humans directly via air and water pollution. Fossil fuel combustion, responsible for the majority of GHG emissions, also leads to ocean acidification, impacting marine biodiversity and disrupting carbon cycles.

Primary activities such as agricultural expansion, fossil fuel extraction, and deforestation not only modify vast landscapes but also release considerable quantities of GHGs and other pollutants into the atmosphere, soils, rivers, and oceans. For example, over half a century of oil spills in regions like Bayelsa in Nigeria have left dangerous levels of toxic chemicals in the environment, impacting soil, water, crops, animals, and human health (Bayelsa State Oil and Environmental Commission, 2023). The expansion of croplands for commodities such as soya bean or palm oil, or the expansion of cattle farming, has been a major driver of deforestation, not only destroying habitats but also releasing carbon stored in forests and soils into the atmosphere.

Climate change usually undermines biodiversity and biodiversity loss usually contributes to climate change. Climate change accelerates the depletion of natural capital and ecosystem services by altering major geophysical conditions such as temperatures, precipitation patterns, and ocean acidity

Figure 2.16: Common indirect and direct drivers of climate change and biodiversity loss

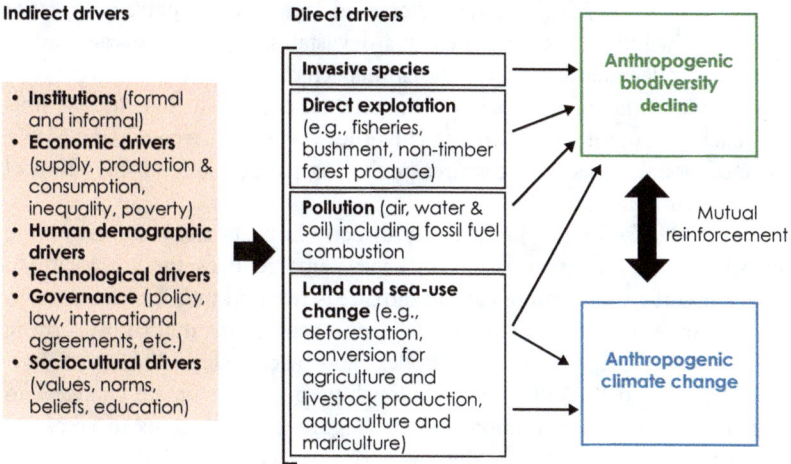

Indirect drivers

Direct drivers

- **Institutions** (formal and informal)
- **Economic drivers** (supply, production & consumption, inequality, poverty)
- **Human demographic drivers**
- **Technological drivers**
- **Governance** (policy, law, international agreements, etc.)
- **Sociocultural drivers** (values, norms, beliefs, education)

Invasive species

Direct explotation (e.g., fisheries, bushment, non-timber forest produce)

Pollution (air, water & soil) including fossil fuel combustion

Land and sea-use change (e.g., deforestation, conversion for agriculture and livestock production, aquaculture and mariculture)

Anthropogenic biodiversity decline

Mutual reinforcement

Anthropogenic climate change

Source: Author's elaboration drawing on Pörtner et al. (2021).

Figure 2.17: Mutual reinforcement between climate change and biodiversity loss

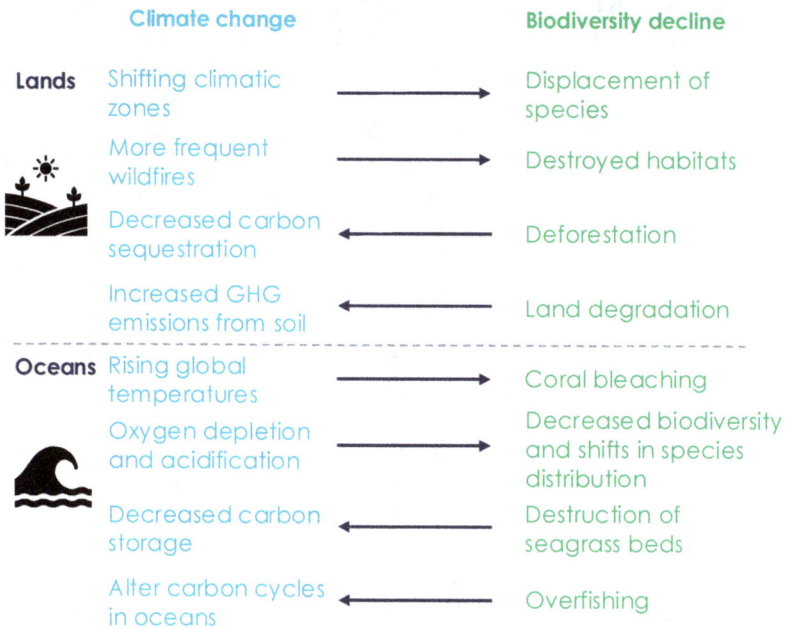

	Climate change		**Biodiversity decline**
Lands	Shifting climatic zones	→	Displacement of species
	More frequent wildfires	→	Destroyed habitats
	Decreased carbon sequestration	←	Deforestation
	Increased GHG emissions from soil	←	Land degradation
Oceans	Rising global temperatures	→	Coral bleaching
	Oxygen depletion and acidification	→	Decreased biodiversity and shifts in species distribution
	Decreased carbon storage	←	Destruction of seagrass beds
	Alter carbon cycles in oceans	←	Overfishing

Source: Author's elaboration drawing on Pörtner et al. (2021).

(Burkett et al., 2005; Woetzel et al., 2020). These changes can occur at a pace which exceeds the adaptive capacity of many species and ecosystems, leading to biodiversity loss. For example, increased frequency of extreme weather events like wildfires and severe storms devastates habitats and can lead to catastrophic declines in biodiversity. Climate change and biodiversity loss can lead to migrations of species and fundamental change in their interactions with each other and with humans. Thus they are also interconnected with a range of global crises and pressures, including pandemics and migration of diseases in animals and humans.

Conversely, the loss of biodiversity directly impacts climate change. The degradation of forests and marine ecosystems reduces their capacity to sequester carbon, thus more carbon dioxide is released into, and remains, in the atmosphere. The degradation of marine ecosystems due to overfishing and pollution impacts carbon cycles and accelerates climate change. Figures 2.16 and 2.17 illustrate relationships between climate change and biodiversity loss. The common drivers (Figure 2.16) point us to areas for intervention that can tackle the two crises together. And the feedbacks between the two (Figure 2.17) imply that tackling one helps tackle the other.

2.5 Concluding remarks: the science is clear and sets the timetable

The science has been built over the last centuries following Fourier's fundamental insight of 1821 that something is trapping the emissions of heat. The identification of GHGs, the first quantitative estimations, and the physical mechanisms at work were established over the next century or so. Since then the science and modelling of climate change have become ever more quantitative and precise as data and analytical techniques have improved. The nature and magnitude of the risks have become ever more clearly established.

Since the first IPCC report of 1992, each subsequent report has become ever more worrying. We now see that we are close to tipping points which could accelerate the pace of climate change. We also see that we will overshoot the 1.5 °C target within a decade and that holding below the Paris target of 'well below 2 °C' will not be possible without strong and urgent action. Much time has been lost since the magnitude of the risks became clear; further delay is ever more dangerous.

The effects of climate change are with us, already on great scale and intensity. That scale and intensity will increase. Adaptation and the building of resilience will be essential and require substantial investment.

Finally, in this chapter we saw that we do not have one crisis of climate, but twin crises of climate and biodiversity. They are intertwined and mutually reinforcing in their causes and dangers. They must be tackled together.

An effective response to the twin crises and the building of sustainable, resilient, and equitable development will require massive investment, powerful innovation, and systemic and structural change. Fostering that investment and change will require sound economic and political insight and analysis. The next chapter examines the changing world, including its opportunities, and sets out an agenda for action.

Notes

1   This trapping process is now known as the greenhouse effect. The sun emits energy which passes through the Earth's atmosphere and warms its surface. When the surface of the Earth absorbs sunlight it reflects some of this energy back as infra-red. Some gases in the atmosphere, those whose molecules oscillate at a similar frequency to infra-red, interfere with the escape of the infra-red energy from the atmosphere. These gases are transparent to shorter wave-length radiation but block the longer wave-length, in particular, the infra-red. In effect, they act as a blanket preventing that energy from escaping and thus cause global warming. Gases which are GHGs are those with these particular frequencies.

2   I am very grateful to leading climate scientists John Schellnhuber and Brian Hoskins for early guidance on the science.

3   I am very grateful for the guidance of scientists Keith Shine and Brian Hoskins on these issues. For references see Martin and Barker (1932) and Shine and Perry (2023).

4   CO_2 equivalent (CO_2e) is generally used to describe emissions concentration and trends. Sometimes CO_2 data are more broadly available, and sometimes it is relevant to mention both, to make a distinction between different CO_2 and CO_2e trends.

5   Historical GHG concentrations are analysed by examining air trapped in ice. Glaciers in Greenland and Antarctica developed over time, allowing scientists to examine concentrations corresponding to up to around 800,000 years ago. Current GHG concentrations are mainly measured using spectroscopy. This entails collecting air samples from various remote locations worldwide and shining electromagnetic radiation through them. By measuring absorbed radiation and examining the wavelength of emitted light, present substances and their quantities are identified. A similar process, which applies the same spectroscopic principles, can be conducted with satellites (Moseman, 2021).

6   The World Weather Attribution, an institution dedicated to understanding how human-caused global warming is linked to these events, described Milton as 'yet another hurricane wetter, windier and more destructive because of climate change' (WWA, 2024).

7   Excessive exposure to heat in the workplace is indeed a dangerous threat, being already responsible for 7.2% of all occupational injuries in Africa (ILO, 2024).

[8] This event was found to be 'virtually impossible' (WWA, 2025b) without human-caused climate change. A blackout in Buenos Aires in January 2022 was found to have been triggered by extreme heat.

[9] We are simplifying here. There will be lags. And some GHGs such as methane decay fairly rapidly causing concentrations to fall.

[10] The definition of heatwaves varies. Wong et al. (2024) defines it as three or more consecutive days on which the maximum temperature is equal to or exceeds the local 90th percentile of daily maximum temperature over a 40-year period (1980–2019).

[11] Arboviruses (e.g. dengue, Zika, West Nile, yellow fever and chikungunya) are transmitted to humans through arthropods.

[12] For example, population displacement projections by 2050 in Latin America, sub-Saharan Africa, and South Asia due to climate change vary, due to different assumptions, between 31 and 143 million people (Cissé et al., 2022).

[13] See Hendrix et al. (2023) for descriptions of debate on climate and conflict.

[14] Given the complications of GHGs having different timescales and radiative forcing, carbon budgets are generally expressed in terms of CO_2.

[15] Both estimates were calculated using a 67% probability (to remain under each temperature limit) at the current emission rate of around 40 $GtCO_2$ per annum.

[16] In short, the cost of removal varies greatly across methods of removal. While natural CDR methods are less costly (such as afforestation), their permanence could be a problem; forests can burn down. Currently the high costs of direct air capture (DAC) hinder its deployment. There may be impacts on biodiversity, from some forms of afforestation or soil carbon sequestration. Public and private support for innovation may eventually drive down these costs. See Chapter 9 for further discussion.

[17] SRM acts by preventing energy from reaching the Earth.

[18] A statement President Reagan made in November 1984 (see Reagan, 1984).

[19] For example, restoring wetlands can, by absorbing excess water, reduce the risk associated with the *hazard* of run-off from heavy rain. It is also important to improve the *vulnerability* to the hazard. For example, improving the drainage systems of a city would mean that it has become less vulnerable to run-off from heavy rain, and therefore flooding of homes and businesses might not occur even when the hazard of intense rain and associated run-off takes place. Lastly, adaptation is also about

acting upon the exposure to climate change risks. For example, modifying land-use planning can reduce *exposure* by avoiding urban sprawl in flood-prone areas (IPCC, 2019).

[20] Many adaptation projects imply large up-front costs with long payback timeframes. And information and understanding are key: often the risks of inaction are underestimated. These factors contribute to strong limitations in private finance flowing to adaptation, a matter that we discuss in Chapter 9.

[21] The IPCC distinguishes hard and soft limits to adaptation. Hard limits occur when no feasible adaptation actions are possible. For example, if a small island is submerged, or continually inundated by storm surges, there is a hard limit in the sense that survival requires moving (IPCC, 2022a). Indeed, that can occur in major regions in bigger countries, such as Bangladesh. Soft limits, on the other hand, can exist where potential adaptation options are theoretically available but practically inaccessible due to 'financial, governance, institutional, and policy constraints' (IPCC, 2022a, p. 26). These soft limits are often exacerbated by inequality and poverty, which disproportionately affect the most vulnerable groups by increasing their exposure to climate impacts (IPCC, 2022a). For example, desertification might in principle be reversed with enough planting and irrigation but that might be way beyond the resources available.

[22] A biome is a geographical area classified according to the species that live in that location.

[23] These ecosystems are complex systems that perform critical functions like energy flux through photosynthesis, nutrient cycling, and organic matter decomposition. The loss of biodiversity reduces the capacity of ecosystems to cope with climate change and hampers their ability to sequester carbon effectively.

[24] For example, 60% of wild coffee species are at risk of extinction in the next decades due to climate change, deforestation, and diseases, including the popular Arabica bean, due to temperature sensitivity (Davis et al., 2019).

3. More fundamentals: politics, economics, ethics

The fundamentals of the science provide lessons which are crystal clear. The climate change and biodiversity crises have already brought the world into great difficulty and the challenges, disruption, and destruction will increase. The difficulties could become unmanageable unless we take urgent and large-scale action on mitigation. These difficulties are already of an intensity, and worse is to follow, that strong adaptation and the building of resilience is essential. Delay on both fronts is dangerous. But we also saw that actions for mitigation, adaptation, and development are intertwined. It is a mistake to insist that there is a horse race between them.

The urgent and large-scale action involves investment in the new, and innovation in our technologies and activities. That innovation will, in large measure, be embedded in the investment. We must invest in the new before we can run down the old. Further, much of the structure and systems of the economy, including around energy, transport, cities, land, and water, must change in fundamental ways. And all this must be done rapidly. The next two or three decades are crucial.

All this action involves, at its core, economic, political, and societal change – at a pace which will involve dislocation and stress. The prize is the avoidance of catastrophe and a new and more attractive path of growth. It will not be easy, but failure would be still more difficult. It is clear that this transformation, at the pace necessary, will not be possible without sound foundations in analyses in the social sciences. Economics, politics, history, geography, and ethics are at centre stage.

Indeed, one can argue that the majority of the challenge now lies in the social sciences. The scientists and engineers, of course, have much more to do, as we saw in Chapter 2, and will see in Chapters 5 and 6. But it is in economy, politics, and society that the biggest obstacles to the changes we must make will be found.

In Section 3.1 we examine the politics, shaped in large measure by history and geography. Economics and ethics are the subject of Section 3.2. I will be critical of where my own subject, economics, has steered the discussion. And much of that discussion has been cavalier with the ethics. There are major ethical issues here and they must be discussed directly. Section 3.3 looks forward to how constructive analysis can and should develop, emphasising sustainability, political economy, and structural and systemic change. The final section is directed to my fellow economists, who are, at last, starting to move in a better direction.

3.1 Politics and its intersections with history and geography

Chapter 2 introduced the basic science which shapes the consequences of our action or inaction. How decision-making actually plays out is in turn shaped by politics. And the politics of climate is influenced by history. The financial, social, institutional, and other crises and frailties that took place in the last two decades (see Figure 3.1) have made political conditions for future progress much more difficult than would have been the case without them. Further, the economic context in which the developed countries became rich – through growth with high emissions – was in some cases also one of colonialism. In the eyes of many in developing countries, this history points to a special responsibility of the developed countries to lead on action and finance in relation to the climate and biodiversity crises. Politics, history, and geography are, of course, intertwined with economics and political economy; these issues are discussed in the next section.

Figure 3.1: Timeline of major global crises and disruptive events (2008–2024)

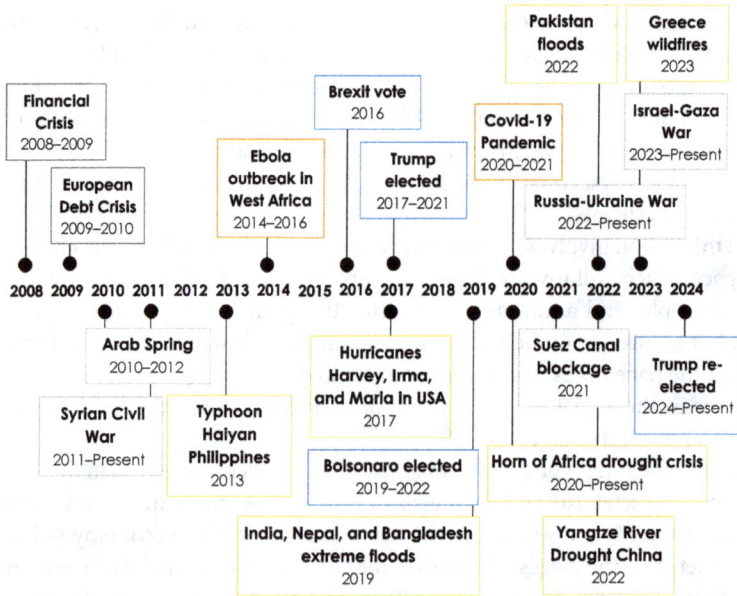

Source: Author's elaboration.

Not only have these recent crises of the last two decades made conditions for action more difficult, they have also consumed bandwidth of politicians and political processes. We have a difficult state of affairs for generating action on climate and biodiversity. We examine briefly some of the international difficulties and show the political friction and mistrust they have generated.

Support for developing countries: failures and eroded promises

Since 2015, the world has experienced persistent policy uncertainty globally, weak trade and investment trends, and rising tides of populism, for example the election of political leaders such as Donald Trump in the USA and Jair Bolsonaro in Brazil, and the Brexit vote in the UK. A common sequence of populist questions has been: Are you angry? Is it the fault of the elite? Is it the fault of the immigrants or foreigners? In skilful hands, such discourse can be very effective. Populist politics has often coincided with climate denial or scepticism. All this has, to put it mildly, not been helpful in the task of fostering the new investment necessary for climate action and a new form of growth.

The Covid-19 pandemic, which had spread to most countries in the world by early 2020, was revealing in many ways of how countries would act in times of crisis and what their priorities would be. Developed countries acted quickly and invested massive amounts of resources in the rapid development of vaccines. However, the reluctance on the part of developed countries to share Covid-19 vaccines or to facilitate their production in developing countries caused great and understandable resentment. Many spoke of vaccine apartheid. It did indeed appear that lives in some countries were seen as much more valuable than others. These moments are tests of trust and the richer countries were seen by the poorer as failing that test.

The war in Ukraine, which started in early 2022, compounded these struggles. Soaring food and energy prices hit developing economies hardest, diverting attention and resources away from climate and economic support. Uncertainties and economic pressures increased caution in financial markets and raised the cost of capital and reduced its availability for poorer countries. The commitment of developed nations to their own green transition was also tested by the energy crisis caused by the outbreak of the war. For example, some EU countries moved back towards coal in the short run.

In the climate change arena, a deep wound came from the failure of developed countries to deliver the promised US$100 billion per year in climate finance by 2020. This commitment, first made at COP15 in the informal Copenhagen Accord (2009), formally agreed at COP16 in Cancún (2010), and reaffirmed at COP21 in Paris (2015), was foundational in building these agreements. Yet, when the deadline passed in 2020 without delivery, many emerging markets and developing countries (EMDCs) saw it not just as a financial shortfall but as a betrayal. Though the funding was finally met – two years late (Organisation for Economic Co-operation and Development [OECD], 2024) – serious damage to international relations had been done. It has signalled a perceived lack of seriousness and commitment from the wealthier nations, particularly since the US$100bn per annum was so far short of what is needed (see, e.g., Bhattacharya et al., 2023, 2024). We will turn to finance in Chapters 5 and 9.

Further, the negotiation processes of international agreements themselves sometimes seem to lack transparency and inclusivity, leading to feelings of disenfranchisement among less influential nations. The structures and direction of agreements are sometimes seen as determined in the G7 or, in more recent times, increasingly in the G20. Large emerging market economies are part of the G20, but the majority of countries are not present at that table. A major determinant of the acrimony in Copenhagen in 2009 was the perception that the G7 was dictating the agenda. The problem may have changed shape and, to some extent, diminished with the increasing role of the G20, but it has not gone away.

All these difficulties provide a troublesome context for the urgent, large-scale, and collaborative climate action which is necessary for the investment and change now required. Restoring trust between nations, ensuring the fulfilment of climate finance commitments, and accelerating progress on the Sustainable Development Goals (SDGs) are not merely desirable policy aspirations but also necessities for the collaboration required to secure a sustainable future. Trust, delivery, and constructive collaboration are mutually reinforcing.

The significance of history and geography for emissions

The challenges of the politics, and the potential areas of action, are also shaped by the sectoral and geographical origins of emissions. In this subsection, we trace briefly the sources and trajectory of emissions, which have, in large measure, emerged from economic history. Indeed, the history of emissions follows closely the economic history of sectors and countries.

Until around 1950, land use, land-use change, and forestry (LULUCF) were the primary sources of CO_2 emissions. Deforestation and agricultural expansion were responsible for the release of significant amounts of carbon stored in forests and soils into the atmosphere. However, over the past 70 or more years, there has been a major shift, with fossil-fuel combustion and industrial processes becoming the main emission sources (Evans, 2021) (see Figure 3.2). This transition mirrors rising energy consumption and the global expansion of industrial activities, together with the reliance on fossil fuels for energy. Indeed, rising emissions reflect changing patterns in economic activity. Among sectors, the energy sector is the largest emitter, responsible for 34% of global CO_2 emissions in 2019, followed by industry (24%), agriculture, forestry, and other land use (AFOLU) (22%), transport (15%), and buildings (5%) (Pathak et al., 2022).

Around 30% of the rise in global temperatures since the Industrial Revolution can be attributed to methane emissions (the primary component of natural gas), which is the second most important greenhouse gas (GHG) emitted by human activities after CO_2 (IEA, 2022). The anthropogenic emissions of this powerful and short-lived gas are mostly generated by the agricultural sector (around 40%), fossil fuels (35%), and waste (20%) (United Nations Environment Programme [UNEP] and Climate and Clean Air Coalition, 2021). There is substantial geographical variation in the importance of methane-emission subsectors by country and region.

Figure 3.2: Global anthropogenic CO_2 emissions

Quantitative information of CH_4 and N_2O emission time series from 1850 to 1970 is limited

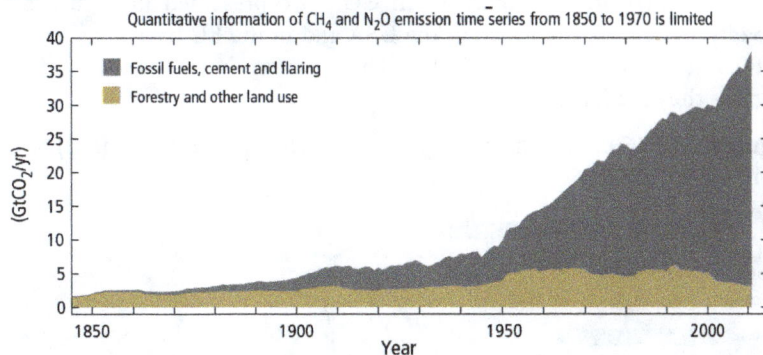

Note: Global anthropogenic CO_2 emissions from forestry and other land use as well as from burning of fossil fuel, cement production, and flaring.
Source: Figure SPM.1(d) in IPCC (2014, p. 3). Copyright 2014 IPCC. Reproduced with permission. Use of IPCC figure(s) is at the User's sole risk. Under no circumstances shall the IPCC, WMO or UNEP be liable for any loss, damage, liability, or expense incurred or suffered that is claimed to have resulted from the use of any IPCC figure(s), without limitation, any fault, error, omission, interruption, or delay with respect thereto. Nothing herein shall constitute or be considered to be a limitation upon or a waiver of the privileges and immunities of WMO or UNEP, which are specifically reserved.

The growth of emissions over the period 1980 to 2010 was striking (see Figure 3.2). Almost two-thirds (62%) of cumulative emissions from 1850 to 2019 occurred since 1970. The powerful growth in the flow of emissions from 1980 to 2000 was driven by rapid industrialisation and urbanisation, particularly in developing economies, and especially China. Emissions are still rising although there is some evidence that the peak is approaching, particularly since China is committed to peaking by 2030, and will likely do so.[1]

Although from 2010 to 2019 emissions were higher than in any previous decade, the rate of increase slowed. This can be attributed to improved energy efficiencies and the gradual shift towards renewable energy sources in many parts of the world, as well as some slowing in growth in China and the world economy as a whole after the global financial crisis of 2008–2010. The global Covid-19 pandemic briefly reduced CO_2 emissions in 2020 by about 5.8% compared to 2019 levels, but the decrease was temporary as emissions rebounded by year-end.

The geographical distribution of GHG emissions per capita, in large measure, reflects the distribution of income per capita. The highest per capita emissions amongst major economies are in the USA and Russia – more than double the world average (UNEP, 2023). China's per capita CO_2e emissions surpassed those of the EU28 in 2014 (Global Carbon Project, 2014). In 2021, per capita emissions in India remained under half the world average but were climbing (UNEP, 2023). Also in 2021, the G20 averaged 7.9 tCO_2e per capita, while least developed countries (LDCs) emitted just 2.2 tCO_2e (UNEP, 2023). Trends (2000–2021) of per capita emissions for different countries in CO_2e are presented in Figure 3.3. Broadly, per capita emissions in the USA and in the EU have been falling in the last few decades whilst China's have risen strongly. For the world as a whole they are flattening.

Figure 3.3: GHG emissions per capita in 2021 for key countries (left) and trends since 2000 (right)

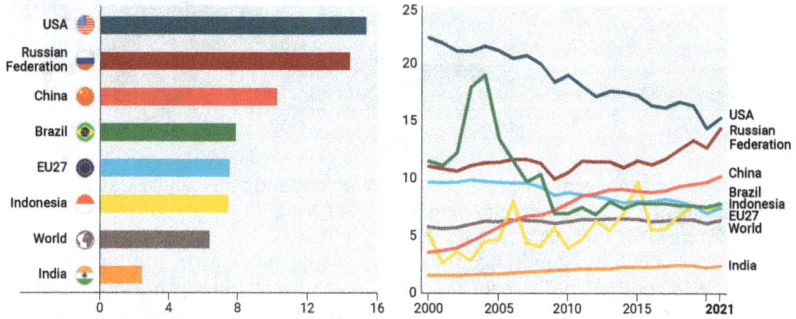

Note: Per capita GHG emissions in 2021 and trend since 2000, including inventory-based LULUCF CO_2 (tCO_2e/capita).
Source: Figure 2.2, bottom panel in UNEP (2023, p. 7). Copyright 2023 UNEP. Reproduced with permission.

It is important to recognise that there are both high and low emitters in rich and poor nations and that there is not an exact correspondence between income per capita and emissions per capita (see Figure 3.4). The link between income and carbon emissions is mediated by factors such as the energy mix, climate, and culture. For example, in 2019, though the poorest half of the USA had half the income of the European middle 40%, it still had similar emission levels to that group.[2] This is substantially explained by the USA having a more carbon-intensive energy mix. Still, it is also related to other factors such as characteristics of infrastructure and device or machine efficiency. For example, not only are cars highly used in the USA as a transport medium, but on average they tend to be larger and less fuel-efficient than those used in Europe (Chancel et al., 2022).

The balance of shares of global emissions has generally shifted from high-income towards low- and middle-income countries in the past two decades, reflecting the higher economic growth rates in the latter. In 2000, high-income nations generated 43% of GHG emissions, compared with 28% in 2021. In 2000, low- and middle-income countries generated 53%, compared with 69% in 2021 (UNEP, 2023) and around 71% in 2023 (Fengler et al., 2023). Though historical emissions have mainly come from high-income nations, the majority of new emissions are being produced by middle-income countries, home to around three-quarters of the people in the world.

Figure 3.4: Countries classified by income per capita and GHG emissions per capita

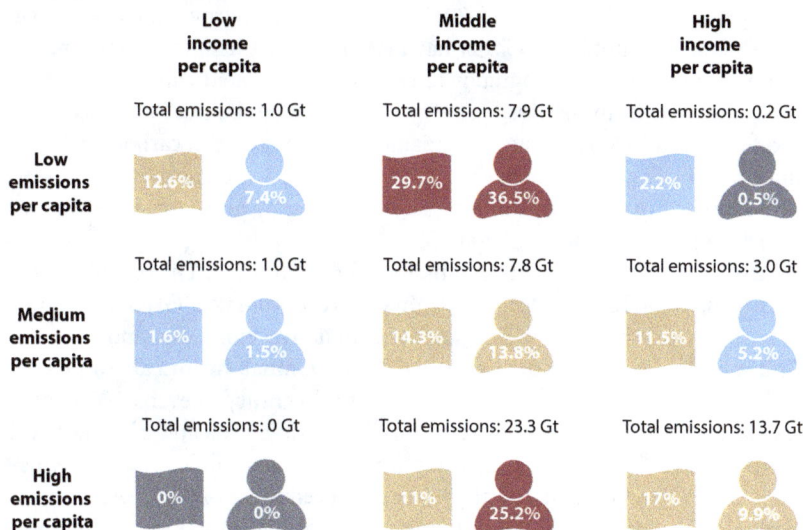

	Low income per capita	Middle income per capita	High income per capita
	Total emissions: 1.0 Gt	Total emissions: 7.9 Gt	Total emissions: 0.2 Gt
Low emissions per capita	12.6% 7.4%	29.7% 36.5%	2.2% 0.5%
	Total emissions: 1.0 Gt	Total emissions: 7.8 Gt	Total emissions: 3.0 Gt
Medium emissions per capita	1.6% 1.5%	14.3% 13.8%	11.5% 5.2%
	Total emissions: 0 Gt	Total emissions: 23.3 Gt	Total emissions: 13.7 Gt
High emissions per capita	0% 0%	11% 25.2%	17% 9.9%

Notes:

- Numbers within flag figures represent the percentage of world nations; and numbers within person figures represent the percentage of world population. For example, 12.6% of world nations, home to 7.4% of the world population, have low emissions per capita and low income per capita, emitting 1 gigatonne (Gt) in total.
- Classification of low-/middle-/high-income countries based on gross national income per capita: Low income ≤ 1,135; Lower-middle income 1,136–4,465; Upper-middle income 4,466–13,845; High income >13,845.
- Classification of emission intensity based on average per capita emissions (given that the current global average of per capita emissions is around 7.4 tons per person per year): low-emission countries <5 tons per capita; medium-emission countries 5 to 10 tons per capita; high-emission countries (HECs) >10 tons per capita.
- Grey indicates cases where the percentage is 0–0.99; light blue where the percentage is 1–9; beige where the percentage is 10–20; and red >20.

Source: Author's elaboration using data and figures from Fengler et al. (2023).

However, emission levels among middle-income countries vary. For example, in terms of GHG emissions per capita, according to the classification presented in Figure 3.4, India is a low emitter (but emissions are growing rapidly), Indonesia a middle emitter, and Brazil a high emitter. Broadly speaking, global objectives (as in the SDGs) would be to increase the number of countries with low emissions per capita and high income per capita (top right corner) (Fengler et al., 2023).[3]

China has been the world's largest emitter of GHGs since 2006 (Liu et al., 2023). And in 2023, it contributed 30% of the global total (CO_2e) (IEA et al., 2024).[4] This substantial share reflects the scale and growth of China's industrial base and its reliance on coal for energy. It became the world's manufacturing hub in the decades 1990 to 2010. Its emissions increased very rapidly in that period of extraordinarily rapid economic growth, which averaged around 10% per annum from 1990 to 2010. The primary energy source was coal. We should also note that China has been manufacturing for consumption in other countries. The geography of emissions associated with consumption looks different from that associated with production, because a reduction in production emissions can be accompanied by an increase in carbon-intensive imports. The distinction between production-based and consumption-based emissions is important in thinking about geography and responsibility, and something we return to in Chapter 8.

Figure 3.5 shows cumulative emissions for 1850–2021. Historically, a few countries, notably the USA and China, have dominated GHG emissions. Whilst the gap between US historical cumulative emissions and those of China is diminishing and some suggest that China's historical cumulative emissions may surpass those of the USA by mid-century (Stevens, 2023), others claim that this assertion is based on 'implausible projections' (Evans and Viisainen, 2024), and that China might never do so.[5] The cumulative emissions of China and Russia together already exceed those of the USA. The G20 is responsible for roughly three-quarters of warming caused up to 2021, and accounts for an even greater share of historical cumulative fossil CO_2 emissions (UNEP, 2023).

Figure 3.5: Countries with largest cumulative GHG emissions (GtCO$_2$e) 1850–2021

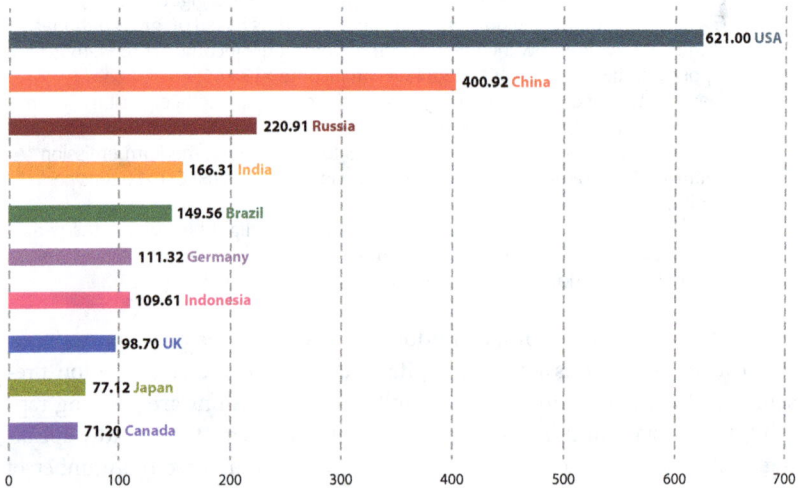

Source: Author's elaboration using Jones et al. (2023).

Figure 3.6 shows the evolution of GHG emissions (CO_2e) from China and other major emitters between 1945 and 2021. Together, the seven major economies included in Figure 3.6 accounted for more than 60% of emissions in 2021.

Figure 3.6: The evolution of GHG emissions (CO$_2$e) from 1945 to 2021

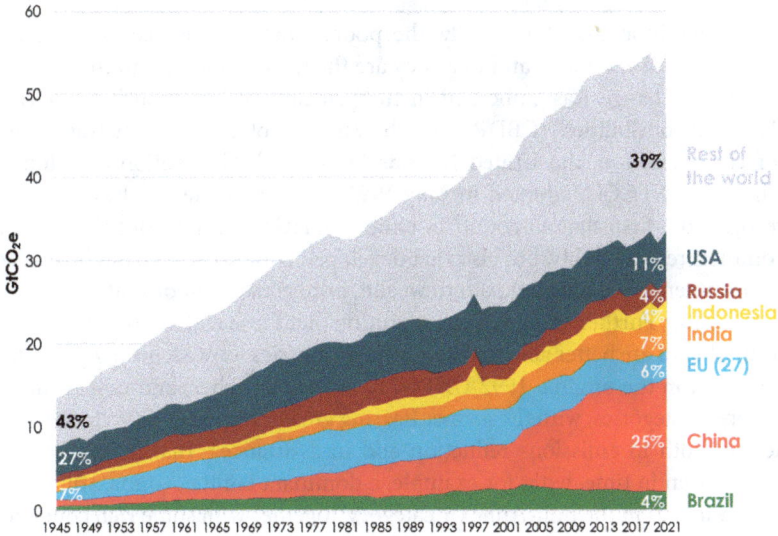

Note: LULUCF are included in GHG emissions.
Source: Author's elaboration with Jones et al. (2024) data.

In contrast, LDCs (a group of 46 nations with low levels of economic and human development) have very low annual contributions to emissions. As shown in Figure 3.7, together they represent only 3% of total GHG emissions in 2021 (UNEP, 2023), although they have 14% of the global population.

Figure 3.7: Current contributions to climate change (% share by countries or regions) compared with population

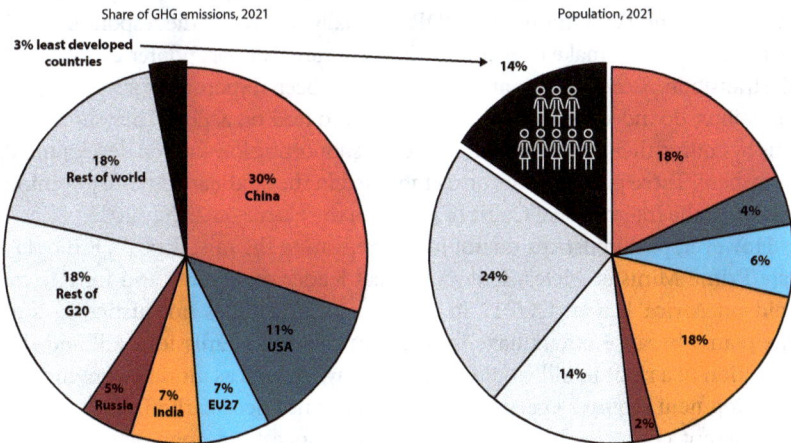

Note: LULUCF CO$_2$ emissions are excluded from GHG emissions.
Source: Author's elaboration with data from UNEP (2023).

Looking to the future: developing countries, growth of emissions, and mutual responsibilities

The recognition that, historically, the poorest nations have contributed the least to global emissions and that they are the most vulnerable to the impacts of climate change was embodied in the principle of 'common but differentiated responsibilities' (CBDR), which was part of the original framework for climate action, the United Nations Framework Convention on Climate Change (UNFCCC, adopted in May 1992). The principle was based on the recognition, first, that all countries emitted GHGs, which meant there was a common responsibility, but also that developed countries, having followed an environmentally damaging path to wealth, bore greater responsibilities for climate action. Further, there was a recognition that those with more resources or technologies are better able to take action. Countries were accordingly divided into a list in the original 1992 UNFCCC: Annex 1 (richer) and non-Annex 1 (poorer) categories, with the former bearing greater responsibilities for climate action, both in emissions reduction and in assistance.[6] That list has largely been frozen in time, with, for example, a dominant emitter, China, still in the 155 countries in the non-Annex 1 category, together with some rich countries such as Singapore, Saudi Arabia, and the United Arab Emirates.

Looking forward, it is clear that EMDCs (other than China) will experience the fastest growth in the coming two or three decades, and their future contributions to global emissions will exceed those of richer countries. Their current annual emissions are already higher; their transition towards the low-carbon economy is of great importance to the future of emissions. Investment is at the heart of this transition, and EMDCs (other than China) will need to increase their investment strongly as a share of income in order to play their part in the transformation which is crucial to the achievement of the Paris Agreement goals (see Chapter 9).

How to foster that investment and finance is at the core of the arguments of Parts II, III, and IV of this book. CBDR originally referred to the responsibility of richer countries to make deeper emissions cuts and support poorer countries in the transition. Richer nations and groups, having been responsible for most of past emissions, do indeed have a moral obligation to lead on action. They should do this by cutting their own emissions as well as supporting low-carbon development elsewhere. These principles are also embodied in the landmark advisory opinion issued by the International Court of Justice in the Hague on 23 July 2025.

However, past injustices do not justify repeating the mistakes. As Ethiopia's late Prime Minister Meles Zenawi, a great leader on climate and for Africa, said on Africa Day at COP17 in Durban in 2011, 'It is not justice to foul the planet because others have fouled it in the past.'[7] Emissions kill and the assertion of a right to kill would not, for many, be seen as morally convincing. Development requires energy, but energy does not need carbon.

Different countries have different circumstances and approaches. We do not attempt comprehensiveness here but, in terms of people and potential future emissions, India and China are especially important.

India is clearly very vulnerable to climate change and is demonstrating in many of its instruments the increasing promise of renewable energy. Prime Minister Modi put sustainability at the core of India's G20 Presidency in 2023. For a long time, including for most of the period pre-Paris 2015, India has insisted on CBDR being narrowly interpreted and that the rich countries should cut emissions much more strongly and deliver much more financial support to poorer countries, before putting pressure on them to cut emissions. Some in India, and other developing countries, have seen such pressure as an intention to hinder its growth. However, India supported the Paris Agreement and has increasingly become a force for stronger climate action.

There are many factors explaining the change but, in my view, the two most important are the increased recognition of the ever-increasing impacts of climate change and of India's great vulnerability. And a rising understanding of the potential in the new growth story and the possibility of leapfrogging dirty stages of development. However, like many countries, the cost and availability of capital, particularly risk capital, is a challenge. And so too, in some cases, the investment climate.

China's approach has evolved from caution in making commitments during the Copenhagen 2009 summit, where it highlighted, often vehemently, the historical emissions of developed nations, to a more proactive stance in recent years. After Copenhagen, China adopted the strategy of 'Ecological Civilization', embedding environmental concerns within its developmental agenda (Stern and Xie, 2021). It became a leading objective of the Communist Party in 2012 (under President Hu Jintao) and a central plank of China's 13th Five-Year Plan in 2016 (under President Xi Jinping).

By the time of the Paris Agreement negotiations, China was advocating for shared global responsibility, notwithstanding its continued commitment to CBDR. It became one of the first major emitters to ratify the Paris Agreement, reflecting its leadership in global climate governance. As described in Chapter 1, in November 2014 in Beijing, Presidents Obama and Xi announced their emissions reduction targets together, one year ahead of Paris, thereby signalling their joint commitment to success at Paris.[8] Mutual trust does matter in creating agreements. Pete Betts, who was negotiating for the EU and the UK at that time, was a much-loved figure and deeply trusted by many across the world. That made a real difference. See Pete's book *The Climate Diplomat: A Personal History of the COP Conferences*.

One of the influences on China's position was consistency of emissions commitments with the five-year plans. The five-year plans are genuine strategies for China; not just aspirational. Achievement of plan targets is one of the key criteria against which leaders are judged. In Copenhagen 2009, the 11th Five-Year Plan (2011–2015) was not ready but in Paris December 2015, the 12th (for 2016–2020) was more or less done – note the six-year gap between Copenhagen and Paris. Thus, in Paris, China could be confident that its COP commitment and five-year plan could be mutually consistent.

The future of emissions will be determined across all countries. But in looking at past and future growth of emissions we have seen that whilst rich countries dominated emissions growth in the 19th and 20th centuries, it will be EMDCs that will dominate the shaping of the future of emissions. Their choices and the support they receive will, in large measure, determine whether the world is able to manage the crises of climate and biodiversity.

3.2 Economics and ethics

It is the science that tells us that we have to act with urgency and scale on climate and that delay is dangerous. Economics is fundamental to guiding the strategies, choices, and investments that will be at the heart of that response. The emission of GHGs damages prospects for others. It constitutes a negative externality and a market failure if uncorrected by policy. The economics of market failures and externalities goes back to the late 19th and early 20th centuries, with the crucial insights of Alfred Marshall and A.C. Pigou (see below). But the economic analysis of climate change from a growth perspective started in the early 1990s, particularly with the work of Bill Nordhaus, as climate change started to emerge in public discussion.

We can now see, and it is a lesson that has emerged ever more strongly in the 20 years since the Stern Review, that those early attempts in the 1990s, whilst understandable first steps, involved analytical approaches, models, and perspectives which were not fit for purpose in relation to the magnitude of risks and the nature, pace, and scale of necessary change. Further, those early discussions of growth and climate were muddled and misleading in their treatment of intertemporal values and discounting.

In this section, we first show the ways in which those early efforts were misleading and narrow, both in relation to the economic modelling and the discounting. Care with the analytics and ethics of intertemporal values is crucial for an adequate treatment of the latter. Indeed, we should look at ethics and moral philosophy more broadly than is standard in economic analyses. We then turn to the broader and deeper issues of how economics can get to grips with the challenges that the science emphasises strongly: immense risks and a response which embodies fundamental technological, structural, and systemic change at pace and scale.

We will put this broader and deeper economics to use in the second part of this book when we describe and analyse the new story of growth and development which can flow from the action necessary in response to the climate and biodiversity crises. This will include the ideas of sustainability and the importance of investing in all four capitals: physical, human, natural, and social. And it will require attention to the difficulties in political economy associated with rapid change.

This discussion of how our economics must broaden and deepen, including in relation to the relevant ethical issues, will underpin the analysis of Part II of this book on the new story of growth and development, including reflections on growth processes in Chapter 5, strategies and policies in Chapter 6, and what is implied for the role of the state in Chapter 7.

Before embarking further on this analysis of economics and ethics, let us begin with emphasising the importance for policy of the concept of market failure, in particular, that associated with the damage arising from the emissions of GHGs. That damage represents an externality in the sense that the activity of one agent affects the welfare of other agents in a way that is not reflected in the price of goods and services. Private and social costs differ, and we have a negative externality. Alfred Marshall, in *Principles of Economics* (1890), highlighted the possibility of a gap between marginal private costs (or benefits) and marginal social costs (or benefits),[9] and the deviation from efficiency that would involve. If agents do not face the full social costs of their actions, they over-indulge in the activity producing the negative externality. In such cases, a reduction of the activity at the margin would produce 'benefits' to those affected by the externality which would be greater than the loss to the agent pursuing the activity. The reason is that, for the final unit consumed ('on the margin' in the language of economics), the person pursuing the activity gets zero benefit because she/he goes on consuming up to the point where the price paid is equal to the benefit perceived. Hence there is zero cost to her/him from reducing by one unit, whereas the person affected has gained from the reduction.

In the early 20th century Arthur Cecil Pigou showed, particularly in his book on the *Economics of Welfare* (1920), that taxation could effectively and efficiently bridge that gap. We call these Pigouvian taxes and the idea has been ever-present in the economics of climate change. And it is an insight and idea of great value. It absolutely should be at the centre of discussion on the policy stage.

Many economists offer versions of the following statement: 'economic theory shows that setting a price for carbon, the Pigouvian tax for example, is the optimum or most efficient policy'.[10] That is a basic analytical mistake because there are many other serious market failures at issue here, which we will discuss in Chapter 6.

To reiterate, the Pigouvian insight is crucial and should be central to policy, even though policy must go way beyond that idea. Economics has been far too narrowly focused on this one instrument. Aside from not tackling many other highly relevant market failures, a carbon price is not by itself powerful or subtle enough to drive systemic change. It does not, by itself, redesign a city, energy, or transport system. A key task for economics should now be the design of incentive, institutional, and behavioural policies, and actions that can tackle the market failures, foster investment, and bring fundamental change. And how to do that at pace: time matters; delay is dangerous.

GHG emissions and climate change are often seen as global public goods or bads. The notion of a public good is another crucial idea in this context.[11] A public good benefits everyone, and no one can be excluded from using it; examples are clean air or a stable climate. Public goods can lead to collective action problems, such as free-riding, where people rely on others to take action and to create the public good, so they can benefit without contributing. Free-riding in the context of public goods is a standard idea in economics and an idea of real significance. It underlines concerns that international agreements on climate may be fragile in the sense that everyone will have a reason to do less than committed, on the assumption – an important one – that others will fulfil their commitments. In other words, one might avoid acting, assuming others will do enough. The idea of a public good will influence much of what follows, including international issues in Part III.

Inadequate assumptions, inadequate models

A second problem in the mainstream economics of climate change, beyond that of excessive focus on just one instrument, the carbon price, has been the way in which models integrating economic growth and the effects of climate change have been constructed and used. Simple models with this integration – called integrated assessment models (IAMs) – were introduced by Bill Nordhaus in 1992. Such integration is important and the models initially provided some insight. But they had and have several problems in the basics of their construction and calibration which have made them very misleading. Modelling should be as simple as possible and will necessarily omit issues of importance. The problem here, however, is twofold: what these IAMs have built in badly distorts the issues; and the central challenge of how to transform the structures and technology of the economy has been essentially excluded.

In part through the use of these distorting IAMs, our subject has perpetuated the idea, all too pervasive in much of public discussion in economics and more broadly, that there is an inevitable trade-off between climate action and economic growth. Bill Nordhaus's early paper in 1991 in many ways cast the discussion in that direction; its title was 'To slow, or not to slow'. Given its model structure and assumptions, it derived the answer to that question (reflected also in much of the following work along these lines), namely that yes, there should be some slowing of growth to help deal with climate change, but not much. And further, the paper concluded that the 'optimum' long-run global temperature increase was around 3.5 °C. A description of that temperature as 'optimum' was a conclusion which seems absurd to those who have looked carefully at the science, but it has been influential in public arguments against strong targets and action.

Nordhaus (2018) has more recently lowered his estimate of long-run 'optimum' temperature but, at 3 °C by 2100, it is still far too high for any sensible interpretation of the dangers indicated by the science. For example, when the Earth was last at that temperature sea levels were 5–25 metres higher than today (IPCC, 2023). Michael Mann (in Carrington, 2023) refers to the dangers in terms of the future of civilisation (see also IPCC, 2018). Substantial parts of the world, including many of those currently heavily populated, would likely be uninhabitable.

The erroneous and misleading conclusions from IAMs follow from inadequate assumptions about the nature of climate change:

1. **Underlying growth impervious to climate change.** Their structure assumes basic underlying aggregate growth, notwithstanding that climate change could thoroughly derail growth and cause decline or collapse.
2. **Underestimated risks.** The risks from climate change are generally trivialised into potential losses of a few percentage points of GDP, even at high temperature increases (e.g., 4 °C or 5 °C), when what is at issue includes potential loss of lives and livelihoods on a massive scale.
3. **Misrepresenting catastrophic change.** The use of expected utility maximisation,[12] a standard tool in these models, as a criterion or objective can be very misleading because of its limited ability to encapsulate catastrophic loss. Seeing losses as potentially unbounded or infinite is surely a possibility when deaths could be in the hundreds of millions. Yet, infinity is not easily incorporated in models. A different approach to risk is necessary when, for many in the world, the stakes are existential.
4. **Large and increasing costs of action.** The models generally assume large costs of action, weak technical progress, and strongly rising marginal costs of action, when we have seen rapidly falling costs of action for many technologies and economies of scale in taking action. Indeed, we should now emphasise still more strongly that we should see the action not in terms of costs but as productive investment in a new form of growth.
5. **High discounting[13] of the lives and livelihoods of future generations.** Future incremental benefits are assigned low value and large long-term impacts can be seen as unimportant because the high discounting in these models assumes them to be so. There is no sound justification for the approach to discounting used, particularly given the scale of damage that could arise. Discounting is discussed in the following subsection; it has been a distinctive theoretical debate in the economics of climate change.

It should surely be clear that with those assumptions, all highly implausible and misleading, the whole structure is tilted against conclusions in favour of strong climate action.

A fundamental problem, as noted above, which is associated with high risk, is the extensive use in these IAMs of the expectation of social utility[14] as a criterion. In other words, the criterion is the weighted average of utilities of possible outcomes, where the weights are the associated probabilities. Utility might be seen as becoming unbounded or move to minus infinity in catastrophic outcomes. That makes the expected utility criterion for deriving formal policy conclusions largely unusable. That is the point which the late Martin Weitzman (2009) made clearly and strongly.[15] He spoke of an apparently infinite social cost of carbon if an extra unit increases the probability of infinite damage. Essentially, the expected utility approach falls apart as a helpful guide in the context of risks which have to take account of the possibility of hundreds of millions of deaths. In those circumstances we need different approaches to evaluation, or ways of understanding policy. In the context of risks of the magnitude we may face, Joseph Stiglitz and I have suggested a 'guardrail approach' (Stern et al., 2022). Guardrails prevent us from getting too close to danger, such as falling off a cliff or under a train. An example would be one that the world has indeed chosen to use: setting an upper target limit or temperature increase (see also discussion on discounting in the next subsection).

It must be recognised that certain adjustments to these kinds of models, IAMs, have been made in partial acknowledgement of some of those basic problems, but some of the faults, particularly around assumptions on underlying growth, the absence of structural change, and the increasing marginal costs of action, are inherent in the approach. As a result, these IAMs largely rule out the new growth story because they essentially assume away the potential drivers of that growth (which we will discuss in Chapter 5). These models have very limited ability to build in fundamental structural and systemic change. The role and strength of technological change has also generally been underrepresented. Yet structural, systemic, and technological change are at the heart of the necessary response and the new growth story.

As we discussed in Chapter 1, economics does make thoughtful and fruitful contributions to our understanding of technological progress and structural change. Joseph Schumpeter's creative destruction, Kenneth Arrow's learning by doing, and Arthur Lewis's story of structural change all carry great insight and potential around these issues. These elements are fundamental building blocks for understanding and unlocking our growth story but all are omitted from standard IAMs. Revitalising these insights and continuing research in these areas around technology, structural, and systemic change is crucial for an economics of climate change which tackles the real problems we face and the transformations which are available to us.

As argued here, much of the economic modelling of climate change, particularly concerning the interaction of growth and climate action, became locked into an approach which turned out to be an unhelpful route to policy analysis. The recognition of the inadequacy of this modelling approach has strengthened, as understanding has deepened of the potential magnitude of the risks; of the scale, pace, and nature of necessary change; and of advances in technology – including those already achieved and those likely to appear. Our focus in this book is on what economic analysis should be doing. The issues for economics now should be about how to foster the large increases in investment and the fundamental structural, systemic, and technological changes which are now necessary and urgent – and on how to think about the management of immense risk.

Too much of the growth–climate modelling of the past, in my view, turned out to be a false start that became locked in. Before we begin the discussion on where economics should go, we must take a careful look at intertemporal values, discounting, and ethics. These values are critical to a discussion of how to act now to tackle fundamental risks in the future.

Flawed analytical and ethical approaches to intertemporal values and discounting

Economics has been far too casual and simplistic about discounting and valuations applied to future impacts and consequences, particularly in the context of the very big potential impacts associated with climate change. All too many economists appeared to think that they could read off such intertemporal values[16] for social decisions, here of immense importance, from rates of interest or return in private transactions on financial markets. That is bad economics and muddled thinking: first, market transactions reflect personal preferences, not moral positions, and, second, the markets in question, here capital markets, are full of imperfections.

Many formulations go straight to the question, 'What is the discount rate being applied?' That is not the place to start. The key idea is a relative valuation between a unit of benefit in the future and one now. That is a shadow relative price or relative social valuation. It is the discount factor, not the discount rate. There is an important technical distinction between discount factors and discount rates. The latter is the proportional rate of fall of the former. For example, if a unit in the next period is worth 0.99 of a unit now, then the discount factor is 0.99 and the discount rate is 1%. The former is logically prior to the latter.[17]

Yet many economists failed to start a discussion of discounting from the basic question of relative valuations and the future increases or decreases in key variables likely to influence those valuations. All too many jumped straight to discount rates. I offer briefly here a slightly technical account, although with minimal mathematics, of the intuitive arguments on discounting just given.

We begin with the question: 'How do we value an extra unit of account, for example consumption, in the future relative to now?' This is a 'normative' issue, a question of values in social decision-making, and we use the adjective 'social' to emphasise this. That relative valuation, or relative shadow value,[18] is the *social discount factor*. *Social discount rates* are the rates of change of valuations, not the valuations themselves.

As soon as we start from first principles in this way, we see that these relative valuations will depend, first, in most ethical frameworks, on how well off, or poor, future generations may be relative to ourselves. In fact, under inaction or weak action on climate they could be much worse off. Second, these relative valuations also depend on how we would value a future life, which differs from a current one only in terms of date of birth.

Note, also, it is clear if we proceed in this way from first principles that we would not expect to find a single discount rate. It will depend on a time period, the good in question, who the good is consumed by, and so on. Relative valuations of the goods concerned in the future relative to now will depend on the date we are discussing, which good, the person at issue, and all the circumstances at that time.

As these valuations depend on how we act now, they cannot be seen as something exogenous, in other words, determined external to our decisions and the system. Most people would value an extra unit going to a richer individual less than one going to a poorer one. That is a value judgement that is probably widely shared. On that basis the social discount factor depends on the state of affairs in the future relative to now, and that in turn depends on our actions now. Social discount factors, and thus also their rate of change, social discount rates, are endogenous. In other words, they depend on what we do and are not something we can 'import' from entirely outside our models; they are not exogenous.

I will argue here that pure-time discounting[19] has little support from most commonly used ethical frameworks. It is a critical ethical assumption often used in economics but with little ethical foundation. Above I referred to value judgements concerning people at different levels of consumption or income. An explicit discussion of values will also be crucial when thinking about how lives in the future, at identical levels of consumption or welfare to now, are valued relative to now. That latter question refers to pure-time discounting, or discrimination by date of birth. It is a different issue from whether or not future generations will be better or worse off and relative valuations in those circumstances.

In my view, it is not easy to find a persuasive ethical argument in favour of applying a value to a future life, identical except for date of birth, which is different from that applied to a life now. To do that would indeed simply be discrimination by date of birth. It would be an extraordinarily strong position to take and would need a strong argument to justify it. It violates principles of symmetric or equal treatment which we often invoke as key features of equality and common humanity. Examples of the use of such principles include equality of right to vote across different ages (subject to a minimum age) and equality before the law. And such notions of equality of individuals in relation to life, liberty, and the pursuit of happiness are embedded in key political and philosophical statements – for example, in the US Declaration of Independence. Further, they are part of many of the great religions of the world.

Finally, we must note that there is some probability that some exogeneous factor, a large asteroid, say, could arrive and obliterate the world; there would be a reason for discounting on those grounds. Such discounting would be logically distinct from pure-time discounting; it does not value the life or utility per se of a future person any less, but simply takes into account the probability that the person might not be there. But such discounting would likely be very small, much smaller than, say, 1%, which would imply that there is a 50% probability of the world being wiped out by an asteroid in the next 70 years.[20] Figure 3.8 sets out the logic of the argument I have just discussed in a simplified way.

Figure 3.8: The ethics and economics of discounting

Decisions now affect lives and livelihoods, and the risks faced, in the future. Intertemporal evaluations are central.

Social discount factor
The relative social valuation of an extra unit in the future, relative to an extra unit now

Social discount rate
The proportional rate of fall of the social discount factor

The valuation of an extra unit at time *t* will depend, for most ethical observers, on:

i. The levels of living of those at time *t* relative to now
High valuation if future generations are likely to be poor; low if they are likely to be rich

ii. The valuation of a future life (or utility) relative to one now
'Pure-time discounting', effectively to discriminate by date of birth. Hard to provide a serious ethical argument in favour of pure-time discounting.

Note: The 'extra unit' could be of consumption or income, or some particular good, depending on the unit of account.
Source: Author's elaboration.

Broader perspectives on ethics

In my view, one of the causes of the weakness of the profession in this area is the absence of discussions of moral philosophy in much of economics education. Further, there has been a line of argument that ethics and moral philosophy should be excluded from the formation of economists and their toolkits and perspectives. These issues, some argue, are not for economists and best left to philosophers, politicians, or religious leaders.[21] Or that values can be 'found on the shelf elsewhere', for example in markets. Whilst there is no single 'correct' ethical position, excluding ethics is to misunderstand and distort policy making. Values should be explicitly discussed, not 'buried' or 'imported' in simple-minded ways.

The crises of climate and biodiversity, and the responses to them, will have profound consequences for this and future generations – indeed, potentially existential for many. And remember that 'inaction' is a choice, even if taken by default, which carries immense consequences. Our ethics and values, whether implicit or explicit, are basic to our choices, including those involving minimal action or inaction. Facing risks of this magnitude forces us to ask ourselves who we are, what our values are, and what we stand for.

To illustrate different perspectives on the relevant ethical issues here, five prominent broad approaches in moral philosophy are considered briefly: Kantian, contractarian, Aristotelian, 'commonsense pluralism', and consequentialism. They can be applied to the deep moral challenges which arise here in relation to the welfare or interests of current and future generations. In my view, they all, and they are very different in moral perspectives, point to strong action on the climate and biodiversity crises.[22] That they point the same way not only helps to build a strong ethical case for action, but also increases the chances of agreement.

One of Immanuel Kant's formulations of his 'categorical imperative' was: 'Act only according to that maxim whereby you can, at the same time, will that it should become universal law.' In other words, act in a way that you would wish all others to do the same. That, for example, would see 'free-riding' through my own inaction on climate change, whilst hoping or assuming that others will act, as immoral in the sense of contradicting the categorical imperative. Kant's second formulation of the categorical imperative was to never treat 'humanity' as an instrument in the sense of seeing others as a means to our own ends (Kant, 1993 [1785]). Again, that points to strong individual responsibilities to act on climate; not to 'free-ride' and not to treat future generations as an instrument for our benefit.

The contractarian approach – as, for example, embraced by Jean-Jacques Rousseau or John Rawls – sees an individual as a member of society and as having a contract with that society. Rawls (1971) examined how such a contract could be based on the idea of what individuals would seek as a social contract if they did not know where or who they would be in a society. That was his idea of the 'original position' being behind a 'veil of ignorance'. If I do not know which generation I will be in from behind that veil of ignorance, I would likely worry about being part of a future generation that could be in catastrophic circumstances as a result of neglect from this generation. That would likely lead to an argument that future generations should have the same weight in social decision-making as current generations. This would constitute an argument against pure-time discounting.

An Aristotelian approach focuses on the idea of moral character or 'virtue', or 'what sort of person should we be?' We have some notion of good behaviour, as we might recognise good playing of the flute or kindness to others. Most people would regard reckless behaviour which potentially causes great damage to others, such as drunken driving, as not virtuous. Similarly, such an approach would likely see neglect of a kind which puts future generations in danger as unvirtuous.

Some have also suggested an approach known as 'commonsense pluralism', which tries to crystallise everyday moral beliefs and treat them as guiding principles.[23] In this approach one adopts the moral behaviours that have evolved, perhaps on the implicit or explicit assumption that they have evolved for a beneficial functional purpose. A problem with this approach in this context is that we have not yet experienced the kind of risks we could be facing. Nevertheless, it is likely that recklessness in relation to the future of our descendants or children would not fit well with such commonsense pluralism.

We should note that many of the religions of the world embody a deep respect and central place for nature, including Hinduism, Buddhism, Taoism, the Abrahamic religions (Islam, Judaism, and Christianity), and many of the belief systems of indigenous communities across the world. An important example in relation to climate was the encyclical (June 2015, six months ahead of Paris COP21) of the late Pope Francis, *Laudato Si'*, on environment, nature, and climate. A succinct expression of one of his key messages was: 'If we destroy creation, then creation will destroy us' (Pope Francis, 2014). Not only a clear position, but also powerfully and simply communicated. 'Love thy neighbour as thyself' is a maxim that also points away from free-riding, away from pure-time discounting, and towards strong climate action.

There is increasing emphasis in public discussion on the notion that the natural world has rights. It has been part of Ecuador's constitution since 2008 and the case has been argued by indigenous communities across the world. The quote from Pope Francis above, and the arguments in *Laudato Si'*, stress the importance of respect for nature as God's creation. Such rights of nature are in the spirit of Rachel Carson's famous book *Silent Spring* (1962). And the idea has recently been argued powerfully by Robert Macfarlane in his book *Is a River Alive?* (2025).[24]

Finally, amongst the broad approaches discussed here we have consequentialism, which is the standard approach in economics. Actions are assessed in terms of the perceived costs and benefits associated with their consequences. Making judgements within consequentialism usually, but not always, requires an aggregate criterion, measure, or index. And if utilities are invoked in making assessments or judgements, they have to be specified; the specification of utilities is itself an ethical position.[25] If the consequences of some set of actions could be catastrophic for large numbers of people, then from a broad range of criteria for evaluation we might regard those actions as unacceptably damaging. Thus, again, consequentialism would likely, in my view, point to strong action on climate change. But that does depend on specifications of consequences and on values used to assess those consequences. The application of the consequentialism approach often takes the form of thinking of an action as a small perturbation to the status quo. That often involves comparing a world with and a world without the action in a static model, and where consequences are small. In this case, however, we have to think of action and inaction as associated with potentially very large consequences, and occurring in systems with complex dynamics.

From now on the analysis in this book is based largely on the consequentialist approach. However, it is important, in my view, that we recognise that this approach is just one of the approaches in moral philosophy of relevance here. How we, as this generation, value the lives of future generations will be critical. If we attach very little weight to their lives and to losses or gains to them, then we are likely to be less willing to take strong action on climate change. We would normally think that interested parties should be part of decision-making but, of course, future generations are not present when decisions are taken.

3.3 Ways forward for constructive analysis in economics and the social sciences

Much of our discussion so far in this section has been highlighting past mis-leading approaches in economics and ethics. They have done real damage and should be avoided in future work. We now turn to the more constructive and creative story of how, in my view, we should be thinking about these great challenges and crises. These ideas will guide us throughout the rest of the book. We examine sustainability and its implications for perspectives on types of capital, political economy, and systemic and structural change.

Sustainability and a broad view of capital and investment

Understanding ecosystems in terms of their functionality helps us appreciate their role as capital goods, essential for the sustainability of our biosphere.[26] Natural capital consists of the world's stocks of natural resources, including soil, water, air, and all living things. This capital yields a flow of services that contribute to human life and livelihoods. As we saw in Chapter 1, the view of sustainability adopted here can be seen as requiring action to maintain or enhance the group of capitals and endowments for future generations, so that opportunities are not diminished over time. These endowments involve all forms of capital: physical, human, natural, and social/cultural.

These four types of capital, and the services they provide, are deeply interwoven. Taken together, they shape future opportunities. Physical capital includes tangible assets such as infrastructure, plant and equipment, housing, and so on. Human capital encompasses education, skills, knowledge, and health. Natural capital includes ecological assets such as forests, flora, soil, fauna, oceans, water, and air. Social/cultural capital refers to the institutions, behaviours, societal beliefs and norms, and networks that foster trust and cooperation within communities.

Power and inequality will be of real significance to social capital. Though harder to define and measure than the other three, social capital is vital to the function-ing of society and the shaping of opportunities. It can be eroded and it can be augmented; the former can happen more quickly than the latter. In some ways it should be seen as social and cultural capital; for example, how we see the notion of responsibility, individual and mutual, is both cultural and social; we sometimes use social, for brevity, as including cultural. The work of Elinor Ostrom (1990, 2000), for example, showed how institutions, rules, and behaviour – essentially what we have called social and cultural capital – could develop in ways that fos-tered care for and sustainable use of natural resources. In this case social/cultural capital is developed in part as a result of a shared need to protect natural capital.

Policies and incentive structures to tackle climate change, biodiversity, and environment together will have a number of key elements. Valuing natural capital and bringing those valuations into economic decision-making is of special importance. We must see natural capital as an over-arching supplier of the services that allow the economy and society to function. In the sense that it facilitates and enables other activities, it is fundamentally infrastructure. If we destroy our natural capital, then our physical capital will be undermined and so too our economies, our health, and our societies. There is a crucial sense in which our economy and society are wholly owned subsidiaries of our environment, our climate and biodiversity, and our natural capital.

The interplay between these capitals creates dynamics that can either enhance or undermine development, sustainability, and resilience. For example, investments in human capital can lead to better health outcomes and longer life expectancy. Healthy populations with a balanced age distribution can, in turn, strengthen social and cultural capital through greater family, community, and societal engagement, including across generations. Environmental degradation, including climate change and biodiversity loss, can harm public health, erode physical assets through climate impacts, reduce social cohesion, undermine livelihoods, and increase the risk of conflict. Investment in, or shock to, one kind of capital can also involve investment in or shocks to other kinds. For example, investing in forests can both enhance natural capital and improve human capital via health and recreation; wildfires can destroy natural capital and undermine physical, human, and social output. There can be virtuous or vicious cycles.

Understanding and managing these capitals cohesively is central to promoting sustainable and resilient development. Their dynamic interplay determines the success or failure of sustainable development initiatives, making it essential to approach actions and investments in relation to each capital as part of an integrated whole rather than in isolation. Investing in all four capitals offers the potential of advances in all of economic growth, social well-being, and environmental health. Considered and effective action in this way can provide real and mutually-reinforcing advance across all development goals.

Political economy

Many of the challenges of acting on climate change are inherently challenges of political economy, including resistance from those who perceive potential loss or dislocation, the challenges of building a just transition, and the processes of institutional and behavioural change. Political economy should be part of the foundations of the economics discipline. Indeed, in the late 18th century a giant figure in the building of economics, Adam Smith, was fundamentally and explicitly concerned with political economy.

Achieving effective climate action and embarking on a new path of growth and development is not merely a matter of scientific and technological advances; it is also, as argued throughout this book, a deeply political, economic, and social challenge. Indeed, many, including myself, would contend that the main obstacles lie within politics, economics, and society. These barriers and difficulties require careful analysis, strategic action, vigorous engagement, good communication, and wise judgement.

Political economy plays a crucial role in either facilitating or hindering progress on climate action and the new growth story. In this context we take political economy as meaning the way political, economic, and social factors influence and shape climate policy decisions and their outcomes, particularly in relation to opposing forces. Analysing the political economy of transition helps understand the factors that can either enable or block changes in action, policy, and practices.[27]

Let us consider the traditional economic approach, which has much merit, to climate change focused largely on market mechanisms. There will be major difficulties in implementation, arising from those who might see their interests as threatened. There are real distributional consequences. There will be very large net gains to effective and timely action on climate change, but there will be some net losers, particularly those receiving rents and profits from fossil fuels. Thus, there will be vested interests and opposition. There will be battles for influence and power, within and across nations. Climate policies have significant distributional impacts across regions, classes, and generations. Understanding the barriers potentially arising from political economy requires careful analytical attention if robust policies to overcome these barriers are to be created. That analysis should, in particular, provide guidance on likely gainers and losers, including over the shorter and longer term.

Hallegatte et al. (2023) provide a useful framework, in the context of climate, which categorises political economy issues into four main components: institutions, interests, ideas, and influence.

Institutions, first, are the established structures, customs, and regulations or rules that influence economic, political, and social behaviour. Institutions, such as climate change legislation and committees, can provide clear direction, fostering confidence and stability essential for long-term planning. However, certain institutional processes, such as spatial planning procedures, can create obstacles to rapid decision-making, for example when determining routes for transmission lines. Regulatory bodies, such as those governing electricity or water, may have a narrow focus, prioritising consumer prices while neglecting necessary emphasis on climate and environmental considerations. Rules for project appraisal in finance ministries and elsewhere can be skewed against the green transition, for example through their approach to discounting or inherent limitations in examining structural, programmatic, and large-scale changes. Overall confidence in institutions and structures can boost investor confidence. In some cases, such as within the EU, regional, or cross-country, institutions can exert a significant influence in creating confidence in the direction of travel. They can, of course, also cause bureaucratic delay.

Second, interests represent what different groups stand to gain or lose from climate action and new approaches to growth. The understanding or perception of how costs and benefits are distributed can strongly influence how various actors respond to climate policies. For example, workers may be concerned about job security as industries evolve, while environmental groups push for action to protect nature. Policy-making will be more successful in gaining support if it works to design climate policies that account for these varied interests by, for example, integrating job opportunities – and, in some cases, protections – into new policies.

Immense forces are at play from the fossil fuel industries, where the stakes, in terms of their profits and rents, are in the trillions of dollars per year (Verbruggen, 2022). These industries can and will vigorously oppose potential threats to their profits and rents. Historically, this opposition has included fostering uncertainty or denial regarding climate science (see, e.g., Oreskes and Conway, 2010).[28] They are also likely to exploit crises, such as those arising from the war in Ukraine, to put forward dubious arguments for slowing the transition. For further discussion of such issues, see Chapter 7.

Third, ideas refer to the beliefs, values, and perspectives people hold – including their views on government 'interference', the virtues of the market economy, attitudes towards climate science, prioritisation of the environment and biodiversity, and notions of fairness and responsibility. These all shape their preferences and their support or otherwise for climate policies and approaches to growth. Trust in science or government, as well as the perceived fairness and effectiveness of policies, play critical roles in gaining public support.

Education can be influential here; science and environmental education in schools have already had a major impact on public understanding of climate issues. Most European students entering universities now would have been taught the basics of the greenhouse effect and climate change at school. The media, including social media, also play a role, sometimes in the promulgation of denial of the importance of climate change.

Finally, influence refers to the fact that different actors in society have varying capacities to use their power, authority, and resources to sway decisions and shape ideas in their favour. Influence concerns who has the power to generate the support that can foster or block changes and shape perspectives. This can involve lobbying by businesses, campaigns by environmental groups, or public protests. For example, large companies might use their influence to slow down policies that would harm their profits, while civic groups might rally public support for stronger climate action or cleaner air in cities. Others might oppose, for example, creating pedestrian zones or limitations on the use of motor vehicles.

Understanding who holds influence and how they wield it can explain why certain climate policies succeed or fail. Influence is not only about shaping specific decisions and policies but also about shaping ideas and attitudes. Here, politicians, religious and cultural leaders, and those skilled in social media, can exert powerful influence.

These four categories are helpful in structuring some aspects of political economy but do not fully cover crises, preferences, and building support. These are examined partly in the following paragraphs and in Chapters 4, 6, and 10.

In the very early days of work on the Stern Review, in early 2006, I spoke to the late Danny Kahneman, a famous psychologist and economist. He argued that 'the messenger' was critical in conveying ideas and pointed to trust in religious leaders. Pope Francis has been of great significance as a leader, with his cyclical *Laudato Si'* in 2015 being an important example. See Section 4.1 in Chapter 4 for a more detailed discussion of the contribution of both these men.

For communities to achieve shared and stable commitments to tackle climate change and to embark on a new path of growth, some shared understandings will be necessary. That does not, of course, mean unanimity or 'group think', but it does mean both the discussions of arguments and values and some cohesion around a sense of direction. Understandings shape both preferences and perspectives, which are then expressed through behaviour, both politically and personally. It is therefore important to analyse how preferences, perspectives, and behaviours are formed in relation to climate, the environment, and growth. The use of 'citizens' assemblies' can help ordinary citizens come together and listen carefully to, and cross-question, evidence and analyses on a subject on which decisions are necessary. They can help develop considered views, rather than the instant reactions of opinion polls.[29]

The UK held a citizens' assembly on climate change in 2020, bringing over a hundred people together. The resulting report sets out the principles the members agreed on and the resulting recommendations, which included a levy for frequent fliers and a ban on the sale of polluting vehicles by 2030–2035 (Climate Assembly UK, 2020). France held a Citizens' Convention on Climate in 2019–2020, bringing 150 randomly selected citizens together. The final proposals included measures such as making ecocide a legal offence, reducing speed limits on highways, and requiring renovations to improve building energy efficiency (Convention Citoyenne pour le Climat, 2021).

This overall set of topics – political economy, institutional and behavioural change, and the formation of values – is increasingly the focus of research by political scientists, social psychologists, economists, and others. Particularly notable is the work of my LSE colleague Tim Besley and his collaborators, who have made important contributions to the political economy of climate and environment, for example shedding light on the role of values in supporting or hindering green transitions (Besley and Persson, 2023). And on how values and behaviour change in response to experience, evidence, and argument. Their research starts from the assumption that demand depends not just on changing prices and taxes but also on values. Besley and Persson (2023) investigate the way in which market failures interact with government failures (such as short-termism) and stand in the way of green transitions, preventing them outright or slowing them down.

The way ideas are shaped and political opinions are formed will be of crucial significance in the future of climate action. So too will be an understanding of the political economy of opposing and supporting forces. This is a vitally important area of research.

Systemic, structural, and technological change; EMDCs; and investment

The growth story, and within it fundamental economic change, should now, in my view, have still stronger emphasis than it did in the Stern Review. This is a much more dynamic story than we could see at the time, even though we rightly said, in the Review, that climate action is pro-growth in the long term. We now see that climate action is a driver of growth in the short, medium, and longer term. And that the devastation resulting from weak action would undermine and reverse growth. The technological understanding and advances and the lessons from experience are telling us that many poorer countries can leapfrog the dirty, destructive phases of development followed by the rich countries to a much more attractive and sustainable path of growth.

The dynamics of systemic, structural, and technological change must now be at the centre of economic analyses. As we saw in Chapter 1, the economics of structural and technological change has a strong history. There are Schumpeterian theories, which are of great significance. Development economics, particularly in the three decades after the Second World War, highlighted changing sectoral composition as countries develop. The works of Arthur Lewis, Albert Hirschman, Paul Rosenstein-Rodan, Ragnar Nurkse, and others are, in their different ways, full of insight around these processes of change (see Chapter 1). The collection of approaches and how they are chosen, and how they are combined into decision-making, will all be crucial to social policy formation. They should include sector studies (e.g., energy), system studies (e.g., cities, land, or water), and technological studies on how to discover and implement lessons from economic history. Overall economy-wide formal models will have some roles, but they should not dominate attempts at understanding policy because they will leave out, in most cases and in large measure, the key structural, systemic, and technological issues.

Aggregate growth models cannot really address these questions, since the modelling structure leaves them out. Thus, they are silent by assumption on the most central action and policy issues. We will need a number of analytical approaches.

In a similar vein on structural change, I would now put the EMDC challenges still more prominently at the top of the research and action list. Their growth, and how they grow, will be the most important driver of emissions. And their development and ability to reduce poverty, and how they develop, will be the most important element of advance in well-being across the world. The majority of people and the incidence of poverty are concentrated in these countries. They are the most vulnerable and they have the greatest opportunities to find new and better paths to development. The biggest challenge for them, and where international action should be focused, is in fostering and financing the large increase in the investment necessary for the new path of development: sustainable, resilient, and inclusive.

3.4 Concluding remarks: to my fellow economists

I close this chapter with an appeal to, and an agenda for, my fellow econo-mists. Our profession has not contributed to this most important challenge of our day to its full potential, although it is now making progress. There are six priorities:

1. The analysis of how fundamental structural, systemic, and technologi-cal change can take place and how it can be promoted; including anal-ysis of its political economy and a just transition.
2. How to foster and finance a major increase in investment, from micro, structural, and macro perspectives.
3. How to understand and promote policy towards the interweaving of the economy, climate, and biodiversity.
4. How to understand the analytics of and policy towards extreme risk.
5. How to understand the role and political economy of institutional and behavioural change in the processes of transition and in establishing the new growth story.
6. How to build a public economics where time matters, urgency is cen-tral, there is real dynamics, and there are multiple market failures.

A broader research agenda, including more detail on these priorities, is pre-sented in Chapter 10.

Creating the understanding of what is necessary to respond to the climate and biodiversity crises and setting out on a new growth story should have been a key task and challenge for economics over the last two decades. In my view it has, in large measure, failed to step up to that challenge with the sense of urgency, depth, and clarity of focus required. Further, much of what has been done has been misleading about the magnitude and nature of the problems, issues, and on the necessary responses. Thus, it has been one of the causes of slowness of action, although arguably not the most important one. Whilst I do recognise that eco-nomics is now moving in a more positive direction, precious time has been lost.

If I were to start my research career and activities in public life again, these are the analytical issues I would want to focus on. And, if I were starting the Stern Review over again, it would be these sets of issues, on action and analyt-ics, that I would make the highest priority.

The last two chapters have provided foundations and principles for the analysis of climate change. The essence of the challenges and crises requires that we examine carefully key guiding principles from science, politics, eco-nomics, and ethics. Such examination is essential to building sound policy. Indeed, building foundations through an understanding of these subjects, principles, and challenges is basic and necessary. This chapter has shown that we need to change not only our economies, but also economics. It is of utmost importance that, when conducting economic analysis, we leave behind inad-equate assumptions and models, incorporate ethical considerations, value all four capitals comprehensively, recognise the role of political economy, and acknowledge the dynamics of systemic, structural, and technological change.

I have been strongly critical of major strands in the economics literature – and, I think, for good reasons. However, economics is, at last, moving in a better direction. It is moving beyond narrow textbook frameworks and unwise attempts to shoehorn new and immense issues into standard models. Some of this is reflected in the discussion of the economics of the new growth story in Chapter 5 and the policies and actions to realise that new growth story in Chapter 6.

The next chapter provides an analysis of the context for action, particularly developments over the last two decades that present opportunities to accelerate action. After two chapters on analytic foundations, often highlighting daunting and complex challenges, let us remind ourselves that we are building towards a positive and hopeful story. But one where there are major obstacles to overcome. They can indeed be overcome, but care with both principles and reality will be crucial to that endeavour.

Notes

[1] Various projections suggest that China's carbon emissions will peak soon, or may have already peaked (You, 2024). Supporting this, a survey by the Centre for Research on Energy and Clean Air involving 33 domestic and 11 international experts found that 44% believe China's emissions have either already peaked or will do so by 2025. This marks a notable increase from 2023, when only 21% held this view (White, 2024).

[2] European emissions were still high relative to global standards in 2019 (Chancel et al., 2022).

[3] Figure 3.4 represents a cross-tabulation between income and emissions per capita across countries. There is a close correlation. Among low-income countries (a group of 26 nations with low income per capita), none of them have high emissions per capita.

[4] This is a marked increase from holding 25% of the share in 2021, see Figure 3.6.

[5] China currently emits around 10 $GtCO_2e$ per annum more than the USA and is a little over 200 $GtCO_2e$ 'behind' in cumulative emissions.

[6] Additionally, the convention assumed that lower-carbon or cleaner versions of economic activities would be unambiguously more costly, raising concerns about who would bear the costs if poorer countries pursued such paths (UNFCCC, 1992, e.g., Art. 4.3). As we mentioned in Chapter 1, the great changes in the costs of action were not foreseen. Developments like these fundamentally change, indeed enhance, the case for action.

[7] I was on the platform with him to hear that important statement.

8 As noted, this was built on the long-standing friendly relationships between Xie Zhenhua on the China side and John Kerry and Todd Stern on the USA side of negotiations.

9 In economics, 'marginal' refers to the impact of a small, incremental change, such as the cost or benefit of producing or consuming one additional unit of a good or service. If marginal private costs (or benefits) differ from marginal social costs (or benefits), it means the true costs or benefits of an activity are not being fully accounted for in the incentives faced by the private decision-makers.

10 See, e.g., a joint letter to the *Wall Street Journal* of 16 January 2019, signed by 3,649 economists from the USA. The first sentence of its first recommendations reads: 'A carbon tax offers the most effective lever to reduce carbon emissions at the scale and speed that is necessary' (*Wall Street Journal*, 2019). And see Chapter 5.

11 Formally speaking, a public good is a special case of an externality where the effect on others works through the sum total of the relevant actions – in this case, emitting GHGs.

12 Expected utility maximisation is a decision-making criterion where agents choose the option that provides the highest average or 'expected' satisfaction (utility), taking into account both the range of possible outcomes and their probabilities, and where the probabilities are used to calculate the average. In this sense the decision-maker looks at averages over possible outcomes.

13 Discounting refers to valuing a unit occurring in the future differently from one now.

14 Social utility is the utility (or satisfaction) from the perspective of the welfare of all individuals in the economy taken together. Many IAMs aim to capture the aggregate effects of climate change by considering an aggregation of the expected utility of representative agents.

15 Marty was a good friend for almost half a century and we discussed these issues at length.

16 Intertemporal values refer to the valuation of effects occurring in the future. These are value judgements and would usually be influenced by ethical perspectives and the circumstances envisaged in the future.

17 In continuous time, if the discount factor is k, then the discount rate is (minus) the rate of change of k, or $-(dk/dt)/k$, where in the notation of simple differential calculus, dk/dt is the rate of change in the discount factor over time.

18 A shadow value refers to the social value of an additional unit of a resource or commodity; it is in contrast to the private value as embodied in market prices. In this case, the valuation of an extra unit of consump-

tion in the future relative to now can be understood as the shadow value of future consumption, captured by the social discount factor.

[19] Pure-time discounting refers to attaching a lower value to utility, welfare, or satisfaction occurring in the future relative to now, simply because it is in the future. This aspect of discounting is separate from and in addition to any discounting of consumption units because we may be better off in the future.

[20] For further discussion of exogenous extinction discounting, see Stern (2015) and Chichilnisky et al. (2020). The extinction–discounting link goes back, at least, to Kenneth Arrow and James Mirrlees in the 1960s (see, e.g., Mirrlees, 1963). Indeed, I recall discussing exactly these issues in the late 1960s with Arrow, Mirrlees, and Solow, respectively. It has also been examined by Partha Dasgupta and Geoffrey Heal (1979), Robert Solow (1991), and Joseph Stiglitz (1982).

[21] That was, in essence an argument put forward, for example, by Lionel Robbins (1932).

[22] This discussion draws on Chapter 6 of Stern (2015) and on two articles in *Economics & Philosophy*, namely Stern (2014a) and Stern (2014b). See those sources for further references.

[23] See, e.g., Fesmire (2020), Light (2002), and Palmer (2014). For an overview of ethical perspectives including commonplace pluralism in the context of climate change, see Jamieson (2007).

[24] See also Philippe Sands' (2025) article in the *Financial Times* reviewing Macfarlane's book on 26/27 April 2025.

[25] Although there do seem to be some who come close to arguing that we can make objective assessments of 'happiness' – see, e.g., Layard (2005).

[26] See, e.g., Dasgupta (2021, Section 13.1).

[27] See, e.g., Worker and Palmer, 2021.

[28] At a conference in Columbia University in November 1959 to celebrate the 100th anniversary of the USA oil industry, Edward Teller, the famous physicist, warned explicitly, analytically, and numerically of the dangers of global warming arising from the CO_2 released from the burning of fossil fuels. He included the statement 'All the coastal cities would be covered' (in Franta, 2018a, paragraph 5). For decades afterwards the oil industry, notwithstanding the fact that its own scientists were verifying these arguments, tried to undermine the science. In 1965, Frank Ikard, the president of the American Petroleum Institute (API), acknowledged at API's annual meeting that climate change was occurring and would bring challenges for the sector (Franta, 2018b).

[28] They usually involve citizens chosen randomly. They have been pioneered in Ireland (Farrell et al., 2019) where they were influential, for example, in shaping decisions on gay marriage and abortion.

4. A changing world: new opportunities and an agenda for action

The foundations for the analysis of climate change in science, politics, economics, and ethics were set out in Chapters 2 and 3. It is a story of challenges, difficulties, and obstacles. We saw that the science pointed strongly to action with urgency and scale. That requires fundamental technological, structural, and systemic change. The intense and interwoven problems of biodiversity add further difficulty; we have twin crises. The political assessment was one of eroding trust between nations in relation to a problem that, in its essence, requires global action – and of pushback from vested interests. We also saw that much of the economics of climate change had been misleading and unhelpful.

This chapter highlights how a changing world, particularly developments in the two decades since the Stern Review, has created forces that provide opportunities for the acceleration of the action that is now urgently needed. This does not mean that the obstacles are easily overcome; however, there are strong tail winds as well as head winds.

The first set of forces for change come from public pressure and legal accountability. They are examined in Section 4.1. The extraordinarily rapid change in technology and the increasing recognition of the opportunities in investment in these technologies and in the new growth story by the private sector will be a second set of drivers of change. These are discussed in Section 4.2.

International institutions are moving climate to centre stage, including most of the multilateral development banks (MDBs), and new coalitions are forming. The major countries and regions, in their different ways, are moving too. The USA was moving quickly under President Joe Biden. Indeed, under the first Trump presidency (2017–2021), investment in renewables for the power sector moved rapidly, for the simple reason that they are highly competitive relative to fossil fuels. The second Trump presidency is a setback, but the forces of change will continue. Further, developing nations are increasingly taking a leading role. International action in a changing world is the subject of Section 4.3.

These changes in international action across the world are creating a new geopolitics of climate change, indeed of development more generally. Geopolitics is also influenced by changing attitudes to trade and international relations which are showing turbulence, particularly in relation to the second Trump administration. The changing geopolitics is examined briefly in Section 4.4.

In Section 4.5, we look forward and set out key elements of the agenda for action. We bring the analysis of the challenges of Chapters 2 and 3 and of the opportunities and forces of Chapter 4 together, setting the framework for the analysis for the rest of the book.

4.1 Forces for change: public pressure and legal accountability

There are major pressures for change from outside government. These come from public perception and opinion, from political action, and from the law. We examine them in turn. The private sector is also a powerful force – and often a voice for change – which we consider in Section 4.2, alongside technology.

Public perception: the climate message matters, as does the messenger

As we saw in Chapter 2, the role of human influence on climate change is almost universally accepted within the science community. The accumulation of evidence, along with public discussion and direct experience, has led to widespread recognition, across most of the world, of the existence of climate change and an understanding of its causes in human action.

At present, global surveys indicate that the general public believes climate change is happening, that humans are responsible, and that strong action is necessary. There is also increasing public recognition of the dimensions and intensity of the danger. Recent research by Vlasceanu et al. (2024), with more than 59,400 participants across 63 countries, found that the average belief in climate change was high (over 85%), with modest variation among nations – see Figure 4.1A. Most participants also supported active climate policies, with the average global score being 72% (Figure 4.1B).

There are many who believe that stronger action should be pursued. For example, research conducted by Andre et al. (2024) surveyed nearly 130,000 individuals across 125 countries, finding that 89% think that their 'national government should do more to fight global warming'.[2] A nationally representative survey (although with only 1,033 people in the sample) (Leiserowitz et al., 2023) found that 72% of the population in the USA think global warming is happening and well over half of Americans (58%) understand that global warming is mostly human caused. Another study found that 74% of Americans support the USA participating in 'international efforts to reduce the effects of climate change' (Tyson et al., 2023).

Further, many surveys show that people are prepared to pay more for goods that are sustainably produced. For example, a recent consumer survey by PwC, which collected more than 20,000 responses across 31 countries, found that 80% of consumers said they were willing to pay more for sustainably produced or sourced goods (PwC, 2023). However, results from such surveys are often context specific. For example, Bain and Co. (2023) found environmental concerns to be highest in fast-growing economies, such as

India, Indonesia, Brazil, and China, which also reported higher willingness to pay premiums on sustainable goods. Consumers from these countries were willing to pay premiums between 15 and 20%, compared to 8 and 11% for consumers in countries like the USA, UK, Italy, and Germany.

Figure 4.1: Country-level means of (A) climate change belief and (B) policy support

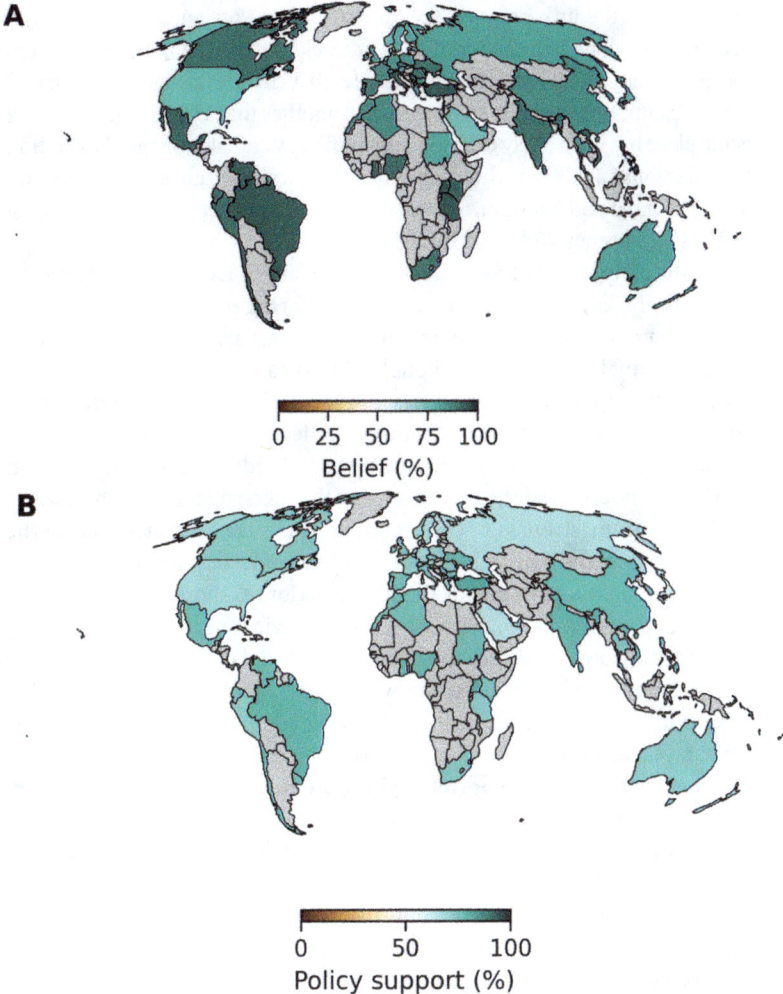

A

B

Note: Countries without available data are shown in grey.[1]
Source: Figure 4 in Vlasceanu et al. (2024, p. 9). Copyright 2024 The Authors, CC BY 4.0.

Increasingly extreme events can be linked directly, via work known as 'attribution science', to climate change, as opposed to a simple statement such as 'climate change makes such events more likely' (McSweeney and Tandon, 2024; van Oldenborgh et al., 2021). Attribution science has helped to shift public opinion by linking the seemingly abstract concept of climate change with personal and tangible experiences. Indeed, improvements in the science, under efforts such as those by the World Weather Attribution initiative (WWA, 2024), have enabled the rapid establishment of the link between specific extreme events and climate change. This can now be done in a matter of days or weeks, before the news about them dissipates. It is one thing to claim, for example, that climate change has increased the overall length of heatwaves in general, and another that a specific heatwave in a specific place has been likely caused or magnified by climate change. Attribution has been demonstrated for a whole range of events, from wildfires in the USA and heatwaves in India and Pakistan, to typhoons in Asia and record-breaking rainfall in the UK (McSweeney and Tandon, 2024).

During the early stages of the Stern Review in 2006, I had the opportunity to discuss, with the late Danny Kahneman, a pioneering psychologist and economist, the issue of how views, values, and behaviour are shaped by scientific evidence, personal experience, and public discussion. Danny, who was one of the world's great pessimists, emphasised the many difficulties, highlighting the forces of inertia, resistance to change, and fear of loss. Whilst he was far from optimistic about our ability to change as individuals, communities, and the world, he underscored the importance of the messenger rather than solely the logic and persuasiveness of the argument itself. He suggested that in the USA, for example, the 'evangelical right' and its leadership could be influential in shaping opinions in favour of climate action. Bishop James Jones, the former Bishop of Liverpool, expressed a similar view, seeing one of his key tasks as reaching out to this group, given his potential credibility with them.[3]

Many religious leaders around the world, across all faiths, have provided leadership on these issues. In June 2015, the late Pope Francis published his second encyclical, *Laudato Si'*, which clearly articulated the importance of the environment, climate, and biodiversity in relation to economic development. This was just a few months ahead of COP21 in Paris. He was very skilful and effective in his use of language and ideas, for example stating on another occasion, 'If we destroy creation, then creation will destroy us' and 'God forgives always, we men forgive sometimes, but [nature] never forgives' (Pope Francis, 2014). Whatever your religious beliefs may be, and I note mine are not similar to those of Pope Francis, these tight, concise phrases convey real insight into the issues at stake. It is powerful communication.

The media, traditional and social, have been very varied in their approach to communication relating to climate change. Climate deniers, sceptics, or minimalists/lukewarmers[4] have commanded substantial attention, particularly on social media. There has been a recent resurgence in the promulgation of ideas that seek to delay climate action by discrediting or playing down climate actions, climate science, social movements, and the negative impacts of global warming.

Denying that global warming is happening or that humans are its cause is increasingly seen for what it is, irrational and unscientific. However, denialism has not gone away; indeed, Donald Trump still speaks of climate change as a hoax. But, on the whole, outright denial has been increasingly replaced by attempts to dismiss the scale of risk and exaggerate the cost of response.[5]

It is important to be aware of, and to try to understand, how disinformation and misinformation evolve. Some public or publicly regulated communication outlets have, for extended periods, operated under the misguided belief that 'balance' required equal time and attention for serious climate science and climate denial. The BBC was guilty of this for a long period and eventually had to apologise (e.g., *Today* Programme, October 2017) (Hickman, 2018). As one scientist friend remarked, 'A balanced discussion of science does not require putting up Coco the Clown against a Nobel Prize winning scientist, just because they hold different views.'

On the other hand, there have been outstanding communicators who have presented the evidence effectively. David Attenborough is an exemplary figure, with his powerful television programmes on the loss of biodiversity and its links to climate change. King Charles III in the UK has also, with persistence over many decades, been influential in promoting understanding of the deterioration of the environment and biodiversity, and the potential consequences. The leaders of small island states, such as Mia Mottley of Barbados, have been very effective. So too has Al Gore. We must also recognise the courageous involvement of journalists, lawyers, indigenous people, academics, communicators, innovators, investors, young people, and more in exposing inaction and driving response. *Standing Up for a Sustainable World* (Henry et al., 2020) highlights many of these transformative efforts.

Both the message, with its logic and conclusions, and the messengers are, as has been argued above, critical. Further, how the message is framed can also shape how it is received. In my view, and that of many others, it is crucial to combine the message of concern and fear over great risk, and thus the imperative to act, with a message of hope. The hope here is that, although it is late, it is not too late to make a profound difference to the prospects for future generations. Wise and purposive action can not only drastically reduce the risks but also lead us to a new growth path that is much more attractive than what went before. There are great communicators who can offer that message of hope, including Barack Obama, John Kerry, Christiana Figueres, and Mary Robinson, among others, alongside the late Pope Francis, David Attenborough, King Charles III, and Al Gore, as already highlighted. The late Prime Minister Meles Zenawi of Ethiopia also spoke and led powerfully on these issues. I have the privilege of being on the Council of Hope, founded in 2017 by the great leader and communicator Jane Goodall. All of these recognise and articulate very clearly the magnitude of the threat, but they also convey and communicate a vision of the promise of a new path of development. In the last century M.K. Gandhi and Nelson Mandela were inspirational on the creation of a new world, offering hope, leadership, and action, and both took the environment very seriously. This book is, I hope, itself a contribution to a spirit of hope.

Political action and youth activism

Activists, in particular the young, can and do keep climate action in the public eye and push the limits of what is politically possible. Generally speaking, the young show a higher concern about climate change than older generations. People Climate Vote[6] found in 2021 that people under 18 are more likely to think climate change is a global emergency than other age groups – 70% of people under 18 think climate change is a global emergency, compared to 65% of those aged 18–35, 66% of those aged 36–59, and 58% of those aged over 60 (United Nations Development Programme [UNDP] and University of Oxford, 2021).

The world has witnessed strong public pressure from young people over the last two decades. Since Greta Thunberg's first protest, aged 15, in 2018 at the Swedish Parliament, she has become an icon of climate activism. Her protest launched the Fridays for Future (FFF) movement, which spread quickly and around the world. The core demand of FFF has been delivery on the Paris Agreement. I was in the audience during her 2019 speech at the UN Climate Action Summit to dozens of world leaders. Her speech remains a pivotal moment in the history of climate awareness and activism:[7]

> For more than 30 years, the science has been crystal clear. How dare you continue to look away and come here saying that you're doing enough, when the politics and solutions needed are still nowhere in sight? (Thunberg, 2019).

There has been a clarity and simplicity in young people's demands, dating back many decades. For example, Severn Cullis-Suzuki (from Canada) was just 12 years old when she spoke out at the 1992 Rio Summit about environmental degradation and the responsibility of older generations to act (Grasso and Giugni, 2022). Veena Balakrishnan and young colleagues from across the world have led the Youth Negotiators Academy at the COPs and it has been a real pleasure to be asked, as a person from an older generation, to join their Wisdom Council and work with them. Their skill and energy are remarkable. Hilda Flavia Nakabuye founded Uganda's FFF movement (see Henry et al., 2020) and has worked to empower more women to join the fight against climate change.

In many cases, the young have moved from protest to proposal. In my own experience as a professor at LSE, I have seen our students move beyond demands for action to detailed and strong work on the elements of an agenda for action. For example, in 2017, LSE created a sustainability strategy after careful consultation across the whole university. Students played a key role in its formulation. Demand for courses with environment, sustainability, and climate elements continues to rise. Many of the best students seek graduate study and research on these issues. They will play a strong role in the Global School of Sustainability, established in January 2025 at LSE with the aid of substantial support from an alumnus, Lei Zhang.

Indeed, more than merely acting as protestors, the youth are now shaping policies and strategies, demonstrating how large-scale change happens. For example, youth groups have initiated climate litigation actions themselves (Nisbett and Spaiser, 2023). As of December 2022, around 34 cases across the world were presented on behalf of children and youth on a human rights basis, relying on evidence of these groups' particular vulnerability to climate harm and including principles of intergenerational equity (United Nations Environment Programme [UNEP], 2023).[8] It was the initiative, and campaigning, of the Pacific Islands Students Fighting Climate Change – work that started in a classroom in Fiji – that eventually led to the landmark International Court of Justice opinion of July 2025 mentioned in the previous chapter.

Legal accountability

Beyond widespread recognition of the serious injustices associated with climate impacts, courts are being used to pursue accountability. As we saw in Chapter 3, wealthier countries and the richer groups are responsible for a high share of current and historical emissions, yet the repercussions are felt earliest and hardest by the poorer nations and the poorer communities within societies. And it is the current generations that are undermining the future opportunities of coming generations. There are major issues of ethics and justice around both intergenerational and international equity (see Chapters 3 and 6). The notion of injustice can be understood (following Sen, 1999, 2009) in terms of the actions of some denying rights to others. Here emissions harm opportunities and rights to development.

Legal action is being pursued along two broad lines. First, lawsuits are being brought against entities, often corporations, for depriving people or communities of rights or for the creation of damage to others. Though it was eventually (May 2025) dismissed, an exemplary case of this is *Lliuya v. RWE*, where a Peruvian farmer, Saúl Luciano Lliuya, filed a lawsuit against the German energy company RWE in 2015, claiming that its contributions to greenhouse gas (GHG) emissions were responsible for the glacial melting threatening his hometown of Huaraz. The case, which was allowed to proceed by the German courts in 2017, argues that the company, as one of the largest CO_2 emitters in the world, should be held accountable for the increased risk of flooding caused by the melting glacier, which puts Lliuya's community in danger. This was the first case in the world that relied on climate attribution science. The dismissal was not on the grounds of absence of liability for damage but because, in this case, the risk to the particular farmer was deemed to be very small.

Second, legal systems also play a growing role in enforcing climate commitments – where legal challenges are brought against entities, often governments, for failing to meet climate commitments. Through legislation and international agreements, countries commit to standards; these can become legally binding and subject to enforcement mechanisms within the nation. These legal frameworks provide a system of accountability for those failing to meet their obligations – they are part of the broader architecture of global climate action. National courts and international tribunals are increasingly holding governments and companies accountable, putting pressure on the translation of promises into action. In total, Setzer and Higham (2024) reported that more than 2,666 climate litigation cases have been registered globally, with around 70% of them filed since 2015.

Landmark cases like *Urgenda v. The State of the Netherlands* (2018) and *Neubauer et al. v. Germany* (2020) forced governments to strengthen emissions targets, given the respective governments' failure to engage with their obligations under the Paris Agreement. In the USA, *Juliana v. United States* (2015), though ultimately unsuccessful, argued that government inaction on climate change violated the constitutional rights of young Americans. Litigation is also being used, for example, to demand the incorporation of climate risk into financial decision-making or by shareholder activists pursuing climate agendas. Many more climate litigation cases have been filed in the Global North than in the Global South, although it is the Global South that will suffer most severely from inaction on climate and although they, through their emissions, have contributed least to the problem (see Setzer and Benjamin, 2019).[9]

4.2 Technology, innovation, and the private sector

As argued in Chapter 1, the Paris Agreement has been a basic strategic foundation and guiding star for much that has followed. The clear sense of direction embodied in the Agreement has built confidence for investment and technological innovation. This 'Paris effect' has created a global shift of investment, innovation, and research and development (R&D) in terms of major technological advances and new forms of private sector investment (Systemiq, 2020). Although this effect has gathered momentum in the private sector, government strategy and policy is essential in continuing to enable progress – as will be discussed in Chapter 6.

The global agreements and national commitments associated with them have advanced markets for low-carbon technologies. Progress in solar, wind, and storage have made the clean become cheaper than the dirty in electricity generation across the world. That is increasingly true for surface transport via electric vehicles. Investments in clean technologies and infrastructure in one region have created global ripple effects, driving down costs and increasing accessibility across the world. Since 2015, progress on low-carbon technologies, investments, and markets has been much faster than anticipated and greater than many realise (Systemiq, 2023).

Transforming energy systems

At the heart of the action agenda is the global transformation of energy systems, with a rapid transition away from fossil fuels. And the rapid development and improvement of clean technology have given us a springboard for this transformation. Further detail on clean investment, positive tipping points in cost and adoption, and other advances is provided in Chapter 5, along with other key dimensions of the new growth story. But it is important to introduce this progress in the first part of the book, because it is a foundation of much that follows.

The clean is already cheaper than the dirty across around 30% of emissions and this proportion is rising quickly (Systemiq, 2020). Wind and solar, including storage, now offer the lowest cost means of generating electricity in most of the world (Bond et al., 2024). Many new technologies, including clean technologies, follow an uptake S-curve. Positive technological tipping points refer to critical cost and technology thresholds after which the technology out-competes incumbents on affordability, attractiveness, and accessibility. Adoption and diffusion accelerate as a result. Through reinforcing feedback loops, such as learning by doing, the technology can become more efficient; economies of scale can reduce costs, attracting new investments and sales. Policies aimed at triggering positive tipping points can have an outsized impact on accelerating progress in the adoption of new and clean technologies and in reaping the associated benefits of these technologies (see this acceleration illustrated in Figure 4.2).

Figure 4.2: Accelerating tipping points for climate technologies fast-tracks impacts

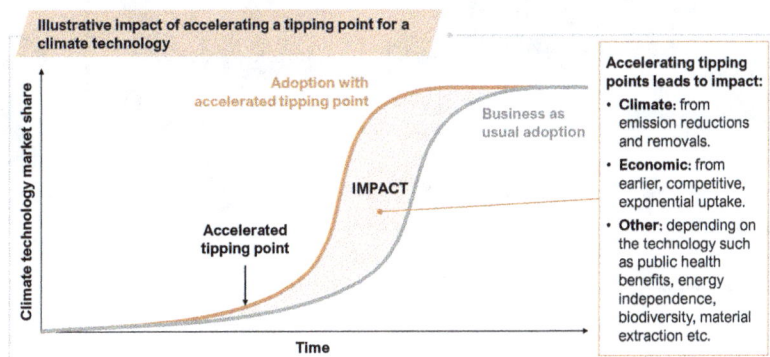

Source: Systemiq (2025, forthcoming). *Accelerating the breakthrough of climate technologies: Driving exponential growth in climate technologies with positive tipping points* [Unpublished report]. Copyright Systemiq. Reproduced with permission.

Such positive tipping points are being reached in more and more technologies; in other sectors or activities, they are close to being reached. For example, by 2018, the levelised cost of electricity (LCOE)[10] for new solar and wind was already lower than that for new coal and gas in many parts of the world. Other tipping points are on the horizon: unsubsidised battery electric vehicles (BEVs) are expected to reach purchase price parity with internal combustion engines (ICEs) by 2026–2027 in major regions (S&P, 2022). Green ammonia and green hydrogen are also gaining traction, with strong policies and major projects expanding rapidly worldwide (Systemiq, 2023). Costs fell particularly rapidly in the decade 2010–2020. They are still falling and will likely continue to do so, given the intensity of R&D, learning-by-doing, economies of scale, and growing competition. The remarkable progress in cost reductions in renewable energy technologies is illustrated in Figures 4.3 and 4.4. The rollout of these technologies is already transforming energy systems around the world, making accelerated climate action more feasible and offering a renewed sense of hope.

Figure 4.3: Renewable power technologies: decreases in LCOE, 2013–2023

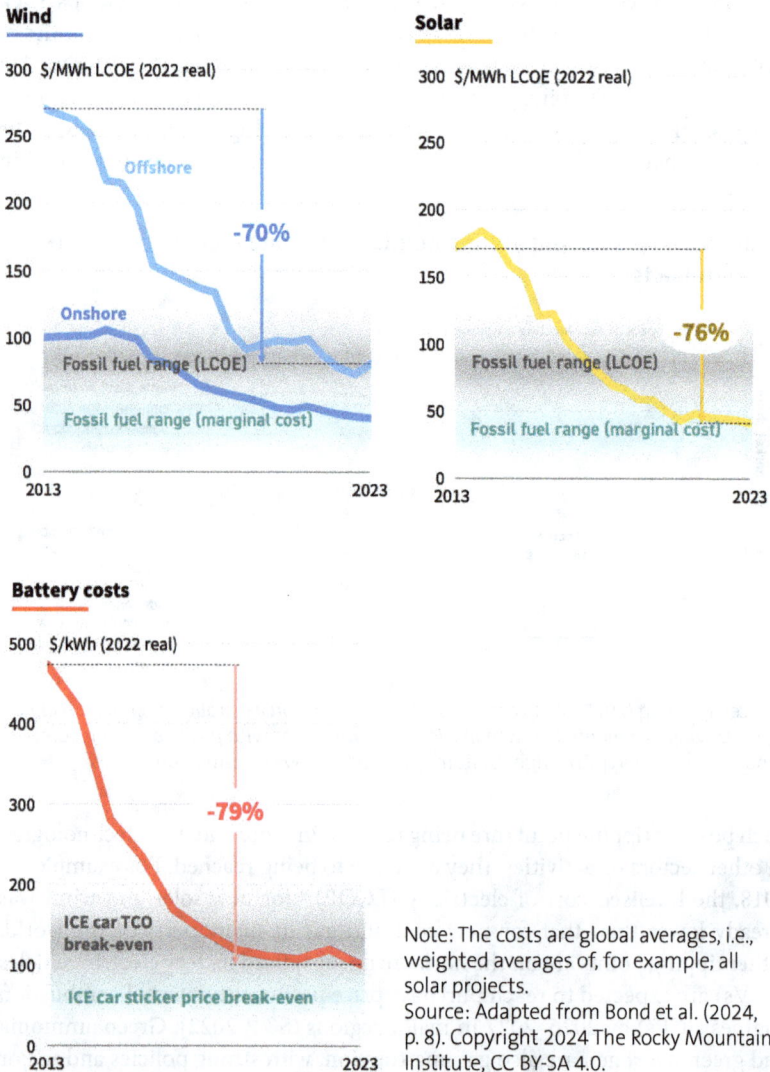

Wind

300 $/MWh LCOE (2022 real)

250

Offshore

200

-70%

150

Onshore

100
Fossil fuel range (LCOE)

50
Fossil fuel range (marginal cost)

0
2013   2023

Solar

300 $/MWh LCOE (2022 real)

250

200

-76%

150

100
Fossil fuel range (LCOE)

50
Fossil fuel range (marginal cost)

0
2013   2023

Battery costs

500 $/kWh (2022 real)

400

300   -79%

200

ICE car TCO
break-even

100

ICE car sticker price break-even

0
2013   2023

Note: The costs are global averages, i.e., weighted averages of, for example, all solar projects.
Source: Adapted from Bond et al. (2024, p. 8). Copyright 2024 The Rocky Mountain Institute, CC BY-SA 4.0.

Figure 4.4: Global levelised cost of energy for renewables

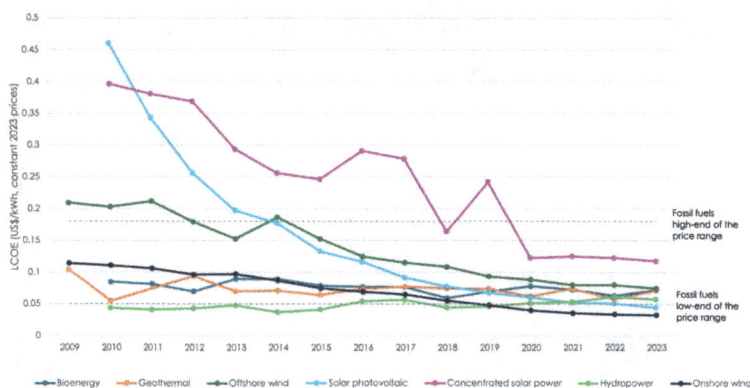

Note: The average cost per unit of energy generated across the lifetime of a new power plant. These data are expressed in US dollars per kilowatt-hour. It is adjusted for inflation but does not account for differences in living costs between countries.
Source: Author's elaboration using IRENA (2024) data – with minor processing by Our World in Data.

In 2022, according to the International Renewable Energy Agency (IRENA, 2023a), the worldwide average cost of electricity from solar photovoltaic (PV) reached US\$0.049/kWh – almost one-third lower than the most economical global fossil fuel alternative. For onshore wind it was US\$0.033/kWh – nearly half the price of the least expensive fossil-fuel-fired option in the same year. 'Round-the-clock' renewables (including storage) are already competitive with fossil fuels, provided capital costs are manageable. Costs of generation and storage continue to fall. Better grids reduce the storage needed (BloombergNEF, 2023; IRENA, 2023a). Investment in grids is crucial to the advancement of renewables on an economy-wide scale. But decentralised solar in villages or small towns can play a powerful role in expanding opportunities for poor people.

Given that technical progress and cost reduction have moved so strongly in the past decade or so, it is striking that estimates of cost reductions and deployment have continually underestimated the pace of change (see Figure 4.5). At last, forecasts seem to be catching up with reality. And the world would be wise to plan for continuing technical progress and cost reduction in new low-carbon technologies.

As a consequence of rapid falls in costs, solar and wind deployment is exploding and global annual investments in renewables are already far higher than in fossil fuels. According to the International Energy Agency (IEA), for 'every USD 1 spent on fossil fuels, USD 1.7 is now spent on clean energy. Five years ago [in 2018], this ratio was 1:1' (2023, p. 12) (see Figure 4.6). In 2025, this ratio is expected to reach 2:1 (IEA, 2025a). There is huge solar deployment potential for the world. China has played a crucial role in driving down

Figure 4.5: Actual versus IEA's projected LCOE of solar PV, 1970–2040

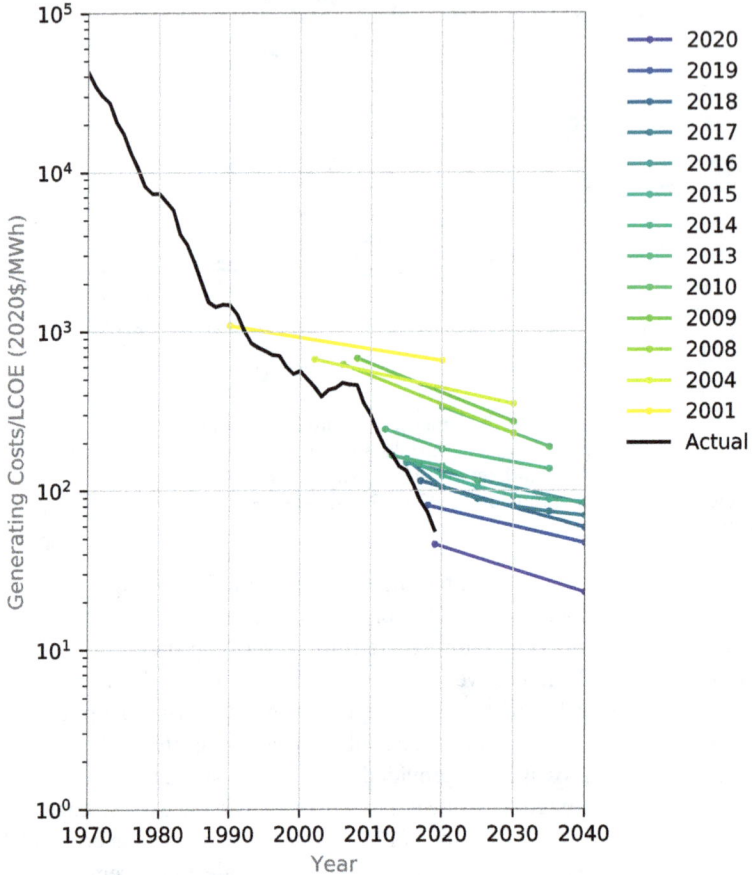

the cost of solar panels and has accounted for the majority of global solar power deployment since 2015. Nevertheless, other countries are picking up quickly – we shall return to this in Chapters 8 and 9. China's very large production capacity for low-cost solar, and for wind power and batteries, embodies great opportunities, particularly for the developing world.

Different countries have different potential in energy technologies. Land scarcity in Bangladesh and Indonesia limits solar potential relative to Africa, which has 60% of the world's best solar resources (IEA, 2022). Some countries, for example those in northern Europe, are well-endowed with wind potential. Other countries such as Mongolia have strong potential in both.

Progress in the new growth story has depended on policy and entrepreneurship. Indeed, promoting clean innovation through sound policy and the fostering of entrepreneurship is a fundamental aspect of

Figure 4.6: Global investment in clean energy and in fossil fuels, 2015–2025e

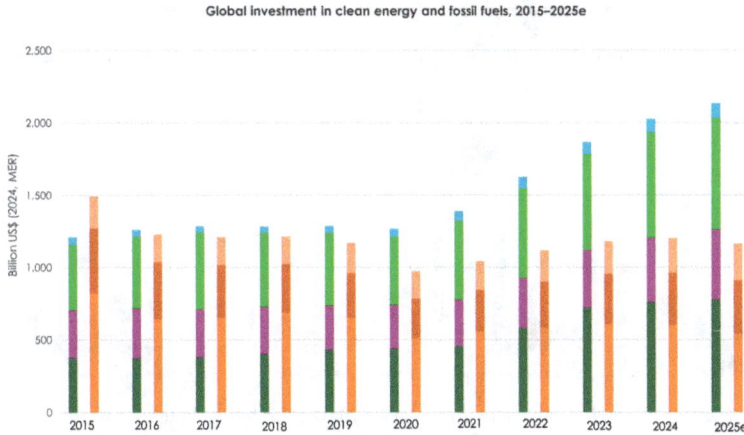

Global investment in clean energy and fossil fuels, 2015–2025e

Note: 2025e = estimated values for 2025. MER = Market Exchange Rates. Other clean fuels = modern bioenergy, low-emissions H_2-based fuels, and CCUS associated with fossil fuels, and also includes direct air capture.
Source: Author's elaboration using IEA (2025a) data.

climate leadership. Two interrelated events that were key to achieving scale and falling costs are the early subsidies in Germany for solar in 2000 under their Energiewende policy (Öko-Institut, 2013; Trancik et al., 2015) and, partially in response to those initiatives, China's industrial development and entrepreneurship. The boom in solar demand experienced in Germany gave an important push to the, at that time, pioneering solar manufacturing firms in China, which reacted by investing aggressively in the sector (Quitzow, 2015). It is remarkable how advance has taken place even with, on the whole, rather modest policies across the world to support renewables. Technological acceleration really is possible. On the other hand, frequent changes of policy can undermine investor confidence. And some policies, for example in support of gas in the USA or coal in Indonesia and Pakistan, may lead to entrepreneurial interests and capacity which constrain future progress in renewables (see Chapter 6). Fossil-fuel subsidies remain substantial (see Chapters 5 and 6).

In 2023, around 90% of global net power capacity additions came from solar and wind – versus 6% from fossil fuels (BloombergNEF, 2024a). In 2024, the majority of the increase in global electricity demand was met by the growing supply of low-emission sources, with 80% of the increase in global electricity generation being produced by renewable and nuclear sources (IEA, 2025b). Further, clean energy technologies and infrastructure investment hit new records in 2024, reaching US$2 trillion (IEA, 2025a).

The high-income countries and China are still playing a dominant role. However, emerging markets attracted more than double the total new renew-

Figure 4.7: Annual energy investment by selected country and region, 2015 and 2025e

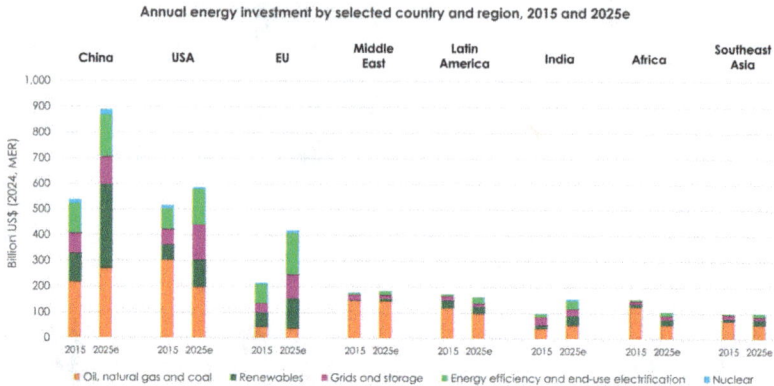

Annual energy investment by selected country and region, 2015 and 2025e

Note: 2025e = estimated values for 2025. MER = Market Exchange Rates.
Source: Author's elaboration using IEA (2025a) data.

ables investment in 2023 than in 2018, and 85% of low- and middle-income economies deployed more renewables capacity relative to fossil fuels in 2023 (Cuming, 2024). The shift has begun in these countries, although the scale remains small in relation to their economies, population, and opportunities, and in relation to the progress seen in high-income countries and China.

Though the progress in clean technology has been remarkable, in many countries major barriers to investment persist. Figure 4.7 illustrates the recent increases in energy transition investment, with emerging markets and developing countries (EMDCs) other than China experiencing, indeed, smaller absolute amounts of investment and slower investment growth. In many cases, there are in these countries powerful barriers in the investment climate and the cost and availability of capital.[11] These barriers will be a major theme for Chapters 8 and 9 in Part III of the book.

Further, notwithstanding strong trends in renewables, it is important to recognise that investments in fossil-fuel extraction and use have continued in most regions worldwide. Globally, governments still plan to produce more than double the amount of fossil fuels in 2030 than would be consistent with the long-term temperature goal of the Paris Agreement (SEI et al., 2023).

In pathways aligned with net zero by 2050, beyond the projects already committed as of 2021, the IEA has argued that there should be no new oil and gas fields approved for development, and no new coal mines or mine extensions (IEA, 2021b). The coal, oil, and gas extracted over the anticipated lifetime of coal mines which are producing or under construction and oil and gas fields as of 2018 would emit more than 3.5 times the carbon budget available to limit warming to 1.5 °C and almost the entire budget available for 2 °C.

Thus, carbon budgets aligned with the long-term temperature goal of the Paris Agreement of well below 2 °C require that much of the existing capital stock will need to be retired early, retrofitted with carbon capture and storage (CCS), and/or operated below capacity (IEA 2023; Intergovernmental Panel on Climate Change [IPCC] 2022; Trout et al., 2022). Globally, this leaves no room for new fossil-fuel infrastructure if we are to stay on track with the goals of the Paris Agreement. Investing in such infrastructure today is, in effect, a bet that the world will fail to meet those targets – or that, even if fossil fuel use declines in line with the Paris Agreement, this particular investment will remain viable while others do not. The latter is a risky assumption – and one that many investors may ultimately find to be mistaken.

The role of artificial intelligence

The advances in green new technologies are occurring simultaneously with extraordinary progress in artificial intelligence (AI) and machine learning (ML, an AI subset). Applications that utilise these technologies can contribute to advancing both adaptation and mitigation action, as illustrated in the examples in Figure 4.8. One notable application is the use of image recognition technology for monitoring and planning purposes. Deep learning, a subset of ML, can be used to enhance the classification and analysis of images provided in real time by satellites. For example, it can be used to classify images of vegetation cover to identify changes over vast areas, such as desertification or deforestation, or to identify emission hotspots, for example of methane, through live satellite imagery (Vinuesa et al., 2020). In these cases, and many others, AI provides useful information to plan necessary biodiversity and climate-associated policies. Importantly, AI can also enhance our understanding of potential climate impacts, supporting more effective adaptation and resilience-building efforts.

Figure 4.8: Examples of AI applications that can help advance adaptation and mitigation action

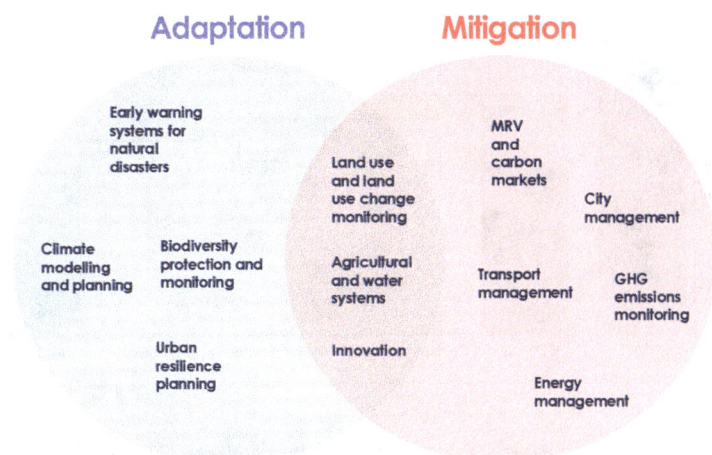

Note: MRV = measurement, reporting, and verification.
Source: Author's elaboration drawing on UNFCCC (2024).

AI applications are likely to have very powerful benefits in the design and management of systems, including cities, energy, transport, land, and water, which are central to driving the climate transition. They will be a powerful force in discovering new technologies, new materials, and the location of critical minerals. We will return to these and other AI applications in Chapter 5.

However, a key challenge that must be highlighted is the substantial electricity demand already associated with AI-related data processing. Those demands will continue to increase rapidly. For example, in the USA, nearly 5% of all energy used is directed toward data centres, a figure that has increased two-fold since 2018 (Dominici, 2025). It is estimated that electricity demand from data centres globally will more than double by 2030, which is equivalent to more than the entire electricity consumption of Japan today (IEA, 2025c). Accommodating this demand and ensuring it has zero-carbon supply will be challenging.

On the other hand, the application of AI itself could help in reducing overall energy demand. For example, Stern et al. (2025) estimate that when looking at AI applications in only three sectors – power, food, and mobility – AI could reduce global business-as-usual (BAU) emissions by 3.2 to 5.4 $GtCO_2e$ annually by 2035 (Figure 4.9). These emissions reductions outweigh the estimated 0.4–1.6 $GtCO_2e$ increase in emissions from the global power consumption of data centres based on all of AI's activities (not just those related to decarbonisation). And it seems likely that AI applications to systems management and discovery in the future will be less energy intensive than the many other applications which require searching over literature, information, and data sources covering most of the world and most of the past.

Figure 4.9: Total emissions and potential emissions savings from AI in 2035 for the sectors in scope (power, meat and dairy, light road vehicles)

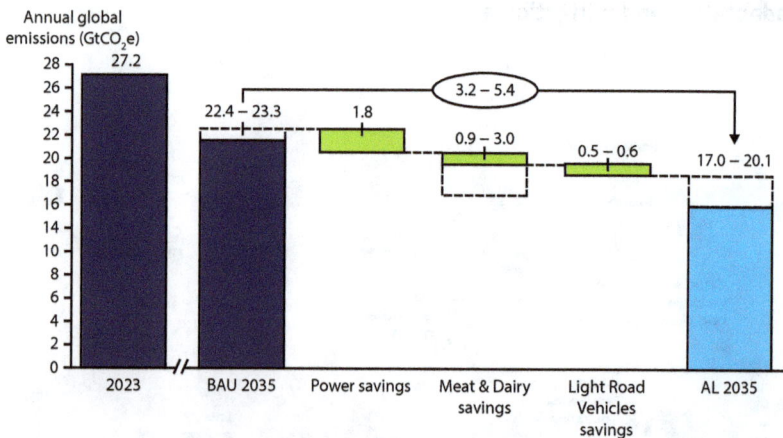

Note: The 2023 bar is the total 2023 $GtCO_2e$ emissions of the power (15.3 $GtCO_2e$), meat and dairy (8.7 $GtCO_2e$), and light road vehicles (3.2 $GtCO_2e$) sectors.
Source: Figure 1 in Stern et al. (2025, p. 4). Copyright 2025 The Authors, CC BY 4.0.

The private sector

Despite setbacks and slowdowns, the direction of travel on climate action over the decade since Paris has been positive, even though sharp acceleration now is required. For example, as Figure 4.10 shows, the number of net zero pledges by nations, states, cities, and companies increased from 2020 to 2024. Mark Carney, former Governor of the Bank of England and elected Prime Minister of Canada in early 2025, has described climate change as 'the greatest commercial opportunity of our time' (2020, p. 2). Indeed, much of the driving down of technology cost has been the result of investment, innovation, and going-to-scale in the private sector. The whole story of the green transition and the new model of growth is driven by investment and the majority of that will be in the private sector. The attraction of the private sector to these great investment opportunities has been growing since Paris COP21 in 2015.

Figure 4.10: Net zero coverage 2020 to 2024 (by number)

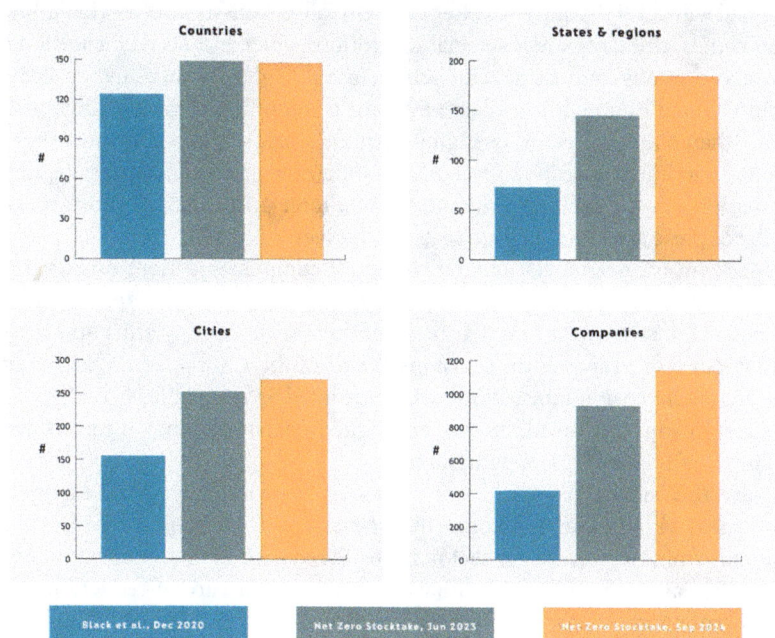

Note: Number of net zero pledges per entity group covered by the Net Zero Tracker database and growth since December 2020. Data for December 2020 and June 2023 are taken from Black et al. (2021) and Net Zero Tracker (2023).
Source: Figure 1 in Net Zero Tracker (2024, p. 14). Copyright 2024 Net Zero Tracker. Reproduced with permission.

The private sector has responded to the overall strategic sense of direction embodied in Paris, to price and subsidy measures, to regulations against polluting activities, to targeted transformation in many countries of economic and particularly energy systems, and to social pressures. A manifestation of their involvement, as mentioned in Chapter 1, is that COPs have now become much more than a meeting place for government officials. They are now a gathering also of private sector investors working in these sectors and of community and social institutions and groupings. Reflecting the increasingly important role of the private sector, the present trend is that national climate plans – the nationally determined contributions (NDCs, see Chapter 1) – now take a shape that helps private investments discern where they can 'step in' and invest.[12] Also, fossil fuel companies now see the importance of their involvement and are present at COPs, although in large measure, it seems, to see if they can slow action.

There has been a wide range of initiatives collecting, monitoring, and facilitating private sector buy-in and commitments. The Science-Based Targets initiative (SBTi) was established to drive ambitious climate action in the private sector by enabling companies to set out GHG reduction targets aligned with a 1.5 °C future. The Race to Zero rallies non-state actors (including companies, cities, regions, financial institutions, and educational institutions) to take rigorous and immediate action to halve global emissions by 2030. And Climate Action 100+, a large investor coalition on climate change with more than 600 investors, is engaging with the world's largest corporate GHG emitters to curb emissions, strengthen climate-related financial disclosures, and improve governance on climate change. Table 4.1 provides a summary of some of these and other private sector initiatives.

Many investors and shareholders interpret companies' actions on sustainability as in their own interest, and some have prosecuted companies for lack of climate action or the misalignment of their sustainability aims and their behaviour. For example, the Pensions Board of the Church of England and the National Pension Fund of Sweden confronted 55 companies from Europe because of their 'misleading and misaligned corporate lobbying practices' (Church of England, 2018, paragraph 4).

Nevertheless, it is also true that during times of difficulty – for example, associated with the war in Ukraine that started in 2022 – some companies, for example from the oil and gas sector, have diluted their climate commitments. In particular, both BP and Shell have weakened their carbon reduction targets, attributing this to future strong gas demand and uncertainty associated with the energy transition (Bousso, 2024). They may also be discovering that they over-estimated their potential advantages in the new low-carbon technologies and are going back to core business.

Retreat has also taken place recently, in early 2025, in the USA under political pressure, particularly in Republican states. Indeed, there have been attempts to force financial institutions or major companies to reduce or reverse net zero commitments. These pressures may continue or increase as the new Trump presidency progresses. Since his re-election in late 2024 we have already seen the six largest American banks – Goldman Sachs, Wells Fargo, Citigroup, Bank of America, Morgan Stanley, and JPMorgan Chase –

Table 4.1: Private sector-oriented initiatives

Name and launch date	Focus	Participation
SBTi (Science-Based Targets initiative), 2015	Defines best practices for science-based GHG targets, provides guidance and validation of targets	10,891 companies with targets or commitments; 8,206 companies with validated targets; 1,880 companies with net zero targets
GFANZ (Glasgow Financial Alliance for Net-Zero), 2021	Unites finance sector's net zero commitments with science-based guidelines, interim targets, transparent reporting	Over 700 institutions across 50 countries representing 40% of global private financial assets
FAST-Infra (Finance to Accelerate the Sustainable Transition Infrastructure), 2020	Creates a labelling system for sustainable infrastructure and market mechanisms to mobilise private capital (especially in EMDCs)	80+ public and private institutions including banks, asset managers, governments, MDBs, national development banks (NDBs)
VCMI (Voluntary Carbon Markets Integrity Initiative), 2021	Ensures credible, net zero-aligned carbon offsets, develops guidance on climate claims and high-integrity voluntary carbon markets (VCMs)	Ongoing global consultations shaping claims guidance and market integrity
Race to Zero, 2020	Mobilises non-state actors (companies, cities, etc.) to halve emissions by 2030, sets rigorous criteria for transparent action	16,000+ members (12,000+ companies, 600+ financial institutions, others)
Climate Action 100+, 2017	Engages top corporate emitters to cut GHGs, strengthen governance, and improve climate-related disclosures	600+ global investors across 30+ markets, 169 focus companies (US$10.3 trillion market cap)
CFLI (Climate Finance Leadership Initiative), 2019	Mobilises private capital for climate solutions; emphasises sustainable infrastructure in emerging markets	Founding members include Allianz, AXA, Enel, Goldman Sachs, HSBC
GISD (Global Investors for Sustainable Development), 2019	Scales long-term finance and investment in sustainable development, especially in EMDCs	30 leaders of major financial institutions and corporations representing US$16 trillion in assets
Sustainable Markets Initiative, 2020	Accelerates the global shift to sustainable markets; flagship initiatives include Terra Carta	250+ private sector CEO members; 22 CEO-led task forces

Note: Figures in the final column reflect data current as of Spring 2025. For further information see SBTi (2025), GFANZ (2024), FAST-Infra (2023), VCMI (2025), Climate High-Level Champions (2024), Climate Action 100+ (2025), Bloomberg (2025), UNDESA (2020), Sustainable Markets Initiative (2024).
Source: Author's elaboration based on Songwe et al. (2022).

depart from the Net Zero Banking Alliance (NZBA). The largest asset management company in the world, BlackRock, has also retreated from the Net Zero Asset Managers (NZAM) initiative (Dennis, 2025).[13] Thus private sector commitment is not necessarily robust in the face of short-term political pressure or energy crises.

Notwithstanding some setbacks, and others will occur, the increase in private sector commitment over the last decade has been remarkable and the longer-term trend seems unlikely to be reversed. The clean is becoming cheaper than the dirty across more and more sectors and positive technological tipping points are being unlocked. Further, as it recognises and embraces opportunities, the private sector can foster change in public policy by highlighting those opportunities and linking them to the new approach to growth. The private sector is key to the credibility of this narrative, as it will be committing resources and taking risks. At the same time, the private sector can help create the conditions for investment and change by working with governments to identify obstacles to investment and the kinds of policies that can create the signals and confidence for investment in the new.

4.3 International action in a changing world

From Paris COP21 onwards, international mechanisms for collaboration have been building beyond the confines of the United Nations Framework Convention on Climate Change (UNFCCC) and the COPs. Of special significance have been the collaborations between the finance ministries and between the central banks, including via the G20. MDBs and international financial institutions (IFIs) have been re-orienting their activities. Major strategic changes have taken place in key countries towards lower carbon and sustainability. In this section we describe and analyse these collaborations and changes. Whilst recognising that the second Trump presidency will be disruptive and create steps backward in the USA, the strategies and commitments have been growing over the past decade. In the world as a whole, they will continue, as they did during the first Trump presidency. No doubt other disruptions will appear on the way to 2050 and beyond, but the sense of direction is clear.

We also emphasise the changes that are taking place in the EMDCs, where the bulk of economic growth will take place in the coming years. This growth – coupled with the fact, as we have seen, that investment in green technologies has been much smaller in these economies (except China) – implies that it is these countries that will see the main growth in emissions and it is here that climate action is of special importance. In large measure, it will be in these economies that the climate and biodiversity future of the world will be determined. And it is these countries that have the opportunity to leapfrog to a new form of development.

International collaboration

The UNFCCC and the COPs have been central to collaboration and agreements on climate. But for the last two decades, there has been increasing recognition that climate and the economy, or climate and growth and development, cannot be seen as separate subjects. Hence, increasingly and particularly post-Paris, it has been recognised that the responsibilities for climate action cannot be confined to the environment ministers, who have been the principal partners in the COP. For example, with Felipe Calderón and working closely with Andrew Steer, then head of the World Resources Institute (WRI), in 2013, we founded[14] the Global Commission on the Economy and the Climate (GCEC; usually termed the New Climate Economy (NCE)) to explore and emphasise the economy-wide importance of climate action. The first report of the GCEC, *Better Economy, Better Climate* in 2014, showed that climate action could drive growth. The idea was gaining increasing traction before Paris 2015, and that in part helped to achieve and underpin the agreement. The NCE assembled evidence on the energy security and air pollution benefits of reducing emissions, possibly influencing China's targets of reaching peak coal consumption by 2020 and peak emissions by 2030 (World Resources Institute [WRI], 2015). And Felipe Calderón's engagement with the Colombian government influenced their decision to integrate climate into their five-year development plan (WRI, 2015).

Recognition of the importance of the interweaving of climate and the economy led to the foundation of the Network for Greening the Financial System, the group of central bank governors, founded in 2017, with over 140 members in 2025 (NGFS, 2025) and the Coalition of Finance Ministers on Climate Action, founded in 2019, with around 100 members in 2025.[15] Among its core

Figure 4.11: Frameworks for sustainable finance

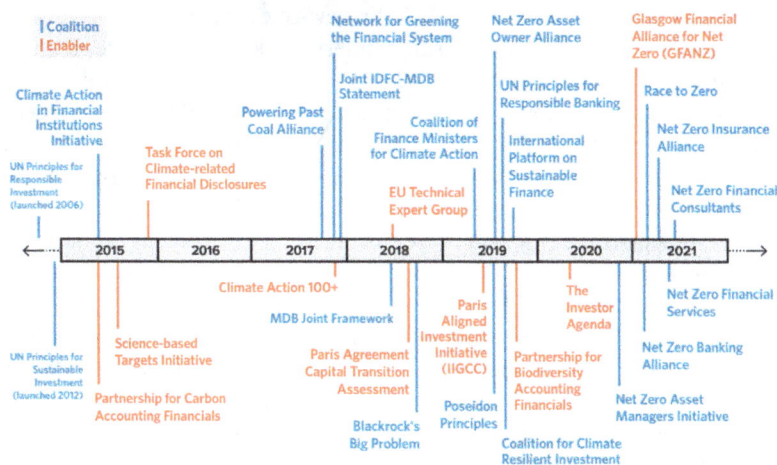

Source: Figure 1 in Climate Policy Initiative (2021, p. 1). Copyright 2021 Climate Policy. Initiative. Reproduced with permission. Courtesy of Climate Policy Initiative.

actions, the Coalition generates a space that enables ministries of finance to become lead agencies in NDC development and implementation. They do so by facilitating knowledge-sharing and peer-to-peer learning across ministries, and through capacity building. For example, the coalition developed a guide which details the climate action options finance ministries usually already have at their disposal, drawing on a wide range of member case studies (see CFMCA, 2024). A summary of these and other initiatives, both public and private, is provided in Figure 4.11.

Climate and sustainability are also central now to the work of the G20. An example of particular significance was the work of the G20 Independent Expert Group (IEG) on MDBs established under the 2023 India G20 Presidency, chaired by N.K. Singh and Larry Summers (I was a member). It published two reports: *The Triple Agenda* for MDBs in June and *A Roadmap for Better, Bolder, and Bigger MDBs* in September 2023 (Singh and Summers, 2023a, 2023b). The first of these emphasised the importance of including global public goods, particularly climate and biodiversity but also pandemics, in addition to fighting poverty and promoting shared prosperity. The second – from the perspective of necessary investment for the Sustainable Development Goals (SDGs) and for climate, nested within the SDG framework – emphasised the importance of expanding (tripling) MDB finance and of MDBs working more with the private sector, including by taking more risk and functioning more effectively overall, including working better as a group. These reports have set the agenda for MDB reform since – placing particular emphasis on the role of MDBs in tackling today's global challenges. We return to the elements of this agenda in Chapter 9.

In 2025, the International Maritime Organization (IMO) sent a clear signal on the role of low-emissions fuels in the future of shipping. A landmark agreement was reached to introduce the first global levy on GHG emissions from ships. Set to take effect in 2027, the new rules will apply to a fleet that accounts for 85% of CO_2 emissions from international shipping (International Maritime Organization [IMO], 2025). A carbon-trading mechanism will accompany the levy, allowing ships to trade emissions credits. The scheme is expected to generate around US$10 billion annually, which will be invested in low-emissions shipping technologies and fuels. However, this falls short of the US$60 billion that many had hoped for (Harvey, 2025). Despite opposition from several petrostates, and the USA exiting the talks before they concluded, the agreement marks a significant shift, with no other industry having made such a commitment at the global level.

The UN leadership itself, through its Secretary-General and Deputy Secretary-General, has long been in the vanguard of the arguments for action on climate. Ban Ki-moon championed the Paris Agreement. António Guterres and Amina Mohammed have constantly warned of the urgency of action and set that action in the context of development and its finance. In 2022, Mohammed proclaimed that 'we are living on borrowed time' (2022, paragraph 3). In 2023, Guterres stated:

We are miles from the goals of the Paris Agreement – and minutes to midnight for the 1.5-degree limit. But it is not too late. We can – you can – prevent planetary crash and burn. We have the technologies to avoid the worst of climate chaos – if we act now. (2023, paragraph 13)

The UNFCCC has been at the heart of international commitment. Since Paris COP21 in 2015, the world's international institutions have tilted strongly towards action on climate and important coalitions have been created, both public and private. Within this context of growing emphasis on climate in international institutions, we now examine the transition in major countries and groups of countries.

The European Union

The EU has, mostly, been at the vanguard of multilateral cooperation in global climate efforts and negotiations. At the international level, the EU has played a leadership role in negotiations throughout the years. For example, in the negotiations for the Kyoto Protocol (1997), it pushed for binding emissions reductions, a contentious issue.[16] It also significantly influenced other nations' climate policies through its own actions. For example, by establishing an emissions trading scheme in 2005, it provided an early model for market-based climate responses. Further, the European Green Deal (of 2019) set the goal of making Europe the first climate-neutral continent by 2050, including the target of reducing GHG emissions by at least 55% by 2030 compared to 1990 levels. With this goal in mind, the Fit for 55 Package (2021) included some key revisions to the Emissions Trading System (ETS)[17] and the introduction of a Carbon Border Adjustment Mechanism (CBAM)[18]. These targets and instruments, together with other policy reforms, contributed to the sharp increase in the price of carbon on the ETS since 2018; the price reached a record level of over €100 per tonne of CO_2 in February 2023 and is currently (Spring 2025) around €70 per tonne of CO_2.[19]

However, the EU has much to do if it is to become a leader of the growth story of the 21st century. Its investment and its growth of productivity has been weak. In 2023, the European Commission tasked Mario Draghi – former Italian Prime Minister (2021–2022) and President of the European Central Bank (ECB) (2011–2019) – with writing a report on the future of competitiveness in the EU, in the context of US and Chinese leadership in growth of productivity. The 400-page report was published in 2024 and titled *The Future of European Competitiveness*. There, Draghi highlights the importance of closing the innovation gap between Europe and the USA and China, noting both that EU companies are specialised in mature technologies where there is limited potential to achieve breakthroughs, and that research and innovation spending in the EU is limited relative to the USA and China. In the last two decades, the sectors of the top three companies in terms of

R&D spending in the USA have changed, from automotive and pharmaceutical industries (2000s) to software and hardware companies (2010s), and then to the digital sector (2020s). This story has been different in the EU, where automotive companies have consistently dominated research and innovation spending. It seems that investment in the EU has remained concentrated on technologies and sectors where the productivity growth rates of larger companies are slowing. And now the European car industry has serious problems, in large measure because the future is in electric vehicles and Chinese firms have a substantial lead.

Draghi argues convincingly that it is critical that the EU unlocks its innovative potential to lead in the new technologies of this century and incorporate their benefits into existing industries. The EU is behind in the breakthrough digital technologies that will drive future growth. For example, since 2017 around 70% of foundational AI models have been developed in the USA. Also, though quantum computing seems to be a promising area of major innovation, none of the top 10 companies globally in terms of investment in quantum technologies are based in the EU – five are based in the USA and four in China (Draghi, 2024a). Trying to take advantage of the potentially powerful links between competitiveness and climate action (see also Chapters 5 and 8) is part of the EU's agenda.

Recent developments in the USA under the second Trump presidency, including in relation to defence and trade, have led to a still stronger focus in the EU on efficiency, productivity, and green technologies, as embodying the growth of the future. To this end, Ursula van der Leyen, the President of the European Commission, emphasised in her World Economic Forum (WEF) Davos January 2025 speech three key messages from the Draghi report: unify financial markets, reduce and harmonise regulation, and foster low and stable energy prices through renewable energy (World Economic Forum [WEF], 2025). This shows the EU's commitment to pursuing competitiveness and productivity and highlights ways forward which align with sustainability.

The United States

The recognition by the US government of the importance of climate action has shifted since the 1990s. It is an illustrative and important example of how the political pendulum can swing dramatically – for better, and for worse. At key moments they have been leaders and on other occasions obstacles to progress.

In the early 1990s, under the administration of President George H.W. Bush, the USA was hesitant to commit to binding emission targets, emphasising concerns over implications for economic growth and energy security. Non-binding targets were adopted in the initial UNFCCC agreement. But it must be recognised that without the strong personal commitment of the first President Bush, the success of the 1992 Rio Earth Summit and the creation of the UNFCCC would not have occurred. And it was another Republican

President, Richard Nixon, who created the US Environmental Protection Agency (EPA) in 1970.

The narrative shifted towards stronger climate action under President Bill Clinton, a Democrat, who supported the Kyoto Protocol (1997), with its binding targets. Vice President Al Gore was a real leader here and played a crucial role in delivery of the Kyoto agreement. However, the subsequent Bush administration (under the second President Bush, George W. Bush, a Republican) withdrew from the protocol, citing the lack of commitments from developing countries and overly stringent targets as major concerns (Depledge, 2005, p. 19).

Under President Barack Obama, a Democrat, there was renewed focus on climate security and promoting a low-carbon economy, and he showed real international leadership in the creation of the Paris Agreement of December 2015 (Zhang et al., 2017).[20] It was the joint announcement of targets by Presidents Xi Jinping and Barack Obama in Beijing one year ahead of Paris that sent the message that both China and the USA sought a strong climate agreement.

This collaborative approach was reversed when President Donald Trump took office in 2017. He rolled back numerous environmental regulations and withdrew the USA from the Paris Agreement, arguing it imposed unfair economic burdens. The response from many other US institutions to Trump's withdrawal was striking and substantial, with states, cities, and businesses across the USA rallying to commit to the Paris goals (with initiatives like the US Climate Alliance for Cities and the 'We are still in' campaign), demonstrating strong domestic support for climate action irrespective of federal policies. I was at COP22 in Marrakech in November 2016 when the US presidential election results came through. The steadfastness of countries, the private sector, and NGOs in relation to the commitment to 'keep going' was remarkable. Under the first Trump presidency the cost advantages of renewables were such that the US solar industry grew by 128% between 2016 and 2020 (SEIA, 2024). The good folk of Texas, where solar power generation increased by more than 100% between December 2016 and December 2017 (Graves and Wright, 2018), may or may not take climate change seriously but they can recognise and act on cost advantages.

The Biden administration in the USA rejoined the Paris Agreement immediately on taking office in 2021 and the Inflation Reduction Act (IRA) of 2022 represents one of the most significant pieces of climate legislation in world history. It set a new course for a clean energy industrial strategy in the USA to drive investment and growth and support the achievement of decarbonisation ambitions, with around US$370 billion allocated to energy and climate investments. Rather than focusing on regulation or emission targets, the IRA sought to achieve decarbonisation through production and investment subsidies, with 43.6% of the planned funding set aside for tax credits on green energy production (Landais et al., 2023).[21]

An estimate from Rhodium Group (August 2024) indicates that support for US$78 billion in federal clean investment – almost entirely via tax credits but also through grants and loan guarantees – occurred in the post-IRA period and that private spending in those technologies over the same period was five to six times larger than public investment. In fact, since the enactment of the IRA, clean investment accounted for more than half of the total US private investment growth. The fastest growth occurred in investment in the manufacturing of clean energy and transportation equipment and technology, which saw, in the year and a half after the IRA, a more than quadrupling of the amount invested relative to the two years prior to the IRA's enactment (Rhodium Group and MIT CEEPR, 2024).

The election of Trump in November 2024 for a second term has brought a further, and very sharp, swing of the pendulum of US policy. Figure 4.12 shows projected emissions, estimated in March 2024 during the presidential race, under a potential Trump presidency versus a potential second Biden (subsequently Kamala Harris) administration. But circumstances have changed greatly since 2016/2017, particularly in the advance of technology,

Figure 4.12: The projected effect of the election of Donald Trump as US president in 2024 – US GHG emissions

A Trump election win could add 4bn tonnes to US emissions by 2030

GHG emissions, billion tonnes of CO_2e

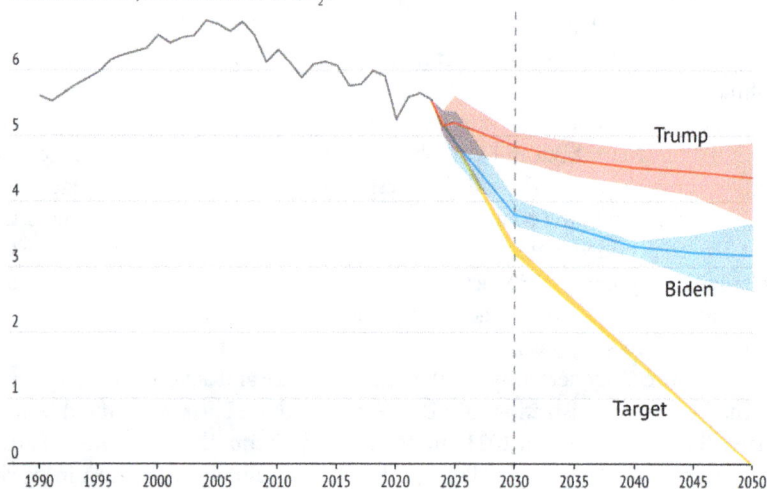

Notes: Black line: Historical US GHG emissions 1990–2022, billions of tonnes of CO_2 equivalent. Red line and area: Projected emissions under the 'Trump' scenario where Biden's key climate policies are eliminated. Blue line and area: Projected emissions under the 'Biden' scenario with the IRA and other key climate policies. Yellow: US climate target trajectory pledged by the Biden administration (50–52% by 2030). The range for each projection corresponds to results from six different models and uncertainty around economic growth, as well as the costs for low-carbon technologies and fossil fuels. This graph was created in March 2024, before Joe Biden withdrew from the presidential race. Source: Evans and Viisainen (2024). Copyright 2024 Carbon Brief. Reproduced with permission.

with the result that today, shortly after the start of Trump's second presidency in 2025, the clean is cheaper than the dirty across a wide range of sectors. The private sector momentum is strong. As noted, the first Trump presidency saw strong increases in investment in renewables and the cost advantages are still stronger now. Nevertheless, it is difficult to say whether US emissions will be as projected by Carbon Brief, as their analysis focuses on the impacts that Trump's rollback of key Biden administration climate policies will have. The analysis does not consider additional measures that Trump could take, for example to increase fossil fuel extraction or undermine the energy transition.

How far and in what way President Trump will, in his second term, reverse and unwind this support remains to be seen. He will, one might suppose, want to take into account the many green jobs created in these new geographies and sectors in the USA. Many of these are in Republican voting states – over 70% of IRA-related investments were in Republican congressional districts (Nilsen and Rigdon, 2024). And he may be concerned about US competitiveness with China where there is a risk of the USA losing ground in the race for the technologies of the future. The private sector momentum in green technologies and investment is now very strong; indeed, the investors of, e.g., Texas may well continue with or accelerate their investments in solar and wind as the cost advantages in relation to fossil fuels widen still further. Nevertheless, President Trump's commitments to undo climate progress are not only very dangerous for the world as a whole but will also directly affect US citizens, through job opportunities and future climate impacts. And they will damage the USA's standing in the world – something we will discuss in the section 'A new geopolitics' and in Chapter 8.

People's Republic of China

China, the world's largest emitter, recognises its major role in world emissions, its own vulnerability to climate change, and the special vulnerability of poorer countries. Moreover, it recognises and is taking the great national and global economic opportunities in the green transition.

Correspondingly, in the last few decades, China, whose approach in the international negotiations we have described in Chapter 3, has adopted increasingly strong climate strategies. The 13th Five-Year Plan for Economic and Social Development of the People's Republic of China (2016–2020), where the country sets its economic and social goals, established targets for reducing carbon intensity (emissions per unit of GDP) by 18% by 2020 compared to 2015 levels and to increase the share of non-fossil fuels in the energy mix to 15%. The following 14th Five-Year Plan (2021–2025) set more ambitious targets, including reducing carbon intensity (emissions per unit of GDP) by 13.5% and increasing the share of non-fossil fuels to around 20% by 2025. It also emphasised innovation in green technologies and the development of a low-carbon economy. Further, in 2020 China announced its 'dual carbon goal' to peak carbon emissions before 2030 and achieve carbon neutrality by 2060. It will peak well before 2030, and indeed may do so in 2025 (Climate

Action Tracker, 2024). And it is now expanding its ETS, while also increasing its penalties for non-compliance in energy-intensive industries.

China has made extraordinary strides in its ability to produce capital equipment for renewables on great scale and at low cost, and in installing that capacity. In 2022, solar power capacity added in China was 45% of the global total installed that year (IRENA, 2023b). By July 2024 China's solar production capacity was more than 1200 GW per annum and rising (Climate Action Tracker, 2024); it is adding 300 GW of renewable capacity internally each year (S&P Global, 2024) – compared to a UK total installed capacity of all forms of around 100 GW (Ember, 2024a) – and can now produce solar panels at a cost of around US$0.10 per watt compared with around US$0.30 per watt elsewhere and costs of around US$2 per watt in the world in 2010 (IEA, 2020; Swanson and Rappeport, 2024). A major global opportunity and challenge will be to use that remarkable production capacity for installations across the world (see Chapter 9).

China also dominates in the production and deployment of electric vehicles (EVs). It is by far the largest EV market and producer. In 2023, more than half of the EVs on the roads of the world were situated in China. In 2024, estimates indicated that 45% of vehicles sold in China were EVs (IEA, 2024a). It also dominates at various points in the supply chain – for example, by 2025 more than 70% of all EV batteries ever produced were made in China (IEA, 2025c). This is driven largely by the low cost at which China can produce batteries – they are cheaper than in Europe and North America by over 30% and 20%, respectively (BloombergNEF, 2024b).

The rapid growth in clean technology production was strongly influenced by Chinese government support (Myllyvirta, 2024), but it is also a result of powerful competition and entrepreneurship in areas which the Chinese industrialists and the government have identified as the technologies and products of the future.[22] These developments in power, EVs, and green growth are illustrated in Figure 4.13, showing that China has made great strides in embracing a new approach to development and seeking to achieve 'harmony between humans and nature'.

At the same time as noting the remarkable advance in renewables and EVs in China, we should also note that it is still opening coal-fired power stations. China began building around 95 GW of new coal-power capacity in 2024, the highest rate since 2015 (Centre for Research on Energy and Clean Air [CREA] and Global Energy Monitor [GEM], 2025). That likely arises from a still strong emphasis on energy security, given turbulence in the Ukraine, in the Middle East, and in US politics. Renewables will, over time, squeeze out coal and plants will retire or run below capacity.

In my view, based on nearly four decades of working on and in China and extensive discussions with policy-makers there, China's rapid economic change towards the development and use of clean technologies has been driven by four considerations. First, the country is very vulnerable to climate change and recognises both that its own emissions are large and that

Figure 4.13: China is embracing a new approach to development under the principle of 'harmony between humans and nature'

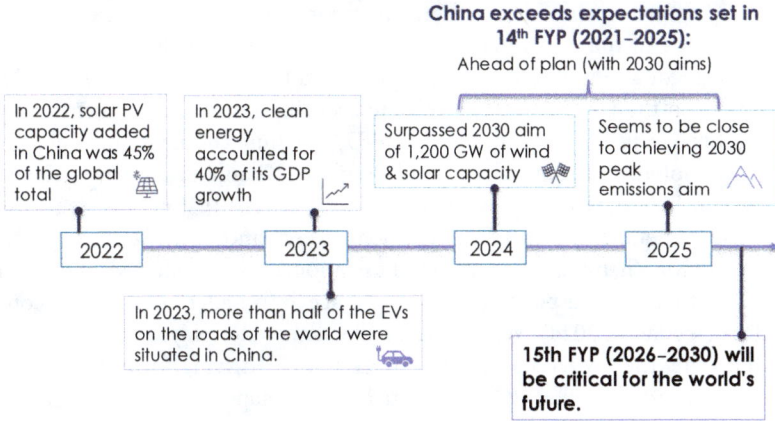

China exceeds expectations set in
14th FYP (2021–2025):
Ahead of plan (with 2030 aims)

In 2022, solar PV capacity added in China was 45% of the global total

In 2023, clean energy accounted for 40% of its GDP growth

Surpassed 2030 aim of 1,200 GW of wind & solar capacity

Seems to be close to achieving 2030 peak emissions aim

2022   2023   2024   2025

In 2023, more than half of the EVs on the roads of the world were situated in China.

15th FYP (2026–2030) will be critical for the world's future.

Sources: Author's elaboration using IRENA (2023b), IEA (2024b), Myllyvirta (2024), and You (2024).

its actions influence others, so that its own activities are major determinants of the world's climate. Second, China has been very worried about air and other pollution in cities, with senior figures making remarks of the following kind, 'What is the point of all this advance in material development if we cannot breathe in our cities?' Third, China recognises the tremendous potential in its next stages of development, which lie in a new growth story based on clean and efficient technologies. Fourth, China sees that by developing new technologies quickly and going to scale it can capture new markets around the world. Taken together, these arguments are powerful and have led to the dramatic changes we are witnessing. These arguments give reasons to believe that these strategies and changes in China are robust and will continue.

The next plan, the 15th (2026–2030), will be of paramount importance in the world. It is likely to accelerate action and further emphasise China's leadership. As planning goes forward, in 2024–2025, for this plan China will revisit its NDC climate pledges. Clean energy and industry are becoming a key part of China's economic and industrial development: in 2023, clean energy was the top growth driver of China's economic expansion, accounting for 40% of GDP growth (Myllyvirta, 2024). Solar was the largest contributor. China's leadership on these fronts will shape global power dynamics and economic geography. It is likely that in the 15th Five-Year Plan, China will emphasise high-end manufacturing (including AI), the expansion of the service sector, and the pursuit of the green transition.

EMDCs: pushing for climate action and growth

The EMDCs have great opportunities to leapfrog the dirty development stages and also face major adaptation challenges. The future of their emissions, given their population and likely growth, will be the single biggest geographical determinant of our climate futures. Progress on these fronts requires strong and urgent investment. There is both an opportunity and an imperative.

For much of the past three decades, developed nations led international climate negotiations in terms of setting the agenda. At COP21 in Paris (2015), Europe and the USA played key roles in securing a strong agreement.[23] The breakthrough in Paris was partly due to USA–China cooperation. When, as we have noted, President Obama and President Xi announced their climate targets in Beijing (2014), they set a collaborative tone that helped shape the final agreement. Key negotiators like John Kerry and Todd Stern from the USA and Xie Zhenhua from China built relationships that allowed China and the USA to align their positions, creating space for broader international consensus (see Stern, 2024). Whilst the small island states played a crucial role, at the time, the EMDCs as a whole were not the driving force behind the climate agenda. The EMDCs did participate in shaping the agreement, but their primary focus was ensuring that rich countries took the lead in cutting emissions and delivering financial support. They saw these as necessary conditions before committing to stronger action themselves.

EMDCs, understandably, continue to exert pressure on rich countries on those two key dimensions, but they have become more flexible, particularly as the opportunities in the new growth story become ever more visible and tangible. And their vulnerability to climate change is ever more obvious and threatening. At COP23 in Bonn (2017), the Fiji COP,[24] they played a central role in the Talanoa Dialogue, a process aimed at increasing climate ambition. At COP26 in Glasgow (2021), they successfully pressed for stronger climate commitments, demanding the long-promised external support of US$100 billion per year be fulfilled. At COP27 in Sharm el-Sheikh (2022), they won a historic victory with the establishment of the Loss and Damage Fund, securing financial assistance for nations hit hardest by climate disasters. Whilst the sums in the Fund are miniscule relative to the problem, a principle was established.

As noted, small island states have long fought for stronger action and were very influential in securing the Paris Agreement and the retention of the 1.5 °C target (see Stern, 2024). Larger EMDCs are also setting the pace. China, India, Brazil, and others are driving investment in renewable energy, clean infrastructure, and other green opportunities, reshaping the global response to climate change. EMDCs are no longer waiting for richer nations to lead. On the other hand, the scale of investment necessary for their transition will require strong external financing flows if it is to move at the pace necessary for the Paris Agreement (see Chapter 9).

EMDCs: the next growth engine

In the coming decades, EMDCs will shape the global economy. Since 2000, EMDCs have contributed an average of 60% of annual global growth –

double their share in 1990s – and this figure is set to rise to 65% by 2035 (Perez-Goropze et al., 2024; World Bank, 2025). Driven by urbanisation, infrastructure investment, and rapid population growth, this expansion will have profound consequences for emissions and climate action.

EMDCs vary widely, from economic powerhouses like India and Brazil to some of the least developed nations. Although China is an upper-middle-income country, it is often analytically excluded from the EMDC grouping because its energy investment and technological advances have been very distinctive and require their own analysis. And China does not need external finance. We have indicated where China has been excluded. China is indeed a special case within the EMDC story. It has the world's largest GHG emissions and has half the world's coal power plants. At the same time, China leads in cleantech innovation and manufacturing as well as electrification. It outperforms the USA and Europe on a number of fronts – in 2023, for example, China added more solar capacity and sold more EVs than the USA in 30 years (Butler-Sloss et al., 2024).

A common challenge faces the EMDCs: delivering better living standards without following the high-carbon path of past industrialisation. While advanced economies bear most historical responsibility for climate change, the future depends on choices EMDCs make now. Their decisions – on energy systems, infrastructure design and build, and patterns of consumption – will lock in emissions trajectories for decades.

In 2024, advanced economies grew at an average rate of 1.7%, whereas EMDCs (other than China) expanded at around 4% (IMF, 2025). This growth is driven by 'catching up' technologically, young populations, growing service and industrial activity, rapid urban expansion, and the associated investments. Young, rapidly expanding populations contribute a potential 'demographic dividend' as more people enter the workforce and stimulate demand for goods and services. The extent of that dividend depends on education, skills, and complementary infrastructure. Cities are being built at pace with roads, housing, and public services demanding energy, land, and materials. As income and output rise, so does energy use both in industry and consumption. Transportation illustrates the challenge: booming car ownership increases pollution and emissions unless cities invest in public transportation and low-carbon alternatives. Most EMDCs are moving through income levels where the income elasticity of demand for energy is strong.[25] Energy demand is surging, driven by growing populations and rising incomes, although energy consumption per capita remains relatively low in many EMDCs (IEA, 2021a). The world's population will grow by 1.9 billion by 2050, with all of that growth taking place in EMDCs (other than China) – especially in sub-Saharan Africa and South (and Southeast) Asia (Gu et al., 2021).

Neither EMDCs nor the world can afford for EMDCs to follow the old model of fossil-fuelled growth, building the dirty first and attempting to 'clean up' later (Dhakal et al., 2022). These countries are among the most vulnerable to climate change, facing rising threats from extreme heat, floods, and other climate-related disruptions. Without decisive action, climate change could stall or even reverse development progress. While advanced economies have historically contributed a much larger share of cumulative emissions, and many EMDCs rightly assert their right to development and the need for international support, the dangers from a 'business-as-usual' approach, in the sense of following past models, are immense. If EMDCs adopt carbon-intensive growth paths, limiting global warming to well below 2 °C will quickly become unattainable, intensifying climate-related risks and undermining development gains.

Yet EMDCs (other than China) also have an unprecedented opportunity. Because much of their infrastructure has yet to be built, many of them can leapfrog straight to cleaner, more efficient technologies. Many are rich in renewable energy – 70% of the world's solar and wind resources and 50% of critical minerals are in the Global South (Singh and Bond, 2024). Investing in renewables can meet the rising energy demand in these economies without deepening fossil fuel dependence.

The economics have shifted in their favour. Solar, wind, and battery costs have plummeted and are still falling, making clean energy the most affordable way to expand electricity access. With millions still lacking power, meeting their needs through renewables would transform lives without driving emissions. They represent an engine for new industries, jobs, and economic leadership. Despite the potential, many EMDCs struggle to break out of low-income status. Progress on this has slowed or stalled since the early 2000s (World Bank, 2025). Given limited budgets and the scale of necessary infrastructure and other investment, without external support the green transition will be difficult. As we shall show in Chapter 9, the green transition, at the pace necessary, will require much larger financial flows from private investors, multilateral institutions, and advanced economies than we have seen in the past or than have yet been committed.

India is beginning to lead on the global stage. Its G20 Presidency in 2023 (see Chapter 9) was an important example, as are several initiatives that it co-founded, including the Coalition for Disaster Resilient Infrastructure and its solar initiatives. Through initiatives like the Leadership Group for Industry Transition, the country also aims to develop industrial applications in high-emitting industries like steel and cement, including the use of green hydrogen. Brazil followed India as G20 President in 2024 and is hosting COP30 in 2025, a unique combination of leadership roles which Brazil is putting to good use.[26] It will be followed in its G20 presidency by South Africa. The trio of India, Brazil, South Africa offers a real opportunity to move EMDCs to centre stage in setting the world agenda. India is likely to host the UNFCCC COP33 in 2028.

India has also emerged as a leader in solar energy. In 2023, it became the world's third-largest solar and wind power generator, behind only China and the USA (Ember, 2024b). Additions of solar PV and onshore wind are expected to more than double by 2028 (compared to the period 2018–2023) (IEA, 2024c). The International Solar Alliance (ISA), headquartered in India, is fostering global cooperation in solar technology. India has proposed the 'One Sun, One World, One Grid' and a 'World Solar Bank', aiming to harness solar energy at a global scale. In the first quarter of 2024, renewables accounted for 71.5% of India's new energy capacity added (Institute for Energy Economics and Financial Analysis [IEEFA], 2024). India's large population and rapid economic growth make it crucial to global climate action. It is India that will experience the world's largest surge in energy demand over the next three decades.

Brazil, for long a leader in hydropower, is accelerating investment in solar, wind, and biofuels. In 2023, renewables supplied 93.1% of the country's electricity, making its grid among the cleanest in the world (Ministério de Minas e Energia, 2024). Hydropower still dominates, but solar and wind are growing rapidly, with PV projects expected to account for 70% of all new capacity in the coming years (IEA, 2024d). Brazil is also a pioneer in second-generation biofuel, with flex-fuel cars covering a large share of its domestic market (IEA, 2024d). At the same time, oil and gas continue to play a major role in Brazil's economy and politics, shaping energy investment decisions even as the country advances its renewable energy ambitions. Following current production projections, by 2030 Brazil is set to become the world's fifth-largest oil producer (Rystad Energy, 2024).

Chile is a model for renewable energy policy and planning. In 2023, around half of its electricity came from renewables, and the government aims to reach 80% by 2030 (IEA, 2024e; Ministry of Energy, Government of Chile, 2022). Chile's transition has been carefully guided by long-term policy frameworks and strong public–private partnerships, making it one of Latin America's most attractive destinations for green investment.

Further, China is helping drive clean energy expansion in partner developing countries. Investments in Malaysia and Vietnam have turned them into major exporters of solar PV technology, with PV products now making up 10% and 5% of their trade surpluses, respectively. Indeed, China's investments in clean infrastructure and manufacturing in developing countries are reshaping global investment in clean energy. Though originally focused on fossil-fuel infrastructure, the Belt and Road Initiative (BRI) pivoted in 2021, pledging to end financing for new coal plants (Bian et al., 2024).[27]

With their huge renewable potential and pressing development needs, investments are urgently needed in EMDCs (other than China). They would be expected to account for nearly 50% of the increase in clean energy investment by 2035 on a global trajectory to meet global climate goals (Bhattacharya et al., 2024) (see Figure 4.14). Though many EMDCs are making progress in the green transition, more than 90% of the clean energy investment increase between 2019 and 2023 occurred in developed nations and China (Bhattacharya et al., 2024) (see also Figures 4.15 and 4.16).

For many poorer countries, finance for renewables remains prohibitively expensive. In Africa, solar power is cheaper than fossil fuel electricity if capital cost is 7–10%, but not if it is the 20% plus faced by much of Africa (see Energy Transitions Commission, 2023). The cost of capital for utility-scale solar PV projects in EMDCs (other than China) is more than twice that in advanced economies (IEA, 2023a). There are also often barriers in public decision-

Figure 4.14: Clean energy investment needs by economic regions for 2030 and 2035

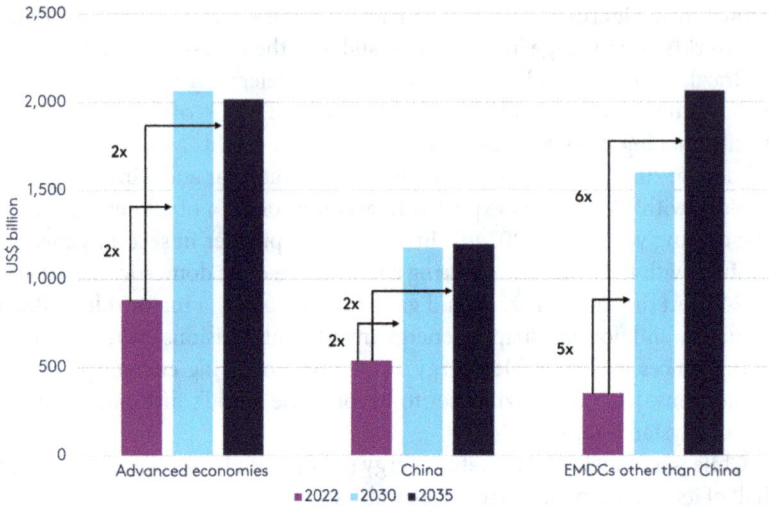

Note: Clean energy investment needs are estimated based on IEA projections, supplemented with additional figures to account for early coal phase-out and just transition-related costs.
Source: Figure 1.2 in Bhattacharya et al. (2024, p. 15). Copyright 2024 The Authors. Reproduced with permission.

Figure 4.15: Increase in annual clean energy investment in selected countries and regions, 2019–2023e, billion US$ (2022)

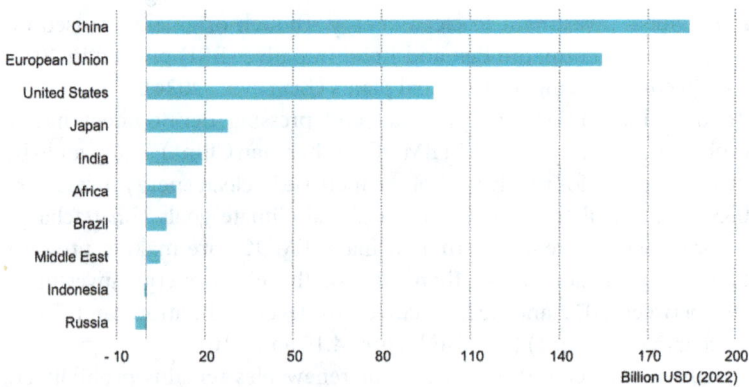

Note: 2023e = estimated values for 2023.
Source: IEA (2023, p. 14). Copyright 2023 IEA, CC BY 4.0.

Figure 4.16: Renewable energy investments in developing and emerging markets, by top countries, 2013–2020

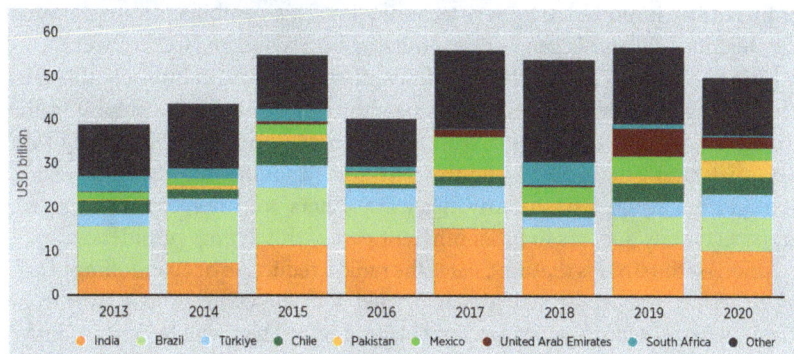

Note: These figures relate to developing and emerging economies, excluding China.
Source: Figure 2.11 in IRENA and CPI (2023, p. 58). Copyright 2023 IRENA, and CPI.
Reprinted with permission.

making, land availability, grid structures, and so on. Investment requires confidence in revenue streams, in the ability to manage costs, and in being able to get things done. Much of that high cost of capital in poorer countries is due to perceived risk. Some of that risk is real, for example exchange rate risk and creditworthiness of utilities. But perceptions of risk are often exaggerated. We return in Chapters 5, 6, and 9 to ways to overcome the financial barriers, increase flows, and lower the cost of capital.

At a broad level the conclusions for the action agenda for EMDCs (other than China) should be clear: improve the investment environment, radically bring down the cost of capital, and help them procure at the cheapest source for supplies.

4.4 A new geopolitics

It is the developed countries and China that have dominated past emissions. And they have, on the whole, up to COP21, dominated climate negotiations. The developed countries, although their emissions remain high relative to the developing world, have led the way in emissions decline. China's emissions will start to decline in the next few years; indeed, they may be around their maximum now.

However, as we have seen, the future of emissions growth will now be in EMDCs (other than China), and these countries are increasingly leading the debate on climate action. They recognise the growth potential in the new model, and they particularly will have to build the resilience in the face of climate forces to which they are especially vulnerable. Leadership is increasingly with these countries, both in terms of where the crucial investments will be located and in climate discussions.

It is in EMDCs (other than China), as we have seen, where the solar potential is concentrated and many of them are endowed with the minerals which will be of increasing importance in new technologies. This combination implies that the weight of economic geography is moving towards them (see Chapter 8).

China and the USA will, in very different ways, be central to how this new economic geography and geopolitics play through. China will respond to the change brought by the second Trump presidency, so let us begin briefly with how that might develop. I write this just a few months into that presidency. Some future strands seem fairly clear. But others are unpredictable. Indeed, unpredictability appears to be an inherent part of this Trump presidency.

Amongst the strands that are clear is the radical reduction or abandoning of aid. Much of the support for activities associated with the US transition away from fossil fuels and reducing emissions will be cut off.[28] There will be a sharp move away from multilateralism. The USA has already announced its withdrawal from the World Health Organization (WHO) and the Paris Agreement. Military support for Europe will be diminished, for Ukraine in particular, but also in terms of the robustness of the US commitment to the North Atlantic Treaty Organization (NATO). Protectionism in relation to trade will rise strongly.

All these examples, and they are of great importance, together constitute a stepping away from, and scaling down of, the constructive involvement of the USA with the rest of the world. These are the embodiment of an America First strategy where the interests of the USA are as interpreted by the Trump administration. Often that interpretation will be, in the view of many, narrow, short-sighted, and transactional.

The USA's withdrawal on the economic, environment, and health fronts comes, however, with some possible expansion on the territorial front, as witnessed by Trump's articulated intentions towards Greenland, the Panama Canal, Canada, and Gaza. The combination of economic and territorial threats, together with the weakening of the USA commitment to NATO, has led other coalitions to move more strongly together on defence. And the EU's commitment to implement the Draghi Report (2024a, 2024b) is likely to become stronger in terms of innovation, productivity growth, and the energy transition. As Ursula von der Leyen put in in her speech at WEF Davos in January 2025, the EU will seek to integrate more closely each of its finance and energy markets and co-ordinate on reductions in regulation (WEF, 2025).

The foreign ministries of China and India, hitherto with a tense relationship, have indicated their desire to work more closely together (Dayal, 2025; Kugelman, 2025). The Brazil, Russia, India, China, and South Africa (BRICS) organisation is widening membership, particularly in response to these new US positions.[29] And Africa, in particular, will be looking to enhance its cooperation with China, and Asia more generally.

As space is vacated, others will step in and look for new alliances. It is clear that China sees these new developments as an opportunity and is looking forward strategically. Its potential leadership on the climate front will become still stronger. First, China will show the powerful example of its own transition, which, as we have seen earlier in this chapter, is gathering pace. Second, it will be working with other countries to create and make use of the technologies in which it has become a world leader. These include renewables, storage of electricity and grids, EVs, and green hydrogen/ammonia. Third, China will work with other countries to make the most of their great expertise in implementation and delivery, particularly in relation to infrastructure. Fourth, it will make greater use of its powerful development banks. All these are predictions we can make with confidence.

China is likely to advance its influence on the trade front. As US-inspired tariffs wars intensify (as of 2025), China will open up trade with the developing world more strongly. And it has already announced zero tariffs for goods from developing countries. The Trump approach to international interactions is therefore likely to intensify existing trends of power and influence moving away from the USA. The Chinese economy has been bigger in purchasing-power parity (PPP) terms than that of the USA for more than a decade. In 2024 the developed countries' share of the world's output (in PPP terms) was around 40%, compared with 58% in 2000 and 64% in 1980 (IMF, 2024). Of the remaining 60%, the EMDCs (other than China) constituted around 40% and China around 20%. The EMDCs will look to each other and China for growth in demand, trade opportunities, technologies, and finance. And they will not be confident, should the need arise, of military support from the USA. Further, Europe, China, and India will be exploring ways in which they can work more closely together.

On the climate front these changes in geopolitics, both the underlying trend and the shock from Trump's second presidency, will see leadership pass towards China and the developing world. Brazil's COP30 in Belém in November 2025 is likely to move more effectively and strongly without the USA. The USA stepping back will be disruptive, but it will not derail climate action. In some ways it will simplify discussions, strengthen resolve, and see others taking the lead.

At the same time, there will be those who might argue that if the USA is moving less quickly, then their own obligation to move is reduced. And the oil and gas companies see an opportunity to loosen or abandon commitments to the transition, as Shell and BP have done. Some financial institutions are already diluting or abandoning commitments as a result of pressure from US Federal institutions and some state governments. US climate science has been a leading force in the Intergovernmental Panel on Climate Change (IPCC) and support for climate science in the USA is being sharply reduced or abandoned.

Thus the geopolitics and the new circumstances do not all point one way on climate action. My own judgement is that the underlying forces for change – in terms of the lived experience of climate change becoming ever more severe, the advance of clean technology becoming ever more rapid, and the realisation of the attractions of the new form of growth – will enable world action on climate change to ride out this storm. And it will see a permanent change in the balance of leadership. Experience will give its own judgement on these predictions before too long.

4.5 Concluding remarks: the agenda

We have seen from the science that without radical changes current policies will put the world on a path of emissions that could result in a global temperature increase of close to 3 °C this century. The science tells us that the potential consequences of such temperatures are so severe that emissions must be cut sharply, starting now, and reach net zero by 2050. And we have seen from the work of the IPCC (2018), comparing 1.5 °C and 2 °C, that 2 °C would expose the people of the planet to great danger, including the possibility of dangerous tipping points.

The agenda for action is determined both by the urgency and by the scale of action necessary to tackle the immense risks and to embrace the opportunities for a new form of growth. Fundamental structural and systemic change will be necessary. And large-scale investment, and the innovation it embodies, will be at centre stage. This investment represents an imperative for climate action and an opportunity in the new form of growth.

The overarching tasks

The 3 °C path we are currently on would be catastrophic, likely existential for many hundreds of millions, possibly billions, of people. Therefore, the first and overarching task on the agenda is cutting emissions sharply and moving to net zero by mid-century.

It now seems likely, given the delays in taking action, that the world will overshoot 1.5 °C. Indeed, 1.5 °C is likely to be exceeded within the next decade. That has important implications for the agenda. First, the challenge of adapting to the climate change which is now unavoidable will demand strong action and investment. The ever-increasing scale of the challenge is illustrated by the latest (2025) data on droughts and floods collected by NASA's GRACE Programme. Since 2019, the intensity or frequency of extreme dry and wet events (defined as beyond one standard deviation from the mean) has doubled globally. In other words, around 2025 the world's water systems, and related activities, have to deal with extreme droughts and floods twice as often relative to 2019.[30] That is an indication of how fast the challenge of adaptation and resilience is intensifying. The pace of change is deeply worrying and the investment required for water infrastructure to cope with these extremes will be very demanding. Second, the overshooting of 1.5 °C will require substantial action for carbon dioxide removal (CDR), and research and innovation on bringing down the costs of CDR are essential. It also raises the possibility of solar radiation management (SRM), although that seems likely to have its own inherent and potentially immense risks (see Chapter 9).

The topics on the agenda then move to how the necessary reduction in emissions and building of resilience, or adaptation, can be achieved. And the context for this is a world that has major development challenges and has poverty across the various dimensions of well-being affecting billions of people.

The SDGs set in 2015 were indeed intended directly and specifically as targets for a sustainable response to the development and poverty challenges. The core of the action agenda therefore must be the overarching task of *combining all three* of the following:

1. *Development* and poverty reduction across all dimensions of well-being.
2. Rapid and large emissions reductions, i.e., *mitigation*.
3. *Adaptation* and the building of resilience.

We have seen in this chapter that there is momentum for change. The action has started. There are some positive tail winds. But we are moving too slowly. Acceleration is crucial. Change on the necessary scale and pace will not be easy. There will be opposition. The new geopolitics has some difficult strands. But climate action, as embodied in the Paris Agreement, is both necessary and possible. And it will deliver development, growth, and poverty reduction.

The key instruments and mechanisms

This rest of the book tells the story of how development, mitigation, and adaptation can indeed be pursued together. The necessary actions to advance on all three fronts constitute the agenda. The key elements will be, on the one hand, investment and innovation and, on the other, systemic, structural, and technological change.

Investment and innovation will drive change across the whole economy. There is an investment imperative from the climate and biodiversity crises and an opportunity in creating a new path to development. The necessary investment is substantial, as we have seen. The conditions for investment and its finance will be critical. Nations and the international community and institutions will have to work individually and together. Clear strategies are needed for the short, medium, and long term.

All of the actions have to keep urgency and scale constantly in mind. Time is of the essence; delay is dangerous. The emissions reductions are not some broad aspiration that can come through whenever we are able to deliver. The science dictates the timelines if the risks are to be manageable and the risks of catastrophic outcomes are to be substantially reduced.

The argument – and in many ways this is the key message of the book – is actually stronger than 'it is possible' to achieve all three together. The investment and innovation as well as the systemic, structural, and technological changes which are essential for mitigation and adaptation will drive a *new and much more attractive form of growth and development*: sustainable, resilient, and inclusive.

The necessary scale and nature of the investments and change were sketched briefly earlier in this chapter. We come back to them in more detail in the rest of the book, particularly Chapters 5 and 9. The necessary investments are large, involving increases of the share of investment in GDP of around 2 to 3 percentage points (ppt) globally and 4 or 5 ppt in many developing countries. Achieving that increase will be a major challenge. But it is a realistic goal: first, because investment rates have in recent times been historically low; second, because the world has potential savings rates which can accommodate that investment; third, because failing to achieve the increase would put the world in great peril. The alternative to carrying through these investments and the necessary systemic and structural change is to risk the worst outcomes. That should be regarded as reckless and, indeed, in this sense the most unrealistic of options.

The investment necessary will be in all forms of capital, particularly physical, human, and natural. Social capital not only plays a key role in relation to the just transition (see below), it would be very damaging to the transition if eroded. The biggest quantitative investment demands, as we have seen and will explore further, are in connection to the energy transition. However, investments in natural capital are critical. These are, of course, of direct significance to reducing emissions. But, and this is of great importance, they can and must contribute to tackling the biodiversity crisis, which is interwoven with that of climate. Natural capital enables, sustains, and is necessary for so many of our activities and needs, including air and water, and should be seen in this crucial sense as part of our basic infrastructure. The climate and biodiversity crises are so interwoven that we must tackle them together. If we fail on one, we fail on the other.

The next items on the agenda follow from the earlier ones. In particular, how to *generate, manage, and finance the investment and change*. They concern first how to create the conditions for investment; innovation; and systemic, structural, and technological change – in terms of incentives, institutions, and behaviour. That is a major challenge, since investment and change are risky and will be inhibited and restricted unless conditions to manage that risk are in place. We will need country platforms[31] and structures to generate a positive investment climate and policies and institutions which can steer investment in the necessary directions.

Finance for investment must be at centre stage. The extra saving that is necessary must match the extra investment and be translated into financial flows if it is to be realised and effective. The first source of finance will be domestic, both public and through the local capital markets and financial system. In most countries the major source of finance for investment is domestic savings. But for many EMDCs, particularly the poorest, external savings or flows, private and public, will be crucial.

The second source is external private flows. These are often restricted by powerful perceptions of major risk. Managing and reducing that risk will be critical to the enhancement of those flows and lowering the cost of capital. Third is external public flows, particularly from MDBs and development finance institutions. These flows can and should be structured in ways that contribute to reducing and managing risk and thus enabling private flows. MDBs will play a vital role in funding public investment too. And MDBs can and should help create a positive investment climate.

The fourth source of finance, critical for certain investments, will be low- or zero-cost funds, with minimal or zero service obligations. These include Official Development Assistance (ODA), highly concessional finance[32] from MDBs, special initiatives such as the use of Special Drawing Rights (SDRs) from the International Monetary Fund (IMF), flows from carbon markets, new forms of international taxation (e.g., on sea or air travel), and philanthropy. Chapter 9 will explain and discuss these instruments. Different combinations of these flows will be appropriate for different projects and programmes. Highly concessional flows would be of special relevance to adaptation, loss and damage, and natural capital. Further work on enhancing all of these concessional sources is of great importance.

If these changes do take place, as they must, there will be the challenge of managing inevitable dislocations. They will be of particular significance to poorer people. Thus, we have on the global agenda, the *creation of a just transition*. This is of significance for all countries. There is a moral duty to protect people who may suffer from rapid structural and technological transformation, pursued for a common good. And a just process is necessary to make the transition politically feasible. A just transition will be of special significance to poorer countries and communities who may be less able to manage dislocations or disruption to lives and livelihoods. The just transition underlines the importance of our fourth class of capital, social and cultural. This capital includes mutual trust, sense of community, institutional strength, recognition of shared values, and so on. It is more difficult to define and measure than other forms of capital, but it is crucial to development and to effective action. It is an asset or stock that can be undermined and destroyed, and it can be enhanced, particularly by working together in the face of a common threat.

This agenda is summarised in Table 4.2. The table also includes references to chapters in this book where the issues are examined. It is an agenda which follows directly from the arguments set out in this chapter and its predecessors. It is rooted in the science, basic moral values, technologies, economics, and politics. In examining how that agenda can and should be pursued, we shall show that an effective response to the challenges and delivery of the agenda will drive the growth and development story of the 21st century – sustainable, resilient, and inclusive. It will yield a much more attractive path of development across all dimensions of well-being than the dirty, destructive paths followed in the past. Not growth forever, but a transformation over the next two or three decades which is full of opportunity and which can set the world on a new and much more attractive path. That opportunity is of special importance for developing nations who can leapfrog the dirty, destructive models of the past.

The first block (I) constitutes the overarching agenda. It is crucial to deepen the understanding both that this is the new growth story, and that strong investment and change are necessary for its delivery. And to be clear that delay is dangerous. Block II contains the instruments for change. Block III concerns international action.

Table 4.2: The agenda for the growth and development story of the 21st century

Agenda	Cross-reference
I. Overarching tasks	
1. Accelerate emissions reductions.	Chapters 2, 5, 6, 7, and 9
2. Strengthen adaptation/resilience	Chapters 2, 5, 9, and 10
3. Combine development and poverty reduction with mitigation, adaptation/resilience, and protecting and enhancing biodiversity.	Chapters 4, 5, 6, and 9
II. Key instruments and mechanisms	
4. Identify strategic investments.	Chapters 5, 6, 7, 8, and 9
5. Foster systemic, structural, and technological change.	Chapters 5, 6, and 8
6. Create conditions for investment; innovation; and systemic, structural, and technological change: incentives, institutions, and behaviour.	Chapters 6, 7, 8, and 9
7. Catalyse the right kind and sources of finance for investments.	Chapters 6, 7, 9, and 10
8. Manage inevitable dislocations: create a just transition.	Chapters 6, 7, 8, and 9
III. International	
9. Continue to build international collaboration around delivery on the Paris Agreement by pursuing the above agenda. Strengthen ambition in all countries. Finance at centre stage.	Chapters 8 and 9

None of this will be easy. On the contrary, as I have taken care to emphasise, change of this magnitude and pace will be difficult. And the greatest difficulties, given the advances in technology and its future potential, lie in economics, politics, and society. That is why the purposive and creative application of the social sciences and humanities is so important to the tackling of the climate and biodiversity crises we face. Economics, as part of the social sciences, has many insights and tools to offer in responding to these challenges. We will continue to draw and build on the analytical insights from economics, and the social sciences more generally, in the rest of the book. The difficulties are not a reason to hesitate and delay. These are problems to be tackled with a sense of urgency. The alternative of not making the changes, or of delaying them, would be far more difficult. Tackling and delivering on this agenda forms the subject of the rest of this book.

Notes

[1] Climate beliefs were measured by the answer of participants to the question 'How accurate do you think these statements are?' from 0 = not at all accurate to 100 = extremely accurate. The statements were: 'Taking action to fight climate change is necessary to avoid a global catastrophe', 'Human activities are causing climate change', 'Climate change poses a serious threat to humanity', and 'Climate change is a global emergency'. The climate policy support variable consisted of the level of agreement of participants from 0 = not at all to 100 = very much so, in relation to the following nine claims: 'I support raising carbon taxes on gas/fossil fuels/coal', 'I support significantly expanding infrastructure for public transportation', 'I support increasing the number of charging stations for electric vehicles', 'I support increasing the use of sustainable energy such as wind and solar energy', 'I support increasing taxes on airline companies to offset carbon emissions', 'I support protecting forested and land areas', 'I support investing more in green jobs and businesses', 'I support introducing laws to keep waterways and oceans clean', and 'I support increasing taxes on carbon intense foods (e.g., meat and dairy)' (Vlasceanu et al., 2024).

[2] These results are consistent with Ipsos's (2023) analysis of more than 21,200 adults in 29 countries. Aside from identifying that people think their nations should do more to combat climate change (global average of 66% agree), they also identified that more people believe that the economic cost of climate change will be higher (42%) than the cost of actions to reduce it (26% believe the latter will be higher). Furthermore, Poushter et al. (2022) analysed public opinion in 19 advanced economies in North America, Europe, Israel, and the Asia-Pacific region. They identified that the percentage of citizens who believe that global climate change is a major threat in 2022 reached an all-time high in 10 countries – with data every year from 2013 to 2022. They found that a median of 75% of people across 19 countries (mostly in Europe, North America, and Asia-Pacific) claim that climate change is a 'major threat to their country' (Poushter et al., 2022).

[3] He shared these thoughts in a discussion.

[4] Lukewarmers try to say that whilst they claim to recognise climate physics, the effects of climate change are likely to be modest and, further, the costs of change in our economies to tackle climate change will be very high.

[5] Examples include spreading the idea that clean energy does not work or that climate policies are harmful. This approach is known as 'new denial' and has been identified as the cause of 70% of YouTube climate denialism in 2023 – a substantial increase from 35% in 2018 (CCDH, 2024).

⁶ People Climate Vote is the largest survey of public opinion on climate change, which encompasses 50 countries.

⁷ And see my chapter in Greta Thunberg's *The Climate Book* (2022). See also 'Coming Generations on the Front Line', Part IV of *Standing up for a Sustainable World*, edited by Henry et al. (2020).

⁸ Covering countries such as Pakistan, India, the USA, Australia, Brazil, Guyana, Spain, and many more.

⁹ Only around 200 climate cases were registered in Global South nations (around 8% of all cases) – 87% were filed in Global North courts and around 5% in international and regional courts (Setzer & Higham, 2024).

¹⁰ The definition of LCOE used by the Department for Energy Security and Net Zero is 'the discounted lifetime cost of building and operating a generation asset, expressed as a cost per unit of electricity generated (£/MWh)' (2023, p. 10).

¹¹ Other barriers include unclear policy frameworks and market design, financially strained utilities, weak grids, high implementation costs, and a push-back from vested interests, in addition to the high costs of capital (IEA, 2023b).

¹² An NDC with these characteristics is called 'investable NDC'; we will return to this in Chapter 6.

¹³ NZBA is a banking-specific coalition; the NZAM is an asset management focused coalition (shown in Figure 4.11). Both are affiliated with GFANZ (discussed in Table 4.1).

¹⁴ See Chapter 1.

¹⁵ The Grantham Research Institute (GRI) on Climate Change and the Environment, together with the World Bank, have worked with the co-chairs of the Coalition in its creation and development. From the GRI, Amar Bhattacharya, Nick Godfrey, Anika Heckwolf, and myself have all been involved.

¹⁶ The agreed reductions were, in principle, binding on the Annex 1 (developed) countries. But the enforcement mechanisms (that if you missed your target in the first period, you had to 'catch up' in the second period) had no traction (see Chapter 1).

¹⁷ The overall cap on emissions has been tightened further, with a reduction rate of 2.2% per year starting in 2021, up from 1.74% in the previous phase. This means fewer allowances are available each year, driving up the price of carbon and incentivising reductions. Moreover, the EU ETS expanded (in January 2024) to include the maritime sector. A separate ETS, referred to as ETS2, will come into effect in 2027 and will include

road transport and buildings. This extension is designed to cover sectors that are harder to decarbonise and where emissions have been stubbornly high.

[18] The CBAM is aimed at 'levelling the playing field' for EU industries that are subject to the ETS by providing border charges on high-carbon imports thereby discouraging 'carbon leakage'. See Chapters 8 and 9.

[19] Other key policy developments include the new renewable energy target of at least 42.5% of energy from renewable sources in the EU's gross final consumption of energy by 2030, but aiming for 45% by 2030. This implies a near doubling, from 2024, of the existing share of renewable energy in the EU. They have also introduced an energy efficiency target of reducing final energy consumption at the EU level by at least 11.7% compared to projections of energy use for 2030 (based on a 2020 reference scenario). In addition, a major portion of the EU's Multiannual Financial Framework (MFF) for 2021–2027, as well as the NextGenerationEU recovery fund, has been designated for climate action, with at least 30% of expenditures allocated to climate-related projects.

[20] 'There are multiple paths and mechanisms by which this country can achieve – efficiently and economically – the targets we embraced in the Paris Agreement. The Paris Agreement itself is based on a nationally determined structure whereby each country sets and updates its own commitments. Regardless of US domestic policies, it would undermine our economic interests to walk away from the opportunity to hold countries representing two-thirds of global emissions – including China, India, Mexico, European Union members, and others – accountable' (Obama, 2017, paragraph 19).

[21] Originally, the IRA allocated US$30 billion in production tax credits to accelerate domestic production of solar PV, wind turbines, and batteries, as well as critical minerals processing; US$10 billion investment tax credit to build clean technology manufacturing facilities (e.g., to make EVs, wind turbines, and solar panels); US$20 billion in loans to build new clean vehicle manufacturing facilities; more than US$9 billion for Federal procurement of clean technologies made in the USA to create a stable clean products market; and US$27 billion clean energy technology accelerator to support the implementation of mitigation technologies (Democrats Senate, 2022).

[22] According to Song et al. (2024), Chinese subsidies to the EV sector in 2018 were around US$17 billion; in 2023 this number was US$45 billion. As costs of production fall strongly, these subsidies are likely to decrease.

[23] The USA pledged significant emissions reductions and committed to the Green Climate Fund, designed to support climate action in developing countries. The EU pushed for the inclusion of a 1.5 °C temperature limit,

working closely with small island states, and committed to financially supporting developing countries in their transition to low-carbon economies and adaptation to climate impacts, which was crucial in encouraging their participation in the agreement. It then included the UK; Pete Betts of the UK's Department of Energy and Climate Change was the lead negotiator for Europe.

24 It was a Fijian presidency, but the UNFCCC took the view that holding the meeting in Fiji itself posed excessive logistical problems.

25 An income elasticity of energy above 1 means that as incomes rise, people's demand for energy tends to grow more than proportionally with income.

26 I am a member of the economic advisory group to the Brazilian COP president and, with Amar Bhattacharya and Vera Songwe, am co-chairing the Independent High Level Expert Group on Climate Finance (IHLEG) for the COP30 presidency.

27 Some concerns remain over debt risks associated with Chinese infrastructure loans, particularly in Sri Lanka and Tajikistan (Jones and Hameiri, 2020; Samaranayake, 2021; van Twillert and Halleck Vega, 2023; World Bank, 2019).

28 There may be some residual support for CCS associated with enhanced oil recovery or for some jobs, for example, in Republican states.

29 As of 2025, members of BRICS are Brazil, Russia, India, China, South Africa (since 2010), Egypt (since 2024), Ethiopia (since 2024), Indonesia (since 2025), Iran (since 2024), and the United Arab Emirates (since 2024). This expanded membership has been informally referred to as BRICS Plus since 2024.

30 Presentation by Christopher Gasson at Paris Water Summit, 13 May 2025.

31 Country platforms concern clear and coherent strategies, reliable institutions, problem-solving abilities, and finance. In other words, the conditions for investment and finance and the conditions for the practicalities of 'getting things done'. See also Chapter 9.

32 International Development Association (IDA) is the 'soft window' of very long-term, low-interest finance at the World Bank. Other MDBs have similar windows.

Part II.

The new growth story: investment, innovation, and fundamental structural change

5. Rising to the challenges: the key elements of a new growth story

The global economy is on an unsustainable path. The climate and biodiversity crises are now so acute that the ways in which the world produces and consumes must change rapidly, and on a great scale. We must break the destructive relationship between production and consumption on the one hand and the environment on the other. This was the conclusion of Part I of this book, leading to the action agenda set out at the end of the last chapter.

The second part of the book, Chapters 5 to 7, shows how we can forge a new path for economic development that does break this destructive relationship. With the right policies and public and community action, this new path can deliver a much more attractive structure for growth and development, with rising living standards across all dimensions. It can deliver sustainable, resilient, and inclusive development. But the next two decades are critical to success. This chapter explains how well-designed climate action constitutes the growth story of the 21st century. Chapter 6 goes deeper into the story, particularly how it can be fostered, examining details of central concepts, strategies, policies, institutions, and incentive structures. Lastly, Chapter 7 emphasises and examines the state's role in shaping the new growth story.

Failure to implement change at the necessary pace and scale will likely result in an environment so hostile that disastrous decline could become highly likely or, indeed, inevitable. There is real hope but also real danger. It will be our actions as a world community over the next two or three decades that will decide which way we will go.

The changes in economic structures and the required increase in investment and innovation are substantial and challenging. Some may argue that these changes are too difficult and 'unrealistic'. However, the consequences of inaction far outweigh the challenges of realising the necessary investment and change needed to embark on the new path. Investment and innovation will be at the heart of the story of creating the new. That will in large measure be from the private sector. Creating the conditions and finance for that investment will be central to public action and policy.

Our first task in this second part of the book is to outline the key elements of the new growth story (Section 5.1). The discussion highlights six key drivers of growth in the transition to a new form of development: rapid innovation, increasing returns to scale, greater resource efficiency, better-functioning systems (such as cities, energy, and transport), improved health, and the increase in investment itself. We must be clear that we are identifying and emphasising the growth and development opportunities over the next two or three decades, driven by investment and structural and systemic change. This is not a story of growth of the kind depicted in the long-run infinite-horizon models of standard growth theory.[1]

In Section 5.2, the analysis focuses on integrated action and the centrality of investment, innovation, and their financing, as well as the scale and urgency required for the entire process of change and growth, for the national economy and across the economies of the world. It also stresses the importance of integrating development, mitigation, and adaptation in our analysis and action. There is a brief discussion of the powerful influence on growth likely to be exerted by the combination of artificial intelligence (AI) and the green transition.

In Section 5.3, we examine and refute two misleading arguments that are sometimes offered as counters to our own: first, that there is a trade-off between climate action and growth when, as explained in Chapter 4 and Section 5.2, such action drives growth and development, and second, that we must reverse growth, sometimes called 'de-growth', to tackle the climate crisis. We also examine some arguments around population. Whilst they have some relevance, the central question for future action is not primarily the number of people, but how they consume and produce. In this section, we show that the investment, innovation, and systemic change necessary for managing the climate and biodiversity crises will drive the creation of a new, attractive, and sustainable growth without having to 'balance' climate or environment on the one hand and growth on the other.

Further details on investment across sectors, geography, and the role of public versus private sectors are set out in Section 5.4. There are significant variations across these dimensions, and coherence within and across dimensions is of critical importance. The implications for growth, development, and poverty reduction in developing countries are drawn together in Section 5.5, emphasising the themes that run through the chapter and the book. There is a great opportunity to leapfrog the dirty, destructive models of the past, but the challenge is to foster and finance the key investments. There is no horse race between action for strong development and climate action. They can drive each other forward.

To meet the challenges, we will need a response encompassing three core elements: investment, innovation, and systemic/structural change. This response must correspond to the urgency and scale of the challenges to avoid the most severe risks of climate change (emissions reductions), to adapt to the inevitable impacts and risks, and to create a new path of growth and development. The new path is the growth story of the 21st century. We know enough now about this path to embark on it confidently, but we must plan to learn and adjust along the way. Delivering change at the pace and scale necessary will not be easy. But the alternative of delay and inaction will create danger and destruction, which would be far more difficult.

For economics to play a constructive role in this new story, it must shift towards the economics of rapid structural, systemic, and technological transformation, moving away from narrow, mostly static general equilibrium perspectives or narrow models of exogenous growth, with often fixed underlying structures and systems.

This is the moment for economics to step up. In doing so, its analysis must be interwoven with politics, finance, law, geography, international relations, history, culture, and, crucially, moral philosophy. It must integrate with social sciences and the humanities while also being informed by and collaborating with the natural sciences, technology, and engineering. Although much more will be needed, science, technology, and engineering have advanced rapidly and will likely continue to do so. Most of the obstacles and difficulties will arise in the realms of economy, society, and politics.

5.1 The drivers of growth

There are six driving forces arising from action to reduce emissions that together can generate a new growth story. These forces operate both individually and in interwoven and complementary ways. As we will see in Section 5.2, effective adaptation and the associated resilience it also creates, protects growth; without it, incomes, lives, and livelihoods would be at risk from damage. The six drivers are:

1. Lower costs, learning-by-doing, and induced innovation in the new cleaner technologies.
2. Increasing returns to scale in new technologies.
3. Greater resource efficiency across the economy.
4. Stronger system productivity, including in energy, transport, cities, land, and water.
5. Improved health, with associated higher productivity.
6. Increase in the share of investment in overall output.

They are clearly conceptually and practically distinct, but powerfully inter-related. They are set out diagrammatically in Figure 5.1.

We should note that most of the processes embodied in these growth drivers, except increased investments, are often excluded from standard macroeconomic or general equilibrium modelling. High on the research agenda for economists should be to better integrate these crucial drivers into their analyses. We examine these six drivers in turn.

Lower costs, learning-by-doing, and induced innovation in clean technologies

The pace of technological advancement and cost reduction in key new technologies has been extraordinarily rapid and has often far exceeded expectations, as we saw in Chapters 1 and 4. Capital costs for renewables

Figure 5.1: The 21st-century growth story: six interwoven, mutually reinforcing drivers

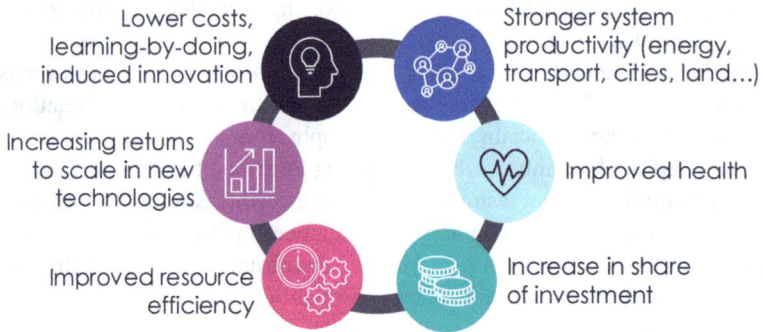

Source: Author's elaboration.

continue to fall much faster than those for conventional technologies. During the work on the Stern Review in 2005/6, we did not anticipate that the cost of solar electricity would drop by 80% over the next decade and continue falling rapidly thereafter (IEA, 2020). Nor did we foresee that by 2024, most car manufacturers would build their planning around the end of the era of the internal combustion engine. As of 2025, solar (with storage costs) is cheaper than coal generation of electricity across major markets, including India, China, Australia, and the USA (Energy Transitions Commission [ETC], 2025a).[2] Outside these major economies, there is greater variation, but with increasing government support schemes and the rapidly declining costs of batteries and solar, cost competitiveness will only increase. By 2035, this cost competitiveness of solar and batteries will be achieved across most markets (ETC, 2025a).

Driving these costs reductions has been (a) research and development (R&D) and innovation, (b) learning-by-doing, and (c) economies of scale. Whilst these overlap and interweave, they are distinct. We treat the first two together as part of technical progress. Economies of scale constitute a logically distinct phenomenon which occurs for given technologies. Innovation plus R&D and learning-by-doing have been very powerful as new approaches and materials have developed and experience with the new technologies has advanced.

Underpinning these cost reductions is the idea that clean technology follows learning curves – that is, production costs decrease as cumulative production increases. This effect comes from learning, from economies of scale, and from technological advancements. Renewable energy follows learning curves, more so than fossil fuels, because they are newer technologies. Broadly speaking, processes for extraction, processing, and use of fossil fuels have remained largely unchanged. Technical progress has occurred, for example fracking, but, on the whole, these are mature technologies where innovation and learning-by-doing are much less strong than for the new, clean technologies. Furthermore, extracting fossil fuels gets increasingly more difficult and expensive as the resource depletes, requiring deeper drilling, lower-quality reserves, and more complex extraction techniques (e.g., fracking). In contrast, renewable energy technologies, such as solar panels and wind turbines, continue to benefit from continuous innovation, manufacturing improvements, and learning-by-experience. These dynamics explain why the price of solar power has dropped so much over the past few decades. As well as showing much slower technological advance, fossil fuel costs remain volatile, driven by geopolitical factors and resource constraints. Uncertainty adds to cost.

With the declining costs of low-carbon technologies, the clean has become cheaper than the dirty for electricity generation across most parts of the world (United Nations Environment Programme [UNEP], 2019), provided the cost of capital is manageable. The cost of capital is a key issue, particularly for poorer countries, and is a core theme for Chapters 6 and 9. Improved grid structures, management, and markets will further enhance the efficient allocation of electricity across space and time. It will become easier to manage variability and to match supplies and demands, especially with AI, lower-cost storage, increasing links to multiple sources of supply, and improvements in power markets.

China's leadership in the manufacture of solar panels and batteries has played a key role in the remarkably rapid fall in costs of solar photovoltaic (PV) power, bringing it down by 85% over the last decade (Cao et al., 2021; International Energy Agency [IEA], 2023a).[3] The falling costs of renewables and storage have transformed what is possible and profitable. And the ability to manage systems, including via AI, has improved strongly at the same time. As costs fall, perceived risk falls too, thus lowering the financing costs. For key climate technologies these financing costs were expected to fall by 40% due to the IRA, an important factor in the rapid scale-up.

Generation, grid and transmission structures, storage, and power markets are complex entities. Different sources for generation play different roles and have to be combined in markets where supply and demand vary over space and time. It is important not to over-simplify. Yet, we can say, as already emphasised in this chapter, that clean is becoming cheaper than dirty across a growing number of sectors worldwide.

As prices fall, new sectors and technologies become commercially viable, generating further rapid change. For example, cost reduction in renewable electricity decreases the cost of green hydrogen (produced by electrolysis), which drives down the cost of green ammonia, opening opportunities for green fertilisers and sea transport. Further, AI can turbocharge the process of innovation itself, across every sector. The pace of technical progress shows little sign of slackening.

The application of AI in emerging markets and developing countries (EMDCs) could greatly enhance and accelerate the process of leapfrogging.

For example, with AI new factories can be equipped in ways which are much more energy efficient, using built-in energy management systems. That is much cheaper than complex retrofitting in older buildings and factories (see IEA, 2023b).

The speed of change has already been remarkable. And we note that the changes have occurred even though government policy has often been weak and problematic. In large measure, they have been driven by a recognition of the opportunities in the new path of growth and that a shift towards a low-carbon future is inevitable. Clearer and more credible strategy, sound government support, and more robust policies could greatly accelerate innovation and investment (see Chapter 6). Digital technologies and AI will not only improve grid management and meeting of demand and supply, but also accelerate the pace and effectiveness with which clean technologies are discovered, improve the geographical identification of key minerals, and accelerate and improve how capital is deployed towards the net zero transition.

Current economy-wide models, mostly general equilibrium models, often struggle to account for these dynamic factors, leading to an underestimation of the pace and potential of technological change. We should recognise, however, that such dynamics are not easily modelled. Economic analysis is more straightforward with simple competitive equilibrium, diminishing or constant returns to scale, and modest changes in technology. But we need analyses and actions which embrace the complexity of the real dynamic forces which are at issue here. These are the forces which will play a powerful role in shaping the transition and the new approach to growth.

As we saw in Chapter 4, the world has already passed many positive technological tipping points and others are close.[4] These achievements represent major opportunities for governments and the private sector to commit to and invest in the global transformation. Further, the costs of decarbonisation of previously 'hard-to-abate' sectors (see Section 5.4) are falling with advances in underlying technologies (such as renewable energy and clean hydrogen production). For example, the costs of decarbonising 70% of global emissions in 2021 was estimated to be 40% lower than it would have been just two years earlier (Systemiq, 2023a). The links among sectors can sometimes imply that when one tipping point is crossed in one sector, it can lead to tipping point cascades, where progress in other sectors is accelerated. For example, falling costs in clean electricity, an output of the energy sector, can lead to lower costs – and thereby a tipping point – in other sectors, such as electric vehicles (EVs) or green hydrogen (Systemiq, 2023a).

Well-designed policies can help initiate and drive forward such cascading. For example, mandating zero-emission vehicles can generate greater EV deployment, raise demand for renewables, and induce innovation, which reduces the cost of batteries, lowering the cost of renewable energy storage (Systemiq, 2023a). Important policies that could generate new cascading tipping points include the recently adopted regulation and pricing of global shipping emissions under the International Maritime Organization (IMO), mentioned in Chapter 4. At the time of writing in early 2025, these policies were moving towards implementation. This will potentially increase the global demand for clean fuels, such as methanol and ammonia made with green hydrogen. Economies of scale in these fuels could yield advances in other applications that rely on the same hydrogen-based technologies, such as green fertilisers.

Figure 5.2 provides an illustrative overview of different stages of technology deployment, focusing on three enabling conditions for tipping points – affordability, attractiveness, and accessibility.[5] In the formation stage, technologies are neither affordable, nor attractive, nor accessible, relative to incumbent technologies, with prototype solutions still not proven at commercial scale. Systemiq (2025) places technologies such as zero-carbon plastics and sustainable, synthetic aviation fuel (eSAF) currently at this stage. Other technologies are further along the S-curve of adoption and deployment – becoming increasingly attractive and accessible, but not yet fully affordable. Electrothermal energy storage (ETES), which will be crucial for electrifying industrial heat processes and storing excess energy as heat, is in the acceleration stage, getting close to the tipping point. When technologies such as solar power pass their tipping points, they go on to grow rapidly, becoming increasingly affordable, attractive, and accessible. Innovation is needed to achieve tipping points. Policies, as mentioned above, can induce innovation, and thus lead, in turn, to tipping points.

Often however, these dynamics of acceleration and growth are local or regional, as they are influenced by, for example, local regulation, clean energy generation potential, and offtake arrangements. It is therefore necessary that analysis on possible policy levers and remaining obstacles takes place at these levels, rather than at the global level only. Doing so sometimes enables the identification of specific levers that are needed to accelerate reaching specific technological tipping points in particular locations.

Realising and capitalising on these clean initiatives will require a substantial increase in investment, both private and public. It is the potential for profitable investment that drives so much of these dynamics, although we should also recognise that the motivation of many of the engineers, technologists, architects, planners, entrepreneurs, and others involved also comes from enthusiasm for and commitment to finding effective responses to the climate and biodiversity crises and building a new approach to growth. That investment will be realised only if it is supported by favourable policies and strong collaboration. This is a constant theme in our argument throughout the book and particularly in Chapter 6.

As highlighted in Chapter 4, market opportunities in the new technologies and an increasing understanding of the imperative to change produced a major shift in global energy investment priorities over the past decade. However, the current global trajectory of clean energy investment falls far short of what is needed to meet growing energy needs in a sustainable way. And investment is concentrated in advanced economies and China (as we

Figure 5.2: Illustrative overview of technological tipping points

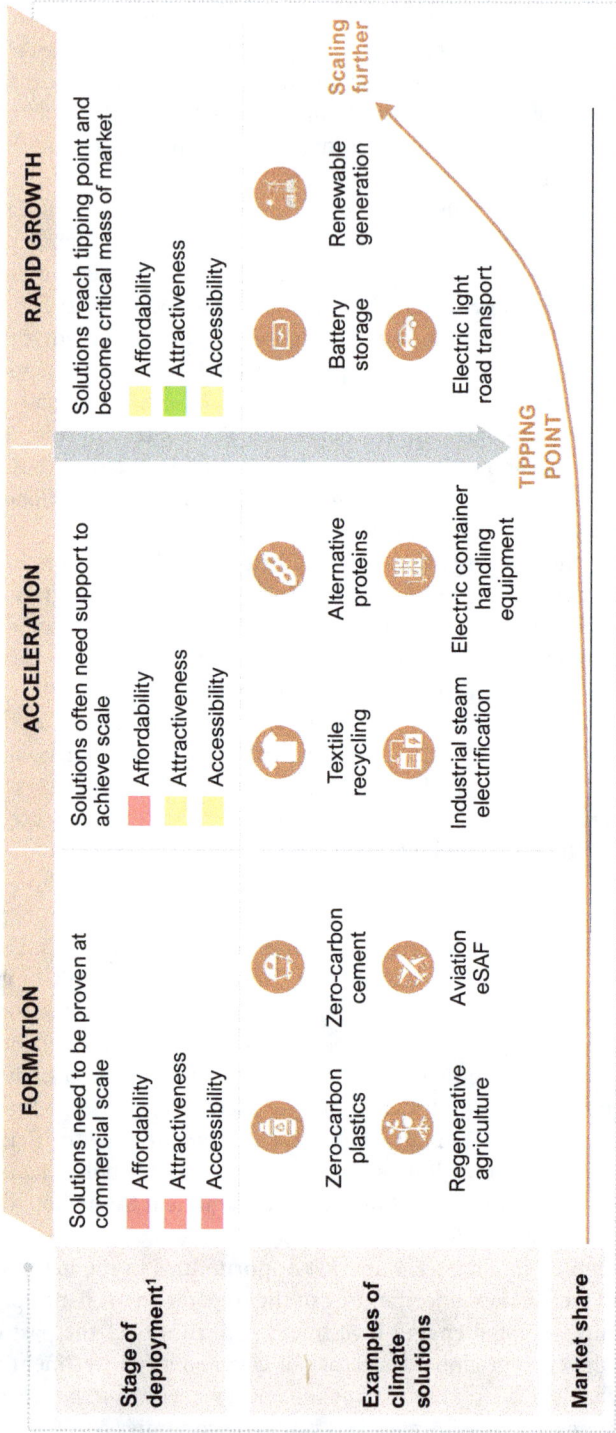

Note: Colours indicate levels of affordability, attractiveness, and/or accessibility: red (low), yellow (medium), and green (high).
Source: Systemiq (2025, forthcoming). *Accelerating the breakthrough of climate technologies: Driving exponential growth in climate technologies with positive tipping points* [Unpublished report]. Copyright Systemiq. Reproduced with permission.

saw in Chapter 4), given the many challenges that hinder investment in EMDCs. Capital investment for fossil fuel energy supply in EMDCs (other than China) exceeds low carbon investment by more than 2:1 (BloombergNEF, 2024).

Increasing returns to scale in new technologies

By definition, increasing returns to scale occur where a proportional increase in the inputs to a production process produce a greater than proportional increase in the output. Increasing returns to scale can emerge from various factors:

- The initial set-up costs involved in building essential equipment and facilities: Once these facilities are established, their costs are spread over larger production runs, reducing average costs as output increases.
- Processes of discovery: For example, strong R&D departments can produce improvements across a whole range of activities.
- Material usage: For example, the volume of a container, such as silos or trucks, could increase faster than the material used to construct it, as seen in storage and transport.
- Information management and the distribution of work and goods: Scale allows for specialisation, shared resources, and lower inventories.
- Critical networks, such as electricity grids, broadband, public transport, or recycling facilities: For example, establishing a network of EV charging stations across the EU would be more cost-effective per unit than installing single chargers; this network, in turn, would promote greater EV adoption, further driving down costs.
- Clustering of activities to enable effective use of supply chains and skills, combined heat and power, and so on: Increasingly we see the rise of clean industrial zones, such as the case of the Tianjin Economic-Technological Development Area (TEDA) in China (see Section 8.3).

These various sources of economies of scale are mutually reinforcing, creating a powerful cycle of cost reduction and scale increases.

Economies of scale are not limited to production in particular firms or clusters; they also play a crucial role in discovery. Some activities, such as particle physics, require large-scale machinery to achieve breakthroughs. In AI, projects such as AlphaFold have shown that looking across a broad waterfront, here analysing all proteins collectively, can yield transformational insights. Ideas are public goods which can be applied repeatedly without being 'used up', a fundamental economy of scale. These forces are highly relevant in the transition to net zero, influencing technologies from solar panels and EVs to recycling and the circular economy.

An illustrative example of production economies of scale can be seen in India's adoption of LED light bulbs. Through large-scale public procurement, the cost of LED bulbs, far more energy efficient than incandescent light bulbs, was reduced by 85% within four years. India now aims to replicate this success with electric buses (Anadon et al., 2022). The government is working on replacing 800,000 diesel buses with electric buses by 2030 using a bulk procurement model to secure lower prices in exchange for large quantities, to then lease to local transport operators on a pay-as-you-go basis (IEA, 2024a). These examples illustrate both the power of economies of scale and the role of organisation and procurement in realising the economies of scale. Scaling up green activities will enable economies to capitalise on these returns to scale.

There are, of course, economies of scale in dirty technologies too. But the ideas and examples above show that they are of particular importance in clean technologies, because substantial set-up costs and growth from small to big appear most strongly in sectors which are currently innovating and starting up.

Greater resource efficiency

Improving resource efficiency means getting more out of the same quantity of resources. Efficiency is productivity is growth. Across most countries and activities, there is great potential for enhanced energy efficiency. So much is wasted through poor design, inadequate insulation, faulty operation, and so on. The shift to renewable sources of energy presents a powerful opportunity for a more efficient energy system. From the basic physics, because zero-carbon energy sources generate electricity directly rather than through combustion or heat engines,[6] there are substantial efficiency gains available across the economy (Eyre, 2021).

The work of the Rocky Mountain Institute, for example, has shown the high returns that can be generated from 'integrative design' (Lovins, 2018). This approach involves designing entire systems – such as homes, industrial processes, or transportation networks – for enhanced efficiency, encompassing everything from heating and water flows to lighting and construction materials. For example, UK houses are notoriously less well insulated and energy efficient than European counterparts further north (Baker et al., 2022). In Denmark, a community district heating revolution started a decade ago; the country considerably reduced its heating emissions for a district by switching two-thirds of households to communal systems powered mainly with renewable energy, instead of having many individual boilers within one neighbourhood. Large and very efficient shared boilers spread hot water through pipe networks to many homes (State of Green, 2024; Whitehead, 2014).

The waste from food is colossal, with food loss and waste (FLW) accounting for around one-third of all food produced. In 2022, 13% of food production was lost through supply chains, from post-harvest up to, but excluding, retail. An additional 19% was wasted at the retail, food service, and household level. One of the first countries in the world to ban supermarkets from throwing away or destroying unsold food was France. It required the supermarkets to instead donate it to charities and food banks (Chrisafis, 2016). Since this law was adopted, other provisions have been enacted, extending action beyond the retail sector to also cover food distribution and setting targets for reduced food waste (Zero Waste Europe [ZWE], 2020). Though storage and transport challenges are particularly important in developing countries (Nature Food, 2024; UNEP, 2024a), recent data suggest that household food waste 'is not just a 'rich-country problem': the average household food waste levels among high-income, upper-middle income, and lower-middle income countries differ annually by only 7 kg/capita' (UNEP, 2024b, p. 1). Reducing food loss is a triple win: it benefits food security, the climate, and the economy. For example, farmers can reduce food loss and increase incomes through improved harvesting techniques and timing and better storage.

The circular economy presents another major opportunity for resource efficiency.[7] In this approach, what might previously have been considered waste is repurposed as an input for new production processes (Pauliuk et al., 2021). For example, materials like plastics, steel, aluminium, and cement can be recycled and reused, reducing the need for new raw materials, lowering CO_2 emissions, and reducing environmental damage. Such reuse does, however, require attention to design of systems, technologies, and incentive structures from the beginning, so that, for example, dismantling is not problematic and components and materials can be reused straightforwardly.

A natural question for economists – indeed, it is common sense – is to ask: Why is this waste happening? If it can be prevented at minimal cost, why is it not already prevented? With loss of energy through poor household insulation, the waste occurs due to lack of information, the high cost of finance, or the basic hassle of getting things done in relation to the necessary investment to make the home more energy efficient. In some cases, such as poor transport facilities in poor countries, waste arises from a weakness in infrastructure, which requires public action to improve. For the circular economy, what is necessary is the coordination, design, and networks that take place at a system level for requiring and organising reuse. Thus, these pervasive inefficiencies are largely associated with market failures, infrastructure, and network issues which require public action. But with such action there are potentially high returns on the related investment.

AI applications can help increase asset use and efficiency, including through industrial automation and robotics, helping to decarbonise manufacturing and processing industries as well as warehousing operations. Higher degrees of industrial robotisation can lead to increased productivity and reduced carbon intensity (Li et al., 2022). AI-powered optimisation systems are similarly able to reduce emissions in supply chain operations. For example, the packaging decision engine developed by Amazon can determine the most efficient type of packaging for each item, helping reduce the number of cardboard boxes, air pillows, tape, and mailers used to send purchases to customers (Amazon, 2024).

Countries and the private sector are moving on resource efficiency, particularly energy efficiency. Within a year of the global energy crisis triggered by the war in Ukraine in 2022, governments representing over 70% of the global economy worldwide had introduced new energy efficiency policies or strengthened existing ones (IEA, 2023c). At COP28 in November 2023, countries committed to doubling the annual rate of improvement in energy efficiency from 2% per year to 4% per year until 2030 (COP28 UAE, 2023). If achieved, this increase in energy efficiency could avoid approximately 80 exajoules (EJ) of fossil fuel demand by 2030, resulting in a reduction of around 5 $GtCO_2e$ to global emissions (ETC, 2023a) – representing about 9% of 2022 global emissions (UNEP, 2023a). Furthermore, between 2011 and 2020, the average annual rate of improvement of global energy intensity (energy per unit of output) was 1.7%, double the rate of the previous decade (0.8%) (IEA, 2023a). Energy intensity has decreased across all regions in different magnitudes since 2010 (see Figure 5.3) (IEA et al., 2023).

Figure 5.3: Growth rate of total energy supply, GDP, and primary energy intensity at a regional level, 2010–2021

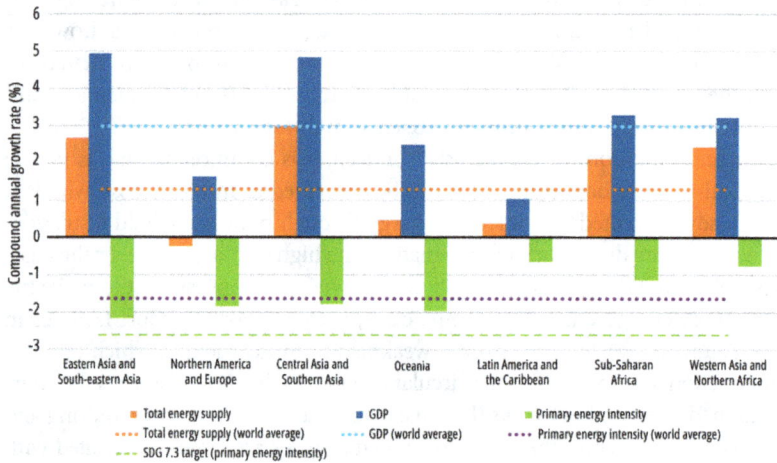

Note: GDP = gross domestic product; SDG = Sustainable Development Goal.
Source: Figure 4.4 in IEA et al. (2024, p. 91). Copyright 2024 World Bank, CC BY-NC 3.0 IGO.

According to work by the Energy Transitions Commission (ETC, 2018), exploiting potential resource efficiencies beyond energy could reduce global CO_2 emissions from the following four 'hard-to-abate' sectors by 40%: plastics, steel, aluminium, and cement. Additionally, improved resource productivity can weaken the link between economic growth and material consumption. Among the G20 countries, resource productivity grew by about 40% between 2000 and 2017 (Organisation for Economic Co-operation and Development [OECD], 2021). A shift in production and consumption patterns towards services also supports a weakening of the relations between aggregate output and emissions. Resource efficiency will be at centre stage in the transition, has much further to go, and will be a powerful driver of productivity increase and of growth.

Stronger system productivity

Energy, cities, land, transport, and water systems can all become more productive by improving how each operates individually and how they interact with one another as systems. Major contributions to emissions reductions and sustainable growth will come from cities that are designed to facilitate cycling and walking, integrate public and private transport facilities more effectively, and provide cleaner air by shutting out vehicles with internal combustion engines. Cities can also benefit from integrated heat and power systems, as well as cleaner homes and industries.

The work of the New Climate Economy (NCE) indicates that more compact, connected, and coordinated cities could generate up to US$17 trillion in economic savings by 2050, while reducing infrastructure capital requirements by over US$3 trillion between 2015 and 2030 (NCE, 2018). Further, more compact cities will not only be more carbon efficient; they will also be more resilient to climate change and disasters (NCE, 2018). Creating clean, liveable, and climate-resilient cities will be an important action area given the urbanisation projected in EMDCs in the coming decades. It is in these countries that the majority of the urban population increase in the world will take place. Around half of the increase in urban population between now and 2050 is projected to take place in eight countries, seven of which are emerging economies (Lwasa et al., 2022). Pakistan, for example, is projected to see 59% of its population reside in urban areas by 2050, compared to 37% in 2020 (World Bank, 2022).

City designs and standards, including for zoning for pedestrians and cyclists, managing waste, and a circular economy, could be major elements in both the drive for net zero and increasing productivity (see Figure 5.4). An example of innovative urban design is the Barcelona Superblock model, where major roads surround superblocks but interior streets are reclaimed entirely for pedestrians and cyclists, reducing pollution and improving public spaces (Mueller et al., 2020). Another example is the Vauban district in Germany, a neighbourhood planned around green transportation, with around 5000 inhabitants, 70% of whom choose to live carless (Thorpe, 2021).

A further example is Bogotá's Bus Rapid Transit (BRT) system, TransMilenio, which introduced dedicated bus lanes, pre-paid boarding, and high-capacity buses, cutting travel times and emissions. It inspired similar systems in Mexico City, Jakarta, and Cape Town (Hidalgo and Gutierrez, 2013). In Lagos, the BRT system has improved urban mobility, and the recent introduction of electric buses aims to further cut fuel costs and reduce air pollution. It set an example for sustainable transport in West Africa (Lagos Metropolitan Transport Authority [LAMATA], 2022). These examples highlight the importance of integrating multiple system-level objectives – urban design, transport efficiency, and sustainability. They illustrate how reducing emissions and improving mobility can also enhance economic opportunities and quality of life.

Figure 5.4: Denser cities produce much lower emissions

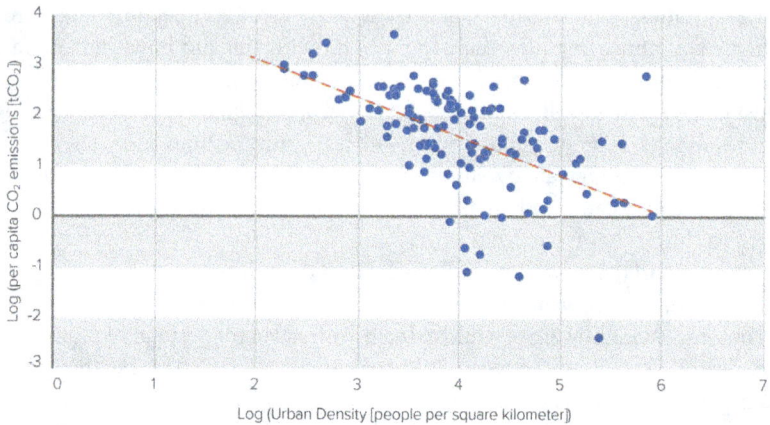

Note: A Pearson's correlation on a dataset of 127 cities found that $r = -0.3383$, with $p < 0.05$.
Source: Figure 9 in New Climate Economy (2018, p. 69). Copyright 2018 New Climate Economy. Reproduced with permission.

Similarly, sustainable approaches to and systemic management of land use can prevent the degradation of land, the destruction of forests, and the poisoning of rivers. This, in turn, prevents damage to further systems, such as for water (Intergovernmental Oceanographic Commission [IOC-UNESCO], 2022; Intergovernmental Science-Policy Platform on Biodiversity and Ecosystem Services [IPBES], 2019).

The potential of information technology (IT) and AI to enhance efficiency, integration, and system management is enormous. Advanced forecasting and control systems in electricity markets, together with good grids and effective market structures, can better accommodate variable electricity sources and flexible demand, while improved transport modelling can enhance infrastructure planning and performance. AI's role extends well beyond optimising existing systems; it will be crucial in designing new systems and managing the processes of systemic change.

The role of entrepreneurship in system development will become increasingly important. We have already seen creative entrepreneurship in systems design and delivery in telecoms, digital banking, broadband linkage, satellite observation and GPS. Given the rapid changes in technologies, systems, and AI, we need not only entrepreneurship within existing systems but also entrepreneurship that drives the development of new systems.

There are important systemic opportunities via international cooperation. For example, intermittency in power supply is more easily handled with a grid structure which can move electricity across geographies. Connecting regions and geographies through energy grids can improve the security of energy supply and reduce costs. In the UK, the electrical connection between

Scotland and England is set to be strengthened, with recently approved plans for a subsea coastal low-carbon electric 'superhighway'. The UK also collaborates with neighbouring countries like Norway (mostly wind), France (mostly nuclear), and soon Morocco (mostly solar). The potential of this type of initiative is illustrated by the fact that the Democratic Republic of the Congo is endowed with sufficient hydropower potential to supply all of sub-Saharan Africa's energy consumption (excluding South Africa) (World Bank, 2023a).

Improved health

Burning fossil fuels is the major contributor to air pollution. Air pollution constitutes a severe global health crisis, contributing to 10–15% of annual global deaths – between 6 and 10 million deaths each year worldwide (Institute for Health Metrics and Evaluation [IHME], 2021; Vohra et al., 2021). In Pakistan, for example, air pollution shortens the average person's life expectancy by around four years (Air Quality Life Index [AQLI], 2023). In the UK air pollution kills around 35,000 a year, 20 times the deaths from road accidents (Kelly, 2018). Air pollution damages health through multiple routes including cardiovascular disease, respiratory illness, and cancer. Children are particularly vulnerable, as they breathe more rapidly, leading to long-term health consequences that carry major costs for the future (Royal College of Paediatrics and Child Health [RCPCH], 2023). The financial burden of pollution-related health issues is

Figure 5.5: Cost of health damage from PM2.5 exposure in 2019 by region, per cent equivalent of GDP (purchasing power parity)

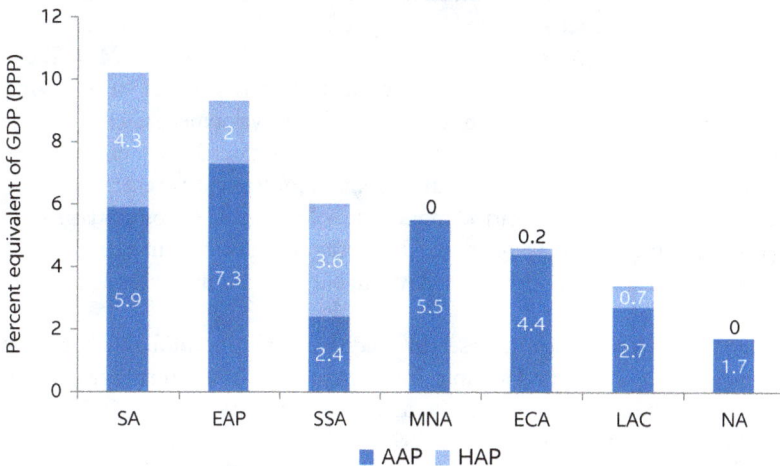

Note: EAP = East Asia and Pacific; ECA = Europe and Central Asia; LAC = Latin America and Caribbean; MNA = Middle East and North Africa; NA = North America; SA = South Asia; SSA = sub-Saharan Africa. Numbers may not add up due to rounding. AAP = ambient air pollution; HAP = household air pollution.
Source: Figure ES.1, top panel in Awe et al. (2022, p. xiv). Copyright 2022 World Bank, CC BY 3.0 IGO.

immense: in 2019, the global annual health cost of mortality and morbidity caused by exposure to PM2.5 particles was estimated at 2.5–6% of global GDP (Awe et al., 2022) – see Figure 5.5. Improving health is a major development outcome in its own right as well as making a strong contribution to productivity. The move towards a new form of growth and away from fossil fuels improves health, and improved health improves productivity and drives growth.

Moreover, many actions aimed at reducing emissions, such as promoting cycling and walking, offer additional health and productivity benefits. Increasing active transport modes like cycling and walking not only reduces emissions but also improves health by reducing the prevalence of obesity, diabetes, and heart disease (Woodcock et al., 2009).

Improvement in health is a core objective of development. It also enhances productivity. Again we see that climate action, growth, and development are closely intertwined.

Increase in share of investments

Climate action requires substantial investment across all countries and sectors. Investment is central to the new growth story. The scale of these investment needs will be discussed in greater detail in Section 5.4, and particularly Chapter 9. In developed countries, investment rates (the fraction of investment in national output) will have to increase by 1–2 percentage points (ppt). In EMDCs, this increase should be around 3–5 ppt, except in China, where investment rates are high; however, a reallocation of investment will be necessary. These figures are based on extensive research aimed at understanding the investment requirements for transitioning to a net zero path and building resilience (see, e.g., Stern et al., 2021; Songwe et al., 2022; Bhattacharya et al., 2022, 2024, 2025; IEA, 2024b). These analyses consistently show that a big investment push is essential for achieving the Paris targets, in the context of the Sustainable Development Goals (SDGs); especially in EMDCs, where investment needs are considerably larger, and must be sustained over time. As we saw in Chapter 1, investment is central to many theories of growth – in the Harrod model, the Solow model, endogenous growth theories, Schumpeterian theories, and Keynesian perspectives. Investment is at the core of most stories of growth. And within our story it is needed in abundance and at speed.

Realising these strong increases in investment will be difficult, but such increases are necessary and feasible. Investment rates, as a percentage of GDP, have been low or declining in both advanced economies and EMDCs since the global financial crisis of 2008–2009 and, for the most part, they have not returned to the levels seen in 2000. Tackling this shortfall in investment would boost aggregate demand and would, in turn, strengthen supply, together a powerful potential spur to productivity and growth. Moreover, new capital goods would generally embody more productive technologies.

Taking these together, we can see that the investments required for the green transition can have a powerful positive impact on growth. The simple

Harrod–Domar condition[8] set out in Chapter 1 would point to a possible increase in world growth rates of one-third to a half percentage point from a two-percentage-point increase in investment.[9] And the innovations, efficiencies, and returns to scale described in the previous drivers could lower the incremental capital output ratio (ICOR), in that sense increasing the productivity of investment and delivering a further increase in growth rate.

AI can also help, in addition to the system and discovery effects described above, by making financial markets more efficient. Such techniques can shape the allocation of funds to superior investment opportunities. Investment decisions could be supported by applying AI to better predict the risks and returns of investments. Leading investors have already trained AI models based on an extensive range of financial data to support their investment processes (Black-Rock, 2024). The potential could be especially significant for EMDCs, where the needs for investment are particularly acute, where there are often great difficulties with the affordability, availability, and accessibility of finance, and where the difficulties are often associated with perceived risk (Bhattacharya et al., 2023).

For many countries, beyond direct productivity, efficiency, system effects, and demand–supply improvements, there are important positive medium-term macro effects from these new investments through the easing of balance of payments pressure by switching to solar and wind rather than imported fossil fuels. For example, in 2021, India spent over 4% of GDP on fossil-fuel imports (*The Economist*, 2022).

Further climate actions fostering growth

These six drivers of growth are largely associated with actions and investments focused on mitigation, the creation of new, clean infrastructure, technical change, and efficiency. Adaptation, by building resilience to climate impacts, avoids damaging decreases in income and thus raises growth relative to output and income in its absence. See, for example, the work of the Global Commission on Adaptation (2019), which shows returns from US$1 invested in adaptation resulting in up to US$10 in net economic benefits. Investments in natural capital can have powerful returns in terms of losses and disasters avoided, such as those associated with deforestation causing flooding and landslides, as hillsides are weakened and hold less water. Many such investments are vital to water systems as a whole, and thus to agriculture and health. Such investments improve development and productivity, avert disasters, and foster growth.

There is no doubt that delivering this new form of growth and the necessary investments will be challenging and will require strong, purposeful, and effective leadership on policy and action (see Chapter 6) as well as international collaboration (see Chapter 9). In summary, the following actions will be at the core of realising this new form of growth. They are critical to the investment and innovation which lie at the heart of the growth story:

- **Fostering sectoral change and driving innovation**: Supporting new ideas, offering some protection against risk, providing public infrastructure, and lowering the cost of capital will all be relevant here. As with past industrial revolutions, there will be both risks of failure and speculation.
- **Implementing macro policies**: Good macroeconomic policies are essential to manage the increase in investment demand. History, particularly the growth of East Asia, has shown that 'macro with high investment' can be successfully managed, provided there is a strong supply response.
- **Managing public finances**: Public finances will need careful management, partly because revenues from fossil fuel taxation will decline. Policy options include reducing harmful subsidies, taxing vehicle use, and implementing a carbon tax during the transition. Advances in digital information and AI can also enhance public revenue collection.
- **Tackling economic dislocations**: The transition will inevitably cause dislocations for consumers and workers, which must be managed, in large measure via investment in people and places. The cost of capital, particularly for poorer populations, will be a key issue not only in fostering investment but also in managing these dislocations.
- **Managing the political economy**: Vested interests will push back. Leadership will be necessary but also an understanding that changes will require some cooperation with those interests.
- **Working closely with the private sector**: Close cooperation with the private sector is essential to manage implementation, identify obstacles, reduce adjustment costs, instil confidence in the long-term viability and profitability of sustainable investments, and realise the benefits of creativity and entrepreneurship.
- **Recasting the role of the state**: These tasks and the necessary policies will imply that major aspects of the state's role will need to be redefined to support this transition effectively – a matter that will be discussed in detail in Chapter 7.

At the end of Chapter 4 we set out the agenda. The six core drivers of growth and the actions just described form the core of the response to that agenda. That is the essence of the new story and how it can be delivered. Further details of policies and actions for delivery are examined in the next chapter.

5.2 Economy-wide integrated action

A change of economic trajectory at the scale and pace required to deliver the new growth story and necessary cuts in emissions must involve strong economy-wide action; as we have seen, many or most sectors and actors will be involved. In considering that action and how it can be shared strategically, there are three basic guiding perspectives. First, we must recognise that the necessary increase in overall investment is large and we must understand why investment action has fallen far below what is required. Second, economy-wide climate action can and should orient investments towards, and integrate them with, all three of development, mitigation, and adaptation. As with action on mitigation, we must pay close attention to the consequences of failure to act on adaptation for poverty and development. Third, the investments must embody innovation and be a part of, and drive forward, systemic, structural, and technological change. We have already set out the centrality of systemic, structural, and technological change in describing the six drivers. In this third element we will focus on the role of AI, in shorthand how to construct 'green and intelligent growth'.

The criticality of investment, innovation, and finance

Many EMDCs, particularly those at early stages of development, have, as a result of the advantages in clean technologies, the opportunity to leapfrog over outdated, polluting technologies by building new infrastructure and economic activities that are clean and sustainable. And we have seen the imperative of changing technology in response to the climate and biodiversity crises. This transition to a sustainable future will require a major push on investment across the world, but particularly in EMDCs. And that is where the potential for growth is also the greatest. If well executed, the increase in investment in EMDCs and elsewhere will yield high returns in terms of productivity, create new opportunities, and enhance and protect the environment. It will be through this investment that the new growth story will be realised. As we have consistently argued, failure to take bold, internationally coherent action on investment will leave us facing a deeply dangerous world.

In the short run, immense investment opportunities already exist for new clean activities and technologies. Sound policy and credible strategies can instil confidence that the economy is transitioning in a purposeful and sustainable way, thus encouraging investment in these activities (see Chapter 6). In the medium run, we will witness a Schumpeterian process of discovery, innovation, and investment. Indeed, the process is already underway in some sectors and beginning in others. The opportunities in low-carbon investment will spur creativity and discovery and unleash new waves of innovation and investment. In the longer run, low carbon is the only viable form of growth and development. Any attempt at a global scale to maintain high-carbon growth will lead inexorably to self-destruction, as it will create an environment hostile to economic activity and to lives and livelihoods. The transition is the growth story, but substantial investment is necessary to drive it forward.

The magnitude of the task requires a strong jump in growth rates of investment. In the last two decades overall investment growth has not risen; in many countries, it has fallen. During the 2010s, EMDCs (other than China) decreased their real investment growth rate from 9% per annum (2010) to 0.9% per annum (2019). Even including China, growth rates in EMDC investment dropped from nearly 11% (2010) to 3.4% (2019) (Kose and Ohnsorge, 2024). In advanced economies, investment growth was more stable but relatively low: during 2010–2019 it remained at only around 2% per annum, similar to its long-term average. Across the world, growth in investment is still low (Kose and Ohnsorge, 2024).

For much of the last 15 years, there has been a global deficiency, in Keynesian terms, of planned investment relative to planned saving, as evidenced by low real interest rates (up until 2021–2022), low investment levels, and sluggish productivity growth, all alongside low inflation. The Covid-19 pandemic and the Ukraine war brought inflation and a rise in interest rates. Investment growth has remained low, and inflation is falling back. This suggests that, looking ahead over the next 15 years, as the world economy goes past these two disruptors, the global aggregate demand and supply balances could accommodate the necessary investment,[10] potentially driving the world out of stagnation or weak growth and fostering a new kind of growth that is strong, resilient, and sustainable. It is not yet clear, at the time of writing in April 2025, how the disruptions associated with the policies and actions of Donald Trump's second presidency will play through. For some countries over the medium term, it could increase strategic investment in economic, energy, and physical security. In the short run, many investors may 'hold off to wait and see'.

The urgency of tackling climate change while simultaneously advancing development has never been more apparent, yet we are far behind on global climate action. Current global investment in new, clean technologies falls short of what is needed to meet critical targets for mitigation, adaptation, and conserving and enhancing natural capital. Meanwhile, too much investment continues to be channelled towards the outdated and polluting fossil fuel economy. Transitioning away from the dirty requires building up the clean.

Investment in the new, clean, and resilient has been far too low for several reasons, including:

- The perception of a trade-off between growth/development and climate action, which has resulted in weak policy (see Section 5.3).
- A failure to think through or bother with ethical considerations, leading to the undervaluation of the well-being of future generations (see Chapter 3).
- The high cost of capital, particularly in EMDCs, due to perceived risks and severe difficulties of debt and the public finance post-Covid-19 (see Chapter 6).
- Political difficulties in managing dislocation and adjustment costs (see Chapter 6).
- Strong resistance from vested interests (see Chapters 7, 8, and 10).
- International tensions, partly rooted in mistrust between developed and developing countries over finance, exacerbated by the behaviour of developed countries during the Covid-19 pandemic as well as by ongoing conflicts and energy security issues, such as those related to the war in Ukraine (see Chapters 4 and 9).

As discussed in Chapter 4, EMDCs will account for the vast majority of new physical capital over the coming three decades, driving a global doubling of infrastructure over the next two decades. If this new infrastructure replicates the old, the impact on climate will be devastating. Fortunately, for much of this infrastructure, clean options are already cheaper and more efficient than dirty ones. And the strong expansion of clean power will be vital to transport, heating and cooling, and much of industry.

The realisation of the necessary investment requires sound policy and the right kind of finance, on the right scale, at the right time. While the majority of investment will come from the private sector, public investment must play a key role, particularly in the early stages and in relation to sustainable infrastructure, including in public transport, city design, power grids, and the management of land. We will return to the details of these investment needs in Chapters 8 and 9.

Integrating development, mitigation, and adaptation and resilience: a unified approach

In thinking about the impacts of climate change and the importance of adaptation for development, we must first understand some of the complexities and dynamics in the interactions between climate change and poverty and development. In so doing, the integration of adaptation with development and mitigation will run through most of the argument and examples. Inaction on climate will stall or reverse development, increasing poverty. Indeed, climate change can be a key factor in generating poverty–environment traps that may increase chronic poverty. Climate change can create a hostile environment that both pushes people into poverty and increases the difficulty in escaping from poverty. For rural populations in particular, climate change undermines agriculture, threatens food and nutrition, disrupts access to ecosystem services, and depletes habitats (Birkmann et al., 2022). Poor people in urban areas are more likely to live in more vulnerable places such as flood plains or areas with weak drainage. They are disproportionately affected by natural hazards and disasters (Hallegatte et al., 2020).

But climate change also increases the risk of individuals who are not currently poor falling into poverty and undermines the ability of people to escape poverty. Understanding these processes can guide good climate action on tackling poverty (Bangalore et al., 2014). Figure 5.6 is a simplified diagram that shows only some of the many ways in which climate change can impact work, reinforcing existing poverty traps. Further, policies such as resilient housing and energy efficiency improvement, adaptation, mitigation, and development, can have cascading positive impact on poor communities, for example by simultaneously improving their personal finances, health, and work performance.

Figure 5.6: The poverty–environment trap can increase chronic poverty (illustrative example)

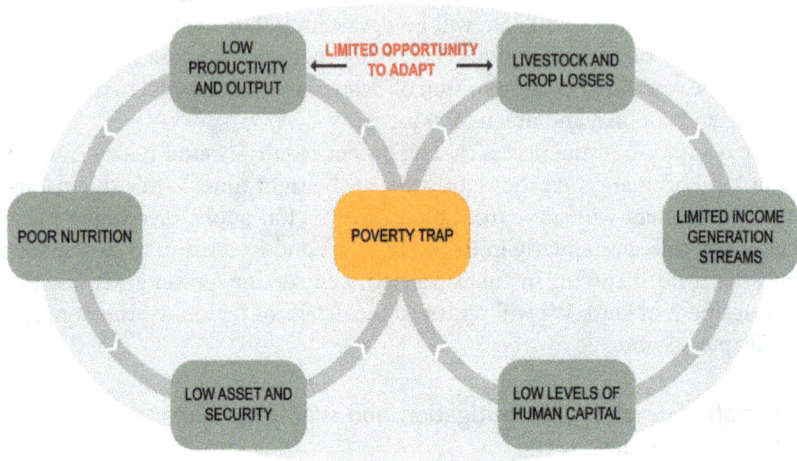

The poverty–environment traps operate through all of health, assets, and incomes, food, and personal security, in both rural and urban areas:

- **Health:** Income loss can result from work disruptions due to hostile weather, increased malnutrition from impaired crop yield and water scarcity, and exposure to air pollution from fossil fuel combustion, to which poorer people are more vulnerable as they are more likely to be close to the pollution sources, both outdoors and indoors, and increased vector-driven diseases.
- **Physical damage:** Poorer households are more likely to live in vulnerable areas and take longer to recover from disasters. It is estimated that between 32 and 132 million people could fall into extreme poverty by 2030 due to the impacts of climate change (Birkmann et al., 2022, p. 1201); with the passage of time and greater severity of impacts, the numbers could be much higher.
- **Food systems:** Reduced agricultural incomes from crop and livestock losses harm poor farmers and workers. High and volatile food prices disproportionately impact people in poverty, who spend a larger share of their income on food.

- **Migration and conflict:** Climate change intensifies droughts, which have been linked to migration and increased armed conflict, including in West Asia and North Africa (Abel et al., 2019). In Syria, climate-induced drought contributed to mass migration from rural to urban areas, which in turn contributed to the causes of civil war (Kelley et al., 2015).

Climate-related investments are crucial for simultaneously tackling the above – as well as further challenges related to health, education, unemployment, growth, inequality, social cohesion, and biodiversity loss. All these issues are integral to the overall effort to achieve the SDGs. Thus, climate-related investments should be understood as a core part of development investments and strategies and not separately. They are nested within and part of strategies to achieve the SDGs. Climate action is central to sustainable, resilient, and inclusive development.

Indeed, the central thesis of this book is that action on climate change can be a powerful driver of development, and not merely that development and climate action can be made consistent or 'balanced'. Similarly, action and investments for development, mitigation, and adaptation/resilience should not be seen as necessarily competitive. On the contrary, in much of the economy, these three elements are strongly interwoven and mutually supportive; see Figure 5.7.

Figure 5.7: Places where development and climate action meet

Restoring degraded land

Public transport

Decentralised renewable energy

Mangroves

Source: Author's elaboration.

For example, mangroves act as strong barriers to storm surges, sequester carbon more effectively than many other plants or trees, and support fisheries, biodiversity (including tigers), and tourism. Restoring degraded land enhances resilience, captures carbon in the soil, and increases income. Public transport protects livelihoods, expands work and education opportunities, and reduces emissions. Decentralised solar power makes people less dependent on fragile grids, reduces use of fossil fuels and wood and thus air pollution, and increases energy access – for example, by being used to power remote homes, shops, and hospitals, or by decreasing energy bills. Such investments foster development, reduce emissions, and promote resilience.

Good adaptation is always good development (see Figure 5.8). It makes no sense to build infrastructure, construct houses, or pursue agriculture activities that are not resilient to the climate impacts already occurring and those that are on the way (see also Chapter 2, Box 2.1, and 2.2).

Greater security, achieved by tackling climate risk, reduces the cost of capital and fosters investment, particularly for small and medium enterprises (SMEs) and poorer households (Global Commission on Adaptation [GCA], 2019), and thus increases living standards. For example, in Uganda renewable energy generation increases have been associated with higher firm productivity by making the supply of energy more stable (Probst et al., 2021).

Careful design and appraisal, including from an economy-wide perspective, is crucial to the quality and effectiveness of climate action. If poorly designed or badly thought through, some measures could increase vulnerability and

Figure 5.8: Relationship between adaptation, mitigation, and development

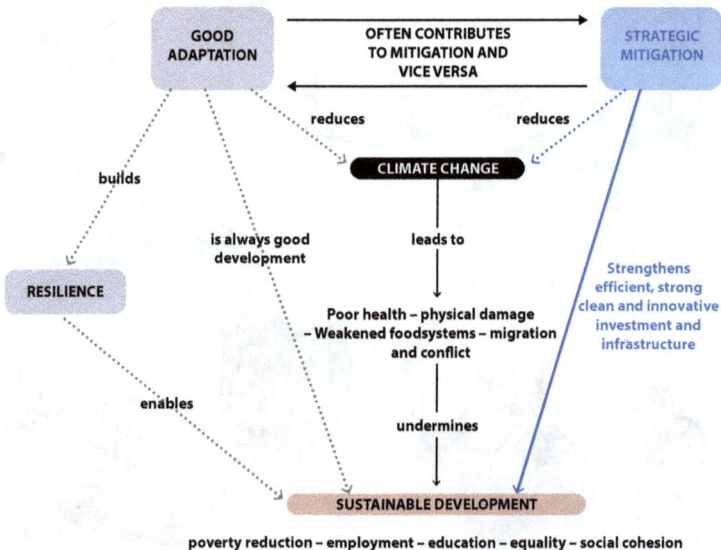

Source: Author's elaboration drawing on OECD (2023).

obstruct development (Noble et al., 2014). For example, Brazil's main source of electricity is hydropower, with an 83% share of power generation in 2021 (World Bank, 2023b). Yet its reliability is under threat as climate change, growing water demand, and land degradation (due to deforestation) are putting stress on water availability, leading to increasingly frequent and intense water crises already being experienced (World Bank, 2023b). This water scarcity has repercussions for climate goals and development alike. It has led the government to consider expanding gas-fired power generation, which would undo progress in clean generation, use capital in assets that might get stranded, and contribute to undermining the water availability that their citizens, energy systems, and agricultural systems need, as climate change keeps worsening.

On the other hand, given water stress, diversifying the energy mix by investing in solar and wind could be a sound adaptation strategy, aligned with both mitigation and national development objectives. The intersection of water, energy, and food systems is complex. In the case of Brazil, a bigger share of solar and wind at an earlier stage would have been a wiser strategy. When developing strategies for climate action, countries must carefully consider the specific energy and related natural resources needs of their systems, their clean energy endowments, and future climate risks; the country context is crucial.

The poverty–environment traps discussed earlier in this section and the constraints and insecurities they embody, together with the difficulties of designing revenue streams to capture the benefits of adaptation, imply that it is often very difficult to raise substantial private capital for investments that are predominantly adaptation. Private capital requires returns that can be captured by the investor and that may not be easy in the absence of directly related development returns. That underscores the importance of sources of low-cost capital for adaptation – see Chapter 9. It also emphasises the importance of integrating the three strands of development, mitigation, and adaptation, because a strong development element can facilitate finance. An example would be the replacing of mud houses (*kaccha* in Hindi) with brick (*pukka*), as occurred in the village of Palanpur in India, which I, with collaborators, have been studying for more than half a century (Himanshu et al., 2018). A second is decentralised solar, which again encompasses all three strands: development, mitigation, and adaptation (see Box 5.1).

Fundamental systemic change: investment and innovation

Effective action on climate and development will require rapid structural transformation across all key systems – energy, transport, cities, land, and water – to achieve the necessary scale of emissions reductions and deliver all three of development, mitigation, and adaptation. These systems interact and form larger, increasingly complex systems, with newer, cleaner technologies at their core. For example, decarbonising supply chains requires rethinking sourcing locations for energy and inputs to reduce emissions, transforming

Box 5.1: Decentralised solar in Africa: development, mitigation, and adaptation

Renewable energy presents an opportunity to increase access to electricity and do so in a clean way. Worldwide, in 2023, there were around 750 million people without access to electricity, with 600 million of these being in sub-Saharan Africa (IEA, 2024c). Access to electricity has wide-ranging development impacts, from allowing students to study when it is dark, to improving industry and agriculture, food preservation, and heating. And it enables access to the internet and mobile phones, both of great importance for development and climate action, including adaptation. For example, these technologies enable people to receive early warnings of extreme weather events and to access information on how to manage agriculture in the face of a changing climate. And it gives them the ability to seek help.

Around 60% of the world's best solar resources are in Africa, but the continent has had only 2% of the world's clean energy investment (IEA, 2022a, 2023e). One of the advantages of solar is its capacity to be run on a decentralised basis: villages which are off-grid can have their own source of power under their own control. Indeed, decentralised solar energy is springing up in parts of rural Africa, particularly in East African countries like Kenya, Rwanda, and Tanzania. The extension of connectivity to grids can be a lengthy and expensive process. And grids can be fragile.

Several decentralised energy service companies (DESCOs) – for example, M-Kopa, Mobisol, and Lumos – provide households, communities, and small businesses with off-grid solar home systems or localised microgrids of varying capacity from which they can power their homes, neighbourhoods, and operations (Wakeford, 2018). These companies have come up with an innovative form of end-user financing: offering these decentralised solar systems under a pay-as-you-go (PAYG) model, breaking down the large upfront costs of solar panel instalment into smaller payments and leveraging the fact that mobile payment systems are increasingly widespread. In exchange for the solar PV systems, customers arrange regular payments through a mobile payment system, on a PAYG basis, meaning customers pay a small upfront fee and then regular payments based on the energy used. The companies often offer flexible payment terms and lease length can vary according to needs. The costs of such schemes are often in line with what households would otherwise spend on kerosene, candles, and other less efficient, less reliable, and more unhealthy energy sources.

Such initiatives both reduce energy poverty and move towards cleaner and cheaper renewable sources of energy. They help protect against uncertainties. They constitute an important example of technological leapfrogging that builds on existing technological advances – using mobile, solar PV, and mobile payment technologies. According to IRENA (2020), around eight million people gained access to energy under PAYG schemes between 2015 and 2020. But all this is still on a very small scale relative to the population and needs of Africa. For further discussion of the potential of major solar PV initiatives for Africa, see Chapter 9.

production methods and switching fuels, electrifying transport and heating/cooling, increasing overall resource efficiency through innovation and collaboration, and reforming degrading, polluting, and toxic structures in agriculture.

Redesigning systems and managing these systems can be, as we have emphasised, greatly facilitated by AI. Indeed, creating and managing the interactions within systems – whether in the circular economy, electricity grid systems with greater flexibility in demand and supply across space and time, transport interactions, or city design – will likely involve the strategic deployment of AI tools (see Chapter 4 and below on AI). These have great potential for design and management of new system structures, more efficient usage of resources, and rapid discovery of new technologies and methods.

At the heart of systemic change are investments and innovations. The necessary investments, innovations, and systemic changes require dynamic and strategic choices from both governments and the private sector. Successfully managing the macroeconomic, public finance, and financial aspects of this investment expansion, along with the innovation that accompanies it, will be critical to success. Much of this involves reducing, managing, and sharing risk to lower the cost of capital. Multilateral development banks (MDBs) can play a fundamental role in fostering investment, providing affordable finance and managing risk, thus enabling private investment to scale up (see Chapter 9). The choices made now on infrastructure and capital will either lock us into high emissions or set us on a low-carbon growth path that is sustainable, resilient, and inclusive. It is the collaboration between public and private sectors around the investments, policy instruments, institutional structures, and their financing that will determine whether the necessary investments are realised (see Chapter 6). All this involves much more than the technical or engineering aspects of systems, important though they are. The integration with policies and institutions is crucial.

To illustrate, while prices are critical to incentive structures in market economies, they alone will not deliver the major systemic and structural changes required. Some economists have argued that all we need to do is raise the price of carbon to levels necessary to yield the required emissions reductions and to overcome the market failure associated with the damage the emissions cause. But this view ignores numerous other market failures. Recognising and overcoming these other market failures can simultaneously reduce emissions and foster better and stronger growth. Market failures inhibit the effective functioning of price mechanisms in guiding decisions and resource allocation and can lead to major inefficiencies, both static and dynamic, and obstruct investment and growth. However, if these failures can be tackled, at least partially, through price or tax/subsidy and institutional mechanisms, markets will play a powerful role in realising the new story of growth and tackling climate change. Chapter 6 examines how policies and institutions can help tackle these market failures. Chapter 7 will examine the implications of this analysis for the role of the state.

The scale and pace of the required investments will present substantial challenges, in terms of dislocation of work and livelihoods for some and difficult changes in relative prices for some others. Many would regard it as unjust to fail to assist those who are affected in finding a response to the dislocation and in realising the opportunities of the new. Delivering a 'just transition' across nations, communities, and individuals will be fundamental to preventing 'push-back' from those who stand to lose from these changes. Such action is not only just but also essential for successful implementation. We return to the just transition in Chapter 6 on policies and strategies underpinning the new growth story.

AI as agent of green and intelligent growth

The six drivers of growth set out in Section 5.1 above, taken together, embody fundamental structural systemic and technological change. And we have just emphasised the centrality of investment and innovation to these changes. Here our focus is on the powerful role AI and digitisation can and should play in guiding and fostering change. An element of good fortune has entered here, as AI has arrived just when it is critically needed to respond to the crises of climate and biodiversity. AI can and must be put to good use in creating green and intelligent growth. We draw here on Stern and Romani (2023) and Stern et al. (2025), who identified five key areas through which AI can drive the climate transition and growth. These are introduced in Table 5.1.

The first key area is the promotion of change in the systems of energy, cities, transport, land, and water, which are at the heart of the growth story. As the energy transition advances, as our energy generation points and sources diversify, and as the electrification of activities surges forward, energy systems are becoming more complex. The role of AI, as the brain of energy systems, becomes more and more important (IEA, 2023f). In the energy sector, AI can help design, optimise, and run grid structures, and how they work, together with power markets, so that demand and supply are anticipated, responded to, and co-ordinated over geography and time. For example, smart-grid technologies powered by AI could speedily interpret real-time information

Table 5.1: AI's potential contribution to the climate transition and growth

1. Transforming complex systems	2. Innovating technology discovery and resource efficiency	3. Behavioural change	4. Modelling climate systems and policy interventions	5. Managing adaptation and resilience
• Integrated management of energy systems, multimodal transport, and the urban ecosystem • Simulations of inter-systemic flows and cross-system interaction through AI-powered digital twins • Prediction of investment risks in low-carbon projects	• Acceleration of scientific discovery and incubation of green tech innovation at scale • Generation of sustainable design options • Maximisation of asset use and efficiency over lifetime	• Modelling of social behaviour, pattern analysis, and prediction • Facilitation of pro-environmental behaviour through advanced data analytics and AI-powered assistants	• Forecasting of extreme weather and climate change scenarios • Modelling the effects of climate change and the effectiveness of different policy scenarios	• Forecasting of climate impacts and early warning systems • Management of financial and human climate risk and impact towards more resilient systems • Strategic planning on climate adaptation

Source: Table 1 in Stern et al. (2025, p. 2). Copyright 2025 The Authors, CC 4.0.

to match the supply and demand of electricity, therefore reducing costs and building stability and reliability into energy access.[11] That would allow for a high portion of renewables in the power mix and contribute to a positive environment for renewable investments, thus accelerating the transition. Further, it would reduce the cost and improve the efficiency of energy use, key drivers of growth. Taking these effects together, it would foster and enable the major expansion of electricity in the energy mix, which will be crucial for the transition. Although some speculate that fully automated systems could be developed in the future, at present AI is being leveraged primarily to provide better insights for human decision-making (Massachusetts Institute of Technology [MIT], 2023). For example, in 2019, the National Grid Electricity System Operator (ESO) partnered with the Alan Turing Institute to enhance renewable energy forecasts, leveraging AI. Using data analysis and ML techniques, for example, they developed solar forecasts that are 33% more accurate (AI Innovation for Decarbonisation's Virtual Centre of Excellence [ADViCE], 2023).

Similarly, AI can play a crucial role in designing, managing, and running urban transport systems with a shift to electrified public transport with strong intermodal connectivity – that is, between mainline trains, metros, buses, trams, cycling, and walking. This would contribute strongly to the quality of life and productivity in cities. Indeed, AI can also be used to help transport run smoothly, from optimising delivery routes to adjusting traffic light timing to reduce congestion (Abdulijabbar et al., 2019). AI could also strengthen commerce and trade systems by managing unexpected transport disruptions due to unexpected events. For example, Unilever uses an AI tool to identify suppliers when needed on short notice. The software creates a list of potential suppliers by scraping websites for data on a variety of the supplier's characteristics, from customer ratings to sustainability scorecards (Van Hoek and Lacity, 2023).

Second, AI can also drive the discovery of new materials and technologies, as it did with AlphaFold (a protein structure database), and vaccine and drug development. For example, Google's DeepMind has used an AI tool to discover 2.2 million new and viable inorganic crystal structures. These new materials have the potential to be used in fields from solar cells to superconductors (Peel, 2023). Not only does this cut years of trial-and-error scientific work, but it can also directly feed into the improvement of green technologies, which are central in the new growth story. AI can accelerate the technological and scientific aspects of the transition. It can enable the much faster geographical discovery of the location of the minerals needed in new technologies, for example batteries. And, as we have seen, acceleration on all fronts of action is paramount to avoid acceleration in climate impacts.

There is likely to be, or has already been, rapid progress associated with AI in nuclear fusion, quantum chemistry, alternative protein design, materials, and many areas of medicine. Indeed, in 2024 two of the Nobel Prizes in science were closely associated with AI. As Demis Hassabis, a leading figure in AI (founder of Google DeepMind) and a pioneer of the characterisation of proteins (Nobel Prize for Chemistry 2024), has put it, AI can give us scientific advance at digital speed (oral communication).

Third, AI can help provide timely information, overcome psychological barriers, promote changes in behaviour, and tailor intervention. An example of an initiative helping to generate change in behaviour through better communication and information is Farmer.Chat. This is a mobile phone app created in partnership with OpenAI by a non-profit organisation called Digital Green, a private initiative. It assists small-scale farmers, some in remote communities, in increasing their profits and their resilience to climate change. Farmer.Chat provides support to local farmers who are on the frontline dealing with the impacts of climate change, answering either written or verbal questions on their farming needs and problems, directly to them or through 'agriculture extension workers' (The Rockefeller Foundation, 2024). In India, for example, the Ministry of Agriculture validates all documents included in its knowledge base, to increase the reliability of Farmer.Chat (Open AI, n.d.). The initiative is linked to VISTAAR (Virtually Integrated Systems to Access Agricultural Resources), a public information repository and network of the Government of India for the agricultural community. As digitisation and information systems advance, the potential of examples like this will become ever stronger. Openness in data, both public and private, will facilitate initiatives such as Farmer.Chat.[12]

Fourth, climate and weather scenarios require understanding of the functioning of global climate systems. Much of that is associated with detailed climate modelling and other data-intensive approaches, for which AI can be a valuable instrument. It can help identify potential non-linearities, tipping points, and extremes. Indeed, AI applications can enhance weather forecasting and climate modelling, providing more accurate predictions of extreme weather events with strong local resolution and enabling better preparedness (Jain et al., 2023). Accurate and timely climate forecasting is crucial for designing and implementing effective climate policies and scenarios. For example, AI has been instrumental in enhancing weather forecasting and climate prediction models. The British Antarctic Survey and the Alan Turing Institute developed IceNet, an AI-powered tool that uses satellite data to forecast sea ice levels at a higher accuracy than state-of-the-art dynamical models (ECMWF SEAS5) (IceNet, 2024). This level of precision can improve climate projections, thereby contributing to better-informed policy decisions in relation to sound mitigation, resilient adaptation, and stronger development. The strengths of old and novel methodologies can also be combined, for example as in Google's new NeuralGCM weather prediction model, which blends traditional physics modelling and forecasting techniques with AI – helping overcome the limitations of both.

Fifth, AI can play a major role in identifying and understanding risks from climate change and thus helping design and manage adaptation and resilience. Building resilience depends, in large measure, on the integration of place-specific analysis of impacts and other details of the place itself. AI can be a powerful tool in learning about and fine-tuning that integration. Levels of data definition and modelling should be as local as possible and the fine-tuning can be greatly enhanced by AI and strong computing power. For example, flooding results in US$50 billion of economic damages each year, with 1.5 billion people globally exposed to these risks (Matias, 2024). Google's Flood-Hub uses ML models to forecast flooding events up to five days in advance; detailed alerts are issued in more than 80 countries, allowing such damages to be reduced or avoided (Matias, 2024).

To cover extreme weather events more extensively, digital twins, such as NVIDIA's Earth-2 and the European Space Agency's (ESA's) DestinE, are being developed. These simulations combine traditional physics-based models with AI to forecast weather in unprecedented detail, thereby improving disaster alert systems and allowing for dynamic adaptation measures (Hoffmann et al., 2023). Another example is that of Microsoft working with the Ministry of Irrigation and Lowlands in Ethiopia and the Ethiopian AI Institute to identify communities that are at high risk of natural disasters by combining satellite imagery, AI, predictive modelling, and local expertise (United Nations Framework Convention on Climate Change [UNFCCC], 2023). Precision and high local resolution are crucial in this context, often requiring substantial increases in computing capacity as well as data generation and availability. That links directly to the energy demand of AI, as discussed in Chapter 4.

As summarised in Table 5.1 above, there is a deep and powerful connection between AI and digitisation, the net zero transition, and broader economic development and growth. While the challenges of climate change and the transition are complex and often daunting, one encouraging development is the timely emergence of AI as a transformative tool in this critical moment.

5.3 Errors in common counterarguments

The central thesis of this book is that strong, well-designed climate action can drive a new attractive form of growth that is sustainable, resilient, and inclusive. There is no inevitable trade-off between growth and climate action. It is not a horse race between the two. Whilst our focus is on the direct and positive case, it is also important to examine the arguments of those who claim that an inevitable trade-off does exist and to demonstrate where their reasoning falls short. We begin with standard trade-off arguments, then discuss degrowth arguments, and finally examine issues around population growth.

Trade-off arguments

Table 5.2 summarises four different concerns or perspectives that could form the basis for the argument of an inevitable trade-off, along with explanations of why each argument is flawed.

Table 5.2: Climate action versus growth

Concern	Argument	Counter-argument
Concern 1	In an efficient world, where market prices accurately reflect all relevant social costs, introducing an additional criterion (here, future state of the climate) must involve a reduction on some other dimension, such as growth or poverty reduction.	This position is not valid in a world with significant market failures. Well-designed climate action can and should help overcome or reduce the many market failures and inefficiencies of relevance to climate change and climate action. The challenge of climate change provides extra motivation for addressing these failures (Stern and Stiglitz, 2023).
Concern 2	Development needs energy, and energy needs fossil fuels; therefore, development must involve increased GHG emissions.	While development generally needs energy, energy does not need fossil fuels. Zero- or low-carbon sources are now cheaper than fossil fuels in many sectors and regions.
Concern 3	Using resources for climate action will reduce those going to growth, via physical or human capital, which would have reduced poverty and increased resilience.	Well-designed climate action fosters investment and innovation in physical and human capital in cleaner and better ways, in turn driving growth, job creation, and resilience. For example, investment in renewable energy infrastructure creates new work opportunities, reduces energy costs over time, and creates new opportunities for innovation.
Concern 4	Climate action involves policies around pricing, technologies, and phasing out of fossil fuel extraction and use, which could increase costs and reduce opportunities for poor people.	Well-designed policies can mitigate such effects and produce better outcomes. For example, reducing inefficient and toxic subsidies frees up resources to invest for better growth and to compensate the poorest. Similarly, revenue from carbon pricing can be used for clean investment and compensation of the poorest. Clean technology can be more inclusive, e.g., decentralised solar can empower small businesses and women working from home. Clean energy is increasingly cheaper than dirty energy.

Source: Author's elaboration.

The first concern or argument suggests that if an economy is already operating efficiently, introducing a new criterion such as climate and sustainability would inevitably lead to a reduction in some other objective, such as output or growth. This idea is rooted in the basic definition of efficiency, which states that a situation is efficient if it is impossible to increase one good or goal without decreasing another.

The flaw in this argument lies in the assumption that the economy's 'starting point', before introducing the climate criterion, is efficient. In reality, it is thoroughly implausible to argue that our current economic state is efficient, particularly given the numerous market failures of strong relevance to climate change. Table 5.3 highlights six major market failures, each important. By working to overcome or reduce these failures, it is therefore entirely possible to improve climate outcomes without sacrificing growth. Here, the focus is on their implications for action, and the horse race between climate action and growth and development. In Chapter 6, the discussion will focus on policy options to tackle these market failures.

Realising potential improvements requires tackling these market failures. Table 5.3, third column, sketches how public policies and institutional responses can reduce the force of these failures. Most cannot be eliminated but their distorting effects can be substantially reduced. For example, the market failure associated with emissions can be reduced through a carbon price or tax combined with standards and regulation. The failure related to research, development, and deployment (R,D&D), given its nature as a public good (everyone can in principle use an idea), can be mitigated by public support for R,D&D. Failures in capital markets can be addressed by sharing risk through wise regulation for transparency and through the activities of development banks, among other measures.

Tackling market failures through public action is not only theoretically possible but also practically achievable. We see this in policies being implemented in many countries, including measures to reduce emissions, increase research and development in clean technologies, and reorient development banking towards climate and sustainability. The recognition of the central importance of tackling climate change can provide a political and economic impetus to work to overcome these failures – a motivating drive which would not be there, or at least not with sufficient force, without that recognition. The issues in Table 5.3 are of such importance to our argument and to policy that we shall return to them in Chapter 6, where we discuss policy issues in more detail.

The second concern in Table 5.2 revolves around the role of energy in development. Economic history shows that improvements in living standards have indeed often been linked to increased energy use, with, for example, coal and steam playing central roles in the Industrial Revolution (Smil, 2004). However, while the connection between energy use and development is indeed strong, the relationship is not rigid or fixed. Increased energy efficiency can reduce energy consumption, conserve resources, and still lead to higher living standards. Moreover, as incomes rise and consumption patterns shift towards services, the energy intensity per unit of output can decrease, potentially reducing overall energy consumption.

Table 5.3: Market failures related to climate and sustainable development

Market failure	Description	Policy options
GHGs	Negative externality because of the damage that emissions inflict on others.	Carbon tax/cap-and-trade/regulation of GHG emissions (standards). Do not subsidise the toxic.
R,D&D	Ideas are public goods, and here their use (climate action) is also a public good. Without public action, the creators of these ideas do not capture the full value to society from their creation.	Supporting research, innovation, and dissemination. Tax breaks, support for demonstration/deployment, including via procurement, publicly funded research. Coherent standards to focus research and innovation.
Risk/capital markets	Imperfect and asymmetric information and assessment of risks. Problems of collateral.	Risk-sharing/reduction through financial structures, e.g., equity, guarantees. Long-term contracts, e.g., for off-take for power. Convening power of development banks on investment climate; transparency (e.g., Task Force on Climate-related Financial Disclosures [TCFD], 2017).
Networks	There are problems of coordination between actors and agents within networks and across networks in key sectors. Examples in energy, transport, cities are important.	Investment in infrastructure to support integration of new technologies in electricity grids, public transport, broadband, recycling. Management of grids. Planning of cities.
Information	Lack of awareness of technologies, actions or support, or product content.	Promoting awareness of options. Labelling and information requirements on cars, domestic appliances, and products more generally.
Co-benefits	Benefits beyond market rewards are often ignored or downplayed.	Recognising and acting on impacts on health. Valuing ecosystems and biodiversity.

Note: The first column describes the area within which the market failure occurs. The second column briefly describes the nature of the problem. The policy options are discussed in the next chapter.
Source: Author's elaboration.

Nevertheless, the link between living standards and energy use is important, particularly in developing countries. At low incomes, the promotion of growth and development is heavily reliant on the increased use of energy. In other words, in the jargon of economics, at these levels of income, the income and output demand elasticities for energy are high. However, energy consumption does not require CO_2 emissions. By this stage in our argument, that should be crystal clear. And it is important in this argument that zero-carbon alternatives to fossil fuels for energy are not only available, but are also, in many cases, cheaper.

Thus, the argument that development necessitates emissions is flawed. As emphasised in Chapter 3, the argument that there is a right to development is strong. But the argument that this implies the right to emit could, I hope, be regarded as morally wrong. Emissions kill, maim, and damage livelihoods for others and are not required for the energy necessary for development. Thus, it is very hard to see any logical or ethical basis to insist on a right to emit.

The third perspective in Table 5.2, closely related to the first, argues that climate action demands investment and resources that could otherwise be used for growth. The counterarguments presented in the first perspective apply here as well. Specifically, the six drivers of growth outlined in the previous section demonstrate how climate action, along with the associated investment and innovation, directly drives growth. It might be argued that investment focused solely on growth, without regard for climate, could yield even higher growth. However, this overlooks the fact that climate action is often centred on overcoming market failures and inefficiencies. Because it carries an additional, widely shared purpose, climate-oriented policies may have a greater chance of tackling these failures than a strategy focused purely on growth. In this sense, well-designed climate and growth policies could, in principle, achieve higher growth than a growth-only strategy.

The fourth perspective concerns the argument that the impact of climate action and associated policies on poor people will make them worse off. Part of the response to this argument is that relating to the third: climate action can drive growth. But that is not a full answer, as some policies could, for example, increase prices of goods used by poor people. The full response to this concern can and should include a package of measures to protect poorer people. The revenues from well-designed climate policies – or the funds made available by reducing subsidies for dirty fuels – can provide more than enough revenue to fund such packages (see Lankes et al., 2023). Given the importance of policy to align climate action and poverty, we return to this issue in a little more detail in Section 5.5.

Throughout all these discussions, the quality and sustainability of growth should be at the forefront. A strategy that integrates growth with climate action can deliver higher-quality growth – growth that is sustainable, resilient, and inclusive. This approach leads to better quality, more robust, and longer-lasting outcomes. There is no inevitable trade-off between climate action and economic growth.

Degrowth arguments

There is a viewpoint that achieving sustainability and net zero emissions necessitates an absolute reduction in consumption, particularly in wealthy countries and communities. This perspective is often summarised under the concept of degrowth.

At a basic level, the argument that degrowth is essential for achieving net zero is flawed, particularly if it claims that emissions cannot be reduced unless consumption is also reduced. Such an argument would imply that net zero emissions can only be achieved if consumption is reduced to zero, unless a massive CO_2 removal programme is implemented. We know that there are many ways to alter our consumption and production patterns to reduce emissions, so the idea that reducing consumption is a necessary condition for reducing emissions makes little sense. Similarly, no one suggests, I hope, that we should achieve net zero by reducing the population to zero.

Beyond these logical and empirical issues, there is also the politics. Policies aimed at drastically reducing consumption in the name of emissions reduction and sustainability are unlikely to gain political traction. Such priorities could face significant pushback, potentially derailing broader climate action and sustainability plans. This resistance would be particularly intense, and understandably so, if it involved wealthier nations trying to impose limits on growth for poorer countries – invoking, implicitly or explicitly, the argument that the carbon budget has already been used up by the rich.

There are hundreds of millions (or billions, depending on the metric) (World Bank, 2024) of very poor people in the world, and it will be impossible for them to achieve and sustain high levels of human development and escape from poverty without substantial economic growth in the communities where they live and work. This is a crucial point that many commentators do not seem to recognise or accept when they advocate degrowth.[13]

A 'softer' version of the degrowth argument does not insist that consumption reduction is a necessary condition for emission reduction. Instead, it suggests that reducing consumption could or should help reduce emissions as part of a broader strategy. While this argument is not entirely wrong, it is weak and it diverts attention from the big issues. We need to drastically reduce emissions – from nearly 60 billion tonnes of CO_2 equivalent (CO_2e) per year today to net zero by 2050. A modest reduction in consumption in some parts of the world would only make a limited contribution to the 100% reduction needed. The real challenge is to break the link between consumption and production on one hand, and emissions and environmental damage on the other. That is where the major changes must occur. The key lies in producing and consuming in ways fundamentally different from the past. Even if pursued, a modest reduction in consumption would be, at best, a small part of the solution. Worse, insisting on it could generate hostility to the entire climate action agenda; it could be powerfully counterproductive. The 'softer' version of the degrowth argument bears similarities to the flawed argument about the inevitability of a trade-off between environment and development, which we have already shown to be mistaken.

Our first conclusion, therefore, is to be very cautious of simplistic versions of the degrowth argument. Some of those who advocate for degrowth have, on the way, offered some thoughtful ideas, which we briefly discuss below. But these ideas do not require the concept of degrowth as their foundation. They can be examined and discussed from the perspectives of sustainability, social cohesion, and the reduction of waste without invoking degrowth.

Some of those arguing for degrowth have begun by challenging 'growth-centric ideologies' and 'consumerism'. Some offer a vision of a society that values mutual care, autonomy, self-sufficiency, and the environment over perpetual accumulation. They paint a different picture of how to enhance human well-being and social equity. Often policies like universal basic income, maximum income caps, and progressive taxation are seen as part of this vision (e.g., Hickel, 2017; Jackson, 2021). Such policies, or versions of them, may have merits, but those merits can be argued without reference to degrowth. The reduction of harmful activities or waste is indeed needed, but that does not require aiming for overall economic decline. Such decline could hamper the perceived demand side for investment for change (van den Bergh, 2011) and uniform support for the changes that really matter for the path to net zero. The more sensible route would be through supporting the growth of the clean that displaces the dirty and arguing in favour of policies that promote equity, reduce waste, and embody a broad view of development.

Those making degrowth arguments often call for a shift away from using GDP as the primary measure of success, and for focusing instead on broader notions of human well-being and happiness. That is indeed a well-founded argument – see, for example, Chapter 4. Those making the case for degrowth often support innovative technologies, efficient resource use, the circular economy, and the scaling down of harmful industries. Advocates for degrowth often call for investments in sustainable infrastructure, such as renewable energy and circular economy practices. Much of this agenda does indeed overlap with the new approach to growth we have described and advocated. However, these actions for sustainability do not require the concept of degrowth for their support.

Population arguments

Population dynamics sometimes emerges as a contentious issue in discussions of sustainable development and climate change. Population growth in wealthier parts of the world is already low or negative. It is negative in China and much of East Asia. In India, and some other Asian developing countries, the fertility rate is already below replacement, although population will rise for a while. Narratives advocating population management or control tend to focus on the poorest parts of the world, particularly Africa. Such arguments, if they come from the rich world, can sometimes, understandably, be seen as condescending, hypocritical, or indeed neo-colonial. They could thus be met with resentment or pushback, especially given that the poorest countries and peoples have contributed relatively little to overall emissions compared to wealthier countries.

The first and most important issue is not simply the number of people, but how resources are consumed and produced. Consumption patterns and the resulting emissions are closely tied to wealth and economic structures (Figure 5.9). Thus, in the shorter term the biggest potential emissions reduction, by changing how goods are produced and consumed, will be associated with richer countries and groups.

Figure 5.9: Global carbon inequality (2019): group contribution to world emissions (%)

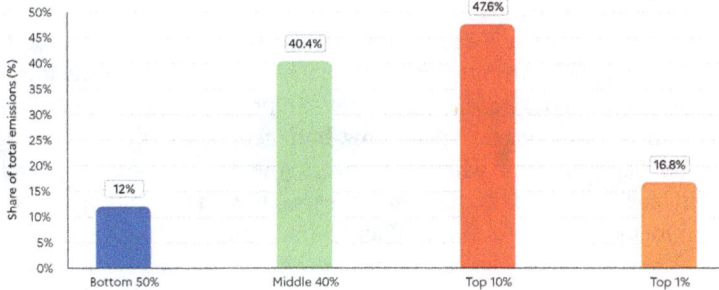

Note: Personal carbon footprints include emissions from domestic consumption, public and private investments, as well as imports and exports of carbon embedded in goods and services traded with the rest of the world. Modelled estimates are based on the systematic combination of tax data, household surveys, and input–output tables. Emissions are split equally within households. Percentiles refer to global income and distribution.
Source: Figure 6.5b in Chancel et al. (2022, p. 122). Copyright 2022 The Authors, CC BY 4.0.

However, as we have seen earlier in this chapter (see also Chapters 8 and 9), the most rapid growth of emissions will be in middle-income countries, particularly in Asia. Thus, their course of development will be critical to the management of emissions over the next couple of decades. But looking ahead, the weight of population in Africa (likely two billion people by mid-century) – and, we hope, rising living standards – implies that the choices of development paths for Africa, starting now, will also be crucial to the course of emissions. And their emissions would rise less fast, or fall more rapidly, if population growth were lower.

Of special importance here are arguments grounded in the rights and empowerment of women, which are compelling in their own right. Rights-based approaches that prioritise empowerment, education, and choice for women and girls are not only matters of justice and equality, but also powerful drivers of sustainable development. Greater access to education, paid work, and leadership opportunities improves living standards for women, their families, and their communities (United Nations Women [UNW], 2018). Expanding access to voluntary family planning and comprehensive reproductive health services increases the ability of women and couples to make informed choices about their lives. These actions stand on their own as vital investments in dignity, rights, and well-being. If, as a result of such empowerment and expanded opportunities, fertility rates and population growth decline, this can be seen as a secondary benefit that may also reduce pressure on the environment.[14]

5.4 Investment across sectors and geographies

It is the sum total of emissions over all countries and over time which will shape our climate future. China, as the biggest emitter (around 30% of total CO_2e GHG emissions in 2021), will be crucial (UNEP, 2023a). So too the developed countries, currently with emissions of nearly 30% of the total in 2022 (Beynon and Wickstead, 2024). However, policy should focus not only on where these emissions take place now, both geographically and in which sectors, but particularly on where in the world the strongest growth in emissions will likely take place in the coming years. Thus, in this section we place a special focus on EMDCs (other than China). It is vitally important that emissions reductions accelerate in China and the developed world. But it is outside these countries that the most rapid growth in emissions is likely to take place in the absence of strong climate action. And it is these countries, EMDCs (other than China), that have the great opportunities to pursue and integrate climate action and development if the necessary investment can be fostered and financed.

Similarly, there are some sectors of special significance to the transformation. First and foremost, it is the energy sector, but agriculture and some other hard-to-abate sectors will also be central to the story of reducing emissions and combining those reductions with stronger and better development. The private and public sectors play distinct but intertwined roles, but we must recognise that the bulk of the necessary investment across the world will be from the private sector, although the public–private proportions will vary across countries.

The centrality of EMDCs (other than China)

In terms of potential future growth of emissions, the EMDCs (other than China) are pivotal to global climate action. There is a special opportunity for EMDCs to leapfrog traditional carbon-intensive industrialisation. This involves the fundamental transformation of key systems, including energy, transport, cities, land, and water. Such a transition promises not only environmental benefits but also, as I have argued, powerful development benefits, including economic growth, quality job creation, reduced energy costs, improved public health outcomes, liveable and productive cities, and ecosystems which are much more resilient and more fruitful.

Investment is at the core of this new growth story and of reaching the goals of the Paris Agreement. Our work for the Independent High Level Expert

Group (IHLEG) on Climate Finance (Bhattacharya et al., 2024) indicates that to achieve the Paris goals, global climate investment must increase and be sustained at levels above US$6 trillion annually by 2030 and beyond.[15] This is necessary to grow and transform physical, natural, and human capital in line with Paris Agreement targets, with a major portion directed towards energy transition. EMDCs (other than China) will require approximately $2.4 trillion of investment annually by 2030 to meet their climate action and sustainable development objectives in ways consistent with the Paris Agreement (Songwe et al., 2022). In this context we defined climate investment using five elements: energy transition, adaptation and resilience, loss and damage, natural capital and sustainable agriculture, and a just transition. That would involve multiplying investments in these areas by a factor of four or five from current levels of around $500 billion. Ways to achieve this will be the topic of Chapter 8.

Despite a growing global commitment, climate action remains uneven across geographies and sectors. In 2023, combined investment in low-emissions generation and batteries outperformed fossil fuel generation by 12:1 in advanced economies, 6:1 in China, and 2:1 in other EMDCs (Climate Policy Initiative [CPI], 2025). The majority of climate finance[16] is concentrated in advanced economies and China: from 2018 to 2022, advanced economies accounted for 45%, China 36%, EMDCs (excluding least developed countries [LDCs] and China) 16%, and LDCs 3% (CPI, 2024). Further, mitigation efforts, particularly in the energy and transport sectors, received over 90% of climate finance between 2017 and 2020 globally, while adaptation investments were much lower (Kreibiehl et al., 2022).[17] Thus, there is a critical need in the coming years not only to advance climate finance across the board but also to tackle geographical disparities and strongly increase adaptation finance. A little caution in interpretation here is important, since projects can be, and often are, both mitigation and adaptation. Nevertheless, the point stands: adaptation is under-invested.

Investment priorities for climate action

As we emphasised in our discussion of EMDCs (other than China), the next development phases of those countries will require strong investment in infrastructure. And they have a great opportunity to build that infrastructure in less carbon-intensive and more efficient ways than in the past. Low-carbon investments tend to involve capital upfront, which economises on fossil fuel inputs in operations down the track. The categories in this section reflect the key areas of climate investment that underpin our working definition of climate finance in the studies we carried out for the IHLEG on Climate Finance, commissioned by the presidencies of COP26, 27, 28, 29, and 30, and co-chaired by Amar Bhattacharya, Vera Songwe, and myself (Éléonore Soubeyran was a key co-author). As mentioned above, there are five elements of investment, which we consider in turn: the energy transition; adaptation and resilience to climate impacts; funding for loss and damage; preserving, restoring, and investing in nature, including the transformation of agriculture; and investing in a just transition. Climate finance, in this context, refers to the flows of funds directed toward these investment areas.[18]

First, amongst climate investments the *energy transition* is central, with a strong emphasis on decarbonising the power sector. It requires significant investment in both new infrastructure and the upgrading of existing systems. Clean electricity is the backbone of the net zero transition, projected to provide more than 60% of total energy consumption by 2050 – a dramatic increase from today's 24% (IRENA, 2018). The global electricity system needs to expand by three to five times with both growing energy demand and a shift to electricity in transport, heating and cooling, and much of industry. The role of wind and solar power will need to grow dramatically, covering 75–90% of power generation, up from 15% in 2024 (ETC, 2023b; Ember, 2025). Achieving net zero emissions will require a surge in energy investment to around $4.5 trillion a year by the early 2030s (IEA, 2023e), primarily allocated to power generation, networks, and storage. Alongside this expansion, enhanced energy efficiency will be of great importance.

Second, *adaptation and resilience* to climate impacts remain a critical part of climate action. These needs will rise rapidly as temperatures keep increasing over the next two decades. Yet mitigation finance far outweighs adaptation. Adaptation finance reached a record US$76 billion in 2022 and thereafter decreased in 2023 – see Chapter 9 (CPI, 2025). EMDCs alone need on average US$212 billion annually between 2024 and 2030 (CPI, 2024). This gap reflects the urgent need to increase funding to build resilience in vulnerable regions, particularly in Africa, which receives a disproportionately low share of adaptation funds relative to its vulnerability to climate change effects (GCA and CPI, 2023). In 2022, LDCs and small island and developing states (SIDS) – groups that include the economies that are the most vulnerable to climate impacts globally – received only 19% and 2% of adaptation finance respectively (CPI, 2024). And it must be noted that the needs of EMDCs might be much higher than indicated here, given the uncertain impacts of climate change and the increasing costs of inaction (CPI, 2024).

As we argued (Section 5.3), it is important to recognise that good adaptation is always positive for development – it makes no sense to ignore potentially destructive change. And we should continue to remind ourselves that many actions from restoring degraded land to public transport embody all three of mitigation, adaptation, and development.

Third, *funding for loss and damage* is already critical to the future of countries in the frontline of impacts such as sea-level rise, extreme weather events, and biodiversity loss (see Chapter 2). However, financing in this area remains inadequate and inconsistent. The potential scale of these losses is very large, damaging infrastructure, livelihoods, and ecosystems. Whilst efforts at COP27 in 2022, including the agreement to establish a fund specifically for loss and damage, marked a significant shift towards recognising the need for dedicated financing in this area, the pledged funds remain far below the required levels.

The initial pledges made at COP28 in late 2023 total nearly US$700 million, falling way short of the needs of EMDCs (other than China) – projected to be at least US$250 billion per year by 2030 and above US$400 billion per year by 2035 (Bhattacharya et al., 2024).

Fourth, *preserving, restoring, and investing in nature* is integral to climate action, particularly in EMDCs where such investment opportunities and needs exist on great scale. These investments are essential for preserving and restoring ecosystems that provide crucial services such as carbon sequestration, water purification, and biodiversity preservation. Investing in natural capital supports climate adaptation and mitigation efforts while enhancing local communities' resilience and livelihoods; it is a prime example of adaptation, mitigation, and development coming together.

The *transformation of agriculture* will play a central role in nurturing biodiversity, in reducing emissions, in enhancing resilience, and in driving development. Changing past damaging policies and strategies offers great potential in improving both quality and quantity of development and, crucially, releasing resources associated with misdirected subsidies. That will involve major policy reform. Agriculture is responsible for 22% of global GHG emissions, directly and indirectly through land use change (Intergovernmental Panel on Climate Change [IPCC], 2022b). Methane is of particular importance.

Amongst past damaging policies, distorted and often toxic subsidy systems are of special importance. Agriculture subsidies are nearly US$700 billion per year,[19] but outcomes are all too often degraded land, deforestation, poisoned rivers, and polluted oceans (Damania et al., 2023). There is substantial scope for redesign of these subsidies with great benefits to the economy and the environment. Two recent World Bank publications, *Detox Development* (Damania et al., 2023) and *Recipe for a Liveable Planet* (Sutton et al., 2024), have set out the issues clearly and there has been valuable work from the Food and Land Use Coalition (Food and Land Use Coalition [FOLU], 2019). By repurposing subsidies for meat and dairy towards low-emission foods, such as fruits, vegetables, pulses, and poultry, significant reductions in emission could be achieved (World Bank, 2024). Effects on diets would likely improve health.[22]

According to the Intergovernmental Panel on Climate Change (IPCC), most mitigation measures here are 'available and ready to deploy' (Nabuurs et al., 2022, p. 751). For example, efficiency can be increased through sustainable agricultural intensification, shifting diets, and reducing waste. In turn, these measures would reduce the quantity of agricultural land needed, enabling the greater deployment of reforestation and restoration. Other supply-side measures include 'cropland and grassland soil carbon management, agroforestry, use of biochar, improved rice cultivation, and livestock and nutrient management' (Nabuurs et al., 2022, p. 750). Though still emerging, there are also experimental technologies with emissions reduction potential, such as vaccines to reduce methane emissions from cows. Nevertheless, given the likely continuation of cattle and sheep farming for meat, milk, wool, and leather at some level, it is unlikely that zero emissions from pastoral activities could be achieved. That is a key reason why CO_2 removal will be necessary.

In agriculture, AI applications are revolutionising water use and crop management through precision-farming techniques that enhance productivity while conserving resources. They also provide small farmers with crucial information and guidance, particularly regarding weather conditions, soil health, and market trends. Similarly, in water management, AI algorithms and approaches can enhance treatment processes and predict water demand fluctuations and trends, enabling sustainable usage patterns crucial in water-stressed regions. As with electricity, managing demand and supply across space and time is vital (De Baerdemaeker et al., 2023).

We should emphasise again that transforming agriculture along these lines will bring mitigation, adaptation, and development. A key example is restoring degraded land, which then captures carbon, makes incomes more resilient, and increases productivity and growth. A second is the system of root intensification for rice, which avoids flooding paddy fields. This reduces methane emissions, increases productivity, reduces costs of energy and water, and is more resilient.

Finally, investing in a just transition means working to make the shift towards a green economy inclusive and equitable. A just transition aims to manage dislocation in an inclusive way that gives opportunities to participate in the new growth story and protects those whose livelihoods might be disrupted. This involves, for example, supporting workers and communities affected by the phase-out of carbon-intensive industries through retraining programmes, economic diversification strategies, local infrastructure, and social safety nets where re-employment is difficult. In other words, it means investing in people and places and, where necessary, offering social protection. A just transition tackles social inequalities and promotes broader social and economic resilience. Failure to ensure a just transition is not only unjust but also risks creating an opposition that could derail the transition.

To tackle effectively the complex challenges of climate change, an integrated approach to development and investment that encompasses all these elements is essential. These elements are interconnected and collectively support the reduction of GHG emissions, the enhancement of economic and social resilience, and the strengthening of the ecosystems upon which many depend.

The scale of finance necessary arises from the scale of investment necessary, which is a result of both the scale of the task and past delay in action. If the world, both developed countries and EMDCs, had started this transformation two decades ago, the required annual increase in investment flows would have been much lower. Financing investment at this scale is itself an immense challenge, which we will return to in Chapter 8.

Hard-to-abate sectors

Reducing and ultimately eliminating emissions from the sectors once labelled 'hard to abate', which in 2024 accounted for nearly 30% of total annual GHG emissions (agriculture not included) (WEF, 2024), is crucial for achieving climate goals.[21] Agriculture, also a hard-to-abate sector, has been discussed above. We focus here on industry (cement, steel, and chemicals) and transport (heavy duty road transport, shipping, and aviation) (ETC, 2018).

Decarbonising these sectors will require a range of strategies – from rapidly scaling green technologies to replacing legacy fuels and creating novel incentive and business models that reward innovation and challenge incumbency (Lovins, 2021). These sectors will need significant structural transformations, including reconfiguring the constitutive elements of existing production processes, such as replacing coal in steel manufacturing. Additionally, decarbonising will require extensive changes across value chains, from upstream activities such as coal mining to downstream practices such as material efficiency in construction.

While comprehensive structural transformations are essential, incremental measures such as energy efficiency also remain vital. For example, in the industrial sector, energy efficiency can reduce emissions from long-lived assets that have still not reached the end of their operational life. Carbon capture can, in some cases, enable the continuation of existing industrial processes, such as cement, without structural transformations, but its scalability depends on further and significant investments.

For industry as a whole, the ETC (2018) highlights three types of action as particularly important in tackling emissions: deploying decarbonisation technologies, limiting demand growth, and improving energy efficiency. There are three main decarbonisation technologies aside from carbon capture to incorporate in the industrial and transportation sectors: zero- or near-zero-carbon hydrogen (to be used as a heat source or a reduction agent), electrification (to decarbonise high-heat temperature generation), and biomass (also as a reduction agent or as feedstock on plastics production). Which technologies among these will be most cost-effective depends on the specific locations, and particularly on the availability of resources.

Although 95% of manufactured products rely in some way on chemicals, credible transition plans in this sector remain scarce. Emissions from chemical production stem, in approximately equal parts, from the use of fossil fuels as an input for chemical reactions (feedstock) and to generate heat, steam, and power for industrial processes such as compression or cooling. Therefore, the chemical industry should find ways to transition away from fossil fuel use in both feedstock and energy needs. Essential to achieving this is the use of green hydrogen and electrification. Therefore, particularly relevant to the feasibility of decarbonising this industry is the increasing competitiveness of hydrogen and clean energy relative to their fossil-fuel-based counterparts.[22]

Steel demand generally increases along the development trajectory, particularly in early industrialisation (IRENA, 2023a). It also features as an input into the energy transition, used in EVs, wind turbines, and solar PV structures (IRENA, 2023a). Demand management for steel can be achieved through increased material efficiency and circularity, particularly by decreasing materials required in value chains and increasing recycling and reuse. It is estimated that 40% reductions of emissions from plastics, steel, aluminium, and cement can be achieved globally by deploying these two measures (ETC, 2018). The role of governments will be key in decreasing the demand for materials, as this type of measure requires strong collaboration between the public sector and companies all along value chains. For example, this is often the case with strategies to decrease material demand, minimise waste, and prolong the lifetime of buildings, which could altogether reduce cement and concrete demand by between 30% and 35% by 2050 (ETC, 2025b). In addition to strategies around material and process efficiency, steel production will need to shift towards using renewable energy, in particular green hydrogen (IRENA, 2023b).

In the case of transport, possibilities to contain demand growth are more limited, but measures such as improving logistics efficiency and changing modalities (e.g., from trucks to rail transport) can reduce up to 20% of emissions from the harder-to-abate sectors in transport by 2050 (ETC, 2018). EVs are now well advanced and cost competitive. Long-range EVs (more than 300 miles, or about 480 km) in the USA market are now cheaper than the average petrol-powered car (Randall, 2024). In the case of heavy transport, electric drivetrains and electric trucks are examples of technologies expected to play an increasingly important role. Further, though electric engines with hydrogen-based batteries are expected to be of some relevance for both shipping and aviation, new technological developments are required to deploy the latter technology for long distances; battery energy density needs to improve. Other possibilities for long-distance travel include bio-jet fuel for aviation and ammonia for shipping.

AI is playing an increasingly instrumental role in these sectors, improving resource use and boosting productivity. For example, under the initiative Carbon Re, founded jointly by University College London (UCL) and the University of Cambridge, AI is being used to conduct research on how to reduce emissions from energy-intensive industries, such as steel or cement. Specifically, their AI technology processes complex manufacturing data to enhance plant efficiency, reducing energy use while ensuring high material quality.

The role of private investment

Private investment will be at the core of the transition, particularly in the development and deployment of renewable energy technologies. Between 2013 and 2020, private financiers were responsible for funding 75% of global investments in renewables. This is especially pronounced in technologies like solar PV, where 83% of total investment in 2020 came from the private sector. Similarly, in 2020 wind energy saw 65% of its investments funded privately (IRENA and CPI, 2023). This predominance of private financing is largely due to the low costs associated with these technologies, together with advances in the building of grids, the creation of markets, and the presence of purchasers of some reliability. Where these features are not present, the profitability of renewable energy supply comes under question and the cost of capital can then increase. The financial viability of power supply via renewables is particularly sensitive to the cost of capital, given that costs are focused on upfront investment and savings on future inputs.

However, the role of private investment is not the same across all renewable technologies. For example, geothermal and hydropower projects often require substantial upfront investments and carry significant community and social risks, including via their potential effects on large geographical areas and populations. These factors make them less attractive to private investors. In many cases, public sector investment will be necessary if they are to be pursued.

Looking forward, the private sector is expected to remain the major driver of climate investment and finance, particularly in mature and commercially viable technologies. As the market for clean technologies expands and as policy and economic environments become more supportive, private investments are likely to increase strongly.

Strategic focus of the public sector and the role of companies in supporting change

The public sector will contribute a smaller share than the private sector to overall climate finance. Its focus will be on areas of strategic importance that are less attractive to private investors. And it will be needed on a major scale. The relevant areas where public investment is likely to appear strongly include financing for power grids, adaptation and resilience, loss and damage, natural capital, and measures aimed at fostering a just transition. These investments often require substantial capital, involve higher risks, or have economic returns which are difficult for the individual private investor to capture, because they are spread across a large area or community, such as flood defences. Adaptation finance remains predominantly the responsibility of public finance. Between 2019 and 2022, less than 3% of global adaptation funding came from the private sector, highlighting the critical role of public funds as not just complementary but a primary source of support for adaptation initiatives (Global Centre on Adaptation and CPI, 2023).

In nature conservation, public capital is often the main source of finance. Unlike energy systems, where private investment can play a substantial role, nature finance saw, in 2022, only 17% of its flows coming from private capital (UNEP, 2022). In the early stages of conservation projects, grant and concessional financing are crucial, but as these projects mature, de-risking and the introduction of pure commercial capital become more relevant. Some private finance can also be generated through regulation and standards applied to private sector investments, so that they are required to consider potential environmental impacts.

A just transition will require concerted efforts from public and private sectors and from communities. Public funds are especially important in regions heavily impacted by the transition from fossil fuels, supporting initiatives such as re-skilling workers and promoting economic diversification. The EU's Just Transition Mechanism is an example of how targeted public finance can help manage and reduce the socioeconomic impacts of transitioning economies, creating new and better opportunities in the process (European Commission [EC], 2024). Additionally, the private sector, including large institutional investors, is increasingly recognising its role and responsibilities in the net zero transition, including in supporting a just transition.

Two cases where investors had a key role in the development and implementation of the companies' just transition strategies and programmes are Scottish and Southern Energy (SSE) and Ayana Renewable Power.[23] SSE, a UK-based energy company, was one of the first companies globally to release a just transition strategy. Published in 2020, this strategy outlines 20 principles to promote a fair and just transition for workers, consumers, and communities. According to the company, by 2024, 29% of its employees had transitioned from high-carbon roles, facilitated by initiatives such as engineering conversion programmes, skills courses, and the integration of just transition principles into its core business strategies (Scottish and Southern Energy [SSE], 2024).

Ayana Renewable Power is a renewable energy development platform established in India. Although the national government created a Skills Council for Green Jobs in 2015, which has already trained around 500,000 workers, many more are expected to be needed, creating a challenge for growing companies in India. Indeed, around one million clean energy workers are projected to be needed in the country by 2030, a significant growth relative to the present workforce, composed now of more than 160,000 workers. This challenge of building 'clean skills' is especially acute in rural areas. And these areas are economically convenient places to advance with clean energy projects due to the low cost of land (relative to urban) and the frequently pressing need for local economic opportunities. In this context, Ayana developed a just transition programme focused on creating the green skills that are needed to support its growth and contribute to the development of local communities. After several skill programmes, which initially depended on grants, the company is now funding the programmes itself, showing 'a progression from grant funding through corporate social responsibility spending to core business investment in developing human resources' (Just Transition Finance Lab, 2024a, p. 9).

The distribution of public and private finance will continue to vary substantially across different geographies and sectors. Developed economies, such as those in the USA and Western Europe, are expected to attract a higher concentration of private investments, particularly in mitigation technologies like renewable energy and energy efficiency improvements. These regions benefit from mature markets structures, established infrastructure, and some support from regulatory environments, all of which make them more appealing to private financiers.

In contrast, regions such as sub-Saharan Africa will rely more heavily on public and international finance to meet their climate goals. These areas often face greater challenges, including less developed infrastructure, higher risks, limited access to private capital, and severe pressures on the public finances, including in relation to debt. As a result, public funds and international support become crucial for supporting climate initiatives, from adaptation projects to renewable energy, enabling these regions to transition towards a sustainable future. In this context the role of MDBs is central – as we will see in Chapter 9.

5.5 Development, poverty reduction, and climate action

The argument that there is an inevitable trade-off between climate action and growth was considered in Section 5.3, where we showed that it was misguided. Similar arguments have been used to claim that there is an inevitable trade-off between climate action and poverty reduction (see Table 5.2). The issue is of great importance and we therefore examine here the concerns set out in Table 5.2 on climate action and growth directly in terms of links to poverty reduction in further detail. There is an inevitable overlap between this discussion and the one in Section 5.3, because growth itself is central to development and poverty reduction. However, poverty reduction is so important that, at the risk of repetition, we examine it directly here.

Concern 1: In an efficient world, introducing an additional criterion – the future state of the climate – must involve some reduction on another dimension.

This claim assumes that the existing equilibrium is efficient, implying that progress toward a new criterion must come at the expense of other objectives, including poverty reduction. As we saw in Table 5.3, however, the real world is characterised by multiple market failures and inefficiencies beyond the externalities of GHG emissions. Climate action policies provide an opportunity to address these inefficiencies while delivering economic and social benefits to vulnerable groups. For example, tackling urban congestion reduces household costs and pollution, leading to improved health and productivity. Public transportation and air pollution are of particular relevance for poorer people. So too is decentralised solar (with batteries): it can create job opportunities for poorer people through, for example, small enterprises; it enables children to study in the evening; and it can reduce time (and danger) associated with women searching for firewood.

Concern 2: Development requires energy and energy requires fossil fuels, and therefore development must involve more GHG emissions.

This claim assumes fossil fuels will remain cheaper than low-carbon alternatives, overlooking the rapid pace of technological change and innovation in clean energy. As discussed throughout the book, clean energy is already cheaper in many contexts, undermining the argument's premise. In most of the developing world, solar and wind offer vast potential for increasing energy access. Scaling up solar PV in Africa (see Chapter 9) would bring electricity to many poor people for the first time, unlocking opportunities for businesses, connectivity, and agriculture, and driving both development and poverty reduction.

Concern 3: Using resources for climate action will reduce capital for growth needed to reduce poverty and increase resilience.

There is no inherent trade-off between allocating resources for climate action and growth. On the contrary, as we have argued, climate action can drive growth through increased investment, enhanced resource efficiency, accelerated innovation, stronger international collaboration, improved health, and more. All these crucial aspects of climate action are drivers of growth. Investing in key sectors such as mitigation, adaptation, biodiversity, and natural capital can generate substantial economic benefits while advancing climate objectives. That growth opens up new opportunities, many of which would be available to poorer people.

Adaptation is of particular importance to poorer people, who in general are the most vulnerable to climate change. Every US$1 invested in adaptation can result in up to US$10 in net economic benefits (GCA, 2019). In the agricultural sector, of special relevance to poorer people in developing countries, adaptation measures are strategic investments to improve crop production, reduce farmers' vulnerability, and enable sustainable sectoral growth in the face of shifting climate conditions.

Moreover, though the effect of transitions on employment varies among regions and timeframes, evidence shows that climate-friendly investments create more jobs per US$1 million compared to unsustainable investments – and many of those jobs can benefit poor populations. Wu et al. (2024) found that growth in sub-Saharan Africa has in the past translated into less poverty reduction than has growth in other regions. This is in part because this growth has been underpinned by commodities – for example, oil and gas – many of which create few jobs. A greener, sustainable approach to growth would enhance the poverty-reducing effects of growth. More generally, for climate action to foster a growth path which brings about poverty reduction, policy should incorporate poverty concerns and manage potential adverse impacts or displacements of job opportunities across regions and sectors; this is related to the fourth and last concern.

Concern 4: Beyond diverting investment, climate action can involve a series of policies (e.g., pricing and fossil fuel phase-out) that could increase costs and reduce opportunities for poor people.

The perception that climate action burdens poor populations originates largely from past failures to incorporate poverty concerns in policy design and implementation, and the lack of accompanying social measures. Overall, leveraging climate action for poverty reduction requires appropriate policy mixes tailored to local contexts. For example, revenues from carbon taxes can be used for social programmes that support low-income groups (revenue recycling), reducing inequality and poverty. However, designing and deploying carbon pricing and redistribution policies can be challenging for EMDCs due to low institutional capacities (Budolfson et al., 2021; Lankes et al., 2023).

Revenue from reforms of fossil fuel subsidies can be reallocated for social protection and inequality reduction, as occurred in Jordan (2012), where the poverty impacts of their reforms were addressed through multiple social safety net measures, such as cash transfers and food subsidies. Another example is the case of Iran (2010), where fossil fuel subsidy reform included electronic cash transfers accounting for 50% of projected savings from the reforms. This had positive effects on inequality reduction, and the remaining savings were used to tackle the reforms' impact on the private and public sector (World Bank, 2019).

Opportunities for sustainable development with strong poverty reduction are available across the world.[24] Some measures impact poverty directly, such as the case mentioned from Iran. Another example is Bolsa Verde, an initiative for payment-for-environmental-services in Brazil which aims to reduce extreme poverty while improving ecosystem conservation: beneficiaries receive a payment for using natural resources sustainably and for monitoring areas to facilitate their protection (Schwarzer et al., 2016). Other policies impact poverty indirectly, for example by generating changes in economic growth (Barbier, 2014).

Key actions that can enable the effective linking of climate action to poverty reduction are, first, including affected communities in designing interventions, and, second, unfolding transitions in a careful and planned manner. Abrupt shifts in employment can be difficult, particularly in low-income areas where weak state capacity and limited access to social and economic data make managing just transitions challenging. Planning will work much better if the people who will be affected are involved.

5.6 Concluding remarks: the new growth and development story

This chapter serves as the heart of the book. Its aim has been to substantiate the idea that a well-structured transition to a cleaner, more sustainable, resilient, and inclusive economy delivers on two crucial purposes together. First, it achieves the reductions in emissions and the protection of biodiversity and environment that are crucial to respond to the twin crises of climate and biodiversity and protect our future from catastrophe. Second, it sets the world on a much more attractive and prosperous path of growth and development. The positive case for the growth story, particularly through the six drivers of growth, is strong and sound both in terms of its theoretical foundations and in the practicalities of the new opportunities, technologies, innovations, and efficiencies.

Beyond the positive case, it is important also to dispose of the arguments that claim there is an inevitable trade-off between climate action and economic growth. These arguments are often built on shaky or mistaken assumptions that underestimate or overlook market failures and overestimate the costs of transition.

The EMDCs will play a pivotal role in the global climate transition. With their significant growth potential and the vast infrastructure needs anticipated in the coming years, these regions must be at the forefront of the new growth story. However, it is clear that in EMDCs (other than China), clean investment is progressing much too slowly to meet the scale of the climate challenge and to foster development at a pace necessary for the SDGs.

At the centre of this new growth story are investment, innovation, and structural, systemic, and technological change. To transition successfully to this new economy, the share of investment in GDP must increase by at least 2 ppt globally, and by a substantially higher margin in EMDCs (other than China) (Stern, 2022). The specific policies and investments required to achieve this shift will be discussed in Chapter 6, while the international actions necessary to support this transition are the focus of Chapters 8 and 9. Following from the analyses of Chapters 5 and 6, it is clear that the transition will require a fundamental reorientation of the role of the state, which is the subject of Chapter 7.

An examination of the investment, innovation, and change necessary makes it clear that the transition will not be easy. There are many obstacles and challenges along the way. But these can and must be overcome and we have shown – and will show – how that can be achieved. The consequences of failure and delay pose far more difficult problems, indeed existential for many. As the first fossil-fuel-free power systems emerge, other countries will learn from them. There will be many challenges, but what matters is that we address problems effectively and continue to develop the solutions we need, from technologies to business models and social policies. Learning is part of the strategy.

In summary, the new growth and development story is not just about avoiding the catastrophic consequences of climate change and biodiversity loss but about seizing the opportunity to create a better, more equitable, and sustainable world for all. This chapter has tried to lay the groundwork for understanding how this can be achieved, setting the stage for the detailed discussions that follow in the subsequent chapters of how it can be carried through, nationally and internationally.

Notes

[1] See, e.g., the description of the Solow model in Chapter 1. And note Greta Thunberg speaking at the UN in 2019 'all you can talk about is money and fairy tales of eternal economic growth. How dare you!' (Thunberg, 2019, paragraph 4).

[2] Estimates are made on a levelised cost of electricity (LCOE) basis. Estimates for the USA take into account IRA funding.

[3] Every time global capacity doubled, prices have fallen by approximately 20% (Ritchie, 2024). China can now produce solar PV panels at 10 US cents a watt or less; in 2015, global solar costs were 80 cents per watt (IEA, 2020; Swanson and Rappeport, 2024). The capital cost for a gas-powered power station is between 60 US cents and US$1.30 per watt, although this varies a great deal by technology, scale, and location (IEA, 2024d; Lazard, 2024). And, in addition, running them requires the purchase of gas. That fall in the cost of solar power, in large measure from the remarkable fall in costs of panels, has transformed the economics of power away from gas. Renewables need accompanying storage, but that cost has fallen very rapidly too. The clean is becoming cheaper than the dirty across the world. In other words, the so-called green 'premium' is becoming negative.

[4] In this chapter we identify tipping points as when the clean becomes cheaper and more competitive than the dirty on the dimensions of affordability, attractiveness, and accessibility.

[5] Affordability refers to whether the cost of the new technology is lower than that of alternatives. Attractiveness refers to whether the technology outperforms alternatives (along a given dimension, e.g., efficiency, taste, or quality) and whether the technology is socially desirable to stakeholders. And finally, accessibility captures both whether the technology can be easily accessed by stakeholders and whether the stakeholders hold adequate information in order to use the technology (Systemiq, 2025). As noted in the preface, I currently chair the board of Systemiq.

[6] There are just a few exceptions, including concentrated solar, which heats water for a steam process.

[7] According to the IPCC (2022a, p. 1797), the circular economy can be defined as a 'system with minimal input and operational losses of materials and energy through extensive reduce, reuse, recycling, and recovery activities.'

[8] The Harrod model described in Chapter 1 is often called the Harrod–Domar model as it was developed independently by each of these two scholars (Harrod in 1939, Domar in 1946).

[9] The growth rate in the Harrod model is given by the equation $g = s / v$, where the growth rate (g) is the investment (or savings) rate, S, divided by the capital–output ratio, (or, as it is sometimes, and more accurately, called, the incremental capital output ratio, ICOR – a measure of how much additional capital is needed to produce one additional unit of output). I assumed a capital output ratio of 4 to 6.

[10] Some countries, such as the UK, which have seen a slow recovery on the supply side from Covid-19, might have to manage the demand side, including by increasing saving, to accommodate the extra investment.

[11] AI here can help by interpreting data generated, for example, by the many smart meters and predicting what might happen next.

[12] Data protection issues are, of course, relevant here but it should be possible to handle these, for example, by anonymising.

[13] On the connections and relations between growth and poverty reduction across its various dimensions, see, e.g., Bourguignon (2004), Dollar and Kraay (2002), and Easterly (1999).

[14] See, e.g., Boserup (1970), Mehra (1997), Duflo (2012), and Balasubramanian et al. (2024).

[15] See also Songwe et al. (2022).

[16] Though there is no universally agreed definition on climate finance, according to the UNFCCC (2018, p. 22) it refers to financial 'flows whose expected effect is to reduce net GHG emissions and/or enhance resilience to the impacts of climate variability and projected climate change'.

[17] In 2022 and 2023 climate finance continued to be directed predominantly to mitigation projects (90%), and so too particularly to the energy and transport sectors (CPI, 2024, 2025).

[18] Sometimes we combine adaptation, resilience, and loss and damage so that there are four elements overall. The detailed investment needs in monetary terms we leave for Chapter 9.

[19] This number accounts for countries with available data. The true global figure likely exceeds US$1 trillion (Damania et al., 2023).

[20] For further detail see Damania et al. (2023), World Bank (2024), and FOLU (2019).

[21] Distribution of GHG emissions by key hard-to-abate sectors: steel (7%), cement (6%), trucking (5%), aviation (3%), primary chemicals

(3%), shipping (2%), and aluminium (2%) (WEF, 2024). Also note that WEF (2024) includes the oil and gas sector as a hard-to-abate one; if we include that sector, it would account for 40% of GHG emissions.

[22] See ShareAction (2021) for further information.

[23] See Just Transition Finance Lab (2024a, 2024b).

[24] For examples, see UN (2020).

6. Perspectives, policies, institutions: actions for rapid structural transformation and sustainable growth

The investments and transformations necessary to tackle the climate and bio-diversity crises can create new paths for development and growth that are sustainable, resilient, and inclusive. This is the central argument of this book. These new pathways will be far more attractive than the destructive, high-carbon paths of the past. However, the scale, urgency, and complexity of realising the necessary investments, innovations, and systemic transformations present formidable challenges. This chapter examines the key policies, actions, and institutions that can drive the necessary transformation at the required pace and scale.

A sound analysis and informed public discussion of strategies and action should be grounded in a clear understanding of the concepts and perspectives that are embodied in and which underpin this new path. Section 6.1 explores these foundational ideas, including sustainability, innovation, and systems. New forms of investment are the core of the story. The creation of conditions to foster the required investments is the focus of Section 6.2. Strategies and policies must be clear and credible if they are to provide the confidence necessary for investment. Section 6.3 discusses strategies and incentive structures within the context of market failures, learning processes, increasing returns to scale, uncertainty, networks, and more. This discussion lays the foundations in two main ways, before we get into the details of policy. The first is to remind us of and emphasise the key outcomes that are sought and the importance of a systems approach as we chart the future transition; the second is to distil and assemble lessons from past transitions. Both of these generate and frame approaches to the challenge of creating a new path for development.

While the details of policy are crucial, so too is the need for flexibility – albeit flexibility that is predictable and grounded in a clear strategy. To shape and guide investment effectively, policies and institutional frameworks must be designed to manage and reduce risks, particularly government policy risks, which can undermine the willingness to invest. The need for diverse forms of finance, tailored to the specific needs of different investments, is the subject of Section 6.4.

The following three sections discuss issues that, if mismanaged, could derail the transition. The scale and nature of the transformation will inevitably lead to dislocation and disruption. Policy and action must, as a matter of jus-tice and practicality, involve those affected, foster their participation in new opportunities, and protect the poorest and most vulnerable. The concept of a

'just' transition was introduced in earlier chapters and some of its practicalities are explored in Section 6.5. Social action, public discussion, leadership,
and the law all have crucial roles to play. The macroeconomic management
of a significant increase in investment, alongside rapid structural change, will
not be straightforward. Finance ministries and central banks often present
arguments for caution and against 'rushing', yet in this case, the urgency is
dictated by science, not impatience. The macroeconomic challenges must be
addressed head-on, as discussed in Section 6.6.

Finally, Section 6.7 reflects on the challenges and decisions involved in fostering change at the necessary scale and pace. Strategic choices and difficult
decisions will be unavoidable. However, in the context of an inefficient starting point, not all decisions will involve trade-offs, as we saw in Chapter 5,
though some will. As we confront these tough decisions, we must remember
that the riskiest and most unrealistic approach is to delay and dither.

6.1 Concepts and perspectives: technologies and systems

Effective strategies and policies should be grounded in objectives – here,
achieving sustainable development. That in turn requires a clear understanding of what sustainability means. This section examines sustainable development and the central role of investing in the four capitals – physical, human,
natural, and social/cultural – in strategies and actions. These conceptual
frameworks are supplemented by lessons from history. Past structural and
technological transformations offer valuable insights for navigating the challenges of today.

Sustainability: objectives and interconnected systems

The concept of sustainability, introduced in Chapter 1, embodies the idea that
future generations should have, in terms of well-being, opportunities that are
at least as good as those of the current generation, assuming they behave in
a similar way to those who follow. Further, Chapter 2 explained that all four
types of capital (physical, human, natural, and social/cultural) are vital to sustainability; they represent the resources and endowments inherited by future
generations. Effectively managing these capitals is essential for achieving sustainable development, in the sense of progress across all development goals.
A useful summary way of specifying those goals is through the dimensions of
the Sustainable Development Goals (SDGs).

As we explore the concepts of sustainability and new forms of growth and
development, it becomes evident that some traditional economic concepts
and paradigms, with their relatively fixed structures and narrowly defined
metrics, require major overhaul. Policies and strategies should be set in relation to objectives. How these objectives are expressed is determined in part by
an understanding of the structures that shape them, alongside an understanding of the notions of well-being, development, and welfare.

For example, conventional economic indicators like GDP, while useful for measuring certain aspects of economic activity,[1] do not capture the full scope of welfare and well-being, particularly in the context of environmental sustainability. This recognition, as we set out in Chapter 1, led to the adoption of the Millennium Development Goals (MDGs) in 2000 and the SDGs in 2015 by the UN.[2]

The use of metrics to measure progress helps to both foster progress towards the goals and deepen understanding of concepts and perspectives. Efforts such as the UNEP's Inclusive Wealth Index and the World Bank's Changing Wealth of Nations (and its System of Environmental-Economic Accounting) aim to create metrics that value wealth from the perspective of physical, human, natural, and social capitals (Bhattacharya et al., 2021; Managi et al., 2023; World Bank, 2021). Creating and using these metrics helps us understand wealth and the four capitals, as well as guide the processes to enhance them.

As argued in Chapter 5, a key part of the growth story depends on the recasting of key systems – energy, transport, cities, land, and water. All these systems rely on different forms of capital as inputs while also influencing the four capitals in varying ways (see Figure 6.1). For example, urban development needs physical infrastructure. How that infrastructure is constructed and managed can help to preserve, or can damage, natural capital. The challenge in a period of rapid urbanisation is to develop and design, or re-design, cities to accommodate growing populations while radically reducing environmental footprints. That requires purposeful and integrated urban planning. As noted in Chapter 5, compact and connected cities can bring about economic growth and job creation while decreasing urban sprawl, poverty, emissions, investment costs, and traffic congestion, therefore improving productivity, health, and the quality of life in general (New Climate Economy, 2015).

Figure 6.1: Key systems (outer circles) and the four capitals (inner circles)

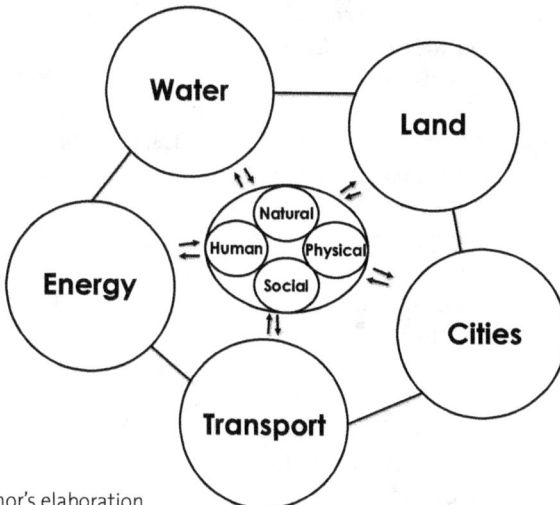

Source: Author's elaboration.

The need for energy crosses every system. Thus, its transformation opens possibilities to do things differently in many sectors. For example, decentralised electricity grids enable internet connectivity and create opportunities for those in rural areas to work with smartphones. This creates opportunities to deploy precision agriculture methods in remote areas. The rise of digital jobs enables teleworking, reducing the costs of travel and commuting.

Artificial intelligence (AI) accelerates innovation and the implementation of many sustainable activities and systems. In energy systems, AI can improve grid management and advance the use and efficiency of renewable energy. In transportation, it can aid in developing smart infrastructure that reduces congestion and emissions. In urban planning, AI-driven tools can help design, model, and analyse urban development scenarios, thereby predicting potential future challenges and finding ways to reduce or manage negative impacts, so that both its residents and physical infrastructure are protected and supported.

As we pursue the new growth story, we must constantly keep in mind the multidimensionality of the objectives being pursued and the interconnectivity of the systems that must be transformed. As the examples demonstrate clearly, the potential in positive feedback loops to boost innovation and accelerate progress is immense. The systemic nature of the changes involved is fundamental to the new story of growth.

Lessons from history for fundamental structural and technological transformation

The transformation to the new form of growth discussed in this book – with its rapid changes in energy, technology, sectoral and systemic structures, AI and digitalisation, and advances in biological and material sciences – can surely be described as an industrial revolution. Due to the scope and magnitude of the changes, it might be considered as multiple revolutions. However, for simplicity, we refer to it as the sixth industrial revolution.

Economic historians categorise industrial revolutions in different ways, but I find the structure identified by Chris Freeman[3] (Freeman and Louçã, 2001) and Carlota Perez (2002) particularly insightful. The preceding five industrial revolutions are, first, the mechanisation of production through factories and mills after the mid-18th century; second, rail and steam in the 19th century; third, electricity, the proliferation of low-cost steel, and heavy engineering from the late 19th century; fourth, automobiles and mass production from the beginning of the 20th century; and fifth, computers and the internet from the late 20th century (Perez, 2010).

Each of these revolutions profoundly affected how people worked, consumed, travelled, and lived. They also had intense effects on the distribution of income and power. The first of these transformed particularly how and where people worked and lived, and their relationship with capital. The second transformed energy and its use in production and consumption, and also geographical connectivity. The third was a 'fast energy' revolution which also transformed the organisation of new products and technologies. The fourth involves both a revolution in transport and a whole new approach to mass production. And the fifth transformed connectivity (again) and the whole service sector. Energy is pervasive throughout, and so too are radical transformations in people's lives.

Other authors count the industrial revolutions differently, for example by merging the first and second or third and fourth (e.g., Schwab, 2016).[4] However, the precise counting is less important here; what matters are the lessons we can learn from these past technological and industrial transformations to guide the significant changes we now need to foster.

Energy is everywhere in development, and changes in its sources have profound effects. In the late 19th century, steam was the dominant force powering factories, transportation, and urban infrastructure. Its replacement of wind and water energy, from the mid-19th century onwards, provided new locations and more intense heat. The introduction of electricity set in motion a gradual yet profound transformation. Initially, its use was confined to niche applications, such as lighting in affluent homes and public spaces. However, advances in electrical engineering, declining generation costs, and the expansion of standardised grids enabled factories to replace centralised steam engines with more efficient electric motors.

These transitions do not happen instantaneously but through a sequence of cumulative changes that, together, redefine industry and economic structures. This dynamic is central to Schumpeter's (1942) notion of creative destruction – where old technologies and business models give way to new, more productive systems, reshaping economies in the process. After the burst of creative destruction, the process of diffusion takes over and the new technologies become dominant across the economy.

A similar process of transformation is now underway as the world moves into the sixth industrial revolution, driven by the imperative for sustainable and resilient economic growth. Indeed, the new energy transition – marked by the rise of renewables, electrification, and resource efficiency – follows a familiar pattern. Initially, these technologies established themselves in high-value applications or growing new opportunities, such as electric vehicles (EVs) in premium markets and solar energy in remote areas. As costs fall and infrastructure improves, their deployment scales up, displacing fossil-fuel-based systems. These processes of diffusion are critical to going to scale. Given the urgency dictated by climate science, accelerating diffusion will be a central part of the challenge.

Just as electrification in the 20th century unlocked new modes of production and economic organisation, today's transition is enabling digitalisation, smart grids, and circular economies. However, the urgency of action on climate change demands that the transition in this industrial revolution moves much faster, even though in historical terms, those processes moved quickly in the 18th, 19th, and 20th centuries. The costs of inaction are immense, while the opportunities for innovation, growth, and sustainable prosperity are unparalleled.

Figure 6.2 provides a pictorial summary of these industrial revolutions, with flags highlighting the nations in the vanguard. Country leadership in the sixth industrial revolution will be complex. We are now in a multi-polar world. There is no doubt, however, that China will be one of those in the vanguard. It is already dominating in renewables.

Figure 6.2: The sixth industrial revolution in a historical perspective

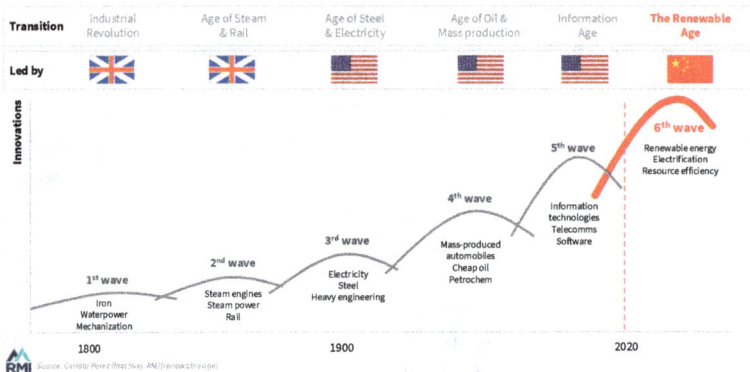

Source: Exhibit 22 in RMI (2024, p. 21). Copyright 2024 RMI, CC BY-SA 4.0.

Five broad areas, rich in historical lessons, stand out from past revolutions: innovation and entrepreneurship; government policies, institutional structures, and investment frameworks; management of dislocation; failures and recoveries; and finance. While this is not the place for an exhaustive treatise on past industrial revolutions, some lessons for guiding the new revolution are briefly highlighted below. The five areas and their lessons are interwoven.

The first lesson highlights the importance of *public action to foster innovation and entrepreneurship.* Chris Freeman was a pioneering researcher in this field, emphasising the key tasks of building skills in science and technology and setting clear missions for technological advancement. However, in Freeman's view, creativity does not stem from science and technology alone. He identified five main streams in history that shape innovation: science, technology, economy, politics, and culture. He later added ecology. A useful summary of his analysis is provided by Louçã (2020). Despite listing science and technology as the first two streams, Freeman never embraced technological determinism; the other streams were also critical and interwoven in shaping the history of innovation and technological change.[5] For example, European universities were probably in the lead in science up to the middle of the 20th century, while the US culture of entrepreneurship was likely a critical factor in its economic leadership. The rise of East Asia (South Korea, Taiwan, Japan) in the three or four decades after the Second World War was likely linked to politics, culture, and economic strategy.

Most serious authors writing on innovation have been profoundly influenced by Joseph Schumpeter. In 1942 (in *Capitalism, Socialism and Democracy*), he developed several key ideas highlighting the importance of innovation as a driver for economic growth, in particular creative destruction. In 1992, Aghion and Howitt formulated a growth model based primarily on three ideas from Schumpeter (see also Aghion and Howitt, 2023). The first emphasises the significance of innovating and diffusing knowledge, which accumulates progressively over time. This process enables innovators to build upon the ideas of their predecessors. The continuing process delivers long-term growth. The second idea concerns the incentives required for innovation. Entrepreneurs' decisions to invest depend on potential 'innovation rents' associated with the distinctiveness of their technology and methods. It is thus important for innovation and associated investment that these be secured, including through property right protection. The third insight is that growth involves disruptive processes, where 'the new' makes 'the old' obsolete. It is the perspective and language of creative destruction for which Schumpeter is the most famous. The words do indeed encapsulate an idea of dynamic, episodic, and dislocating processes.

Other authors, such as Segerstrom et al. (1990) or, more recently, Acemoglu et al. (2014), have also developed growth models based on the idea of creative destruction. In the last decade or so, this theory has evolved into a framework used to understand issues which go beyond overall growth, including microeconomic questions such as the distribution of gains and losses from innovation (Aghion et al., 2014). More recently, Aghion et al. (2021) argued that through governmental and civil society actions it is possible to foster green innovation, overcoming vested interests and obstructionism. This perspective integrates growth processes, political economy, and public policy. The Schumpeterian process of growth highlights the importance of property rights to capture the returns to innovation. Incumbents will work the political processes to protect these. But if they are protected too strongly, diffusion will be slowed, as future innovation will be stifled (the political economy). Public policy will be necessary so that the protection of property rights is neither too weak nor too strong.[6]

While technological invention and innovation, alongside entrepreneurship, were at the heart of these changes, so too was *government intervention*, the second area carrying broad historical lessons. Strategic policy frameworks and substantial public investments nurtured nascent industries and technologies. For example, during the early phases of industrialisation, countries like the USA and England implemented policies supporting infrastructure development and subsidising key industries. They provided legal frameworks, such as limited liability for joint-stock companies (Acts of 1844 and 1856 in England). This historical perspective underscores the importance of proactive government involvement in today's context, particularly in the role of policy and institutional arrangements catalysing and supporting innovation, the deployment of green technologies, and sustainable practices. On innovation and discovery, other countries – such as the UK, with the creation in 2023 of the Advanced Research and Invention Agency (ARIA) – are following the US example of the Defense Advanced Research Projects Agency (DARPA), which has been instrumental in the creation of the internet, micro-electronics, robotics, GPS, and more.[7]

These examples highlight the potential of the right kind of institutions in driving fundamental change. Japan's Ministry of International Trade and Industry (MITI) was created in 1949 and was described as a 'pilot agency' that concentrated key powers to protect and support industrial development into one place in the government structures. Authors such as Johnson (1982) argue that MITI played a key role in the emergence of Japan as an economic power in the 1960s. MITI selected key industries that were then 'nurtured' (*Ikusei*) through tailored measures – including tax breaks, low-interest loans, and access to foreign technology – to achieve their rapid development. It ensured that Japan's industrial interests were prioritised, for example by managing the approval of imported technology and exerting pressure so that patent rights were obtained with beneficial conditions for the whole industry. MITI was part of what Johnson (1982) calls a 'capitalist developmental state', where controlled competition was maintained as ownership and management was left in private hands, but the state led economic growth through planned industrial policies, establishing a cooperative relation among the state and private enterprises (Johnson, 1982, 1999). The role of MITI does not meet with universal acclaim,[8] but at the time it was indeed seen by many, both inside and outside Japan, as a key strategic driver.

Third, one of the most pertinent lessons from past industrial revolutions is the importance of *managing economic dislocations*. As new technologies emerged, some sectors and regions thrived while others declined, often leading to economic and social upheavals. In the first industrial revolution, handicraft suffered as spinning and weaving moved into factories. Those managing and supplying horses and horse-drawn travel were displaced as the automobile arrived. Today, as we transition away from fossil fuels towards renewable energy sources, similar patterns are emerging. The green transition is set to disrupt traditional industries and labour markets, necessitating policies – and, in some cases, social protection – to manage processes of change and offer new opportunities where lives are disrupted. These policies include investing in education and retraining programmes to equip workers with the skills needed for emerging green industries.

Fourth, past revolutions teach us about *the inevitability of failures and the resilience required to overcome them*. The paths to industrialisation were not simple linear progressions; they were marked by periods of boom and bust, experimentation, and adaptation. Railroads in the USA, for example, went bankrupt in the 19th and 20th centuries (Association of American Railroads [AAR], 2024). More recently, some renewable energy companies or battery makers have faced similar fates (British Broadcasting Corporation [BBC], 2016; Christie, 2017; *Washington Post*, 2011). Thus, the journey to a sustainable future will likely involve setbacks and challenges that must be met with perseverance and adaptability, and sometimes with support. Learning from past failures and fostering an environment that encourages innovation and tolerates risk are essential for making sustainable progress. Institutional financial structures which share and reduce risk have an important role to play, for example with joint-stock companies, equity, development banks, guarantees, and so on (see also Chapter 9). And the provision of bankruptcy to limit liability and provide exit.

There will be an extended period where the old and the new co-exist, often in tension. Some have referred to that as 'mid-transition' (Espagne et al., 2023; Grubert and Hastings-Simon, 2022). Each energy system, both the old and the emerging one, imposes operational constraints on the other. This coexistence means that the emerging clean energy system faces system constraints, for example grid structures tilted towards the old. There will be substantial uncertainty around how the tensions are resolved. That uncertainty can inhibit investment and slow innovation. Figure 6.3 illustrates this.

Figure 6.3: Representation of the unstable mid-transition period

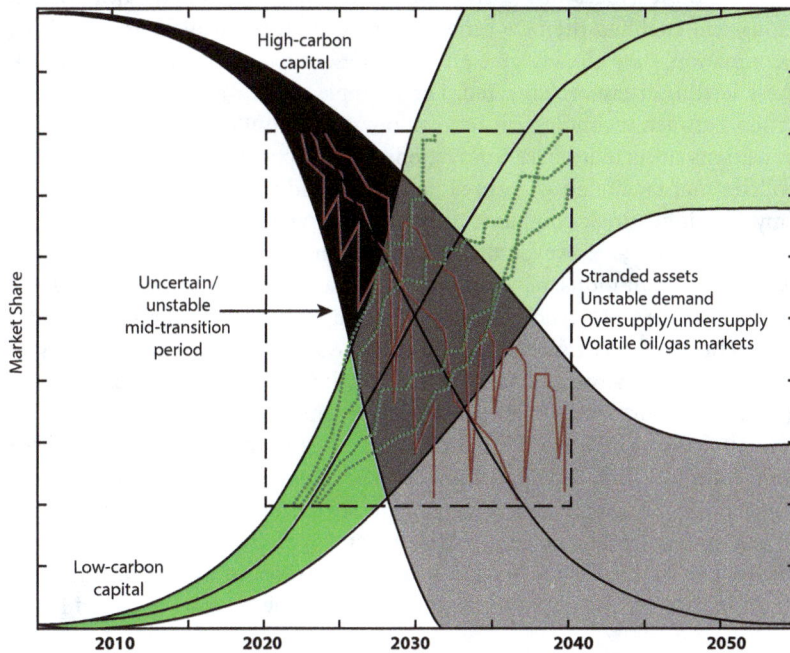

Note: Stylised view of the possible volatility and instability of paths for market shares of high- and low-carbon capital through the middle phase of a low-carbon economic transformation, as low-carbon industries rise rapidly concurrently to a high-carbon industry decline.
Source: Figure 1 in Espagne et al. (2023, p. 5). Copyright 2023 IMF. Reproduced with permission. Font and colours adjusted.

The fifth lesson to be gleaned from history is that *investment requires finance.* Some economic historians have argued that during the first industrial revolution, the sums involved in capital equipment were relatively small (Deane, 1965). Much of the finance was for trade credit, working capital, and inventories. However, subsequent industrial revolutions saw larger requirements for capital equipment, leading to a more prominent role for the banking sector, capital markets, and public finance. By the time of the fifth industrial revolution and the early stages of computing, venture capital became increasingly important. The lesson here is that as the sixth industrial revolution unfolds, different combinations and different types of capital will be needed. However, for this coming industrial revolution the overall sums involved are likely to be large, as argued in Chapters 5 and 9, to support the substantial increase in investment required. Risk, scale, and urgency, together with pervasive market failures in capital markets (see below), point to a central role for government in managing, reducing, and sharing risk. That in turn will bring down the cost of capital.

Whilst these five lessons warrant careful further study, and the tasks of fostering and enabling the new growth story will greatly benefit from such an examination, they already give guidance for action. We must recognise that these lessons are interconnected. For example, there can be powerful interaction between technological innovation and institutional change. Effective transitions occurred where innovations were supported by institutions and policies that facilitated widespread adoption and integration into the economy. The joint-stock company, bankruptcy arrangements, and capital markets are examples. However, different technologies require different kinds of supporting institutions, as we can see when comparing automobiles, where road networks were crucial, with early computers, where the internet and intellectual property rights played key roles. As we advance with technologies like renewable energy and smart grids, enabling institutional frameworks to evolve to support these technologies will be central to success.

In summary, the lessons from economic history emphasise the need for strategic leadership, thoughtful policy intervention, supporting institutions, and an inclusive approach to managing the socioeconomic impacts of technological and systemic change. By studying and applying these lessons, we can understand better how to navigate the complexities of the green transition more effectively. The implications for the role of the state are explored in the next chapter.

In the first part of Section 6.1, we have offered a distillation of core concepts around goals and processes. This allowed us to present lessons, in the second part of the section, from the economic history of fast transformations in ways which provide insights into the challenges of the sixth industrial revolution. In the rest of the chapter, we put those insights to work drawing on economic theory and analysis, particularly theories of public policy, or public economics.

6.2 Fostering investment: strategies, systems, and platforms

Commitment by investors is inextricably linked to confidence in the future growth and trajectory of the economy. John Maynard Keynes, Friedrich von Hayek, and many others have for long emphasised the importance of expectations in determining the level and nature of investment. Governments, therefore, will need to establish strategic and institutional frameworks that indicate clear and credible directions for the economy if investment is to flourish. This is of particular importance in the context of the investment imperative and the urgency of the transition. In this section, we focus on strategies and frameworks. Some detail of policies is offered in Section 6.3.

For a government to be credible, it must demonstrate that it understands the challenges and opportunities involved, and it must provide plausible assurances that its strategic direction will remain stable over time. While no government can guarantee absolute certainty in policy and institutional frameworks over the coming decades, they can, and should, manage uncertainty. In particular, they must recognise that their own behaviour can foster or undermine the confidence necessary for investment. Confidence is undermined if they frequently or unexpectedly alter their stance or if a new government with vastly different priorities looms on the horizon. Domestic legal and institutional frameworks, as well as international agreements and law, can bolster this confidence. In this section, in examining confidence in and credibility of strategy and policy, we focus on strategic commitment, systemic and structural change, the investment climate, and country platforms.

Strategic commitment

A key element of strategic commitment can be setting a national target for net zero emissions. Such targets have gained prominence since the Paris Agreement in 2015, where target emissions paths are part of the nationally determined contributions (NDCs) which countries have pledged, under Paris, to establish. In 2019, the UK became the first G7 and G20 country to legislate a commitment to net zero emissions, under the leadership of Prime Minister Theresa May (Climate Change Committee [CCC], 2020). The EU followed suit in 2021. By the time of the United Nations Framework Convention on Climate Change (UNFCCC) COP26 in Glasgow at the end of 2021, around 140 countries had committed to or were considering net zero targets, with the majority aiming for 2050. China set a target for 2060, and India for 2070 (Climate Action Tracker, 2021). These commitments, particularly within the framework of UNFCCC international agreements, signal powerfully to investors which way the economy and technologies will move.

For many investors and policy-makers, 2050 might seem distant; thus, a net zero target should be accompanied by a well-defined strategy for emissions reduction over time, starting immediately. That will be crucial for shaping the scale and pace of investment. In the UK, forward-looking targets and commitments are set by the Climate Change Committee (CCC) under the climate legislation of 2008, supported by all parties. It provides clear targets 10 years ahead, together with the identification of key measures and investments. It provides specifics on the sense of direction in the central industries and activities.[9] Denmark is another strong example. In 2019 it created the legal mandate to reduce greenhouse gas (GHG) emissions by 70% by 2030 (compared to 1990 levels). To assist in attaining this goal, it established 14 public–private partnerships which cover the key sectors of the Danish economy (The Danish Government's Climate Partnerships, 2022).[10]

There are also powerful examples in Costa Rica and the Philippines. In its 2019 National Decarbonisation Plan (the country's long-term decarbonisation strategy), Costa Rica committed to achieving net zero emissions by 2050 and outlined three distinct phases for transition, along with measures to be taken in each one: Foundations (2018–2022), Inflection (2023–2030), and Massive Deployment (2031–2050). The plan has successfully endured a major political shift in 2022, as the new administration recognised the great value of a robust decarbonisation plan – it has already attracted large investments from multilateral development banks (MDBs), and support from the International Monetary Fund (IMF) (Elliott et al., 2024).

The Philippines has adopted a whole-of-government approach to climate change, developing climate change initiatives across sectors to achieve its NDC of reducing GHG emissions by 75% by 2030 (compared with a business-as-usual scenario). For example, to shift investments to clean energy sources and green technologies, the government has declared a moratorium on endorsements for greenfield coal power plants and liberalised foreign investments in geothermal energy. And so too, through an interagency collaboration called 'Green Force', led by the Department of Finance and the Central Bank of the Philippines, the country is establishing a sustainable finance ecosystem to foster public and private investments in green projects (Department of Finance Philippines, 2023).

Figure 6.4 shows several recommendations helpful for nations when updating their NDCs. The package was created by the Institutional Investors Group on Climate Change (IIGCC), which began in the UK but now has international members.[11] For NDCs to foster private investments, they need to constitute credible and comprehensive plans with sufficient detail on technological and sectoral investment opportunities, and contain information on the policy landscape that could help investors understand the country's direction. An NDC with these characteristics is sometimes called an 'investable NDC'. Long-term plans such as the National Decarbonisation Plan of Costa Rica also have the potential to evolve into investment plans by quantifying needs and presenting local and international investment opportunities aligned with decarbonisation aims (Elliott et al., 2024).

Creating NDCs in alignment with the IIGCC recommendations in Figure 6.4 is a strategy of mutual opportunity. Countries can attract private climate finance necessary for climate and developmental ambitions, and private investors can act with more information and confidence (IIGCC, 2024).

Figure 6.4: Recommendations for creating investable NDCs

Investor views on how to develop 'investable' NDCs
1 Provide more granular detail on the **sectoral pathways and underlying macroeconomic context in a country.**
2 **Quantify investment needs** and prepare financing strategies alongside NDCs, to help investors identify long investment opportunities.
3 Set out **supporting policy and regulatory frameworks** to achieve NDC targets, to help give investors confidence that a decarbonisation plan is credible.
4 **Strengthen governance and the stakeholder engagement process** around NDC development and implementation, to ensure a regular and open dialogue with key stakeholders involved.
5 Enhance global **harmonisation and consistency across NDCs** to facilitate the collection and analysis of consistent information for investment decision-making.

Source: IIGCC (2024, p. 2). Copyright 2024 IIGCC. Text reproduced with permission.

However, creating conditions such as those set in Figure 6.4 will not be easy for countries with limited institutional capacity. The World Bank has carried out several valuable studies of the challenges, country by country, in combining development and climate action, called Country Climate and Development Reports (CCDRs). One of their primary purposes is to indicate ways in which the institutional capacity needs to be and can be strengthened to enable the implementation of climate action policies, such as those discussed in Table 5.3 in Chapter 5.

Confidence in national targets can be further bolstered when key actors within the country also make their own commitments. This is one advantage of the concept of net zero: if the collective emissions of all stakeholders are net zero and the amount of negative emissions is small, nearly everyone must be at or close to net zero. We have seen communities, cities, sectors, firms, and universities making net zero commitments. In this way, making NDCs as 'investable' as possible can garner commitments from other actors, which in turn further bolsters the original NDCs.

Structured targets on the way to 2050 play a vital role in shaping investor confidence in planning. For example, setting a date after which sales of internal combustion engine vehicles will be banned is a significant marker. Similarly, establishing a deadline for decarbonising the electricity system can guide investment strategies. Typically, the target for decarbonising the electricity system should be around 15 years before the overall net zero target, since so many other sectors depend on zero-carbon electricity for their decarbonisation (Energy Transitions Commission [ETC], 2024).

A further source of clarity can come from legal systems. In December 2024 the International Court of Justice (ICJ) began public hearings, at the request of the UN General Assembly (March 2023), on the obligations of countries under international law to ensure the protection of the climate system and environment from the emissions of GHGs. The clear and strong ruling on these obligations came in July 2025 and will have far-reaching consequences. Supreme Courts in countries like India, the UK, and the Netherlands have shown that the legal system can require consistency with the Paris targets or uphold rights to clean air. International examples are described in Chapter 9.

However, confidence in strategic direction can be shaken if commitments are questioned during crises. For example, the surge in fossil fuel prices following the start of the war in Ukraine led to some wavering, and the fossil fuel industry attempted to argue that medium-term energy security would be best served by further investment in fossil fuels. This argument is both unfounded and self-serving. For most countries, energy security is best achieved by moving away from fossil fuels. Whatever the source, any apparent wavering of commitment can undermine investor confidence.

Similarly, if a major country such as the USA is seen to retreat from the Paris Agreement, as President Donald Trump did in 2021 and again in 2025, it can create uncertainty. Interestingly, in 2021 other countries continued to hold their commitments, and many US states, cities, and firms insisted that 'we are still in'. Further, the momentum for investment in the new, generated by the decreasing cost of renewables, continued to build in the USA and globally even after Trump's withdrawal in 2021. However, in the first half of 2025, there have been significant retreats on commitments, particularly in the financial sector and in oil/gas companies. We have yet (Spring 2025) to see how all this will play through. It would be financially unwise to invest in coal-fired power stations based on the short-term policies of a single administration, especially when there is strong global momentum towards sustainability and where the lifespan of such an investment could extend for several decades. Nevertheless, in a world that urgently needs rapid and large-scale clean investment, such political shifts can be damaging.

Policies, confidence, and systemic change

While overarching targets are critical, so too is the way in which governments embed these commitments in their approach to policy for achieving systemic and structural change. Investors need to see commitments translated into tangible actions. These actions will partly be reflected in specific policies (discussed in Section 6.3), partly in their own investments, partly in the promotion of skills, and partly in the facilitation or provision of finance. The overall approach of governments must also be clear and credible, with a minimum of internal contradictions. At the heart of this approach will be strategies, policies, and actions related to key systems and to the development of technology and skills. Finance will be a critical factor, particularly in managing the risks which will be part of the transition. Here, we briefly discuss examples of sectoral policies and actions that can help foster the necessary confidence as well as sound incentives (see Section 6.3).

Those investing in the production of EVs, or those who are thinking of purchasing one, will need confidence in the regulations and legislation that promote EVs and phase out internal combustion engines. They will be concerned about the availability and functionality of charging facilities. All of this will require close public–private partnerships. For example, adding electric charging points to existing petrol stations, which are largely privately owned, could be mandated or subsidised, but that is a different proposition from installing charging facilities on street lighting,[12] which is usually managed by public authorities. In the same way, investors in urban areas will require as much clarity as possible on how city spatial planning and transport systems will function. This includes metro systems, inter-modal links, facilities for cyclists and pedestrians, zoning, industrial parks, and financial clusters. And, in rural areas, stakeholders in farming, construction, and industry will seek guidance on where different activities can take place and on sustainability requirements. They will need to understand how the supporting infrastructure will develop.

Electrification, with green electricity, is at the heart of the transition. The power system represents the largest investment need and is foundational to achieving the most significant emissions reductions elsewhere in the economy, including buildings, transport, and industry. Power systems, and energy systems more generally, should be flexible enough to accommodate change and innovation, and pay close attention to the flexibility of supply and demand across space and time. A strong and flexible power system should seek a sizeable expansion in national and international links (e.g., transmitting wind energy from Scotland to England or solar energy from Morocco). Grids that were originally designed to connect coal-fired power stations to consumers will need to be reconfigured to accommodate new sources of supply, including off-shore wind in the UK and large-scale solar in, for example, South Africa and India. The success of these initiatives depends on public–private partnerships, speedy spatial planning decisions, and active community participation, which must be purposeful and consultative.

In particular, investment in and use of the grid will work much better if, over time, markets for electricity and capacity are established. For example, in the case of the UK grid system, many small decentralised renewable power generators are facing waiting times of 5 to 10 years, or more, to get connected to the grid infrastructure (Octopus Energy Generation, 2023). Many countries use regional pricing systems, where localised markets can respond to local demand and supply. This can improve system efficiency by encouraging capacity to be built where it is most needed, while decreasing electricity bills (Octopus Energy, 2024). One of the problems in China is that whilst renewable capacity has expanded enormously, the functioning of the grid system, for political economy reasons of vested interests, is tilted in favour of the use of coal. Similar problems arise in Indonesia, where there is great renewable potential but the grid system is dominated by the interests of coal.

None of the above flexibility and openness to innovation can be achieved with a passive state or one that refuses to engage with investors and markets. All of this requires an active state. Tackling market failures and creating confidence in future directions are crucial. There will be more on the role of the state in Chapter 7.

It is important to clarify that a more active state absolutely does not mean that the government should take complete control. The design of public–private interactions will remain critical, and most of the investment and innovation will come from the private sector. Therefore, it is essential for the private sector and communities to be central to the creation of decarbonisation policies. This involvement will help to create confidence in the commitment to policy.

Encouraging and facilitating entrepreneurship should be a cornerstone of the new economy. The private sector will generally understand the obstacles and often understand better how to overcome them than civil servants. Further, entrepreneurship in system creation and design will be a key feature. This is not simply about government regulation. For example, the creativity of private sector firms, particularly in new approaches to electricity supply and generation, is extraordinary. In many cases, these firms have a deeper understanding than the government of what is involved in building new structures. They must be integral to 'system design teams'. 'Co-creation' can not only bring better policies, but also provide confidence in both their implementation and their stability.

Governments can also support the private innovation and entrepreneurship necessary to achieve systemic change through direct government procurement – which accounted for around 13% of GDP in Organisation for Economic Co-operation and Development (OECD) countries in 2021 (OECD, n.d.). Serving as 'the first customer' of maturing technologies, through procurement, public actors can pull 'the supply side of the innovation economy down the learning curve to low-cost, reliable production' (Janeway, 2021, p. 9).

Indeed, aligning procurement with broader innovation goals enables governments to encourage research and development (R&D), investment, competition, and technological adoption (Hlács et al., 2024). To support both firms that apply existing best practices and those that develop breakthrough innovation, procurement policies can adopt a two-tiered system which sets a feasible minimum environmental criterion for a portion of total procurement (eliminating negligent firms) and reserves another percentage for companies that comply with ambitious standards. Another alternative is rewarding companies that comply with ambitious criteria through a discount applied to the project price, increasing their competitiveness (Hasanbeigi and Bhadbhade, 2023). Again, the structure of policy can build confidence in policy.

Transforming public procurement for innovative and entrepreneurial ends requires a change in orientation – from a narrowly interpreted least-cost to a potential-based perspective. To achieve this, governments can adopt two-step procurement processes that include demonstration or an incubation programme. They can create engagement spaces to understand the concerns of potential suppliers and communicate to them specific innovation needs, so suppliers can identify innovative solutions. Or even create a public procurement agency to guide businesses in procurement processes and aid policy-makers in being effective innovation customers (C40 Knowledge Hub, 2024).

A key aspect of procurement which can provide both scale and confidence is a number of cities, regions, or countries acting together. Coalitions of cities have provided powerful leadership here. One example is the specification of new electric buses by a number of municipalities or cities, benefiting from economies of scale, improving technical capacities, and having a stronger voice when negotiating with suppliers (C40 Knowledge Hub, 2024).

To instil investor confidence in the clean transition, it is vital that government policy embodies both predictability and adaptability. Investors need policies that are clear, credible, and predictable over the long term. However, the transition is and will continue to be a dynamic process through which new technologies, lessons, and challenges emerge. For policy to be effective in driving momentum, and not hampering it, it must be *predictably flexible*. It should be designed with the expectation that it will evolve in response to learning, whilst maintaining a long-term direction that investors and the private sector can trust.

The investment climate and country platforms

Investors seek confidence in revenue streams, the ability to manage costs, practicalities of timely execution, well-functioning infrastructure, and the avoidance of undue 'interference'. This is what constitutes a favourable investment climate.[13] A stable and clear policy environment and reliable institutions, as previously emphasised, are key components. Clear, credible policy signals reduce perceived risk.

As Chief Economist of first the European Bank for Reconstruction and Development (EBRD) and then the World Bank, I worked to make improving the investment climate central to the strategy of those institutions. As part of that work, we established metrics for measuring improvements in the investment climate.[14] Those metrics themselves improved focus and performance. For example, the metrics identified the extent of monopolisation particularly associated with privatisation of utilities. That helped EBRD bankers to insist on regulation for openness in subsequent privatisations.[15]

A country platform[16] is a collection of institutional structures, practices, partnerships, and policies. It should provide a favourable investment climate, a clear sense of direction, and mechanisms through which various stakeholders can work together. The key stakeholders are government, the private sector, financial institutions, and communities. Such platforms can help integrate the planning of necessary systemic changes and coordinate transition plans across government departments, cities, and regions. These platforms, in principle, act as collaborative spaces where stakeholders can align investment strategies, share resources, solve problems, and coordinate the implementation of large-scale projects. In doing so, they can facilitate the channelling of finance into clean investment opportunities and mitigate some of the risks associated with large, transformative projects. Increasingly, MDBs are recognising the importance of prioritising these platforms in their work to support countries and foster private and public investment.

Just Energy Transition Partnerships (JETPs) constitute an example of a country platform approach where developed and developing economies collaborate to deliver climate finance for decarbonisation in an integrated manner. The idea of JETPs is to tackle several issues simultaneously in a 'package'. A package is necessary for two reasons. First, investment requires complementary units and facilities such as transport links or finance. Second, economy-wide change involves many parties and the effects on the welfare and interest of each party or group will be part of social objectives and will influence political acceptability. This approach to mobilising investment and climate finance is led by the country. It is within the country that cooperative transition processes can be constructed. This is where public and private investments are coordinated and complemented to achieve coherent and investable decarbonisation pathways and growth strategies of transitioning nations which they themselves define. At the same time, though these strategies are led by the transitioning country, outside investors and MDBs can play a role in helping to identify issues and draw on experience and analyses from elsewhere.

Such structures can help create mixtures of no-cost grants and low-cost concessional funds with commercially priced capital, resulting in 'blended financing' (grants, concessional loans, guarantees, commercial loans, etc., as discussed in Section 6.4) (Hauber, 2023). These JETP mechanisms have only recently started to be implemented; South Africa signed the first one in 2021. Other examples of JETPs include the cases of Indonesia and Vietnam. Each JETP confronts distinct planning and implementation challenges, with the funding packages agreed for each partnership varying. JETPs are just one route to country platforms. The key purpose here is the establishment of confidence in the investment climate, the sense of direction, and that problems can be solved.

6.3 Incentive structures for the new economy: tackling market failures

A successful journey towards a sustainable and inclusive economy requires a comprehensive understanding of the challenges and opportunities that lie ahead. Our discussion in this chapter began with an exploration of objectives, particularly sustainability, alongside perspectives on the nature of the problems being confronted and the systems that need to be transformed. As nations and the world confront the climate crisis and seize the opportunity to create a new growth and development story, investment and structural change are at centre stage. The critical question is how to generate and foster the investment and transformation necessary, given the scale and urgency of the task. We started to answer this question by discussing the overarching strategies and structures that shape investor intention and confidence. In this section, we examine more closely the details of policies and incentive structures, focusing on the market failures that must be tackled and the systemic and structural changes required. As emphasised in the previous section, policy must be *predictably flexible* to combine the need for long-term credibility and investor confidence with the capacity to adapt to new circumstances, information, and technologies as the transition unfolds.

To design the right incentives, we must start by identifying the relevant market failures. These were outlined in Table 5.3 in Chapter 5, where we argued that the inefficiencies resulting from these failures imply that we can advance climate action and promote growth simultaneously, in part through action to overcome the market failures. It is misleading to portray decision-making as inevitably involving a trade-off, or 'horse race', between the two. Understanding the origins and implications of these failures directly informs the creation of policies and measures to overcome or significantly reduce them. Here we will focus on the policy options available to correct these market failures, bearing in mind that they are interwoven and that some measures or policies can address multiple failures simultaneously.

For convenience, Table 6.1 repeats the information provided in Table 5.3 of Chapter 5. The information in this table is absolutely central to the substance and organisation of our discussion of policy. The table was introduced in Chapter 5 to demonstrate that there need not be a trade-off between climate and growth. In this chapter, the policies to tackle the market failures, in the third column, are examined more closely.

In this context, three points are crucial. First, as noted, policies to tackle the six market failures are interwoven and should be mutually supportive. Second, the ability to apply policies will be influenced by the political economy of vested interests and institutional capacity. Third, it may not be straightforward to proceed on all six fronts at the same time, for reasons of capacity or politics. Sequencing will be an important issue. In some cases, the choice of sequencing will be influenced by the necessary timing of investments or impacts, in others by the importance of building capacity. Sequencing could also be influenced by challenges in building coalitions.

Policies to tackle the GHG externality

The first and most significant failure listed in Table 6.1 is the GHG externality. In the Stern Review, we described this externality as 'the greatest market failure the world has ever seen' (Stern, 2007, p. xviii). Almost everyone contributes to emissions, almost everyone will suffer the consequences, and the

Table 6.1: Tackling market failures: policy options for action on climate and promotion of sustainable development

Market failure	Description	Policy options
GHGs	Negative externality because of the damage that emissions inflict on others.	Carbon tax/cap-and-trade/regulation of GHG emissions (standards). Do not subsidise the toxic.
Research, development, and deployment (R,D&D)	Ideas are public goods, and here their use (climate action) is also a public good. Without public action, the creators of these ideas do not capture the full value to society from their creation.	Supporting research, innovation, and dissemination. Tax breaks, support for demonstration/deployment, including via procurement, publicly funded research. Coherent standards to focus research and innovation.
Risk/capital markets	Imperfect and asymmetric information and assessment of risks. Problems of collateral.	Risk-sharing/reduction through financial structures, e.g., equity, guarantees. Long-term contracts, e.g., for off-take for power. Convening power of development banks; transparency (e.g., Task Force on Climate-related Financial Disclosures [TCFD], 2017).
Networks	There are problems of coordination between actors and agents within networks and across networks in key sectors. Examples in energy, transport, cities are important.	Investment in infrastructure to support integration of new technologies in electricity grids, public transport, broadband, recycling. Management of grids. Planning of cities.
Information	Lack of awareness of technologies, actions or support, or product content.	Promoting awareness of options. Labelling and information requirements on cars, domestic appliances, and products more generally.
Co-benefits	Benefits beyond market rewards are often ignored or downplayed.	Recognising and acting on impacts on health. Valuing ecosystems and biodiversity.

Note: The first column describes the area within which the market failure occurs. The second column briefly describes the nature of the problem. The existence and nature of these market failures was an important element in the discussion of the last chapter (see Table 5.3). The focus here is on policy options.
Source: Author's elaboration.

potential impacts are vast and catastrophic. Tackling this externality can be through the pricing of emissions (including market mechanisms and taxes) and through regulating or setting standards for emissions. Each policy has its strengths and weaknesses, and they can be effectively combined.

Economists, following Alfred Marshall,[17] often start with the price mechanism when they begin to think about policy in relation to market failure. Intuitively, and in the language of Marshall or Arthur Pigou, if consumers or producers face a price for an activity that is below its social marginal cost, they will continue that activity up to a point where there is net social damage from the marginal units.[18] This happens because they pursue the activity as long as the private benefit from the marginal unit exceeds the price they pay; that price reflects their private cost, which in turn reflects the cost of supply of the good. Thus, on the margin, there is zero net benefit to them from an extra unit, but there is damage to others. In simple terms, social welfare would increase if they, on the margin, decreased the activity.[19]

This line of reasoning has led some economists and policy-makers to argue that economics dictates that putting a price on the GHG externality is unambiguously the most efficient or optimum policy for tackling climate change. For example, a 2019 letter to the *Wall Street Journal*, signed by 3,649 senior and distinguished economists, stated: 'A carbon tax offers the most cost-effective lever to reduce carbon emissions at the scale and speed that is necessary' (*Wall Street Journal*, 2019). However, such statements are misleading for two important reasons. First, they overlook the many other externalities, particularly the other five in Table 6.1. Relying on carbon pricing alone cannot be claimed as optimal if other failures remain unaddressed. Second, they ignore the effects of increasing returns to scale and of uncertainty. When these factors are present, relying solely on prices could risk leading to excessive pursuit of the activities, especially if there are sharply rising damages. Martin Weitzman (1974) explored these issues in his seminal study of 'prices vs quantities'. Consider the example of a severe road accident with lives at risk. Would you hold an auction or set prices to see which supplier could respond most quickly and cheaply, or would you require emergency services, who are on standby, to respond immediately?

In such cases, where the consequences of error are high, standards and regulations can be more effective. They can also lower costs for those supplying new clean products, such as EVs, by providing clarity on when to build and how much to plan for in a context where substantial fixed costs and investments are involved. Thus, the simplistic statement that 'economics says that prices are clearly the most efficient way to tackle climate change' is simply wrong.[20]

Nevertheless, the arguments in favour of using the price mechanism are of great importance and do offer valuable guidance. When individuals face the social cost of their actions (at the margin), they are incentivised to take account of the damage they inflict on others – this is the Marshall–Pigou argument. They are well placed to judge how costly it will be to them to adjust their behaviour on the margin. The analytical and policy error lies in insisting that carbon prices are sufficient, in the sense that they can deliver efficiency by themselves, or that they can fulfil the entire policy mandate. Despite these criticisms of commonly offered arguments, it is important to emphasise that prices should indeed play a significant role. I want to be very clear on this. Carbon pricing should play an absolutely central role in policy. But we will need much more in the way of policy and actions in a risky world, where other important market failures are present, and so much of activity involves increasing returns to scale. The basic theorems of welfare economics (usually studied early on in undergraduate economics), which highlight the efficiency of outcomes in a market economy, assume the presence of a full set of markets for all goods and services, that all markets are perfectly competitive, and that there are no increasing returns to scale.[21] That is not the world we live in and where we have to make policy.

One clear policy lesson, with very broad application, from emphasising the importance of prices in this context is that countries should stop subsidising harmful activities. The World Bank (Damania et al., 2023), IMF (Black et al., 2023), and OECD (2024) have conducted valuable research on calculating the magnitude of existing subsidies for damaging activities, particularly around fossil fuels. Explicit global subsidies for fossil fuels exceed a trillion US dollars per year (around US$1.3 trillion) (Black et al., 2023). When implicit subsidies (such as not paying for damages or 'bads') are included, the figure increases more than five-fold. In total, according to the IMF, in 2022 fossil fuel subsidies (explicit and implicit) amounted to around US$7 trillion, or more than 7% of global GDP. Of the total, 18% is attributed to explicit subsidies and 82% to implicit subsidies (Black et al., 2023). The World Bank's *Detox Development* (Damania et al., 2023) extends this approach to agricultural subsidies, which are often structured in ways that encourage damaging, wasteful, and polluting activities. Abolishing these harmful or toxic subsidies not only reduces environmental damages but also frees up valuable revenue for investment in new, clean activities and for supporting people and places that face challenges during the transition away from dirty activities.

The price mechanism can work through official compulsory carbon markets or via carbon taxes. Carbon taxes are generally easier to administer and can be implemented or approximated without the need for detailed emissions measurement on a firm-by-firm basis, for example by taxing coal, oil, or natural gas. Official carbon markets, on the other hand, focus directly on achieving fixed amounts of reduction, the intended goal. And they can be structured by initially allocating carbon allowances, so that net payments are small at the outset; this approach can reduce opposition when such markets are first introduced.

In an economy with all market failures tackled and perfect markets, the optimum carbon tax would be equal both to the marginal damage from carbon and to the marginal cost of reducing carbon. But those assumptions never apply. We live in an economy with all sorts of imperfections. Thus that equality of marginal damage and marginal cost at the optimal level of activities *does not apply* when other market failures are not fully tackled.[22] This is a fundamental insight of public economics – at least since Meade (1952) – and a defining characteristic of that subject.[23]

Thus, Joseph Stiglitz and I argue that prices to guide choices of government projects and carbon-pricing policies should be based on the marginal costs of action to reduce carbon along the carbon reduction path chosen rather than marginal damages. This is because the world and nations have articulated their policy in terms of a carbon reduction path that manages risk. Prices based on marginal costs are designed to guide us towards that path. Prices based on marginal damages are not guaranteed to do that in an imperfect economy. Our suggestion does indeed lean on the price mechanism, but does so in a way that reflects modern public economy and its analysis of policy in an imperfect world (Stiglitz and Stern, 2017).

Prices can also improve outcomes by operating through voluntary carbon markets (VCMs). For example, in regions or sectors without an official carbon market, firms aiming to achieve self-imposed targets can contribute to emissions reductions by paying entities elsewhere, such as in developing countries, to cut emissions more than they otherwise would have been ready to do. While there are challenges in defining the counterfactual 'otherwise would have been ready to do', such mechanisms can both reduce emissions in ways which take account of relative costs of action and, importantly, generate revenue for poorer countries or regions to finance the necessary investments for a new growth path, alongside actual emissions reductions. If these transfers contribute to a significant and specific investment programme, the difficulties in specifying counterfactuals are less severe. For instance, if a country has a comprehensive plan to transform its entire power sector, verifying whether that plan has delivered overall emissions reductions is more straightforward than doing so on a project-by-project basis. On a project-by-project basis we may not be able to feel confident that a carbon reduction in project A is not replaced by an increase in project B. For example, if deforestation is prevented in place A it might move to place B. We will revisit issues of climate finance briefly in Section 6.4 and in more detail in Chapter 9.

Tackling the other key market failures

The discussion of the five other market failures in Table 6.1 will be more concise, but all five are of real importance. The second market failure is that the market, left to its own devices, generates too little R&D. Much of the action in tackling climate change requires the discovery of new zero-carbon methods. This process could involve searching for new materials, methods for green hydrogen and steel, sustainable marine and air fuels, innovative fertiliser production, different farming techniques, new electricity storage methods, or innovative ways of managing urban systems. It encompasses both R&D and innovation. A discovery is, in large part, an idea. Unlike physical goods, ideas can be used by multiple people without diminishing their value to others. This characteristic implies that ideas are public goods. Without public action, the creators of these ideas do not capture the full value to society from their creation. Therefore, there are compelling reasons for the state to subsidise and support R&D. Many such activities are indeed subsidised or supported by the state, including at universities and research institutes.

The argument for public support of ideas as public goods is not unique to climate change, but it is particularly relevant in this context for two reasons. First, the use of a public good which is an idea to enable the reduction of CO_2 emissions benefits everyone when it is actually used to reduce emissions – one might say it is a 'public good squared'. Second, the urgency of the climate crisis creates a strong motivation to accelerate innovation; the faster the idea is developed and used, the stronger the benefit to all. Public policy should act on the pace of delivery, not just delivery per se.

It is important to recognise that those who generate ideas often respond powerfully to social priorities, and not simply to market incentives. During the Covid-19 crisis, for example, medical researchers moved very rapidly. While potential financial rewards played a role, many were motivated by a sense of social duty and a desire to contribute to global well-being in a shared crisis. Similarly, we are witnessing a growing sense of obligation among scientists, engineers, architects, and indeed some economists to develop cleaner and safer ways of consuming, producing, and living.

The third market failure in Table 6.1 relates to capital markets. As we have seen, effective climate action requires a major step up in investment, much of which will be risky and long term. Capital markets have well-known limitations in managing risk, particularly over the long term. In a perfect market, by definition, agents act as if they can buy or sell as much as they like at the given price they see in the market. However, this is not realistically possible for all, or indeed most, transactions in capital markets. Lenders cannot lend trillions on fixed terms without taking on considerable risk. They worry about the borrower's ability to repay, about how repayments might be enforced or guaranteed, collateral, and more. In a risky world like ours, a key function of financial markets is to allocate and reduce risk. Their ability to do so is substantial and important but it is limited. Here, governments, public development banks, and MDBs can play a crucial role in overcoming some of the limitations and in managing and reducing risk. These capital market interventions are important elements of the public action needed to support the scale of investment required. Further discussion of capital markets is in Chapter 9.

The fourth area of market failure in Table 6.1 concerns networks. Much of the necessary action for climate mitigation and adaptation takes place through networks, including electricity grids, internet connections, transport systems, city infrastructure, reusing/recycling, and circular economies. In a network, the actions of one player affect how the system functions for others. For example, if I drive my vehicle on a road, I might increase congestion for others.

Conversely, if I acquire a telephone, the value of someone else's telephone increases because they can now call me.

Governments will have to provide frameworks, including legislation, for the effective functioning of networks, for example access rules for grids or telephone networks. In some cases, such as city metros, the government may own key elements of the network. As with R&D, these arguments apply to many areas beyond climate change, but they are particularly important in this context, given the centrality of networks to so much of necessary action in the transition.

Fifth, there is the issue of information. As a consumer wishing to reduce my carbon footprint, I need to know the carbon content of the products I buy. If I want to insulate my home, I need to discover what options are available. Similarly, producers and firms face numerous decisions that require access to information around what may be new activities for them. When there is asymmetric information between buyers and sellers, markets do not function efficiently.[24] In many contexts, governments can and should take action to mandate the disclosure of information. Examples include the sugar and salt content of food, whether or not a carpet was made with child labour, ways in which machinery might be dangerous, carbon content, and so on. Again, while these issues extend beyond climate action, they are of special importance in the transition, when many agents will be taking actions which are different from the familiar past.

In this context of changing producer and consumer behaviour where there is limited information or lack of familiarity with new goods or services, ideas from behavioural psychology can become important. Examples directly oriented towards changing behaviour include labelling in hotels or supermarkets or default options in pension plans (see Chapter 7).

Sixth, Table 6.1 identifies so-called 'co-benefits', a term that somewhat understates the great advantages of climate action beyond the fundamental purpose of reducing climate risk. As shown in Chapter 5, the death toll from air pollution is staggering, with most of it stemming from the burning of fossil fuels. Climate action that reduces fossil-fuel combustion can, therefore, have substantial and immediate health benefits. This perspective on co-benefits provides a powerful additional argument for robust climate action, including using both price mechanisms and standards/regulation to limit fossil fuel use.

The analysis of market failures in this section provides a strong foundation for detailed policy development. In Section 6.2, we discussed how overall strategy and policy can help guide systemic and structural change. By combining these strategic approaches with the policy issues just examined, we arrive at what might be termed the 'gardener's craft' or the 'head chef's task' in strategic policy-making. The head gardener or head chef is charged with putting all the elements, ingredients, and actions together. For climate action, which must be economy wide, this coordination should occur at the highest levels of government, typically within the offices of finance ministers or prime ministers/presidents.

6.4 Financial structures for the new economy

The first part of our discussion on the specifics of the challenges of delivering sustainable growth (Section 6.2) focused on creating the conditions necessary to catalyse the investment that will drive this new form of growth. Central to the delivery of investment is the financing of investment. Potential investors will not pursue investment opportunities if they do not perceive financing options that make the investment profitable relative to the risks involved. The cost of capital and effective risk management are therefore crucial to making investment both viable and attractive. In previous sections, we have high-lighted the importance of managing risk in relation to the costs and returns of investment, particularly through the investment climate and the credibility of public policy. In this section, we turn our attention to the financing itself, with a particular focus on risk management. While the issues discussed are rele-vant to all countries, the challenges are particularly acute in emerging markets and developing countries (EMDCs), where international support and action will be crucial. And these are the countries where the new instruments are of special importance, nationally and globally. These challenges are explored in more quantitative detail in Chapter 9.

Risk and the cost of capital

Investment inherently involves risk. Financial structures are typically designed to manage, reduce, and share this risk. Finance is most effective when it allo-cates risk to those who are best equipped to bear it, and to those who can actively manage and mitigate risk through their own actions.

Different financial instruments embody varying levels and types of risk, and they serve different functions. These instruments include, among others, loans of different durations and flexibilities, bonds with different conditions, equity, guarantees, and first-loss agreements. Moreover, these instruments can be blended to create financial packages that allocate and balance risk and return in ways that fit best with the different abilities and 'risk appetites' of different players. Investors who take on greater risk typically expect higher returns.

When considering the terms and conditions of financing, potential finan-ciers will evaluate different sources of risk. At the project level, these risks might include the costs and challenges associated with land acquisition, con-struction, the functioning of local infrastructure, revenue streams, and the reliability of counterparties and collaborators. Such risks are influenced by the stability and reliability of governments, input suppliers, customers, and other stakeholders, such as electricity purchasers. Sectoral issues, such as the operation of grid structures and anticipated future demands and supplies, may also play a role. Additionally, there are regional or city-specific risks as well as macroeconomic factors, including country risk, related to the liquid-ity and solvency of governments. All of these will affect the availability and cost of finance. Regulatory issues, which could influence the types of instruments and risks that financial institutions can undertake, are also relevant. The examples in this paragraph make it clear that government-induced policy risk will be a critical element in shaping investor confidence, or the lack of it.

The cost of capital and the associated risks are crucial for all investments, but they are especially significant for green or sustainable investments, for three reasons. First, investments that avoid using fossil fuels require substantial upfront capital, which reduces future operating costs by eliminating the need for fuel purchases. Second, these investments are often innovative or novel, meaning there may be less experience with similar projects and less data, leading to a perception of higher risk. Third, such investments are frequently long term, particularly in the case of infrastructure, which means there is a longer period during which potential problems could arise around revenues, costs, and the ability to function effectively.

For an example of how important the cost of capital is, consider how it affects the cost of solar power compared to that of fossil-fuel-based electricity generation. In much of Africa, for example, solar power can be cheaper than fossil fuels (even without subsidies or carbon pricing) if the cost of capital is around 7% or so (real rates of interest). However, if the cost of capital exceeds 20% – as is often the case, particularly due to perceived country risk – solar power becomes less competitive. According to the International Energy Agency (IEA) (2024), utility-scale solar photovoltaic (PV) projects in EMDCs face more than twice the cost of capital compared to advanced economies. The Energy Transitions Commission (ETC) (2023) suggests that the higher capital costs in middle- and low-income nations make additional finance and innovative strategies by MDBs necessary, as well as national financial policies and actions. That is also the conclusion of the reports of the Independent High Level Expert Group on Climate Finance (IHLEG) (Bhattacharya et al, 2023, 2024; Songwe et al, 2022) and the G20 Independent Expert Group on MDBs (IEG) (Singh and Summers, 2023a, 2023b), both of which I have been actively involved in. Overall actions to reduce the cost of capital through risk management are critical.

Some of the risk is perceived rather than actual, particularly in EMDCs, discouraging investments relative to developed nations (Blended Finance Taskforce, 2023; International Finance Corporation [IFC], 2024). Factors such as local market familiarity and information asymmetries contribute to this gap (Climate Policy Initiative [CPI], 2022). The presence of partners such as MDBs can reduce that perception. And, indeed, they can also reduce government-induced policy risk, since a government may be less likely to 'interfere' if an MDB is involved. Better data, or greater familiarity with existing data, can reduce risk. But many risks are real, including foreign exchange risk and other aspects of country risk. There is work to do in reducing those risks (see Chapter 9).

Blended finance

The first and logically prior task in private–public collaboration around fostering the necessary investment is on the investment climate – that is, creating conditions where investments can be carried through and the risks around execution, cost control, tax liabilities, possible predation, and revenue flow can be managed. That was the subject of Section 6.2. But managing risk through financing structures is also of great importance.

Different financial actors can manage and take on risk in various ways, and different parts of climate-related investment and associated finance involve distinct risk patterns. The blending of different sources of finance will therefore be central to creating the necessary scale and nature of investment. Broadly speaking, there are four main sources of finance: government, private, development banks and finance institutions, and highly concessional or grant funding. Each of the four sources includes various strands. Government finance can come from different levels and departments, and it may be offered on a range of terms. Private finance might come from retained earnings, banks, capital markets (including venture capitalists, private equity, pension funds, and asset managers), and other sources. It can be national or international. Development banks and finance institutions can be national (such as BNDES in Brazil or the UK Infrastructure Bank, now the National Wealth Fund), regional (such as the European Investment Bank), or global (such as the World Bank). The final category, which is particularly important for EMDCs, includes overseas aid, voluntary carbon markets, earmarked revenues from official carbon markets, initiatives such as Special Drawing Rights at the IMF, dedicated international taxes (e.g., as under discussion on maritime or air transport), and philanthropy.

In terms of Figure 6.5, it must be noted, first, that public and philanthropic sources often think of themselves as mobilising private resources, but the causation can go both ways. Second, funding from development banks is usually classified as concessional, but the term can be misleading. Many MDBs borrow from markets at close to interbank interest rates and lend at rates a little higher. But generally they do not make losses. In that sense, whilst the rate of interest to the borrower is lower than elsewhere on the market, it is not really 'subsidised' or 'concessional'. The two types of lenders (MDB and private) face different risks.

These different sources can offer finance on very different terms, and the necessary components in any financial blend will depend on the specific projects or programmes involved. For example, a renewable electricity generation project in a well-developed power market might rely almost entirely on private sector financing. In contrast, major investments in reconfiguring electricity grids in many countries might require substantial public sector finance (from government or development banks), given the involvement of multiple independent players and the risks associated with public and community decisions or the credibility or reliability of the electricity companies in the system. Financing for adaptation finance or natural capital projects, which may struggle to generate sufficient

Figure 6.5: Blended finance

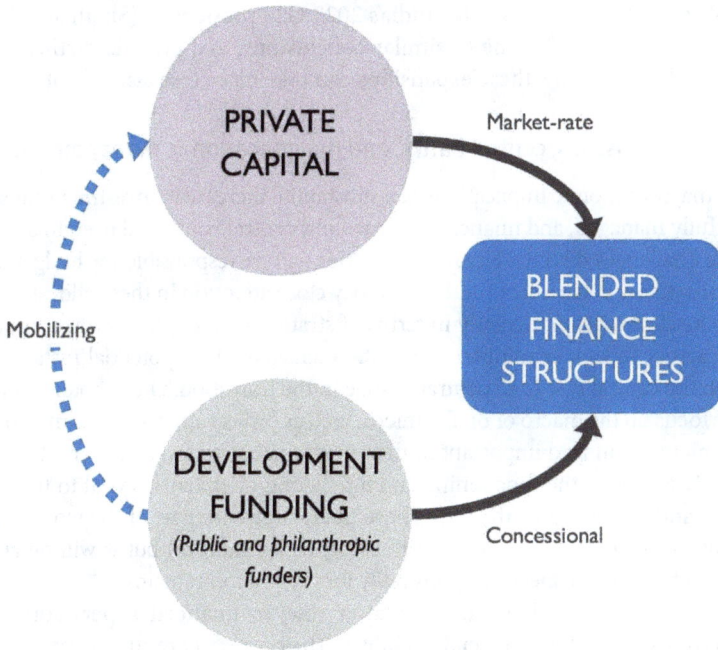

Source: Convergence Blended Finance (n.d.). Copyright Convergence. Reproduced with permission.

revenue streams to service debt, might require significant amounts of low- or zero-cost capital or government grants. Many projects will require blending these financial sources in varying proportions, depending on their characteristics. An important example is the Tropical Forests Forever Facility, where various financial actors, public and private, are collaborating to invest in forests (see Section 10.4). There are also initiatives that seek to attract private finance flows for biodiversity conservation directly, such as biodiversity credits, a growing area (NatureFinance and World Bioeconomy Forum, 2023).

The Blended Finance Taskforce (launched in 2017) notes that MDBs 'mobilise' on average less than 30 cents of private capital for every public dollar invested in climate change in the sense of the amount of private finance that accompanies their dollar. The taskforce argues that it is possible to do much better. Using blended finance as a structuring approach, public funds can unlock private investment by mitigating investment risks through currency hedging, first-loss structures, guarantees, and technical assistance for preparing projects (Blended Finance Taskforce, 2023). Through blended finance, different organisations can invest together so that each one forwards its differentiated aims, such as financial return, environmental impact, or both (Convergence Blended Finance, 2024). The reports of the IHLEG for COPs 27, 28, and 29 (Songwe et al., 2022; Bhattacharya et al., 2023; Bhattacharya et al., 2024) and those of the IEG for India's 2023 G20 presidency (Singh and Summers, 2023a, 2023b) come to similar conclusions, and provide further ideas and evidence on how these expansions can take place (see also Chapter 9).

Finance ministries, central banks, and macroeconomic management

The macroeconomic implications of a substantial increase in investment must be carefully managed, and finance ministries and central banks will need to address these challenges directly. Finance ministries will be responsible for budget allocations across departments and should pay close attention in these allocations to structural change, particularly in terms of strategies for clean growth, coordination across sectors or ministries, and the management of potential instabilities, given the central role of structural change in the transition. Often finance ministries focus on the macro or on the micro, such as project approval. The macro and the micro are indeed important in the transition, but so too is the structural. As already explored, the functioning and interacting of systems is vital to the transition and the new growth story. Those interactions are part of the work of line ministries (e.g., transport, energy, housing, or agriculture) but it will be at the aggregate level that special responsibility for integration remains.

Central banks will inevitably play a role as financial supervisors with a responsibility for financial stability. They must carefully consider the implications of their actions on the allocation of finance during periods of rapid change, in the face of macroeconomic pressures, and around potential instability. Mark Carney, then the Governor of the Bank of England, highlighted these issues in his pioneering speech on 'the tragedy of the horizon' in September 2015, just three months before COP21 in Paris.[25] He outlined compelling reasons why central banks should pay close attention to how financial markets operate concerning climate finance. These reasons include financial and price stability. The involvement of central banks in these issues is discussed further in the next chapter on the role of the state.

In my view, any major public entity, including central banks, with a special responsibility for price stability and some responsibility for economic growth and financial and economic stability should play its part, often proactive, in managing a major transition which has such profound implications for both economic stability and future well-being. Current actions, particularly on investment, will have significant short-term and long-term consequences. Notwithstanding some resistance from 'institutional fundamentalists' – a term also used in Chapter 7 – who advocate a narrow focus on price stability, the Network on Greening the Financial System (NGFS), a network of central banks, was established in 2017. As of Spring 2025 it had 145 members and 23 observers. This indicates that many central banks are indeed prepared to address responsibilities concerning climate and sustainability, even though some critics continue to grumble in the background.

Finance ministries will need to tackle the fiscal and macroeconomic management challenges associated with the substantial increase in investment required. The 'fiscal rules' established by these ministries should carefully consider these investment needs. It is likely that (as a global average) about two-thirds of the necessary investment will come from the private sector, with a higher proportion in wealthier countries and a lower proportion in poorer ones. This investment will impact aggregate demand. Many countries have been 'demand-constrained' – in Keynesian terms – over the last one or two decades, particularly in the period from the global financial crisis of 2008–2009 up to the Covid-19 pandemic in 2020. These countries may be able to accommodate such an increase in demand through supply response and economic expansion. However, this will not be true for all countries; in some cases, fiscal management will be needed to accommodate the additional investment to avoid excessive pressure on domestic demand. Domestic and foreign savings will both contribute to the financing of this investment. Central banks will also play a role in managing demand.

Government investment will be critical in many aspects of this investment expansion. This public investment will generally yield substantial economic returns, as argued in Chapter 5, and will help to enable and leverage private investment. In my view, fiscal rules should be designed to facilitate and accommodate this investment expansion, particularly during this critical period in world history. Borrowing for the medium or long term, with long-term finance at moderate costs – available to many governments – for investment with strong returns, is *fiscally responsible*. Over time, the returns on these investments will reduce debt-to-GDP ratios, even if there is a short-term increase (see Zenghelis et al., 2024). Not to do so would be fiscally irresponsible as well as, in this case, profoundly irresponsible in relation to the well-being of future generations. In some cases, the domestic savings rate will have to rise. Foreign borrowing can have its role to play, depending on cost and availability. And in some cases, output expansion can be a key element in generating savings. In 2024 the incoming Labour government in the UK reoriented its fiscal rules in this direction (see budget of October 2024).

Some government borrowing can be undertaken via green bonds, which signal to markets the direction of investments. The issuance of green bonds has grown rapidly worldwide, from a total stock of US$2.6 billion in 2012 to nearly US$600 billion in annual issuance by 2023 (Bloomberg, 2024; European Commission, 2016). Whether or not green bonds are used, the key principle here is that facilitating and accommodating the required investment – both public and private – demands the attention and action of finance ministries and central banks.

Regulation and asset classes

For financial systems to function effectively and contribute to the efficient and productive allocation of resources, particularly for investment and for economic and financial stability, regulation is essential. Without regulation, the transparency which is necessary for sound economic and financial decision-making would be lacking, there could be a higher risk of fraud, and investments would be discouraged. Moreover, without regulation, financial instability becomes a real danger. Many would argue that the global crisis of 2008–2009 was exacerbated by excessive enthusiasm for financial deregulation in the 1990s, particularly in the USA and the UK (Crotty, 2009; Krugman, 2009; Shiller, 2008; Stiglitz, 2010).

However, financial regulation should not focus solely on learning from past mistakes. Instead, it should also anticipate future risks and economic structures, particularly in a period where rapid change is critical for a sustainable future. If regulation is overly backward-looking, focusing only on past crises, it could inadvertently increase the risks and instability of the future. It is, therefore, time to review financial regulation from the perspective of its potential impact on the expansion of green investment. Regulation should, of course, seek to discourage recklessness or destabilising investment, but it should not arbitrarily impede investments that are essential for creating a more stable and sustainable future.

This is not the place for an exhaustive analysis of these issues, which can be technically quite complex – detailed discussion is generally confined to 'policy wonks'. But the idea is important to understand. Two examples suffice to illustrate the point that details of regulation and supervision really do matter.

First, financial institutions holding debt from very large companies are often viewed by regulators as carrying less risk than those holding debt from smaller companies. Due to past history, fossil-fuel-based energy companies are disproportionately represented among these large companies. As a result, there is a regulatory bias in favour of fossil fuel companies. Over the next two decades, these companies must shift decisively away from fossil fuels. If they do not, they will not only cause significant harm but also become increasingly risky investments, as they will be betting on the declining technologies of the past rather than embracing those of the future. Some of these companies may attempt to delay international climate action, but, in doing so, they will make the world a more dangerous and unstable place for everyone.

The second example concerns the treatment of finance for long-term infrastructure relative to shorter-term financial assets. Many of the new green investments that are urgently needed involve long-term infrastructure in energy, transport, cities, and so on. Regulators often view such assets as illiquid and, therefore, risky. However, if a financial investor, such as a pension fund, needs to provide pension payments several decades into the future, investing in long-term assets that are central to the future growth story may be a wise decision, both in terms of long-term returns and matching assets to liabilities.

The Basel rules, as currently structured, could hinder the shifts in investment structures that are necessary for future growth and stability.[26] Currently, climate and environmental (C&E) financial risks have been unevenly integrated into the pillars of the Basel Framework; policy-makers left the task of making sure that these risks are correctly priced to market participants and private banks (Dikau et al., 2024). There has been limited regulatory action to incorporate C&E risks into Pillar I (regulatory capital requirements). Supervisors and regulators are taking the first steps to incorporate C&E risks in their supervisory procedures (Basel Pillar II) and the Basel Committee is still working on climate financial risk disclosure requirements (Basel Pillar III) (Dikau et al., 2024; Kammourieh and Songwe, 2024). Continued efforts to address and incorporate climate risks are needed. Moreover, the 2023 Basel III finalisation proposal, put forward by US banking regulators and currently under discussion, might slow down the energy transition by requiring banks in the USA to hold significantly more capital against projects critical to the transition. This regulation would make tax equity[27] (an important financial mechanism in the funding of some renewable energy projects) less profitable, leading to fewer renewable energy investments by major banks that have historically funded renewable energy projects this way (Smith et al., 2024).[28]

On the contrary, in the pursuit of financial stability, central banks should take climate change explicitly into account. The climate crisis will generate new risk events, such as 'green swans' – unprecedented or unanticipated events associated with climate change (Bolton et al., 2020). Both physical and transitional climate-associated risks can, individually or in combination, have a dramatic impact on the financial and economic activities of businesses, households, and governments. As physical risks worsen, losses that are not insured can undermine solvency, while the scale and frequency of insured losses might put insurers in a financially difficult situation. Transition risks also arise as economies decarbonise, for example, due to shifts in market demand, reputational impacts, and the emergence of stranded assets (Bolton et al., 2020).

Bolton et al. (2020) suggest that there is a need to integrate climate risks into central banks' financial stability mandates and offer two risk management 'epistemological breaks' to be incorporated. First, since climate-associated risks are highly uncertain and non-linear, they do not tend to follow a normal distribution and, second, the probability of their occurrence cannot be extrapolated from historical data. Therefore, central banks should incorporate forward-looking risk assessment methodologies grounded in climate scenario analysis while also embracing new tools such as non-equilibrium models. The incorporation of these ideas is already underway: central banks are now beginning to redefine risk in the context of climate change. For example, the Network for Greening the Financial System (NGFS) consists of central banks and supervisors that are voluntarily contributing to the development of environment and climate risk management practices in the financial sector. Incorporating climate risks is critical not only to ensure that the financial system is resilient to them, but also to increase financial investments in green assets. Climate risks could help frame a carbon shadow price, increasing the cost of polluting assets and contributing to tackling the 'tragedy of the horizon' (Bolton et al., 2020; Carney, 2015).[29]

Given the limits of risk-based approaches and deep climate uncertainties, central banks 'may inevitably be led into uncharted waters' (Bolton et al., 2020, p. 1). Therefore, it is central that they do not restrict their action solely to creating scenarios and managing risk. To fulfil their mandates of price and financial stability over the long term, central banks should pursue proactive actions to address climate risks. If central banks just wait for other government agencies to act, they would not be delivering their mandate for financial stability. Proactive policies could include better coordination of fiscal, monetary, and prudential policies; more international cooperation on environmental issues among monetary and financial authorities; and integrating sustainability criteria more systematically within the financial sector (Bolton et al., 2020).

Further, and importantly, many central banks have output and growth as part of their mandate (Dikau and Volz, 2021). For this reason, given that climate action is the new growth story, they should pay close attention in their analyses, scenarios, and modelling to the growth effects of policies on the financial regulation and monetary issues which foster that growth.

6.5 Distribution and a just transition

Economies undergo continuous change and transformation. Sometimes transformations are driven by long-term trends such as demographic changes or shifts in economic structure towards services. Medium-term changes, like the rapid expansion of China's exports between 1990 and 2010, can also substantially shift the international division of labour. Additionally, technology can transform jobs, such as in the case of automated supermarket checkouts. And major events like the global financial crisis of 2008–2009, Covid-19, and the war in Ukraine can suddenly disrupt lives and livelihoods.

The transition to a zero-carbon economy will also bring dislocations that are part of these broader economic and social changes and that interact with them. But what is distinctive about the climate-driven shifts is the combination of all of, scale and urgency, medium and long-term processes, and economy-wide impacts. The rapid and deliberate nature of the green transition necessitates special focus. It is a set of changes consciously driven by societal decisions about how to address collective dangers and achieve collective goals. Therefore, there is a collective responsibility to manage this transition in an equitable and just manner.

While the mechanisms and policies required to manage the dislocation arising from the green transition can also apply to disruptions from other sources, there are distinctive features of climate action. It is economy-wide, both in mitigation and adaptation. And it is crucial to recognise that the poorest people are often the most vulnerable to such dislocations. This is particularly relevant when considering adaptation and resilience to the growing impacts of climate change, as well as the job market disruptions, price changes, and costs associated with the green transition.

In this section, we highlight two main aspects: first, the key concepts and insights that should underpin policy and, second, the range of policies needed to manage dislocations while enabling access to new opportunities and jobs. Since the decision to change is a collective one, and the effects of the changes impinge on so many, it is important that individuals and communities participate actively in the decision-making processes. Moreover, it is crucial to try to anticipate potential problems and develop actions for an equitable and just transition from the outset. Whilst the green transition is new, there is much that can be learned about potential problems by looking at past economic dislocations or industrial revolutions (see Section 6.1 for examples).

Before we embark on some of the details of policies for a just transition, we should emphasise, as noted, that dislocation and distribution issues should be central aspects of policy. They are always with us. If climate action brings them to the centre stage in a more effective way than before, then that is indeed a benefit of climate action. That is similar to climate action having the benefit of making us more attentive to existing market failures around R&D, capital markets, networks, and so on.

Justice, equity, and social capital

Effective policy-making relies on an understanding of objectives and of associated concepts together with an understanding of the processes at work, as emphasised in Section 6.1. In discussing the distributional aspects of the transition, we focus here on justice, equity, and social capital. These concepts should be seen as both ends and means in the policy-making process. For example, justice is an objective, but a perceived lack of justice in the transition could lead to opposition, potentially derailing the entire process; thus, justice is also a means.

Social and cultural capital (social capital for short) includes trust, respect, cohesion, and cooperation within communities. It is crucial for sustainability. If social capital erodes, opportunities can be undermined, as communities with lower levels of trust tend to function less effectively. Social cohesion is therefore both an instrumental means to achieve a just transition and an end in itself.

Amartya Sen's approach to the concept of justice, as outlined in his book *The Idea of Justice* (2009), is particularly relevant here. Sen argues that it is often easier to define injustice than justice, framing injustice as the denial of a right or entitlement. That shifts the analytical focus to rights and entitlements. In the context of the green transition, the core right would be 'the right to development', as discussed in Chapter 1. Internationally the COP processes have incorporated the notion of justice (see, for example, the preamble to the Paris Agreement). The *Sharm el-Sheikh Guidebook for Just Financing* was developed for COP27 under the Egyptian presidency.[30] It defines just financing as 'financing that accounts for historical responsibility for climate change while ensuring equitable access to quality and quantity climate financing that supports resilient development pathways, leaving no one behind' (Ministry of International Cooperation, Arab Republic of Egypt, 2022, p. 64).

Inequities, in this context, refer to disparities between people and their opportunities that lack ethical foundation. Given our emphasis on sustainability – defined as offering future generations opportunities that are at least as good as those available to the current generation – our analytical focus would be on actions and processes that generate opportunities for all, and on identifying and avoiding those that generate arbitrary deprivation of opportunity. Social capital plays a vital role in fostering a sense of community and mutual responsibility in ways that can promote opportunity and offer some social protection.

Injustices and inequities in the transition process

With these understandings, it becomes clear that poorly managed dislocations or disruptions to livelihoods during the transition could be seen as unjust or inequitable. Such perceptions could erode social capital and undermine the sense of community. Coal miners who lose their jobs may see this loss as a violation of their right to development, as their primary means of earning a living is taken away. A transition could also be seen as inequitable if it results in losses of opportunity stemming from earlier career choices made before the relevant environmental issues were well understood. A just and equitable transition therefore involves supporting or protecting those who lose opportunities or livelihoods. This often requires targeted investment in both people and places, as the impact of dislocation is frequently concentrated in specific communities.

This issue extends beyond mining to sectors like manufacturing and agriculture. For example, while a major car company might be able to redeploy workers from vehicles with internal combustion engines to the manufacturing of EVs, the knock-on effects on the supply chain could be severe. A company producing pistons to supply a car manufacturer, for example, would not necessarily transition easily to making batteries. Similarly, in agriculture, practices dependent on heavy use of fertilisers or scarce water may need to change rapidly, potentially causing losses during the transition period.

Consumers, too, will feel the effects, as prices for transportation fuels, certain foods, and air travel could rise. Households required to switch from gas boilers to heat pumps might struggle with the upfront costs, particularly as poorer individuals often face higher borrowing costs than wealthier ones.

These examples highlight the many ways in which injustices and inequities can arise during the transition. Ignoring these issues is not only unethical but also risks jeopardising the transition itself. This leads us to the importance of implementing actions and policies aimed at ensuring a just and equitable transition.

Building a just and equitable transition

A broad range of actions can be taken to overcome dislocations, create new opportunities, and foster a just and equitable transition. However, it is important to recognise that not everyone will benefit equally, and some may inevitably lose out. For example, while some coal miners may be able to retrain and find new opportunities, others may not be so fortunate, particularly if they are older or have health issues or lack an educational background. The actions we discuss in this subsection relate to local skills and investment, lifelong learning, relocation of public services, cost and availability of capital, tackling the effects of price changes, and social protection.

The first aspect to consider is *local skills and investment* – investing in people and places. Job losses can have a geographically focused and cumulative impact, especially in specific locations where industries like coal mining or steel production are concentrated. Policies in these areas should focus on concentrated investments in infrastructure and skills, along with incentives for industries to relocate. While these processes take time, they can be accelerated. For example, some older industrial cities in richer countries have built civic universities of real excellence, which are valuable existing assets which could help in moving quickly to drive the expansion of service sectors and create new opportunities.

It is important that structural changes to the economy, with their associated dislocation, be supported by policy. For example, Spain's plan to phase out coal by 2025 has been devised with region-specific just transition policies. It included, in 2018, a deal for €250 million to be invested in mining regions. The government agreed to support early retirement schemes, local re-employment in environmental restoration work, and reskilling programmes in green industries (World Resources Institute [WRI], 2021).

In poorer countries, external aid at low or zero cost may be necessary for such investments. However, with mass local unemployment, the opportunity cost of labour could be very low, making economic returns on investment potentially higher than financial returns. Thus, investing in those places and in the skills and opportunities for the local population makes economic sense, and outside help with finance could be very valuable. Nevertheless, some individuals may choose to relocate in search of better opportunities.

A society that integrates *lifelong learning* into its education system will have a more flexible workforce, capable of switching occupations as needed. While building such systems can be challenging, particularly in poorer countries, local institutions that provide skills training are feasible in a wide range of contexts. The benefits of lifelong learning extend beyond the climate transition, offering resilience in the face of rapid technological and structural change. Again, if climate action encourages a stronger focus on issues like this, which should have been developed more strongly earlier, it carries a direct benefit.

There are numerous examples of poorly managed industrial declines. In the UK during the 1980s, under the government of Margaret Thatcher, the closure of the coal mines, in this case due to rising costs and competition from gas, was handled in a socially divisive and brutal manner, resulting in long-lasting depression and unemployment. The seeds of mistrust sown during that period have persisted for generations. The green transition must learn from these mistakes.

As far as the *relocation of public services* is concerned, it must be emphasised that certain public services, particularly administrative functions, are not necessarily tied to specific locations. Social security administration began moving from southern England to Newcastle in the late 1960s (Hansard, 1968), and the move was boosted in the 1980s. It continues. This was a period of decline of shipbuilding and the coal industry in the region. In today's digitally interconnected world, there is increasing scope for such relocations in all countries.

The *cost and availability of capital* is another important factor. The transition will require significant investment from both public and private sectors, not only in major industries and infrastructure but also in households and small businesses. However, poorer individuals and smaller firms often face higher costs of capital, which can be a barrier to investment. For example, the installation of heat pumps or solar panels might be difficult to finance for lower-income households.

A just transition would involve providing direct assistance to manage the cost of capital. Governments, which can often borrow at low costs, could offer on-lending programmes with modest interest rate mark-ups to cover their costs without requiring subsidies. Costs can be lowered by delivering change on scale throughout a locality (e.g., insulation or heat pumps) and integrating it with support that lowers the cost of capital. Action to encourage local skills for installation could be another element in a package which facilitates and reduces cost.

In terms of *tackling the effects of price changes*, removing subsidies on diesel for farmers or imposing a carbon tax on fuels could leave some people worse off unless other policies are in place. These measures can provoke strong reactions, as seen in the farmers' demonstrations in Germany in 2023–2024 against the removal of diesel subsidies and the *gilets jaunes* protests in France in 2018–2019 against tax increases on petrol and diesel. In such cases, it is important to implement comprehensive packages of measures within which the negative impacts are mitigated. If subsidies on polluting fuels are reduced or removed, as in the German example, other support for more socially productive benefits of farming could be included. In the case of the *gilets jaunes*, one possibility could have been support for the purchase of more efficient vehicles. These changes should be discussed well in advance, allowing time to design and phase in the necessary support. As the policies were presented, in each of these two examples they were seen as the last straw in a series of actions and processes which the protesters saw as having undermined their living standards. A package could have done so much better in each case.

Some of those affected by the transition may not be able to participate in new opportunities, notwithstanding best efforts to foster their participation. In these cases, *social protection* and transfers would be necessary. The first priority should, however, be the development of new opportunities. It is common to argue, and with some justification, that the best social protection is a secure job. The better the creation of job opportunities, the smaller the number of those receiving social protection transfers – and thus the greater the possibility of stronger social benefits.

There is much that can be done to foster a just transition. However, the necessary policies must be integrated into the overall transition strategy from the beginning. Ensuring that the transition is just is not only an ethical priority; it can also accelerate the transition and enhance its benefits. And, as we have noted, it can encourage the tackling of broader issues in society which have received insufficient attention.

In conclusion, while the challenges of achieving a just transition are substantial, they should not be used as an excuse for inaction. As emphasised throughout this book, delay is the most damaging and difficult path of all.

6.6 Macroeconomic challenges

Climate is macrocritical, both in its impacts and in the task of creating an effective response. In both cases, growth, living standards, and stability are at stake. Consequently, climate must be treated as a central macroeconomic issue by finance ministries and be firmly on the agenda for central banks, both in relation to incentive structures and in relation to the economic implication of climate risks and to income distribution. Some macro issues around finance and fiscal rules were examined briefly in Section 6.4 but, given their importance, a more explicit examination of the macroeconomic challenges is provided here.

The impacts of climate change are already having major macroeconomic effects in many countries. Hurricanes in the Caribbean and typhoons in the Pacific island states can devastate major infrastructure. Desertification can lead to population movements and conflicts between pastoralists and agriculturalists with real macroeconomic consequences, as seen in Sudan and Syria. In Nigeria, desertification has led herders and their livestock to move south, increasing competition over land and contributing to conflict among herders and farmers (Lenshie et al., 2021). Wildfires in Southern Europe have had very large impacts, including on tourism. Disruption of major staple foods can trigger inflation, as observed in India, where the Reserve Bank recognised the weather shocks due to climate change as a key risk to the nation's inflation prospects (Reserve Bank of India [RBI], 2024). The areas around the Himalaya, sometimes referred to as the Third Pole, are especially vulnerable. Many of the major rivers of the world rise in these mountains: including the Yangtze, Yellow, Mekong, Ganges, Brahmaputra, and Indus. Millions depend on the flows of water from the snows and glaciers. Floods are increasing in intensity. For example, the 2022 floods in Pakistan submerged one-third of the country and displaced more than 30 million people. We know that much more climate change is inevitable in the coming decades, even if we act decisively to reduce emissions. These examples from the past, powerful as they are at the macro level, are likely to seem modest compared to the impacts we will face in the next two or three decades.

As we have already discussed in Chapter 5, action on climate change will necessitate substantial increases in investment; as 1–2 percentage points (ppt) of GDP in developed countries and 3–5 ppt in EMDCs (other than China) (Bhattacharya et al, 2024; Stern et al, 2021). Increases of investment on this scale will require careful macroeconomic management. Macro issues related to these increases in investment were raised in Chapter 5 on growth and Section 6.4 on finance, and we will revisit some aspects of the macro picture when we examine international finance in Chapter 9. Our discussion here will therefore be brief.

If investment is to increase by a few ppt of GDP, this will result in a significant increase in aggregate demand. If there is Keynesian slack – unemployment or underemployment – in the economy, output will expand through the multiplier effect, thereby generating the necessary extra savings to finance the additional investment. Consumption will also increase along with the extra output and income.

However, many economies will not have Keynesian slack. While there were many that did in the decade following the global financial crisis, this was less evident following Covid-19, the energy crisis linked to the Ukraine war, and the eventual inflationary impact of prolonged expansionary monetary policies in several G7 and other countries. In such cases, extra investment must be accompanied by a reduction in consumption or by resource inflows from abroad. Consumption can be reduced through a combination of higher taxation, incentives to save, or increased domestic government borrowing, though this will undoubtedly carry political challenges. It should be recognised, however, that reducing consumption to increase savings, in order to finance investment with strong returns, is a set of actions that increases welfare, just as a farmer might save to invest in irrigation facilities that make her or him better off despite entailing smaller consumption in the short term.

Some countries may be able to borrow more from abroad at reasonable cost of capital, but others, particularly heavily indebted poorer countries, will not. For these latter cases the international financial institutions and systems will play a crucial role, as will be discussed in Chapter 9. When additional borrowing from abroad occurs to finance the increases in investment, the current account deficit will rise as a counterpart to capital account inflows. This will lead to relative price effects, with the price of non-tradable goods increasing relative to tradables.

All these effects will require careful macroeconomic attention. While they can be managed, they must be addressed specifically and directly. For highly indebted poorer countries, some form of mechanism for debt relief may be necessary, along with access to low-cost or concessionary capital.[31]

The finance ministry and the cabinet office of the prime minister or president should bear central responsibility for economy-wide coordination, as emphasised above. Each line ministry – whether it be transport, energy, health, housing, or agriculture – will have its role to play, but the transition will require these different sectors and activities to move in a coherent manner.

For example, the electrification of transport and heating will necessitate a prior expansion of electricity supply. It is important to build the new and clean infrastructure before phasing down the old and dirty systems; solar and wind installations and grids must be coordinated with land-use planning. The finance ministry and office of the prime minister or president will both have a major role in overall coordination and strategic direction.

The existence of serious macro management issues does not mean that 'realism' or 'macro risk' should be used as an excuse to postpone action. As emphasised many times in this book, delaying action would be the most risky and unrealistic path to follow.[32]

6.7 Concluding remarks: opportunities, choices, trade-offs, and commitment

In concluding our chapter on policies and measures delivering climate action and creating the new growth story, we must recognise and emphasise the opportunities and rewards that can arise from achieving the major changes in our economies which are now an urgent necessity to tackle climate change. These rewards are way beyond those associated with climate action narrowly conceived. Most economies suffer from serious inefficiencies and market failures, particularly around those six identified in Chapter 5 and this chapter (see Tables 5.3 and 6.1). In a highly inefficient world, it is possible to make progress both on climate action, including mitigation and adaptation, and on advancing growth and development. Indeed, our argument is precisely that climate action can drive the new growth and development story.

The realisation that we must make radical changes to respond effectively to the climate and biodiversity crises can help us overcome inertia, thorny market failures, distributional challenges, and vested interests that have been difficult to tackle in the past. For example, the new Labour government (2024–) in the UK is seeking to reduce the inherent delays of the spatial planning system to help accelerate the recasting and building of grids and renewables. This change in spatial planning can also accelerate house building, whereas arguments based solely on the importance of housing might not have been sufficient to overcome the inertia in reforming the planning system.

The organisational changes necessary to drastically reduce waste, including ideas around the circular economy, might gain sufficient momentum if the climate and biodiversity arguments are added. Providing affordable capital to poorer households to insulate their houses, of great value to them, could be facilitated by linking the effort to energy efficiency and climate action. In these ways, the climate crisis can help us overcome inefficiencies, market failures, distributional challenges, vested interests, and other obstacles that have previously proved 'too difficult' to address.

All of these points to understanding our responses to the climate and biodiversity crises as strategic choices rather than inevitable trade-offs. No doubt there will be difficult decisions. Choices will need to be made, and significant investment will be required. No one should pretend that this will be easy. There will be dislocations, reservations, and opposition, all of which must be taken seriously, examined, and discussed. There will be many problems associated with change of this pace and magnitude, but they can be overcome, as I have tried to demonstrate.

We conclude this chapter by emphasising the building of commitment. Throughout this book, an argument of opportunity has been presented, but where investment must be undertaken and dislocation must be managed. That requires commitment.

Building commitment and carrying through its content requires making the positive case clearly and strongly, understanding reasons for resistance, and combating misinformation. Overcoming inertia, short time horizons, and entrenched vested interests is crucial. Industries that protect their profits, such as those in fossil fuels, can exert significant influence over political figures, attempting to maintain the status quo as much as possible or slow down change. This resistance is compounded by fears in the workforce of job loss and economic dislocation, or among consumers of higher prices or the inconvenience of change, which lobbying groups can exploit to hinder transitions (Denton et al., 2022). As a result, some politicians may shy away from strong actions on sustainability or use possible hesitation in the electorate as an argument against such action. This was effectively the approach of the Conservative government of Rishi Sunak in the run-up to the UK elections of July 2024. It did not succeed. But it may be continued by the Conservatives in opposition, partially because they have lost support to Reform UK, which has been hostile to action on climate. Similar phenomena have occurred across Europe, for example in the resistance of the German car industry or reactions to the requirement to replace gas boilers with heat pumps.

To respond to these challenges, marshalling arguments for change may not be sufficient. Strategies will need packages to directly respond to concerns. Policy sequencing will matter, as will care in governance processes. Effective climate governance involves adapting institutional architectures, creating inclusive frameworks, and involving various stakeholders so that the governance process reconciles different perspectives, recognises potential winners and losers, and works to reduce losses and dislocations. Examples include local community climate action groups (there are a number where I live in West Sussex, UK) and finding ways for those who object to solar farms or new electricity pylons to share in the overall gains. Young people can be especially motivated and effective.

A crucial starting point and constant thread is the development of a compelling national vision and narrative. This narrative should integrate domestic advantages and global responsibilities, framing climate action not only as a necessity but also as an opportunity for economic and social development. This involves communicating the effectiveness, fairness, and benefits of climate policies in ways that can resonate with the public's values and expectations. As emphasised, the narrative must offer hope as well as concern. That is why understanding that climate action delivers the growth story of the 21st century is so important.

Notes

[1] For Kuznets, Stone, Meade, and Keynes in the 1930s and 1940s, the idea of GDP was critical to understanding and managing employment and unemployment. And it is still useful from the perspective of capturing and measuring economic activity. But that should not be confused with 'well-being'. Economic activity is seen as the aggregate of value-added in the different sectors of the economy, and employment is associated with these activities.

[2] See, e.g., Sachs et al., 2024.

[3] For transparency, Chris Freeman, an extraordinarily distinguished economic historian and analyst of technical change, was a close family friend. He was at LSE with my mother, Marion Swann, in 1939–1940 when it was evacuated to Cambridge in war time in the UK.

[4] Hence, Klaus Schwab had three preceding the current one (four in total), rather than the five predecessors (six in total) of Freeman and Perez.

[5] Other important analytical insights into waves of technological change, in particular from evolutionary perspectives, come from Richard Nelson and Sidney Winter. Chris Freeman interacted with them and spoke enthusiastically of their contributions (personal interaction with Chris Freeman).

[6] The history of antitrust policy in the USA provides one set of examples. Another is the UK telecoms sector after privatisation, when BT was eventually required to relinquish dominance in key areas to enable the entry of alternative providers.

[7] DARPA is an R&D part of the Department of Defense and was created in 1958 by President Eisenhower (after the launch of Sputnik by the USSR in 1957).

[8] See, e.g., Lynn (1994).

[9] Further, the UK's Climate Change Act of 2008 tasked the CCC to recommend targets over time and monitor progress.

[10] The primary objective of the partnerships is to develop sector-specific strategies for reducing emissions. Each partnership is led by a chairperson appointed by the Danish government, typically a leader from a prominent company within the sector. Each partnership is tasked with presenting proposals on how their sector can contribute to emissions

reductions in a just manner, supporting Danish competitiveness, exports, jobs, welfare, and prosperity. These proposals include measures that the sector itself can take and recommendations for government to remove barriers and improve framework conditions to support emissions reductions and green competitiveness (The Danish Government's Climate Partnerships, 2022).

[11] The IIGCC was created in the UK in 2001 by the responsible investment leads at USS (a UK pension scheme) and later, in 2005, became incubated within the Climate Group, under the leadership of Stephanie Pfeifer (current CEO).

[12] In many countries surplus power capacity in an electric lamppost can come from replacing incandescent light bulbs with LED or other low-wattage bulbs.

[13] See, e.g., the work of the EBRD (2023), the IFC, and the World Bank (Sawaqed and Griffin, 2023).

[14] See, e.g., the Transition Reports of the EBRD from the 1990s (EBRD, 1994, 1995, 1996, 1997, 1998, 1999) and the World Development Report 2005 (World Bank, 2004). I was Chief Economist of the EBRD, 1994–1999, and of the World Bank, 2000–2003.

[15] Our publication of data on corruption certainly provoked attention and reaction. The transparency helped public discussion, but in many cases the corruption was so entrenched that it was very difficult to shift.

[16] The language of 'country platform' originated around the mid-2010s (for a discussion see, e.g., Plant, 2020) and has been part of the work of the G20 International Financial Architecture Working Group for a number of years. It has become more prominent with the work around Just Energy Transition Partnerships and the report of the G20 Independent Experts Group on MDBs (Singh and Summers, 2023a, 2023b).

[17] See, e.g., Marshall's *Principles of Economics* (1890) or Pigou's *Economics of Welfare* (1920). Marshall wrote about externalities, but it was his student Pigou who developed the distinction between social cost and private cost and highlighted its relevance for policy. See, e.g., Boudreaux and Meiners (2019) for textual detail.

[18] See Section 3.2.

[19] As explained in Section 3.2, the Pigouvian tax is directly designed to bridge the gap between the market price (without public policy) and the social cost (private cost plus damage to others).

[20] This is why Joseph Stiglitz and I refused to sign this letter. It is not only wrong in its economics; the 'carbon price alone' dogma also distorts policy. See also Section 3.2.

[21] Formally, there are two basic theorems. The first says that a competitive equilibrium is Pareto efficient. The second says that a given Pareto efficient can be decentralised so that it is chosen by private actors in a competitive market. The required assumptions are as in the text. A state of affairs is Pareto efficient if it is such that no one can be made better off without some other person being made worse off. It is often referred to as 'first-best', given the assumptions in the text. Pareto efficiency does not take the distribution of welfare or income into account.

[22] Optimality, in this context, would be defined as maximising some welfare function subject to all the relevant constraints, which could include constraints beyond those of the 'first-best', such as constraints on policy instruments – see preceding endnote.

[23] See, e.g., the *Journal of Public Economics,* of which Tony Atkinson, Joe Stiglitz, and I were co-editors for two decades from its foundation in the 1970s.

[24] See, e.g., Arrow (1963), Akerlof (1970), Spence (1973), Stiglitz (1975), and others in the substantial literature on the economics of asymmetric information.

[25] Jeremy Oppenheim and I had the opportunity to discuss these issues with Mark Carney in the months before the speech.

[26] The Basel rules are established by the Basel Committee on Banking Supervision. They guide banking supervision in many countries. They set international standards for banking for capital requirements, stress tests, liquidity, and regulations to control risks taken by banks, and thus the risk of bank runs or failures. Capital requirements are defined in terms of ratios between different types of capital held (such as equity or reserves) and outstanding loans or 'assets' weighted by their risk.

[27] Tax equity financing is a way for investors to fund renewable energy projects in exchange for tax benefits, like tax credits or tax reductions. This mechanism helps project developers raise money, while investors reduce their tax bills.

[28] The Expanded Simple Risk-Weight Approach (ESRWA) proposal would increase risk weights for tax equity investments, potentially making them too expensive for banks to participate in and reducing the availability of tax equity financing for clean energy projects.

[29] Though the impacts of climate change are increasingly being felt, and potentially will generate huge impacts in the longer term, economic and financial actors usually plan and act with a short timeframe. For example, rating agencies usually assess credit risks with a time horizon of three to five years (Bolton et al., 2020).

[30] The work on this important contribution was led by Rania Al-Mashat, then Egypt's minister of international cooperation.

[31] See, e.g., Kharas and Rivard (2024) for further discussion.

[32] For further discussion on these issues, see Zenghelis et al. (2024), Soubeyran and Stern (2024), and Pisani-Ferry (2021).

7. The role of the state in a changing world

This chapter draws out the implications for the role of the state in creating the new growth story. The argument follows from the more detailed analysis of Chapter 5 on the process of growth and of Chapter 6 on policies to foster the investment and structural change which will drive the new growth. It is deliberately a 'big picture story'; it is important to lift the line of vision above the detail and be clear and strong on what all this means for the role of the state. The challenge requires an active state but not a dominant state. Thus, this chapter is in part a synthesis and a summary of Chapters 5 and 6.

The transition to a net zero economy is a complex and difficult economic, political, and social process. This transformation, driven by necessary structural, systemic, and technological changes, along with major increases in investment and innovation, will in its essence require active state involvement within the context of a global response to an intensely global crisis. Institutions and behaviours will need to be built and modified, rights must be understood and upheld, and powerful pushback is to be expected from groups whose perceived interests might be threatened by these fundamental changes, or from those who are simply resentful of change or perceived interference. These changes are crucial for future generations and, indeed, for this one. But they cannot happen on the scale and urgency necessary without an active state that fosters and enables them.

This new understanding of the role of the state contrasts sharply with the simplistic market fundamentalism prevalent in much of the developed world during the 1980s and 1990s. We must now recognise that transitioning to a sustainable economy, which can create the growth and development story of the 21st century, requires a redefined role of the state, making it fit for the immense challenges of the modern world.

Whilst this redefinition of the role of the state is logically necessary for tackling the immense crisis of climate change effectively, formulating and carrying through the redefinition will not be simple or straightforward. It will require careful thought and analysis, public discussion, community participation, sound and creative economic policies, private sector and civil society recognition and encouragement, and political leadership. The necessary actions of this more purposive state will be difficult, particularly given the urgency, but they are necessary. Inaction is the most unrealistic and dangerous path of all. However, in understanding and thinking about the detail, it will be crucial to carry throughout an overall picture of the magnitude, nature, and necessity of the transformation.

This chapter offers an initial analysis of the key elements of a recasting of the role of the state. While much more thought and analysis are needed, the urgency of the crisis and the necessary response requires simultaneous reflection and action. We will learn along the way and must integrate this learning into our strategies.

The structure of this chapter is as follows. In Section 7.1, I show how the misguided market fundamentalism of the 1980s and 1990s, clearly damaging at the time and with scars and consequences that continue, is particularly misleading and destructive in the face of the climate and biodiversity crises. Section 7.2 examines and emphasises the role and importance of an active state in driving the transition and the responses to the global crises and in fostering rapid systemic, structural, and technological change. This is not a story of just one country; the climate and biodiversity crises are rooted in the loss of or damage to global public goods and require global action. A re-emphasis on, and a recasting of, internationalism must be part of the story of action. In Section 7.3, I set the context for and examine issues around internationalism and global state action.

The great changes involved in this transition will require supportive institutions and changes in behaviour. Section 7.4 offers some ideas on institutions at different levels of governance, and their ability to protect rights and shape our behaviours. I also examine how these changes affect, and are influenced by, individual and community rights and how relevant rights and freedoms can be upheld. The current generation will make choices in markets, in their own lives, and in politics. And in so doing they affect freedoms and choices of those that follow, and, indeed, the prospects of others across communities and nations. Section 7.5 addresses the political economy, recognising the likelihood of active resistance from those who perceive their interests as being at risk. Overall, this chapter is a gathering together of some of the policy and strategy discussions of preceding chapters to show how, taken together, they require and constitute a recasting of the role of the state. Section 7.6 offers concluding remarks.

7.1 The confusions, failures, and dangers of market fundamentalism

The shortcomings of integrated assessment models (IAMs), which attempt to combine economic and scientific perspectives in a particular and narrow way, were examined in Section 3.2. Integrating economics and science into our decisions is of the essence, but these models are unsuited for the analysis of the big questions on climate action, because they usually assume both exogenous underlying growth and economic equilibrium. They incorporate only limited technical progress and do not take into account increasing returns to scale. Further, they have minimal scope for the analysis of structural and systemic changes. They have been highly misleading, not only in terms of these defects but also by portraying potential damages as both modest and far off in the future. These modelling failures have led, as we saw in Section 3.2, to absurd propositions such as 'the optimal long-run temperature rise is 3.5 °C.'

Above all, however, the models fail to address the central strategic and policy issues of fostering and managing structural, systemic, and technological changes, because essentially the models are constructed in a way that leaves out these changes. Thus, the models are largely silent on these issues; they are simply absent.

Our canvas here on the underlying economics is much broader than IAMs and their defects. At issue here are the big questions of strategies for the transition and the role of the state in fostering and guiding it. I argue that the market fundamentalism of the 1980s and 1990s, which posited that growth and development would flourish only if the government played a minimal role – that is, if 'government gets out of the way' – is untenable in the face of the current environmental crisis.

First, even when environmental issues were set aside, the dogma of market fundamentalism was badly flawed and deeply damaging. The pursuit of this approach increased inequality in the UK sharply in the 1980s, and unemployment in the UK was above 10% for much of that decade. The income share of the richest 1% in the USA (before tax) has doubled since 1980, from around 10% in 1980 to nearly 21% in 2023, while in the UK it increased from nearly 7% in 1980 to a peak of 15% in 2007 (Our World in Data, 2024).

Market fundamentalism undermined the sense of community. In its support for minimal government, that overall approach essentially abrogated national responsibility to help those whose employment prospects were rapidly changing as a result of structural or technological change. For example, in the UK, Norman Tebbit (Secretary of State for Employment in the Conservative government) suggested in 1981 that the unemployed should 'get on their bike' (Tebbit, 1981). Margaret Thatcher famously declared, 'There is no such thing as society' (Thatcher, 1987), deriding the idea of mutual responsibility and community belonging. Public service was denigrated, with many in the public sector regarded as having their nose in the public trough and contributing little. Further, civic pride was sneered at, with consequential damage to our architectural heritage and the quality of our towns and cities. The notion of community and mutual responsibility was devalued.

A doctrinal approach to privatisation saw UK public assets sold off, in many cases at ridiculously low prices and with limited social or service responsibilities. Jonathan Portes, who was a civil servant at the time, has described the incompetence around the privatisation of the UK water industry as a clear example: undervaluation of the assets and absence of necessary environmental responsibilities (Portes, 2022). Deregulation of financial markets on both sides of the Atlantic sowed the seeds of the global financial crisis of 2008–2009, with devastating national and international consequences. Social housing policies, including the selling of public housing to tenants at huge discounts, largely benefited wealthier tenants. And those policies directly restricted local councils from using sales revenues for the building of new social housing.

That consequential running down of the stock of social housing has been an important element in the current housing crisis in the UK.[1] Oil and gas assets in the North Sea were frittered away in overvalued exchange rates, which, along with excessively restrictive fiscal policies, led to the heavy unemployment indicated. Norway, in contrast, used the resources far better in creating lasting national wealth funds, which have both responsibilities and the ability to act in the national interest.

There was some increase in labour market flexibility and entrepreneurship in the Thatcher era in the UK – positive outcomes of genuine significance. However, the immense negative effects have all too often been omitted from mainstream historical narratives. The 1980s were profoundly socially divisive, with deep damage to communities, to mutual respect, to standards of individual behaviour, and, in various ways, to all four of physical, human, natural, and social/cultural capital.

Market fundamentalism has weak theoretical foundations in its understanding of how markets work and of their pervasive failures (unless corrected by good policy). It is a profoundly dangerous and misguided ideology. Markets, which are assumed to function well without government interventions, actually suffer from problems associated with risk, asymmetric information, and market power. The problems are especially severe in capital markets; these markets are also tilted towards the short term because it is difficult to cover for long-term risk. These issues disproportionately affect poorer individuals who face higher capital costs because they have limited collaterals. And these market failures are of central importance in relation to the transition we must now pursue.

Market failures result in high economic rents in many areas, including in relation to property. The renowned economist (and an important mentor to me) Robert Solow, who passed away in December 2023 at the age of 99, highlighted the importance of understanding economic rents, how they are created, to whom they accrue, and their dynamics. He argued (personal communication), a few years before he died, that if he was starting his career again, that would be central to his research. Larry Summers, an outstanding academic macroeconomist and former Secretary to the Treasury, said at a public lecture in Delhi in 2023 that if he was starting his career now, he would work less on macroeconomics and more on how to foster structural and technological change (Confederation of Indian Industry, 2023).

At the same time, as we explore the role of the state in driving change in the next section, we must recognise the risks of government failures. These risks should be examined and addressed in the design of strategies and policies, as well as in institutional contexts (see Section 7.5).

All these issues and difficulties with market fundamentalism that we have raised arise before considering environmental and sustainability-related externalities and market failures. Environmental damage, unregulated or untaxed, is a classic and crucial example of externality and market failure. Environmental destruction affects future generations, who are weakly represented in markets, and hits the poorest in any generation earliest and hardest.

Nevertheless, we should recognise that, perhaps from her background as a student of chemistry, Thatcher was one of the first world political leaders to recognise the potential severity of problems from climate change.

Over the past 30 years, there has been an increasing understanding of the scale of the damage caused by unfettered markets or uncorrected market failures. Markets do have an immensely important role to play in driving forward sustainable and resilient economic development, entrepreneurship, and rising living standards. They are powerful forces for opportunity, creativity, and growth. But they cannot drive growth and development, and particularly sustainability, effectively without government policy and supporting institutions. The necessary pace, scale, and nature of change now demands a strong role for government in leading and fostering this transformation. This role is the issue to which we now turn.

7.2 The role of the state in driving change: crisis, urgency, and systemic transformation

Some approaches to market failure, particularly around environmental externalities, suggest, as we saw in Chapters 5 and 6, that problems can be 'solved' simply by pricing the damage associated with these externalities so that decision-makers 'price in' the associated costs. In this language, such policies 'internalise' the externalities. Many individuals may themselves internalise externalities if they understand the consequences of their actions and do not wish to harm others. The key economic insight on pricing externalities dates back to Marshall and Pigou. It remains crucial for policy-making. We argued, however, that this approach, if it is the sole policy instrument, ignores many other market failures of crucial relevance to the environment. The challenges we face, and our understanding of them, have now gone beyond issues of local damage and the use of static models. We face climate and biodiversity crises which are global, have potentially dangerous dynamics, and require urgent, large-scale action. These are crises where time is of the essence, requiring major structural, systemic, and technological change at a rapid pace. We need an economics of public policy which both takes careful account of market failures and recognises that time matters. At the same time, it must recognise potential government failures. Dogmatism and simplistic 'solutions' will not do.

Historical analogies, such as wars and pandemics, can show how large-scale and rapid responses to crises are both necessary and can be pursued effectively. During the Second World War, in the UK, the government took over factories, allocated key resources, and managed labour directly. Women took on roles traditionally held by men, for example working on the land in agriculture and as drivers in transport. Consumption was managed via rationing to prioritise resources for the war effort. In such times, it was clear that the state had to play a leading role in driving change. And it did. Speed was of the essence and the consequences of failure potentially catastrophic. This role of the state, given the crisis, was accepted by the general population. Indeed, with the external threat and collective purpose, there was a strong sense of community and solidarity.

Similarly, the recent Covid-19 pandemic demonstrated the need for state intervention to drive rapid change and crisis response. Industries like hospitality were shut down, while others, including education and transport, were transformed. Although the scale of transformation was much less than in war time, the pandemic did require large-scale state action. There are important arguments as to what action should have taken place, and how, but the necessity of strong action is not in serious dispute.

The transformations necessary for decarbonisation will to a large extent rely on markets and private decision-making. This is a market and entrepreneurial story, but the role of the state will be crucial in shaping incentives, setting expectations, and fostering entrepreneurship. These transformations will occur over two to three decades, longer time horizons than most wars and pandemics. However, delay is dangerous and the urgency requires strong action now. Tackling market failures, combined with risk and urgency, necessitates, as argued in Section 6.3, not only price mechanisms such as carbon taxation but also standards and regulation. For example, setting a date beyond which internal combustion vehicles will no longer be allowed to be sold provides clarity and confidence for private investors and producers to act quickly and on scale.

Systemic elements of the transition will require economy-wide coordination beyond that which the price mechanism alone can provide. Important examples are the integration of changes in electricity, transport, and city systems. The creation of low-carbon transport has, as a prerequisite, a rapid shift to low-carbon electricity and the expansion of its availability. Similarly, transforming homes for efficiency and heat pumps requires creating and deploying building skills rapidly and on a systemic scale. Much of that will come through markets, whereas other elements such as the ramping up of skill creation are likely to require collaboration, including between government and educational establishments. Providing low-cost finance is likely necessary, particularly for poorer households. All of these demonstrate the importance of managing systemic change in ways that the market alone cannot achieve, but also in ways which promote and harness the flexibility and power of markets and entrepreneurship.

Just as war time saw rapid development of aeroplanes and munitions (and the retraining of people), and pandemics spurred vaccine development, decarbonisation requires swift research, development, and innovation, fostered by public action and investment and driven by entrepreneurship. In the fine tradition of Chris Freeman,[2] Mazzucato (2021) emphasises the importance of 'missions' in advancing new technologies. Examples of relevant missions in the UK would be the systemic transformation of the energy system at the pace and scale required, including its decarbonisation within the next five years. Another would be research to accelerate the creation of affordable sustainable maritime and aviation fuels; that is a mission for the world as a whole.

Whilst a framework of government strategy and policy is essential, private sector initiatives, creativity, innovation, and investment are core drivers. The rapidly falling costs of renewables and battery storage demonstrate how quickly the private sector can move when public policy reflects the importance of pace and entrepreneurship, signals are clear, and social priorities are recognised and broadly shared.

A strong emphasis on an activist state must consider potential government failures. Governments can make mistakes, misread situations and challenges, lack expertise, or be captured by vested interests; they can also be ponderous, bureaucratic, inefficient, and slow. They may have limited understanding of the workings and needs of the private sector. Officials and ministers may act in self-interest or be corrupt. Democratic governments balance longer-term policy with shorter-term electoral considerations. These risks are of real substance and should be examined and addressed in the design of strategies and policies, as well as in institutional contexts.

Recognising these challenges and difficulties of public action – and they are real and important – does not detract from the necessity of a rapid transition or the government's active role. It means that political and economic dynamics must be carefully considered and that improving the efficiency, creativity, and pace of public action must be constantly on the agenda. Fast, large-scale action in a situation of imperfect knowledge can accentuate the likelihood of mistakes. Thus, working with communities, conducting thorough analyses, seeking a broad range of perspectives and advice, and incorporating learning into action plans are all core elements in the programme of change. Some of these issues are revisited in Sections 7.4 and 7.5.

7.3 Global public goods and internationalism

Global warming and climate change are driven by the total concentration of greenhouse gases (GHGs) in the atmosphere, regardless of their origin. This global nature of the issue means that public policy and the role of the state in one country must take account of, and be complementary to, those of other countries. If decarbonisation efforts are costly and yield no direct local benefits, countries may be tempted to 'free-ride' on the actions of others. This underscores the importance of international cooperation and action. Through collaboration, countries can share and shape the benefits of action and, where necessary, share the costs. However, as this book argues, we should recognise that much of action should be seen as investment with strong returns over a range of outcomes, rather than as a narrow cost. Nevertheless, the investments are large and must be carried through rapidly, and there will sometimes be a temptation to wait for others.

International analysis on the science of climate change has been facilitated by the Intergovernmental Panel on Climate Change (IPCC), while global action is coordinated through the United Nations Framework Convention on Climate Change (UNFCCC). Despite numerous forces that could divide nations, both institutions have continued their work and, as described in Part I of the book, there has been significant progress, including in the landmark Paris Agreement. Notwithstanding the progress it has fostered, however, the UNFCCC is an unwieldy process, with its nearly 200 negotiating countries. And it is mostly led by the environment ministries of member countries when the big decisions now lie largely with finance ministries and presidents or prime ministers.

The G20 is beginning to take more responsibility here, with presidents and prime ministers and finance ministries more closely involved. Leadership in the G20 is likely to continue to be fundamental to international advance. In terms of the future of international policies in an increasingly fractious world, the role of the G20 will be critical. As the USA steps away from international action (as of early 2025), it will be for other countries to step forward. In relation to the G20, the countries of special importance, in terms of economic and political clout, in this multi-polar world in the next few years will be China, India, and those of the EU. An alliance here could keep internationalism alive.

The economics of climate action have also evolved. As highlighted throughout the book, clean technologies are now cheaper than their dirty counterparts for about a third of emissions without carbon taxes or subsidies, a figure expected to rise sharply over the decade up to 2035. Systemiq (2020) provides a figure of 25% for 2020 and costs of clean technologies have fallen sharply since then. This is one factor behind the argument of previous chapters that have illustrated the dynamic drivers of growth associated with the green transition, including investment, innovation, scale, and systems productivity.

The fact that the clean technologies are already cheaper than the dirty across many sectors – coupled with the strong growth effects associated with innovation, structural, and systemic change – and that they are less polluting and more efficient mitigates the 'free-rider' problem of countries waiting for others to move. Making these new investments is increasingly seen as very different from incurring a private or national cost for a global benefit. These are investments with strong private and national returns beyond the simple reduction of emissions. This fundamentally changes the politics and economics of interactions between states. The focus now shifts to accelerating the investment, innovation, and change – systemic, structural, and technological – that will drive the new growth story. The 'free-rider' problem diminishes when the transition promises substantial productivity and growth effects and high returns in terms of overall well-being. Moreover, it is important to understand that many nations, cities, firms, farms, and individuals recognise their environmental responsibilities and are taking action. Many do not wish to pollute or harm others.

It should now be clear that the model of the action of nations motivated solely by a very narrow interpretation of self-interest is not the whole picture. International collaboration is essential and depends on an understanding of self-interest and of shared interest in new ways – and on a discussion of mutual responsibilities. The politics of international action and climate will increasingly be shaped by this understanding of self-interest, the growth story, and mutual interest. Active leadership and diplomacy will be critical here, but these understandings provide real foundations.

Many challenges in international action persist, however. The recognition that the green transition is the growth story of the 21st century is not yet universally secure, even though the argument is gaining acceptance. This book aims to reinforce this understanding. Additionally, other crises – such as the global financial crisis of 2008–2009, Covid-19 in 2020/2021, and the continuing Russia–Ukraine war from 2022 – can slow climate action by diverting political attention and resources. Although these crises are of great seriousness, they are, over time, smaller in scale than the twin climate and biodiversity crises. The sense of urgency these other crises create can overshadow the immediacy required for climate action, which must accelerate on a large scale and be sustained over the coming decades. Shared threats can bring people together. We should also recognise that constructive international collaboration on climate can enhance the probability of collaboration on other critical international issues.

For all of the above reasons, it is clear that climate action requires a fundamental recasting of the role of the state not only within the nation but also in its interactions with other nations. There are difficulties but there are foundations.

7.4 Institutions, rights, and behaviours

The role of the state in climate action is already becoming embedded in the institutions of our economy, society, and politics. That will deepen. Legal institutions in many countries actively mandate that climate considerations be integrated into the behaviour of national governments, cities, and firms (Setzer and Higham, 2023). As we saw in Chapter 4, such rulings are often based on the rights of future generations and the national commitment to the Paris Agreement. Mitigation measures that can be the direct subject of internal legal structures and legal decisions include measures such as national carbon limits (as seen in France), cap-and-trade systems (as in South Korea), or investing in low-carbon technology (such as in Australia's Technology Investment Roadmap). As we saw in Chapter 5, the high-level July 2025 opinion by the International Court of Justice has underlined obligations to cut future emissions and obligations for current and past emissions. It is likely to have a significant impact on national legal systems across the world. On average, in nations with a strong rule of law, new climate mitigation laws have been shown to reduce annual CO_2 (Eskander and Fankhauser, 2020). The effects are more powerful when combined with sound policies and incentives.

Local community action has seen some cities and towns move more rapidly than others. This progress sometimes stems from leadership at the top, as seen with Anne Hidalgo, the Mayor of Paris. Such leadership is not always rewarded with popularity; many who might favour clean air are still hostile if they are prevented from moving wherever they wish in polluting vehicles. At other times local pressure groups successfully advocate for action through their local councils. And sometimes, as in my own area in West Sussex, UK, green-oriented councillors or Members of Parliament with mandates for action are elected.

The education system plays a crucial role in shaping the understanding of and commitment to climate action among young people. Experiences in schools, colleges, and universities influence their perception of what needs to be, or should be, done, their understanding of rights, and their personal behaviour. Assertion of rights, particularly here the right to sustainable development and a liveable future for young people, will put pressure on political institutions, local and national, and the behaviour of firms and farms.

The understanding of rights to try to build a better life, together with a view of what is ethical, can change both local politics and the understanding of the role of the state at a local level. And there can be national effects too. At the same time there will, as noted, be pushback from those unwilling to accept shorter-term disruption or dislocation. How these tensions and dynamics play through will determine the pace of change. The direction is, I think, now clear, but the pace is critical. Discussion of the respective responsibilities of the individual and state is long-standing, but now it has a new and vital environmental perspective.

This is about rights, freedoms, and choices. Slow action now curtails the rights and freedoms of future generations. The exercise of choice and freedoms of this generation in sustainable directions determines the nature and pace of action. Public discussion, education, and transparency are all part of the choices and freedoms of this generation, including – and this is of special importance – through political processes.

Religious institutions also wield significant influence over behaviour. The late Pope Francis, as emphasised in Chapter 4, among other religious leaders, has been instrumental in highlighting the importance of considering climate and environment in personal, national, and international behaviour. Pope Leo XIV's track record and early statements as Pope suggest that his thoughts on these subjects align with those of Pope Francis. These leaders help shape values and behaviours across different communities.

Norms and values can enhance responsible behaviour, help recognise good citizenship, motivate people to work in roles that generate public benefits, and encourage participation in political activities. Norms fostered at formative ages can turn into internalised values that can shape the behaviour of individuals beyond narrow calculating behaviour (Besley et al., 2023). Besley and Persson (2023) emphasise the importance of considering both extrinsic (e.g., changes in prices) and intrinsic (e.g., consumers caring about the environmental impact of their choices) motivations or incentives. The intrinsic has its own dynamics of change and can, over time, change the realities of politics and thus strategies and decision-making.

One crucial point in discussing institutional roles in relation to the state and society is to avoid 'institutional fundamentalism'. This notion suggests that institutions and their decisions should be entirely driven by a narrow set of criteria, or indeed a single overriding criterion. For example, some argue that a central bank should focus exclusively on inflation, disregarding other factors like climate. A single focus is often, or usually, a serious and dangerous mistake.

An institution can have a primary objective, which can provide clarity and focus, but that does not imply that it should ignore other critical issues. As argued in Chapter 6, climate change can disrupt the economy, leading to instability and inflation. In an imperfectly organised economy where prices are often distorted by market failures, the decisions of institutions like central banks can impact the broader economy beyond inflation and aggregate demand.

The great economist James Meade emphasised that uncorrected or partially corrected market failures in one part of the economy should be considered in decisions and cost–benefit analyses across the whole economy (Meade, 1973). A major failure in one market can imply that market prices may be distorted across the whole economy.[3] In the context of climate change, these 'knock-on' effects are substantial. Meade's insight, which is, or should be, embodied in the whole theory of public economics in imperfect economies, underlines the importance of integrating climate, biodiversity, and environmental issues with choices across the entire economy, and the strong role of government in so doing. That insight guides policies in relation to responsibilities and incentive structures. Further, changing values and behaviours, if expressed in public discussion, can also influence the understanding of the appropriate responsibilities of national institutions.

It is also important, however, in thinking about institutional responsibilities, to remember that governments can make mistakes, sometimes on a grand scale. Excessive control can stifle the entrepreneurship and creativity essential for advancing climate action. Advocating a stronger role for the state in steering the transition does not imply an embrace of central planning or Gosplan,[4] but it does require the creation of policies and institutions that can guide investment, innovation, and change towards the zero-carbon economy.

The private sector and its functioning are core parts of the institutional structure. And they can influence the way in which other institutions are designed, shaped, and function. The creativity and innovation of the private sector are indispensable across the economy, including in designing and implementing new energy and transport systems, and not only in the energy and transport investments and innovation themselves. The private sector has driven down the costs of clean energy and transport, along with significant advancements in battery technology and associated cost reductions. However, governments set the framework, country by country, and the international Paris Agreement of 2015 provided a global sense of direction. Public–private participation in system design is likely to play an increasing role in the future of the transition. It is the private sector that has special insight into obstacles to entrepreneurship, creativity, and investment, and into how they can be overcome. And it is the private sector that can identify opportunities and expand options for action.

The private sector is also instrumental in driving the transformation brought by artificial intelligence (AI). As discussed in Chapters 4 and 5, the potential for AI to accelerate the process of investment and transformation of our economies is immense. Nevertheless, government intervention will be critical to encourage the application of this potential towards accelerating the low-carbon transition. At present the application of AI is dominated by areas where there is a business model that captures revenues. That tends to be dominated by services supplied to consumers rather than application of AI to system design. Many of these services, such as search, have real value. But some of those services – such as interrupting communications or searches, or targeting vulnerabilities or anxieties – may have negative social value. In contrast, we have shown that the potential for AI in system design and management and in discovery will be of immense significance for the transition. And AI together with digitisation will have profound implications for the role and functioning of the state, including in relation to its interactions with households and with firms and their ability to drive the transition forward. Examples range from smart smallholder agriculture to the discovery of new materials.

Thus, an 'active state' can play a crucial role in fostering the development of AI itself. For example, the active state can provide institutional structures (including research centres), financial incentives, and support for research and development in AI technologies in public and private institutions. It can set regulations and standards for the development and use of AI technologies which reduce risks. And it can promote applications in social and econom-ically productive areas, including, for example, in the green transition and in health.

It is possible that AI could exacerbate North–South divides; as of 2025 AI research and application has been largely led by research institutions and corporates in the Global North, although here China and India are playing ever-increasing roles. That is a risk which should be anticipated and managed. Finally, the computational needs of AI are highly energy-intensive, resulting in a high-carbon footprint where that energy is not low carbon. An active state can support the development of approaches to reduce the environmen-tal impact of AI, for example through using more efficient AI processes and energy-efficient specialised hardware, requiring renewable or nuclear energy for data centres, and adopting circular strategies to decrease e-waste and asso-ciated resource consumption. Nevertheless, it should be recognised that the power supply necessary for activities such as system design and management, or the discovery of new materials, is much smaller than that involved, for example, in searches across the whole of existing literature. It is surely now clear not only that AI has tremendous potential in driving the transition for-ward, but also that AI has a major role to play in enabling the state itself to operate more effectively in fostering the transition. AI will shape both the transition and the role of the state.

7.5 Political economy

The final aspect discussed here on the role of the state in shaping the tran-sition is in relation to political opposition. Such opposition often arises for understandable reasons, including possible rises in prices of goods or services important to workers and consumers, for example fuel prices, the require-ment to make a costly investment such as a heat pump or electric vehicle

(EV), or the loss of jobs in industries such as coal mining. Tackling these problems requires a positive approach by the state to dealing with the issues in a practical way. This approach is central, as we saw in Chapter 6, to what is meant by creating and financing 'a just transition'.

There are indeed ways forward, most of which involve packages. For example, when considering raising the costs of fossil fuels for automobiles – a key element of pricing the GHG externalities – packages could include direct transfers to the poorest people affected, helping lower-income individuals acquire EVs, and improving public transport. Successful action requires thinking through political economy as a key element in designing action and taking a broad view of influences on well-being.

There has been and will be vigorous pushback from the fossil fuel industry, where large profits and rents are at stake. Naomi Oreskes and Erik Conway, in their book *Merchants of Doubt*, convincingly document how the fossil fuel industry conspired to undermine confidence in climate science. The industry has been highly effective in lobbying political parties and governments, often providing substantial political funding (Oreskes and Conway, 2010).

While some political opposition will stem from the self-interest of the fossil fuel industries, other elements will come, as we have consistently explained, from poorer families facing dislocation. These groups cannot and should not be treated in the same way as major oil and gas companies. The former are rich and powerful, and active in protecting their rents and their ability to supply toxic materials. Their assertion of some 'right' to make money from toxic activities, and of a claim to protection from change, are very different from a poorer family facing dislocation, for example through the loss of a job.

At the same time, the energy system cannot be transformed without the participation of energy companies. A key part of the role of the state is not only in standing up to vested interests, but also in encouraging them to play a strong role in the process of change. Constructive engagement with energy companies is a core element in the transition. So too is distinguishing, in public discussion, the kind of dislocation they face as major firms from that faced by poor households.

7.6 Concluding remarks: recasting the role of the state

This chapter has assembled a view of the 'big picture' of the role of the state arising from the more detailed and disaggregated analysis of Chapters 5 and 6. In this sense it draws together some key elements of those chapters and closes Part II of this book.

In summary, the current crisis necessitates re-evaluating the state's role in driving change. Coordinated systemic action is essential. Climate action and creating a new path of growth constitute an immense task that requires a deep understanding of the severity of the current crisis and a strong sense of urgency, whilst at the same time respecting present and future individual rights. It requires an approach where state intervention promotes inclusive development, tackles market failures, and supports private innovation, entrepreneurship, and creativity – in other words, fostering the private sector and harnessing the power of markets. This requires avoiding oversimplification of public policy via narrow, crude – and erroneous – assertions of what 'economics says'; taking community and society seriously, nationally and internationally; avoiding the traps and dangers of market and institutional fundamentalisms; thinking carefully about the dynamics of systemic and structural change; and being watchful and thoughtful on potential government failure.

At the national level there are six reasons for this reappraisal that cover objectives, urgency and scale, market failures, and community action:

1. The climate, biodiversity, and environmental goals have moved to the top of the agenda because of the immense crises we now face.
2. The crises require urgent action. Time is of the essence; delay is dangerous. Resources have to be moved quickly, and technologies have to be created and diffused at pace.
3. The action must be economy-wide and on major scale; this is a moment for a big increase in investment.
4. Market failures which have not been faced adequately now have to be tackled with renewed energy if investment is to take place on the scale and urgency required.
5. Institutional change, from revising spatial planning to remits of regulators, will have to be embraced.
6. Communities in cities, towns, and rural areas will have to collaborate in different and stronger ways. Many basic actions, from public transport to recycling, are at the community level. This is social action.

At the international level, the magnitude and geography of the extra investment necessary will require a re-orientation and expansion of international financial institutions (IFIs) if they are to play their necessary role in fostering and supporting investment action across the developing world. Beyond the IFIs, other key vehicles for international action will have to play a different and stronger role if they are to be able to play their part in tackling the new agenda. Examples include the UN (and its agencies), the World Trade Organization (WTO), the World Investment and Trade Organization (WITO), the International Maritime Organization (IMO), the International Civil Aviation Organization (ICAO), and many others. However, action cannot wait for long processes of institutional reform. Embarking on action at the pace and scale required will have to carry with it the institutional changes that are necessary. There will be learning and reform along the way.

In contemplating the changes to the role of the state, nationally and internationally, which are necessary to respond to the climate and biodiversity crises, we should also recognise the potential permanent benefits of these changes in the role and workings of the state. In the UK after the Second World War, the spirit of collaboration and solidarity generated during war time led to the creation of the National Health Service (NHS) and stronger social security. The Great Depression and the Second World War led to the UN, the Bretton Woods Institutions, and the EU. Responding to national and world crises can bring communities and nations together. The ability to change and to collaborate in response to extreme threat can enable working together for the challenges ahead.

These are lessons both from examining the past and from thinking through the nature of the future transition now required. There will be many more lessons as transitions unfold, and change itself must be designed in flexible ways so that learning feeds quickly and effectively into strategies and actions. All of this requires a fundamental reappraisal of the role of the state.

Notes

[1] See Shelter (2024) and Oliver (2024).

[2] And Carlota Pérez (2002, 2010); see, for example, Freeman (1995).

[3] See also Drèze and Stern (1987).

[4] The old rigid, indeed asphyxiating, Soviet planning system.

Part III.
International action

8. Transformation of the international economy: interdependencies, new structures and geographies, and differences across nations

The new story of growth and development set out in Part II is centred on investment, innovation, and fundamental structural change at the national level. Throughout this book, however, we have consistently emphasised the global nature of the climate and biodiversity crises, the importance of mutual support, and the varying circumstances of different countries. In Part III, our attention shifts to international action.

This chapter explores the transformation of the international economy resulting from climate action and climate change. Chapter 9 examines international policies and actions necessary to foster transitions within individual countries and the global economy, recognising the differential challenges, responsibilities, and opportunities faced by various nations. In the language of welfare economics, Chapter 8 leans towards 'positive economics', or consequences of action and where the world economy may move. Chapter 9 focuses more on 'normative economics', or on values and policies and what countries can and should do, working together internationally, to deliver change on the scale and pace necessary.

In Section 8.1, we explore the interdependencies and linkages that lie at the core of the climate challenge and the necessary response. This response, alongside the impacts of climate change itself, will reshape economic geography. Section 8.2 presents the changing roles of different natural endowments and their geographical distribution. Wind and solar are available across the world and these resources are distributed very differently from fossil fuels. Industry and artificial intelligence (AI) may migrate to geographical locations where clean energy sources are cheap and abundant. The green transition will reshape the demand for a range of critical minerals. All this will change the world's economic geography. Section 8.3 focuses on building the capabilities needed to harness the opportunities that lie in new power sources, different types of resources, changing technologies, and restructured supply chains. As the new growth story unfolds, a new dynamic of collaboration and competition will emerge. Changing comparative advantages and new technological demands for various resources will alter the location of economic activities, generate new industries, and shift trade and investment flows. Section 8.4 recognises that we must examine not only where new green economic activities will take place, but also where the impacts of climate change will be particularly strong. This is of special importance in relation to natural capital, in which strong investment will be necessary.

Different nations have their own circumstances, endowments, and histories and will face their own challenges. Perspectives from developed countries and from emerging markets and developing countries (EMDCs) will differ strongly. Within EMDCs, there are significant variations between and amongst poor economies and emerging markets. The giants of China and India are very different from each other. As explored in Section 8.5, all these differences will carry implications for differentiated strategies nationally and internationally, some of which will be elaborated in Chapter 9. While many elements of action will be common across countries, not all will be. Finally, Section 8.6 concludes by emphasising the potential scale of change in the new economic geographies, and both the opportunities and dislocations it could bring.

8.1 An interdependent world

Climate change affects all countries and arises from the greenhouse gas (GHG) emissions of all. Thus, effective action on climate, perhaps more than on any other issue, depends on understanding the global relationships and interdependencies of countries and requires robust involvement from all countries in both mitigation and adaptation. And all countries will seek the integration of these efforts with sustainable growth and development.

Climate action will be tailored to the geographical and historical contexts of different regions, but given the centrality of moving away from fossil fuels and the shared challenges of building resilience, many nations will choose to implement similar measures. On the mitigation front, all countries will need to increase electrification and transition to zero-carbon power sources. On adaptation, many will need to implement flood control measures in vulnerable areas; the expansion of irrigation may be needed elsewhere. Most will need to tackle issues related to forests, land degradation, and the deterioration of rivers, watersheds, and oceans.

However, the global response requires more than simply nations undertaking their own actions, whether similar or not. The reason for this lies in the powerful interdependencies that exist across a range of dimensions. We examine these linking factors briefly in turn: global public goods, investment, technology, migration, finance, and trade. Recognising these and acting together as nations can produce a response that is far greater than the sum of its parts. We then examine how emissions vary according to income levels and the implications for action involving all nations and groups.

Key linking factors

At a fundamental level, reducing emissions in any one country benefits everyone. This creates a strong incentive for joint decision-making and collective action. The United Nations Framework Convention on Climate Change (UNFCCC), the COPs, and particularly the Paris Agreement of COP21 in 2015 have fostered that collective action.

Investing in the protection of our *public goods*, such as natural capital and bio-diversity, offers mutual benefits beyond the direct impact on emissions. Oceans provide oxygen, are a critical protein source for billions, and regulate the climate by absorbing carbon.[1] Forests provide water management, protection from landslides, carbon sequestration, and medicinal resources. They, along with their flora and fauna, form part of the biome or ecosystems upon which lives and livelihoods depend. Climate and biodiversity are indeed global public goods, as well as being of profound direct importance locally and regionally.

The interdependencies surrounding climate action extend far beyond global public goods, vital as they are. They also arise in relation to investment, technology, trade, the movement of people, and the functioning of international financial systems. When potential investors in one country observe strong *investments* in a new sustainable economy elsewhere, their confidence in this global strategic direction increases, they see potential markets, and they learn from the investing experience of others. Such investor confidence is further bolstered by international agreements. If potential investors and producers witness progress in clean investment in another country, they may gain confidence in their potential and come to believe their own products will need to be clean to compete in those markets. The remarkable development of low-cost, high-quality, fast-charging, long-range electric vehicles (EVs) in China has signalled to the world that this is the future of automobiles. The development of high-efficiency LED lightbulbs in Japan and the USA gave India the opportunity to go to scale and bring down costs. The early encouragement of solar electricity in Germany demonstrated the possibilities to the world and encouraged production elsewhere, particularly in China. The introduction of the Carbon Border Adjustment Mechanism (CBAM) in Europe has signalled that high-carbon imports will be penalised.[2]

Technological advancements and cost reductions from innovation and investment in one country can benefit potential innovators and investors elsewhere through cheaper inputs and 'learning-by-watching'. In the 1960s, 1970s, and 1980s, the world benefited from technological advances and low-cost production in countries like South Korea and Japan, particularly in manufacturing, and in the 1990s and 2000s from rapid advances in China. Green products and inputs will feature prominently in the next wave of technological advances and cost reductions, a process already underway, as discussed in Chapter 5. These processes are facilitated by trade, though protectionism can limit such gains – a topic we revisit in Section 8.3 and in Chapter 9.

Migration can play a role in the development of new skills and entrepreneurship. Consider the many tech leaders, entrepreneurs, and innovators in the USA, particularly in California, who were born in India (Ahmed, 2024). Much of the work in making homes and buildings more efficient and cleaner will be undertaken by construction workers, many of whom in Europe and the USA are immigrants. Services essential for or complementary to the green transition, such as AI and virtual interactions, can often be provided by emerging market economies, facilitating so-called 'trade in services'. Thus, the transition can be enhanced by trade and migration.

Immigration policy, a feature of all countries, requires careful study and discussion both nationally and internationally. No doubt there are political sensitivities and perceived costs, but in assessing policy it is important to recognise benefits when they occur, particularly in relation to both highly skilled and less skilled labour in the transition, and in the context of ageing populations in rich countries.

Conversely, migration can be a major consequence of climate inaction, as climate change affects vulnerable nations and communities. We are already witnessing climate-induced migration at a global temperature increase in 2024, close to 1.4 °C compared with 1850–1900, and much more could occur at 1.5 °C, a scenario that is increasingly likely (World Meteorological Organisation [WMO], 2025). Indeed, at present emission levels, this threshold would probably be reached within the next decade (Climate Change Tracker, 2024). There would be even greater migration at 2–3 °C (Pörtner et al., 2022; Xu et al., 2020), with large, densely populated areas potentially becoming uninhabitable in vulnerable and populous countries such as Bangladesh, India, China, others in East Asia, Nigeria and others in West Africa, and Ethiopia and others in East Africa. The potential scale of forced migration could be such as to cause major political instability and conflict. Indeed, as we saw in Chapter 2, those effects are already emerging both within and between countries. At higher temperatures they could be much worse.

Interdependencies arise also in the operation of the *international financial system*. While most countries finance most of their investment through domestic savings, others will face new investment opportunities – and indeed requirements – associated with the green transition and with building resilience, that exceed what domestic savings can support. Increasing domestic savings should be a key element of national policy in most cases, but international finance can and should play a crucial role, including from international financial institutions (IFIs), as examined in Chapter 9. Financing such investments will yield benefits for all, through emissions reductions, global growth, improved efficiency, resource availability, and greater resilience. The decisions of IFIs are heavily influenced by their shareholders, usually countries, underscoring the importance of international cooperation in shaping these institutions' direction and functioning.

Trade has grown powerfully over the last century, as the world economic system has integrated. Exports have grown by around 40 times between 1913 and 2014, whereas the overall economy has grown by a factor of around 20. That has been a central element in globalisation (Ortiz-Ospina et al., 2024). Since nations are interconnected through trade, even when a country manages to reduce emissions associated with its own production, it could be contributing to the increasing of emissions associated with its consumption through imports of goods and services. The distinction between production-based emissions (PBEs) and consumption-based emissions (CBEs) is indeed important in thinking about structures of trade and responsibility for emissions. It has become an issue in international climate policy discussions,

Figure 8.1: Top six emitters of consumption-based (dotted line) and territorial-based (solid line) CO$_2$ emissions, 1990–2018

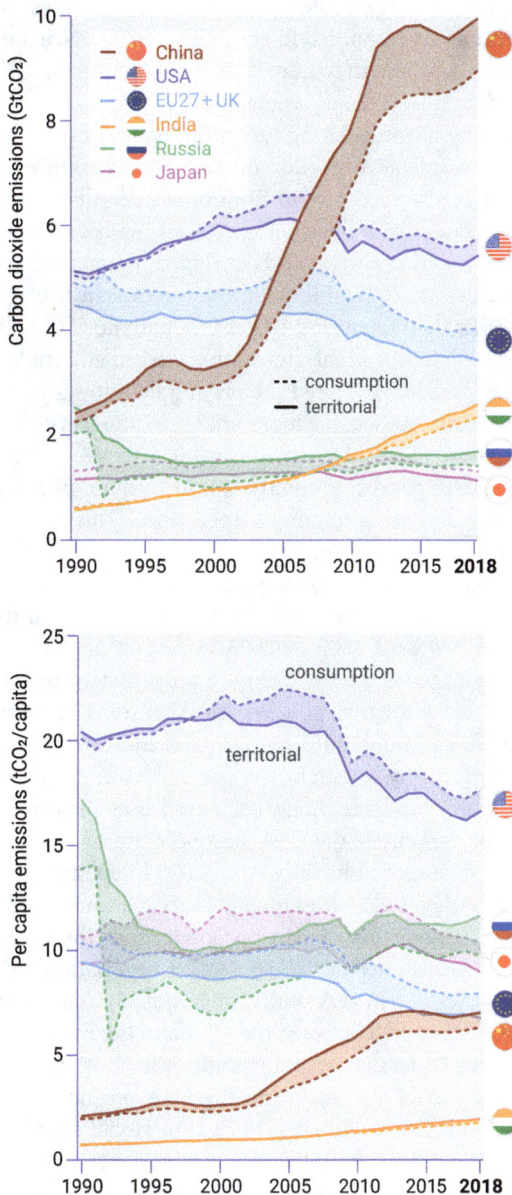

Note: Shading shows the net trade difference for absolute emissions (above) and per capita emissions (below).
Source: Adapted from Figure 2.3 in UNEP (2020, p. 7). Copyright 2020 UNEP. Reproduced with permission.

particularly since the establishment of protocols and agreements such as the Kyoto Protocol and the Paris Agreement and in the context of the shift of low-cost manufacturing out of developed countries.

PBEs account for GHGs generated from domestic activities, whereas CBEs encompass all emissions embodied in goods and services consumed within a country, regardless of where they were produced. Measuring CBEs can highlight the issue of 'carbon leakage', where some countries, particularly developed countries, may have declining territorial emissions but are effectively outsourcing carbon-intensive production to other countries via imports. Developed countries often show lower PBEs than CBEs due to their consumption of imported goods from developing regions, with developing countries often being net emission exporters and developed nations being net emission importers (Dhakal et al., 2022). Figure 8.1 shows that wealthier nations tend to have higher CBEs than territorial-based CO_2 emissions (i.e., PBEs).

Though our world is still highly interconnected, many trade policies and actions in the last decade have tilted the playing field towards fragmentation. The trade environment has become more uncertain and less cooperative. Since the 2008–2009 global financial crisis, geopolitical tensions and other crises, including the Covid-19 pandemic and the war in Ukraine, have led to sceptical narratives on trade and economic interdependence. This is expressed in the increased use of terms such as 're-shoring', 'friend-shoring', and 'decoupling' (World Trade Organization [WTO], 2023). For example, the introduction of new global trade restrictions has been increasing, from around 650 new restrictions in 2017 to over 3000 in 2023 (Seong et al., 2024). The diverse geopolitical drivers impacting trade include a growing tendency to implement protectionist-oriented industrial policies, trade tensions between China, the USA, and the West, and trade shifts by countries and regions, such as the EU stopping certain types of trade with Russia due to the war in Ukraine (Gilbert et al., 2024). Trade policy has increasingly been used as an instrument of political as well as economic pressure in the USA, Europe, China, and elsewhere.

The wave of tariffs imposed in early April 2025 by Donald Trump's second administration has triggered a series of retaliatory measures, with Canada and China implementing their own tariffs on goods from the USA and the EU proposing to impose counter-tariffs, subject to negotiations with the USA. Positions and policies are in flux and change quickly (an outline deal with reciprocal tariffs was reached between the USA and the EU in July 2025). This escalation has raised concerns about a potential global trade war, inflationary pressures, and the risk of a recession in the USA or, indeed, globally. Such tariffs and retaliatory actions threaten to disrupt global supply chains, while the resulting uncertainty undermines investor confidence. These developments could hinder the energy transition by stifling investment, increasing costs, and reinforcing the status quo. The evolving trade patterns and tariffs are coming to represent a third, more structural, disruptor in the global economy, joining the continuing impacts of Covid-19 and the conflict in Ukraine. It is unclear how all this will play through, but the risks of major trade disruption are substantial.

Emissions reductions: global cooperation across the spectrum

In highlighting these interdependencies and their critical role in shaping international action, it is important to remember that whilst all countries can and should embrace the new growth story, the shape that it takes in different countries will be influenced by endowments, incomes, economic structures, and past history. Effective global action requires cooperation across the full range of the spectrum of incomes and backgrounds.

Historically, high-income nations have been the largest emitters of GHGs, and their economies have often been rooted in carbon-intensive industries. These nations possess the financial resources and technological expertise to transition to cleaner energy sources and sustainable practices. And many clean activities and energy sources are now cheaper than their polluting counterparts, as we saw in Chapter 5. In many of the richer countries, emissions have begun to decline, though not yet at a pace consistent with meeting the Paris Agreement goals.

As emphasised throughout this book, EMDCs, both middle- and lower-income, have a particularly important role in determining future emission trends. The current scale of emissions, together with the future potential economic growth in middle-income countries, places them at the forefront of global climate action. Though rich countries still have high emissions per capita, emissions per capita are now higher in China than in Europe. And other middle-income countries will be critical to future emissions over the next two decades, given their expected economic growth. Many middle-income nations, undergoing rapid industrialisation and urbanisation, understandably are focusing their strategy and policy on economic growth and poverty reduction. However, as we have consistently argued, the green transition can drive growth and development forward, particularly in association with infrastructure, the expansion of which is central to development in EMDCs. But realising this new path requires strong investment and structural change. It offers great rewards but it will not be easy.

Lower-income countries, despite having contributed the least to global emissions, both historically and today, must chart a different path from the carbon-intensive development models pursued by high-income and, more recently, many middle-income countries, if there is to be a sustainable future for all. The sheer number of people in poorer countries makes this imperative if the goals of the Paris Agreement are to be achieved. Poorer countries, particularly in Africa, could account for a significant share of potential emissions by mid-century given projected population and income growth – perhaps more than two billion people in Africa by 2050 (Mostefaoui et al., 2024; Rooper, 2024; Stanley, 2023).

Interdependencies are evident in both the effects of emissions and the capacity for change. The poorest countries require support from the international community to build resilience and adapt to increasingly volatile climatic conditions. They also need support in finance and technology to leapfrog traditional, carbon-intensive development paths.

As we have consistently emphasised, the coming decade will bring an unprecedented wave of infrastructure investment, particularly in EMDCs (other than China) and particularly associated with urbanisation and industrialisation. China is now advanced in infrastructure investment and investments are likely to stay strong, particularly around high-tech sectors, AI, and the green transition. However, China will not need much external finance, in contrast to other EMDCs. The choices made today in these regions will shape the future for all, especially for the world's poorest populations. In simple terms, Africa's future will be shaped in large measure by the choices made in Asia's infrastructure development over the next two decades. If Asia's coming infrastructure, so much of which will be built in the two decades or so to mid-century, looks anything like the current, then the world as a whole will be headed towards 3 °C and beyond. All our futures would be bleak, but the situation would be particularly dire for Africa. Mutual vulnerability is at the core of interdependency. The potential and needs of EMDCs (other than China) are discussed in greater detail in Section 8.5.

Tackling climate change is not a simple zero-sum game where some save resources by letting others do the work. Climate action is an opportunity for cooperative progress. By recognising the interconnectedness and interdependences of our economies and ecosystems, we can devise responses and ways forward that offer benefits for all.

8.2 A new global economic geography

As we enter the sixth industrial revolution, as discussed in Chapter 6, we must draw on what we have learned about economic geography from economic history. Industrial revolutions have been shaped by geography and, in turn, have reshaped it. Economic geography is determined by the interplay of technology, natural endowments, physical geography, skills, finance, culture, social structure, and history.

The fossil fuel era: dirty and concentrated

For over a century, fossil fuels – particularly oil and gas – have shaped global economies and politics. Their influence extends beyond energy supply to geopolitical power, economic dependencies, and the distortion of climate science. In many countries, entrenched interests in the production and use of coal in power generation, heating, and heavy industry resist change, further complicating the shift to renewables.

Resource-rich countries, particularly in oil and gas, influence global markets through control over supply chains, pricing, and political alliances and the wealth that these endowments bring. The Organization of the Petroleum Exporting Countries (OPEC), for example, has sought to control oil prices for decades, since its foundation in 1960. Fossil fuel giants wield enormous financial resources and have deep political ties, shaping energy policies. They have become more and more in evidence at COP meetings, with a clear agenda of slowing down change, as I have observed directly. Their influence has been particularly damaging in spreading misinformation about climate science, delaying public understanding of the crisis (Oreskes and Conway, 2010).

It took until COP28 in Dubai (2023) for an international climate agreement to recognise the need to phase out oil and gas. Even then, momentum stalled at COP29 in Baku (2024), where the commitment was not explicitly reaffirmed. Fossil fuel interests continue to work to undermine global agreements and slow the shift to sustainable energy, just as they did via their attempts to throw doubt on climate science.

The war in Ukraine exposed yet again the instability that comes with fossil fuel dependence. It was demonstrated all too strongly with the two oil price crises of the 1970s, which plunged the world into recession and sparked major inflation. The conflict in Ukraine disrupted global energy supplies, driving nations to seek an immediate replacement for Russian gas, triggering a short-lived resurgence in fossil fuel investments. Oil and gas companies seized the moment, pushing a misleading narrative that fossil fuels ensure energy security. In reality, reliance on fossil fuels deepens energy insecurity, tying economies to volatile markets and geopolitical conflicts.

Renewable energy, especially wind and solar, offers a more stable, cost-effective, and secure alternative. Unlike oil and gas, which fluctuate with political tensions and which face rising costs over the longer period associated with exhaustible resources,[3] renewables can enhance domestic energy independence and, as seen in Chapter 5, their cost is increasingly lower. Among their benefits, renewables are available throughout the world, instead of being concentrated in specific sites, and are in this sense a more inclusive source of energy. They do not rely on energy choke points associated with particular places, such as the Persian Gulf; it is harder to disrupt their delivery and they cannot be embargoed; they can be deployed in a decentralised manner, and on most scales have very low operating and maintenance costs. The costs of storage have been falling strongly, with the costs of batteries having declined by 99% between 1990 and 2023 (Walter et al., 2024). The list of benefits goes on and on (International Renewable Energy Agency [IRENA], 2019; Roser, 2020). Yet, despite the growing advantages of renewables, vested interests within the fossil fuel industry sought to exploit the crisis caused by the Ukraine war to justify new exploration and drilling, pushing for expansion under the guise of national security.[4]

As renewable energy grows and fossil fuel demand begins to decline, many oil and gas companies are betting on outlasting their competitors. Their strategy assumes that as the market shrinks and prices fall in the 2030s, they will nevertheless survive. But it will be the lowest-cost producers, mostly in the Middle East and Gulf Region, who will remain profitable the longest. This assumption or belief drives continued exploration, even in high-cost, high-risk areas like the Arctic, deepwater reserves, and fracking sites. Credible plans to limit global warming to well below 2 °C require deep cuts to fossil fuel demand (Energy Transitions Commission [ETC], 2024). Thus, if the world meets its climate goals, most new oil and gas projects will become stranded – expensive ventures with little return. The alternative is the very destructive one of not meeting the climate goals. The oil and gas industry is betting strongly on the failure of the Paris Agreement and working hard to realise that failure.

We should recognise that the transition to clean energy poses significant challenges for developing nations that rely on fossil fuel revenues, especially those that struggle to diversify their export base or develop alternative sources of economic growth speedily.[5] For this reason, as explained in Chapter 6, this shift towards the clean must be as planned, smooth, and just as possible, not shocking, abrupt, and damaging. Unanticipated or abrupt changes in fossil fuel demand and related industries can hit growth and employment strongly, as we saw in the 1970s and, more recently, in 2022–2023 with the war in Ukraine. There are multiplying economic effects and, potentially, political and social disruptions. Financial benefits from fossil fuel price rises generally go to the rich, whereas damages are felt by the whole population, particularly through price rises for energy but also as government and other expenditures are cut.

Global coordination can help mitigate these risks by reducing uncertainty on the path for the global fossil fuel phase-out the world will undertake. For example, increased predictability in fossil fuel demand and in the path of price reductions for producers can help avoid stranded assets by enabling investors to make more measured decisions (Puyo et al., 2024). And, given the many dangerous climate risks described in this book, the timelines for fossil fuel phase-out should be aligned with the Paris Agreement. If some oil and gas development (as opposed to new exploration) takes place in the transition process, it would be better for income distribution and economic justice if this took place in poorer countries. Examples of international initiatives contributing to this purpose are the Fossil Fuel Non-Proliferation Treaty and the Beyond Oil and Gas Alliance (see Pastukhova and Walker, 2024).

The decline of fossil fuels is inevitable. The benefits from alternative energy sources will be immense for all the reasons we have discussed. Economic development needs energy but energy does not require fossil fuels. The challenge is managing the transition equitably and effectively. Some communities are dependent on the production and use of coal, oil, and gas, and they will need support in the transition. That involves economic diversification, worker retraining, and social protections. Governments must engage stakeholders, adjust regulations, and redesign fiscal policies to ease the shift. Done right, the transition can strengthen local economies, create jobs, and build a more resilient, sustainable energy system, as argued in the discussion in Chapter 6 on the just transition. The fossil fuel era is ending. And the Schumpeterian creative destruction process is on its way. Clean and new firms are overturning the dirty and old. The faster the world recognises this reality, and manages it with careful planning and good public policy, the faster, smoother, and fairer the transition will be.

Figure 8.2: The Global South's enormous renewable energy potential

Solar and wind potential as a multiple of current energy demand

Superabundant: >1,000x   Abundant: >100x   Replete: >10x   Stretched: <10x   No data

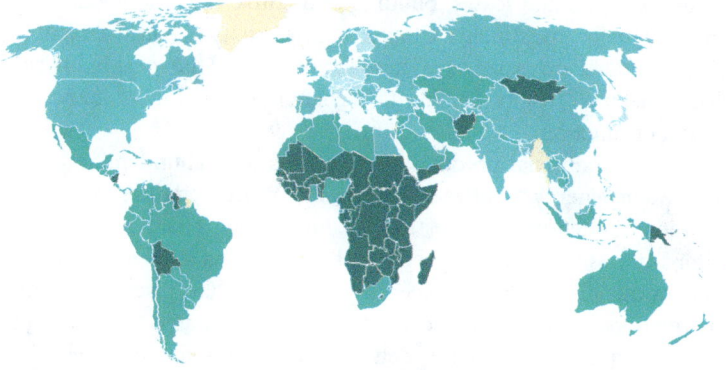

Solar and wind potential as a multiple of fossil production

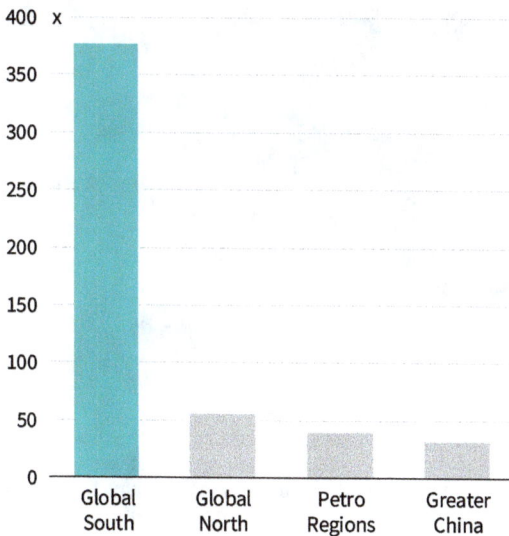

Note: Solar and wind potential, energy demand, and fossil fuel production are measured in terms of energy per annum.
Source: Singh and Bond (2024, p. 8). Copyright 2024 The Rocky Mountain Institute.
CC BY-SA 4.0.

A new global energy landscape: clean, renewable, accessible

We are entering the sixth industrial revolution. The importance of the drive to net zero and advent of new clean technologies is already reshaping the world's economic geography and patterns of trade. Future industrial choices across all countries will place great emphasis on the availability of reliable and low-cost renewable power. Although modern high-voltage DC lines have reduced transmission costs and losses, building and converting the necessary grids will require major investments. Converting electricity to hydrogen and transporting hydrogen are expensive. These factors favour the local industrial use of clean, cheap electricity, leading to a changing global map of energy and industrial production.

The shift of industrial production to regions with abundant and affordable renewable energy is already underway. Many activities will move to where there is plentiful low-cost, clean, low-carbon energy, particularly in emissions-intensive, trade-exposed industries such as steel, aluminium, and base chemicals. This trend is likely to benefit EMDCs, given their renewable energy potential and faster economic growth compared to richer nations, particularly in terms of energy demand and infrastructure development.

Figure 8.3: Renewable market attractiveness

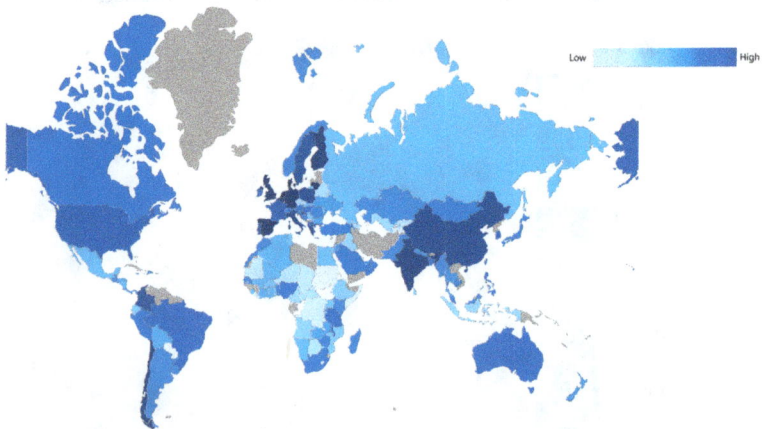

Note: The renewable market attractiveness score is from Climatescope (2022). Climatescope scores markets on a 0–5 scale based on their relative attractiveness for clean power investment and deployment. Scores take into account 163 indicators, or data inputs, which fall under three parameters: fundamentals, opportunities, and experience. The 'fundamentals' parameter encompasses a market's key policies and market structures that help or hamper renewable energy investment or deployment. The 'opportunities' parameter examines a market's potential to grow its supply of renewable power, clean transport, and clean technologies for heating. The 'experience' parameter takes into account markets' experience in deploying renewable power. Source: Adapted from Figure 2 in Stern and Romani (2023, p. 7) with Systemiq analysis, using BloombergNEF's (2022) ClimateScope data. Copyright 2023 The authors. Reproduced with permission.

With the high electricity demands of many applications of AI, there is a similar attraction for data centres to move close to low-cost, clean electricity. They may in turn attract 'digital nomads' who can work anywhere, need excellent and reliable internet and electricity connections, and would be attracted to beautiful locations where living costs are low.

While these shifts in economic geography have great potential for many EMDCs, the investments necessary to accelerate them are constrained for many EMDCs (other than China) by the high cost of capital and their often challenging investment climates. Overall, as can be observed in Figure 8.2, the Global South has enormous renewable energy potential, including 70% of global solar and wind resources. Further, 50% of critical minerals are located within the Global South (Singh and Bond, 2024). Africa and South America are two developing regions particularly abundant in wind and solar power. Realising these renewable potentials across the world will require substantial investments in infrastructure, including strong and effective power grids and storage facilities. Effective power markets, in the context of well-functioning grids, allow flexibility across geography and time to match changing supplies and demands. These markets will play an important role in the design and efficient use of grid and storage systems. Countries like Kyrgyzstan and Canada have extensive hydroelectric power-generation potential, while volcanic regions such as Iceland and Indonesia offer significant geothermal energy opportunities. Mongolia has very strong potential in both solar and wind (and mineral endowments). There are many more examples. We will be seeing real change in the world's economic geography.

Overcoming the constraints of a weak investment climate and high capital costs, and realising the great power and mineral potential across the developing world, will be a central focus of the examination of international policy in Chapter 9. However, despite these difficulties, the pull of the new opportunities is already strong and signs of movement are emerging. Recent outcomes from renewables auctions suggest that many EMDCs other than China are well positioned to become low-cost, zero-emissions power producers; the Middle East and North Africa, for example, are highly competitive (International Energy Agency [IEA], 2019). Figure 8.3 shows the attractiveness of renewable markets across countries. The map shows that some regions, in particular Africa, continue to face significant barriers to attracting renewable energy investment, largely due to high perceived risks and elevated capital costs. This threatens to undermine their natural advantages in solar and wind resources. Chile, Brazil, Saudi Arabia, the United Arab Emirates (UAE), Mauritania, Namibia, and Kazakhstan are all developing landmark projects that could see them become strong players in net-zero industries (Stern and Romani, 2023).

The interaction between AI and the green transition

It has been a constant theme in this book that the intersections between AI and the green transition will constitute powerful drivers of future growth across the world. The geography of these intersections will be shaped by endowments of minerals and resources, together with entrepreneurship, finance, and governance. The *Energy and AI* report from the IEA distils the issues well:

> Countries with a record of reliable and affordable power will be best placed to unlock data centre growth, localise the computing power that is critical to homegrown AI development, and spur the IT industry more generally. Data centres can also be anchors for new low-emissions power projects. However, in regions with frequent power outages or power quality issues, maintaining a data centre can be risky or costly, making overseas hosting more appealing for businesses. […] AI-driven applications may enable some of these economies to sidestep legacy systems and adopt cutting-edge energy management solutions directly. […] For some, attracting data centres through reliable green power could be a catalyst for modernisation; for others, the priority might be smaller-scale digital tools that bolster rural electrification or reduce transmission losses. In all cases, addressing both energy and digital connectivity gaps together is crucial. By fostering local data collection, developing talent and creating robust policy frameworks, emerging market and developing economies can harness AI to drive more inclusive, future-proof growth – growth that integrates renewable energy expansion, meets rapidly rising demand and supports new industries in the process. (IEA, 2025a, pp. 18, 238)

The location of and the pace of change in new geography described in Figures 8.2 and 8.3 can be strongly influenced by the application of AI. It will play a crucial role in the discovery of new minerals. And it will be critical in designing the systems, particularly energy and transport, necessary for the development of both new power and the processing of minerals so that host countries can add value effectively and develop new production and service capacity.

Different countries are likely to follow different strategies here, depending on their skills and endowments. For example, countries such as Chile, Mexico, and Egypt, which are rich in clean energy, will likely seek to attract AI investments. Countries such as India, Brazil, and Thailand, which already have strengths in AI, will seek a rapid build-up of clean energy (Budaragina et al., 2024).

8.3 New opportunities: new resources, new players, competition

The transition to clean energy technologies will depend on the availability and cost of metals and minerals like copper, cobalt, lithium, and rare earth elements, some of which are geographically concentrated, at least in terms of currently known resources. Whilst resource discoveries are occurring rapidly and technology will change demand (ETC, 2023), there is a risk of major demand–supply imbalances for some metals and minerals as the transition accelerates. Diversifying supply chains, developing mining capabilities in multiple resource-rich countries, and investing in recycling technologies are important strategies to reduce and manage these risks. Promoting technological innovations that reduce metal and minerals requirements and encouraging material substitutions can also increase the security and resilience of clean energy technology supply chains and enhance strategic independence.

Both international competition and collaboration are needed to drive the necessary innovation and growth. In past technological changes we have seen that the price mechanism (prices rising with scarcity), entrepreneurship, and creativity in engineering have found new supply sources and possibilities for substitution (Freeman and Perez, 1988; Otojanov et al., 2022). But for those processes to work well and at the pace now necessary, anticipation and planning are crucial. The IEA (2024a) has warned of potential supply shortfalls, particularly around copper and lithium.

Rising use of critical minerals

Many of the needed resources for the clean transformation are concentrated in specific geographical regions. For instance, cobalt, essential for current EV batteries, is predominantly mined in the Democratic Republic of Congo, with 70% of global supply, and refined in China. Similarly, lithium, another key material for EV batteries, is primarily mined in Australia (47%), Chile (30%), and China (15%) (ETC, 2023), with most processing occurring in China (58%) (Bian et al., 2024). So-called rare earths (many of which are not so rare), such as neodymium, dysprosium, and terbium, are essential elements for powerful magnets, hard drives, screens, LED bulbs, and so on. Thus they are crucial for many aspects of high-end manufacturing, digital systems, smartphones, motor vehicles, and so on. At present, processing facilities for these rare earths are also concentrated in China (see below).

Further, as the IEA emphasises:

> Besides the additional electricity demand, a major consideration related to the rapid growth of AI and data centres is the demand for critical minerals. Apart from bulk materials like steel and concrete, the construction of data centres requires sizeable amounts of several minerals and metals, such as copper, aluminium, silicon, gallium, rare earth elements, and battery minerals. There is a significant overlap between the minerals needed for building new data centres and those that are critical to energy technologies. (2025a, p. 213)

Part of the challenge of fostering the contribution of mineral extraction to local development is working to avoid harm to the environment. These activities can be highly damaging, particularly in areas rich in biodiversity, and there are potential social risks related to the treatment of workers in new mining regions. International coordination and standards have important roles to play. For example, the UN's International Seabed Authority is developing norms for deep-sea mining in international waters that are expected to be finalised by 2025 (Ashford et al., 2025).

Here again, we can learn from the past: a rise in mineral requirements occurred in previous industrial revolutions. In the 19th century, the location of industry and populations in the UK was heavily influenced by the availability of key minerals such as iron ore, limestone, and coal, essential for steel production, as well as coal for energy. The technology of the time required these materials in large quantities, and due to their weight and the cost of transport, industries were often located close to these resources. This led to the rise of cities like Manchester in England, Glasgow in Scotland, and Cardiff in Wales. Similar industrial clusters emerged across Europe (e.g., the Ruhr area in Germany) and the USA (e.g., Pennsylvania) during the 19th and early 20th centuries, driven by resource availability and industrial demand.

Historically, force has also played a powerful role. Colonial countries often extracted raw materials from colonised markets through the use of labour with minimal rights and pay and bought these materials at low prices. Processed products were sold to colonies at high prices. That colonial economic geography has echoes to this day (Acemoglu et al., 2001). The history carries lessons. For the new growth story to be sustainable, resilient, and inclusive, nations that have the minerals needed for the transition should be able to benefit from their use.

Processing of critical minerals and manufacturing of clean energy technologies

The new geography of manufacturing operations in clean technologies will, in some cases, be marked by significant concentration, not only because of resource locations but also because of the integrated nature of modern supply chains. For example, in 2022, China accounted for 80% of the global installed manufacturing capacity of solar photovoltaic (PV) modules, dominating every stage of the supply chain. China also held 75% of global battery manufacturing capacity, with the EU and the USA accounting for only 8% and 7%, respectively (Bian et al., 2024).

Manufacturing these goods requires input supplies with reliable quality and timely delivery. Given economies of scale, there will be large demand for local supply of inputs, particularly where transport costs may be high, such as for heavy ores or for energy. Thus, there are forces pointing to local processing or clustering. There may be energy-efficiency possibilities in clustering around heat or energy sources. We see industrial parks arising where clean energy, inputs, and skills are all available in close proximity (Hoicka et al., 2025). And there is real potential for local processing of minerals where cheap clean

Table 8.1: Upstream and midstream dominance: China's role in the mining and refining of main transition-critical minerals (TCMs) for energy generation as a percentage of global figures

Material	% of reserves in China	% of extraction in China	% of processing in China	Needed for
Aluminium			58%	Wind, PV, CSP, hydro, biomass, EVs, grid networks
Cadmium	15%*		42%	PV
Cobalt	1%	1%	65%	EVs, biomass, carbon capture and storage (CCS), electrical storage
Copper	3%	9%	42%	Wind, PV, CSP, hydro, geothermal, biomass, nuclear, grid networks, EVs, electrical storage
Graphite	16%	65%		EVs, electrical storage
Indium			59%	PV, nuclear
Lithium	8%	15%	58%	EVs, electrical storage
Manganese	16%	5%		Hydro, geothermal, nuclear, CCS, EVs, electrical storage
Molybdenum	31%	40%		Wind, hydro, geothermal, nuclear, CCS
Nickel	2%	3%	35%	Wind, hydro, geothermal, biomass, nuclear, CCS, EVs, electrical storage
Rare earth elements (REEs)	34%	70%	87%	Wind, EVs
Selenium	8%		41%	PV
Silicon	37%	70%	68%	PV
Vanadium		70%		Electrical storage

Note *Cadmium reserves are taken as the same as zinc reserves, and all reserves are deposits that are currently financially viable. Less than 20% is highlighted in green, 20–50% is highlighted in blue, and above 50% is highlighted in red. This list includes the nine key TCMs identified in the study by Miller et al. (2023), based on expected demand induced pressures from the net zero transition, as well as those materials required for PV and wind technologies. Data compiled by Bian et al. (2024) based on data from the USA Geological Survey, IEA, and Miller et al. (2023).
Source: Adapted from Table 2.1 in Bian et al. (2024, p. 6). Copyright 2024 The Authors. CC BY-NC 4.0.

Figure 8.4: Installed manufacturing capacity by country/region, 2023 (%)

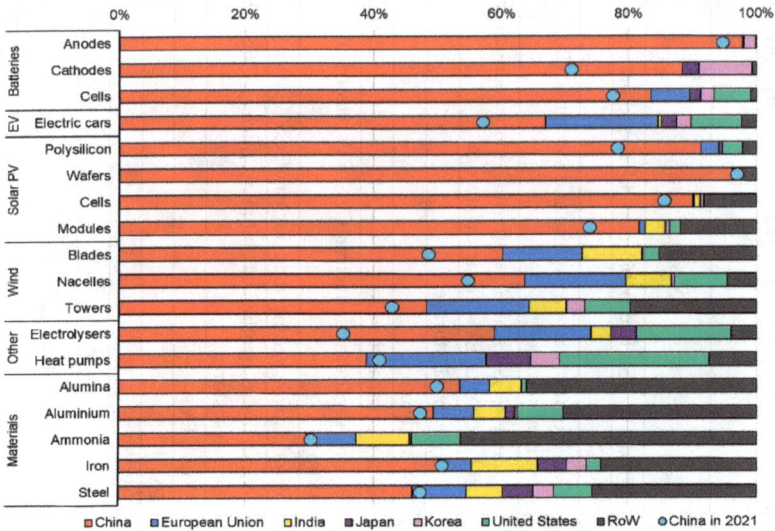

Note: RoW = Rest of World. 'Electric cars' values are calculated based on 2023 production numbers, adjusted according to the utilisation rates of car assembly plants in the region. Source: Figure 1.11 in IEA (2024b, p. 46). Copyright 2024 IEA, CC BY-SA 4.0.

energy is available. At the moment, however, those with processing capacity, particularly China, have been investing heavily in securing access to supplies in resource-rich countries (see Leruth et al., 2022).

As shown in Table 8.1, critical mineral and metal processing capacity is highly concentrated in China: the country owns 58% of both lithium and aluminium processing capacity. And whilst only 1% of global cobalt extraction takes place there, 65% of the global processing capacity of this mineral occurs in China (Bian et al., 2024). Figure 8.4 shows the concentration of manufacturing capacities of key inputs and components of clean technologies, particularly for solar and battery. China exhibits a strong advantage across the different segments of the supply chains as a result of the creation of strong local ecosystems and industrial parks for the production of solar PV and batteries, characterised by high quality, strong reliability, and low costs. Similar developments are occurring in EVs and other key technologies.

China's momentum through research leadership is remarkable, and manufacturing concentrations might intensify still further as a result of this leadership. ASPI (an Australian research unit and think tank) tracks research leadership over 64 critical technologies covering a wide range of activities, including defence, space, environment, energy, computing, and others. From 2003 to 2007, the USA led in 60 of the 64 and China led in 3. More recently, between 2019 and 2023, China led in 57 and the USA led in 7 (Leung et al., 2024). China is no longer 'catching up' in technologies but is in many ways leading the way. This raises issues of how countries of the developed world

Figure 8.5: Levelised cost of energy (LCOE) in the USA before and after the implementation of the IRA[6]

Levelised Cost of Energy (LCOE) pre- and post-tax credits

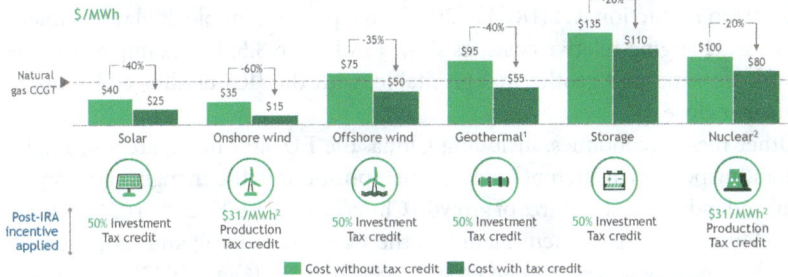

Note: 1. Geothermal values reflect average of traditional flash and EGS technologies. 2. New small modular reactor (SMR). The term 'bonus' refers to additional tax credit percentages available under the IRA for qualifying renewable energy projects. The figure assumes each renewable technology gets the basic tax credit and bonuses for paying fair wages, for using US-made materials, and for being in a disadvantaged or former fossil fuel area. All technologies assume base + prevailing wage bonus + domestic production bonus + energy community bonus. All numbers rounded.
Source: Adapted from BCG (2023, p. 5). Copyright 2023 BCG. Reproduced with permission.

can now catch up. Competition will play its role and many of the developed countries do have strengths in their universities, research institutes, and high-end tech companies. But there is also potential in collaboration. Just as China gained from joint ventures in the 1980s and 1990s (and see discussion on batteries in the next subsection) when it invited foreign investors, there are now real opportunities for developed countries to host Chinese investment. There will, no doubt, be areas such as some aspects of telecoms, or critical infrastructure, where security issues are of relevance, but there will be many others where these issues are minor.

These geographical shifts and the associated opportunities raise important questions regarding the security and resilience of global supply chains for clean technologies and the implications of market concentration. Ensuring a robust, dynamic, and sustainable world economy will require reliable sources of supply, technical progress, and continuing cost reductions. Diversification and competition for key goods and resources, both nationally and regionally, are necessary if the world is to build secure and resilient global supply chains for clean technologies and foster the competition that can promote continuing creativity and rises in productivity.

Collaboration and competition in the transition

There are and will be difficult questions around the role of collaboration and competition in driving the transition forward. Rapid technological advance, innovation, and diffusion will be of great importance. So too the security of supply chains. The fall of output and employment in the increasingly obsolete industries, and the rise in the new clean industries, will both cause tensions.

Countries around the world are now competing to establish dynamic comparative advantages in green technologies and activities. The US$1 trillion US Inflation Reduction Act (IRA) of 2022 was a prime example. It played a major role in changing relative costs, as shown in Figure 8.5. For example, it made offshore wind and geothermal profitable, with the IRA credits, and brought nuclear close.

Other major economies, including China, the EU, and India, are also implementing policies to attempt to secure their positions in the changing geographical and industrial structures of growth. China's 14th Five-Year Plan (2021–2025) on Modern Energy System Planning; the EU's Green Deal; and India's Union Budget, energy access, and green hydrogen policies (Garg, 2022) all combine incentives and finance for green technology and investments with legal, regulatory, and policy support (see Section 9.4). The 15th Five-Year Plan (2026–2030) in China will likely see further strong movements in these directions.

There is also backlash and resistance to change. In his second presidency, Donald Trump seems to want to go back to an era where the car industry in the USA, using the internal combustion engine, was dominant. Some in the steel industry appear to want to retain the use of blast furnaces as long as possible. Trump has been attempting to use very high tariffs as a weapon. The Trump administration is likely to roll back some of the Biden era's IRA support for clean energy. At the same time there is resistance to this attempt both in the courts and in areas (many of them with Republican governors, senators, and congress members) that benefit from IRA measures. We have yet to see how this will play through.

For a discussion of the falling costs of green energy, see Section 5.1. Having four or five major producers of crucial materials, products, and inputs could be sufficient to both provide supply chain resilience and foster the competition that can drive rapid technical progress. Strategic protection designed to foster new globally competitive supplies makes sense. But key sectors which have a chance of 'making it' in global competition need assessment and identification. And across the board protectionism, as economic history has taught us, can lead to prolonged weakness and inability to compete and change. Examples of successful strategic protection were in South Korea and Taiwan in the 1960s and 1970s. There are also many unsuccessful examples, such as in the history of Indian motor cars and in UK motorcycles in the 1960s and 1970s.

As countries attempt to build new production capabilities, there is growing concern about the increasing use of blanket protectionist trade barriers. For example, under the Biden administration (2021–2024), the USA raised tariffs against China's solar PVs, aiming to protect the domestic solar industry and address concerns about unfair trade practices. The EU has previously imposed tariffs on Chinese solar panels, but these tariffs were later lifted. Over the course of 2024, the USA, the EU, and Canada each imposed tariffs on Chinese EVs, citing concerns about unfair competition and subsidies and accusing China of

Figure 8.6: Solar module manufacturing capacity in China versus sales, 2020–2028

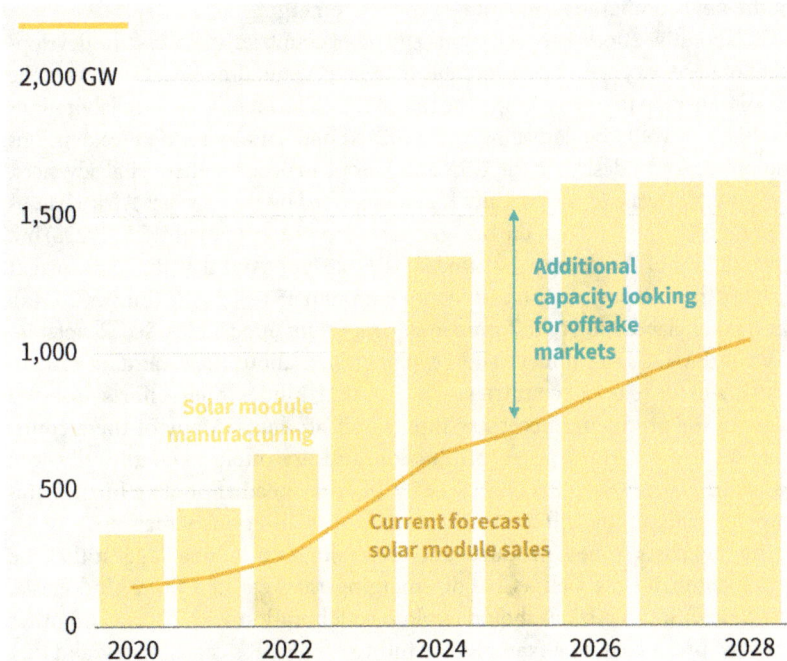

2,000 GW

1,500

Additional capacity looking for offtake markets

1,000

Solar module manufacturing

500

Current forecast solar module sales

0

2020   2022   2024   2026   2028

Source: Singh and Bond (2024, p. 25). Copyright 2024 The Rocky Mountain Institute, CC BY-SA 4.0.

dumping low-carbon technology in their markets at impossibly low prices, due to overcapacity. The protectionist measures of the second Trump presidency are unfolding in 2025 and there have been strong reactions from other countries, particularly China. We have yet to understand how all of this will develop, but it does seem that a serious bout of protectionism may be with us for some time.

Developing alternative supply sources for key inputs will be very important for future competition and reliable supply chains. But in the meantime, with China's low-cost and massive production capacity for solar panels,[7] as shown in Figure 8.6, it makes global economic sense to encourage the export of these panels to poorer countries, particularly given the urgency of climate action. As Figure 8.6 illustrates, China is in a position to export hundreds of gigawatts (GW) of solar panels per annum. Energy is crucial for development. And so much of development now requires digitisation and connectivity, which are very difficult without electricity. The use of this large and low-cost capacity represents a huge development opportunity.

The net zero transition could be an opportunity for EDMCs other than China to realise their development potential by moving up the value chain of new technologies and products. Developed countries and China could and should help build processing and manufacturing capacity, by facilitating the local use of energy, raw materials, and labour in EDMCs to develop and build on the new comparative advantages that are emerging.

Collaborative efforts between China and other countries, including the development of necessary skills and affordable financing, could dramatically advance the global transition to clean energy. The role of IFIs in facilitating such collaborations is critical. It would be damaging and a missed opportunity for political tensions and protectionist desires in the USA and Europe to obstruct these vital advances. The potential scale of opportunity is demonstrated by the fact that Africa has, as mentioned, around 60% of the best global solar power resources (IEA, 2022a) but less than 3% of the energy investment (IEA, 2024c). And that the total current capacity from all power sources in Africa is around 250–300 GW (Ember, 2024a). For further elaboration of this immensely important opportunity, see Chapter 9.

Many EMDCs are understandably concerned about the emerging protectionism in developed countries, which could hinder their efforts to build capacity for producing necessary inputs and products. Many of these countries see the future in green production and are often geographically well positioned to develop this new capacity, but they need affordable finance and confidence that they will not be excluded from rich-country markets.

The low costs of new products and technologies in China suggest that the richer countries, as well as major emerging markets like India, Indonesia, Brazil, and South Africa, should explore collaboration with China and other low-cost producers. For example, the future of the UK's car industry, and that of Europe more broadly, lies in the production of EVs. Batteries make up a high proportion of an EV's cost, generally around 30–40% (IEA, 2022b). EV production cannot be competitive without access to low-cost batteries. Thus, collaboration with China in battery production could be crucial for the car industries of, for example, the UK and the EU. That is of particular importance in the case of the UK, given the need to meet local content requirements for export to the EU under trade agreements with the EU.

Car manufacturers in the UK and the UK government are beginning to recognise these realities. Tata, the Indian owner of Jaguar Land Rover, is collaborating with Chinese firms to produce batteries for its new EV plants in Somerset in southwest England, and plans to use electricity supplied by Électricité de France SA (EDF) from nearby nuclear Hinkley Point. Nissan, and other Japanese car firms, are collaborating with Chinese firms to produce batteries for their EVs being made in Sunderland in northeast England.[8] These examples of international collaboration highlight how transitions can be accelerated, costs reduced, competition and innovation fostered, and robust supply chains built. However, protectionism and isolationism driven by narrow interests could undermine these opportunities, raising costs, stifling innovation, and slowing the transition.

Protectionism is often justified on the grounds that, if given a chance, costs will fall domestically. That is sometimes referred to as the 'infant industry' argument. When protectionism favours established industries, it may aim to delay what could be an inevitable decline, or to provide a window of opportunity for transitioning to newer, more productive methods of production. History suggests that such propositions require rigorous scrutiny. Infant industries may fail to grow up and postponing change can be costly in the case of declining industries.

An interesting development in international competition is the potential importance of integrated industrial parks and local supply ecosystems. Martin Brudermüller, former CEO of BASF, the world's largest chemical firm, has argued that the future of high-temperature processes in the chemical and other industries would rely more on electricity than hydrogen, underscoring the importance of electrification, even in 'hard-to-abate' sectors.[9] He has also emphasised the importance of integrated industrial parks and local ecosystems which allow for the efficient combination of clean electricity, complementary inputs, infrastructure links for distribution and export, and local skills. The Transitioning Industrial Clusters towards Net Zero initiative of the World Economic Forum (WEF), for example, was created at COP26, which now aggregates 35 industrial clusters across 16 nations and 5 continents – representing a combined CO_2 equivalent emission reduction potential comparable to the annual emissions of Saudi Arabia (World Economic Forum [WEF], 2024, 2025a, 2025b).

These ecosystems have been successfully developed in parts of China, providing a real source of competitive advantage. A notable example is the Tianjin Economic-Technological Development Area (TEDA), a mixed industrial park established in 1984 that developed into a complex industrial symbiotic network (Shi et al., 2010). At present, TEDA concentrates a total investment that exceeds US$15 billion, and more than 3,300 foreign-invested firms (United Nations Industrial Development Organisation [UNIDO], 2022). The industrial symbiosis being implemented involves sharing infrastructure and integrated management of solid waste, water, land, and information (UNIDO, 2016). Environmental initiatives at TEDA include the creation of a green development fund, several circular-economy industrial chains, and a real-time air pollution monitoring system (UNIDO, 2020). There are several others of importance in China, including in Guiyang, Shenzhen, Suzhou, and Yunnan. Industrial parks are not, of course, confined to China, with important examples in areas such as Nova Scotia in Canada and Jubail in Saudi Arabia. Brudermüller warned that slow decision-making and local opposition could make it difficult to create such parks in the EU, potentially giving an advantage to countries that can move quickly and effectively.[10]

The example of industrial parks and ecosystems also reminds us that competition within nations can be intense. A key reason why China can supply solar panels, EVs, and other modern clean products so cheaply is the fierce competition among Chinese firms, as mentioned in Chapter 4. Many incorrectly attribute these low costs entirely to subsidies, when in fact they also arise from intense competition, skills ecosystems, and infrastructure (Hove, 2024). Although in some cases Chinese subsidies are indeed large, policy support for clean industries goes beyond traditional forms of subsidies to include strategic signalling on investment priorities, research and development (R&D) spending, and support for developing integrated industrial clusters, resulting in 'an ecosystem of policy support and entrepreneurial drive' (Hove, 2024, p. 1). When we visited the extraordinarily productive Xiaomi EV car factory near Beijing in March 2025, the founder, Lei Jun, told us, in response to a question on state support, that the most important element of support for them was the supply of high-quality engineers as a result of China's investment in its universities. Around half of the employees involved in the car factory were in R&D. Of course, the ability to go to scale is an important element in achieving low cost. China combines fierce competition and scale.

It is, however, important to recognise that China does seem to have greater industrial subsidies (Bickenbach et al., 2024; DiPippo et al., 2022) relative to output than the EU and the USA. But comparisons are difficult because the biggest part is in terms of low-cost credit and it can be argued that this is in part tackling capital market imperfections. Also, many of the subsidies go to state-owned enterprises, whereas the car industry is largely private. Further, China generally works with producer subsidies rather than the consumer subsidies on EVs of the EU. These subsidy comparisons are not straightforward. In my view, whilst China's subsidies are large and important, much EU and US discussion over-emphasises subsidies and under-emphasises competition in explaining low prices of batteries and EVs in China.

The very rapid price falls achieved in China for solar PV, batteries, EVs, and other clean technologies can be of great benefit to the world in achieving sustainable development. Clean technologies have enormous potential in helping low-income countries, particularly in Africa, leapfrog to a new form of development. And their utilisation will reduce emissions to the benefit of us all, particularly the poorest and most vulnerable nations. The challenge is to make best use of the low-cost production capacity in the shorter term and encourage competition so that there will be other low-cost producers across the world in the future.

In some countries, various barriers – including limited fiscal space, high costs of capital, and a shortage of skilled labour – can hinder efforts to build domestic manufacturing capacity. In such cases, an open global trading system can offer lower-cost inputs and machinery produced elsewhere, enabling a more efficient transition. In other instances, some developing countries may find it advantageous to position themselves within the global cleantech supply chain by investing in and supporting local manufacturing capacity. For example, in Latin America, Brazil manufactures wind turbine components and is a leader in biofuel production. Argentina has a mature nuclear energy industry and is working to produce small modular reactors. India is making efforts to position itself as an exporter of solar panels, although it has some way to go to match the low cost of China.

The transition to a cleaner, more sustainable form of economic growth presents an opportunity for developing economies to move up the value chain, rather than remaining at its lower end as simply suppliers of raw materials for technologies manufactured elsewhere. There are real opportunities for developing countries to be integrated into the emerging global cleantech value chains, particularly given their potential in clean energy and the location of raw materials. In all scenarios, international competition can spur innovation and drive down costs, helping to make clean technologies more accessible and affordable worldwide. Sound policies, strong investment, sensible industrial strategies, and openness at the national and regional levels can increase the number of players and foster that international competition.

8.4 Natural capital: investment and impact

Economic geography should change as a result of increased emphasis on biodiversity. As the world comes to better appreciate the value and potential of biodiversity, regions rich in biodiversity could see significant investment in natural capital, including in restoration and conservation. This is an area of research, innovation, action, and investment that has received too little attention.

Much of the world's greatest biodiversity is concentrated in EMDCs. The mountains of the eastern Himalaya region, from southwest China to Nepal and northeast India, form an extraordinary and interdependent ecosystem (Barthakur, 2023), sometimes described as the Third Pole. The hugely rich forest systems of the Amazon Basin, of Indonesia, and of the Congo Basin are examples of immense importance. And there are many more, particularly in the developing world. Yet all of these are now fragile as a result of our failure to understand and act on their fundamental, indeed existential, role in the lives and livelihoods of so many across the world, and as a result of our past patterns of growth.

Our economies are deeply intertwined with biodiversity, ecosystems, and the broader biome. As we learn to invest in these critical structures, we are likely to see investments in new geographical areas across the world (Bromley, 2024); that will be a crucial element of the 'new economic geography'. We should understand these investments also as building and preserving critical infrastructure; this natural capital is indeed infrastructure in the basic sense of supplying services upon which other activities depend for their effective functioning, including in this case clean water and air.

When considering new economic geographies, we must not only examine the future locations of green technologies and activities, but also consider the potential impacts of climate change in relation to natural capital. Some regions could become uninhabitable due to desertification, submergence, or constant threats of flooding. The Sahara is expanding southwards. The melting of glaciers and ice in the Himalayas could destabilise lives and livelihoods across many surrounding countries, causing devastating floods in the rainy season and droughts in the dry season. These changes could drive massive population movements, as discussed in Chapter 2. Many of the world's great cities are built in coastal or flood-plain areas which would be threatened by sea-level rises and storm surges.

The location of economic activities related to natural capital could also be profoundly affected for other climate reasons. For example, if much of southern Europe becomes intolerably hot in the summer, tourism could move to other regions (Rannard, 2023). In agriculturally dependent regions, enhancing climate resilience involves developing drought-resistant crop varieties, water management (including drainage), improving soil health, and implementing sustainable agricultural practices. These measures are of huge importance for promoting and protecting livelihoods and food security. Thus the geography and technology of agricultural production will change. Southern fish populations may also migrate, potentially transforming the geography of the fishing industry (Ojea et al., 2020). As we have seen in Part I of this book, we are on our way past 1.5 °C; even with much stronger mitigation, major adaptation will be required across the world, with its accompanying changes in the location and nature of activities.

8.5 Differences between nations: EMDCs' energy potential and infrastructure needs

In this chapter, we have explored so far the interdependencies of nations and how varying endowments of resources and differing climate impacts could recast global economic geography. A key message from this chapter is that each country's historical development and endowment of natural resources will shape its approach to the climate transition. While scaling up electrification and renewables is a central feature for all countries, the specific paths to climate action and development depend heavily on each country's resources, its history and culture, its infrastructural capabilities and potential, and its socioeconomic context. This is important not only when thinking about pursuing strategic mitigation actions, but also in relation to adaptation and investment and conservation of natural capital. Adaptation strategies must be aligned with the specific vulnerabilities of different regions to manage the impacts of climate change effectively. And ways of conserving and sustainably using natural capital will be specific to geographies.

The following are some examples of how approaches to climate action, including mitigation, adaptation, and development, are likely to vary across countries and regions.

- In nations with abundant critical minerals, such as the Democratic Republic of Congo, which produces around 70% of the world's cobalt (Ritchie and Rosado, 2024), investing in mining capacities is crucial.
- In regions with high solar resources, such as in Africa, which has 60% of the world's best solar resources but less than 3% of the energy investment (IEA, 2024c), increasing clean investments is key to development. There is potential for combination with processing of materials.

Figure 8.7: Share of global population with access to electricity in 2021

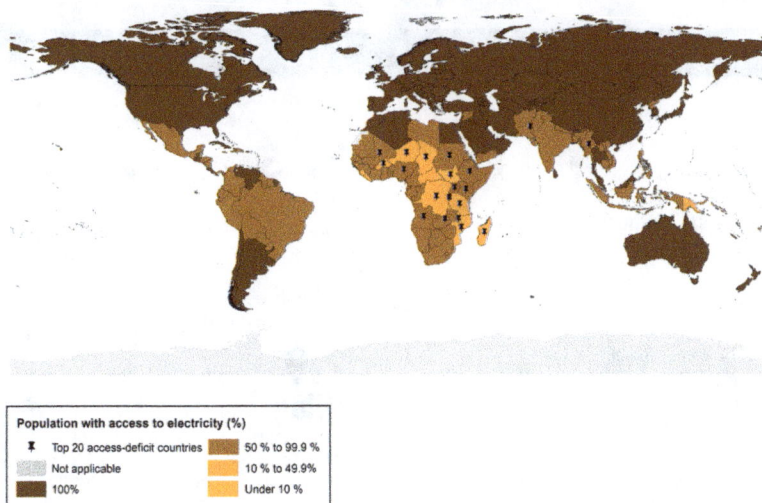

Source: Figure ES.2 in IEA et al. (2023, p. 13). Copyright 2023 International Bank for Reconstruction and Development/The World Bank, CC BY-NC 3.0.

- In forest-rich areas, such as the Amazon, afforestation, combating deforestation, and improving biodiversity protection and management are critical to the world's future.
- In coastal regions, such as the Philippines and small island states, investing in marine conservation and sustainable fisheries is vital to lives and livelihoods, as is protection against rising sea levels and storm surges.
- In regions with high exposure to intense rainfalls, such as Nepal, investing in water management systems is central to managing risk and fostering development.
- In agriculturally dependent regions, such as many nations in Latin America, investing in drought-resistant crop varieties is critical.

In this section, we focus on the differences between nations as they pursue their transitions, particularly focusing on the infrastructure and energy needs of developing countries, which lie at the heart of the new growth story. Indeed, much of the necessary infrastructure in the developing world has yet to be built. That constitutes a huge difference from most developed nations, where the challenge is more in terms of refurbishing and recasting existing infrastructure. But even in developed countries there will be much building of the new.

Across the world, but mostly in developing nations, one billion people live more than 2 kilometres from an all-season road and nearly four billion people lack access to the internet (World Bank, 2024). A substantial proportion of the infrastructure in these countries will be built in the decades to come. For example, nearly 70% of the urban infrastructure likely to exist in India in 2047 will be

Figure 8.8: Global progress toward SDG7 (affordable and clean energy) targets

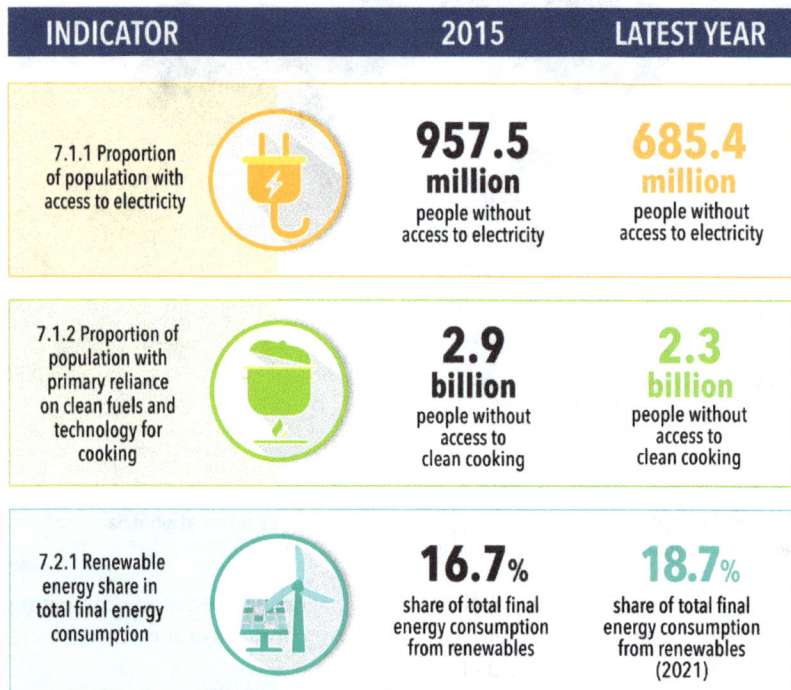

INDICATOR		2015	LATEST YEAR
7.1.1 Proportion of population with access to electricity		**957.5 million** people without access to electricity	**685.4 million** people without access to electricity
7.1.2 Proportion of population with primary reliance on clean fuels and technology for cooking		**2.9 billion** people without access to clean cooking	**2.3 billion** people without access to clean cooking
7.2.1 Renewable energy share in total final energy consumption		**16.7%** share of total final energy consumption from renewables	**18.7%** share of total final energy consumption from renewables (2021)

Source: Figure ES.1 in IEA et al. (2024, p. 10). Copyright 2024 International Bank for Reconstruction and Development/The World Bank. CC BY-NC 3.0.

built between now and then (Kouamé, 2024). And sub-Saharan Africa accounts for about 80% of the people living without electricity (IEA et al., 2023), as shown in Figure 8.7.

By acting now, developing countries can avoid locking in outdated, polluting infrastructure. Doing so requires investment on urgency and scale, and that, in turn, will involve substantial external assistance. The annual investment required for the energy transition in EMDCs (other than China) is estimated at US$1.6 trillion by 2030 (Bhattacharya et al., 2024). That represents a major increase of annual investment in the energy transition in these countries compared with now: by a factor of five by 2030, and six by 2035 (compared to 2022) (Bhattacharya et al., 2024, p. 15; see Figure 4.14 in this volume).[11] These are investments that will be core to the development needs of developing nations with high economic returns beyond reducing emissions, and which will drive growth.

India, a key example, will need to increase its annual wind and solar additions by around five-fold (Ember, 2023) by mid-century to support population and economic growth, higher living standards, and the electrification necessary to power the new growth and reduce emissions. By then, all electricity should be zero carbon. The sources of clean electricity (wind, solar, thermal,

Figure 8.9: Estimated energy investment by type in selected regions, 2024

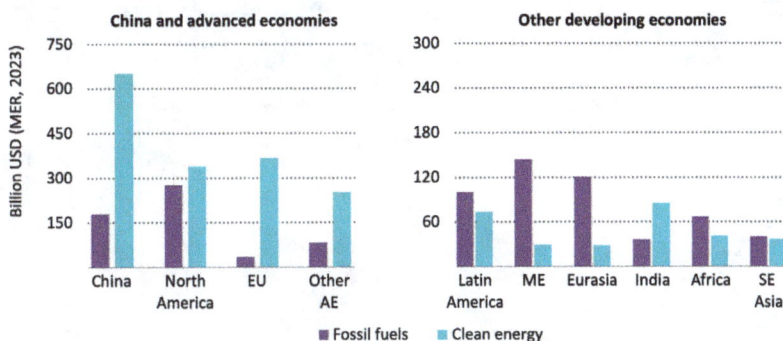

Note: EU = European Union; Other AE = other advanced economies; ME = Middle East; SE Asia = Southeast Asia.
Source: Figure 1.25 in IEA (2024d, p. 62). Copyright 2024 IEA, CC BY 4.0.

hydro, and so on) will vary across countries, but that rapid expansion of clean electricity will be common to all.

As mentioned throughout the book, developing countries have the opportunity to 'leapfrog' to advanced green technologies without the burden of existing outdated systems. However, the cost and availability of capital will be essential in the ability to take these opportunities. At the same time, developed countries, with their established – sometimes ageing – infrastructure and industries, face the challenge of retrofitting or replacing these facilities to accommodate cleaner technologies. For example, Germany, with its long industrial history, is focusing on transitioning its existing industrial bases to integrate renewable energy, increase energy efficiency, change its transport, and so on (Agora Energiewende, 2024). Meanwhile, nations like Kenya are implementing new geothermal and solar projects and leveraging their natural resources, without large legacy power systems. They are building a sustainable energy infrastructure from the ground up (Muiruri, 2024). Financing these changes is the focus of Chapter 9.

The largest share of the necessary infrastructure investment is in the energy systems. Energy, and electricity in particular, is key to unlocking sustainable growth. Ensuring access, affordability, and security in energy supply is crucial for delivering an energy transition that meets development and sustainability goals, and for a just transition. The importance of these advances in energy and electricity is demonstrated by the severe shortcomings and deprivations that many communities and countries faced at the starting point (see 2015 data in Figure 8.8). Even in 2022, 685 million people were still without electricity, and billions had only meagre and unreliable supplies.

Figure 8.10: Solar photovoltaic power potential map

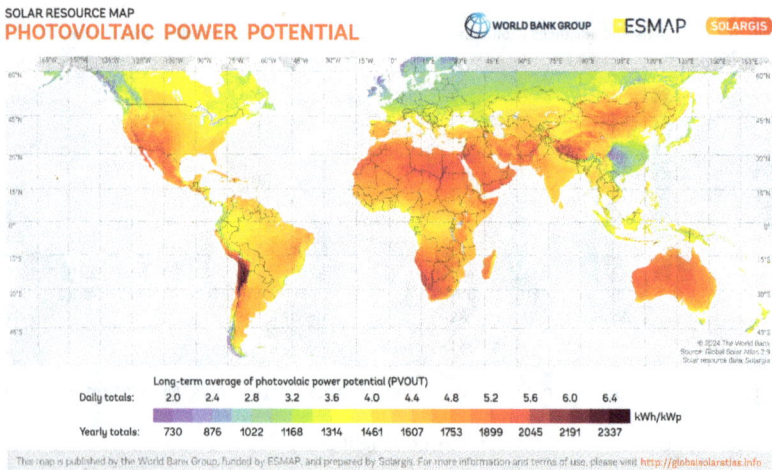

Source: Obtained from the Global Solar Atlas 2.0 (2024), a free, web-based application developed and operated by the company Solargis s.r.o. on behalf of the World Bank Group, utilising Solargis data, with funding provided by the Energy Sector Management Assistance Program (ESMAP). For additional information: https://globalsolaratlas.info.

Figure 8.11: Share of solar in electricity generation, 2010–2020

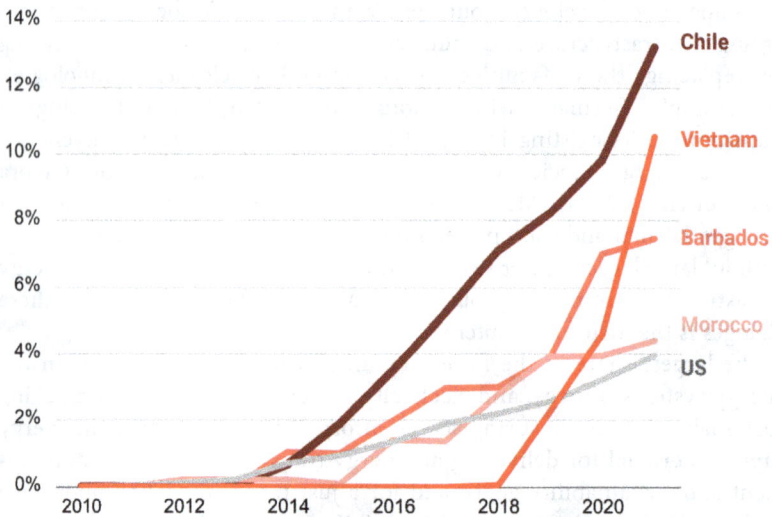

Source: Bond and Butler-Sloss (2023, p. 20). Copyright 2023 The Rocky Mountain Institute, CC BY SA 4.0.

In 2020, the share of renewable energy in total final energy consumption glob-ally was 19.1%. To align with the 1.5 °C target, this share needs to reach 33–38% by 2030, with a share of renewables in electricity generation of 60–65% (IEA et al., 2023). Energy demand is growing fastest in EMDCs, yet, excluding China, these countries accounted for only 15% of global clean energy investment in 2023 (IEA, 2024d). In developing nations, investments in fossil fuels tend to be higher than in clean energy, whereas in advanced countries and China, it is the other way around (see Figure 8.9). However, the Global South has vast renew-able resources (see Figures 8.2 and 8.10) that can be unlocked through invest-ment and taking advantage of the falling costs of clean technologies. However, the realisation of that investment requires the cost of capital to be manageable. The differentials across nations in opportunities are immense but the cost of capital is highest where the opportunities are greatest. That is a key policy ques-tion examined in the next chapter. The risks of delay would be catastrophic; the opportunities for a new and better form of growth are immense.

Many countries in the Global South are already taking advantage of the new technologies and opportunities. For example, Chile, Vietnam, Barbados, and Morocco have already surpassed the USA in terms of share of solar energy in electricity generation. Some EMDCs are also catching up in wind energy. Uru-guay, for example, produces over 30% of its electricity from wind, compared to only 10% in the USA (Ember, 2024b; IEA, 2025b) (see Figure 8.11). Bhutan, Nepal, and Ethiopia are examples of countries exporting clean electricity, particu-larly hydro, with significant economic benefits for them and for the purchasers. Indeed, hydropower largely drives Bhutan's economy: their construction and operation contributed to nearly 18% of its GDP in 2020 (Department of Envi-ronment and Climate Change Bhutan, 2023). Hydropower contributes to 88% of Ethiopia's electricity generation, although 90% of its hydropotential remains untapped and in 2023 only 48% of people had access to electricity (Traide, 2023).

However, as we have stressed throughout, while progress towards green energy is beginning in EMDCs (other than China), it is far too slow to meet the Paris Agreement goals and provide the much-needed boost to growth and development on the scale required. Accelerating this progress requires that poorer countries have access to global capital markets at affordable cost, to technology, and to skills. The difficulties in attracting low-cost capital in some EMDCs (other than China) risk undermining their natural advantages, limiting their ability to leapfrog the dirty stages of development, and slowing their tran-sition to a new, sustainable development path at the pace needed by their popu-lations, the planet, and all its inhabitants. This is the main subject of Chapter 9.

8.6 Concluding remarks: opportunity, international cooperation, and a new economic geography

The narrative around climate and development has evolved from one of bur-den-sharing to one of seizing opportunities. But there are major differences across countries in opportunities and responsibilities. High-income countries, given their historical emissions and wealth, have an obligation to lead by reducing their emissions and providing the financial and technological support necessary for lower-income countries to find new paths of development and build resilience. Middle-income countries can continue to demonstrate that economic growth need not come at the expense of environmental stewardship; with appropriate financing, they can achieve environmental benefits, realise cost savings, and drive a new stage of development. With the right support, lower-income countries can leapfrog in their development, grow sustainably, and build resilience against the impacts of climate change. We can now see, through the potential of the new growth story, that the transition is in the self-interest of each nation and in the shared interests of all.

It is also vitally important to remember that many developing countries are more vulnerable to climate change impacts and that, even with strong adaptation, there will still be substantial loss and damage that cannot be prevented (Intergovernmental Panel on Climate Change [IPCC], 2022). Tackling this will require rapid and substantial international support, both for humanitarian reasons and to mitigate longer-term impacts. In this context, there is a moral responsibility – rooted in historical emissions and the resulting damages – to assist in addressing these dangers. This is in addition to the fundamental imperative of our shared humanity. For many EMDCs, the costs of loss and damage from climate change are already very large. For example, it is estimated that the Vulnerable Twenty (V20) economies suffered losses of more than 20% of their 2019 GDP due to climate impacts over the period 2000–2019 (V20, 2022). The severity of the impacts is rising rapidly.

This basic inequity – that although rich countries are responsible for the bulk of past emissions, poorer countries are hit earliest and hardest by climate change – constitutes a central feature in the differences across countries and a fundamental reason for strong support from rich countries. The very different stages of development across the world, particularly around infrastructure, require different patterns and scales of investment and support.

The transformation of the international economy towards a sustainable future demands a collective effort. It is a shared journey, one that acknowledges the diverse starting points and capabilities of different nations while recognising interdependencies and aiming for a shared future: a world where economic development and environmental sustainability are not only compatible but mutually reinforcing, offering a better future for all. As we work as a world to build that new future, an understanding of both differences and a shared fate will matter. And we must all recognise that the new economic geography that will emerge will bring its own strains as well as new opportunities in new places. We return to these issues of international collaboration in Chapter 9.

Finally, in thinking about differences and new opportunities, we must recognise and understand the new economic geography that has been the main focus of this chapter. The new technologies and the location of potential for renewable energy will be key elements in re-casting global economic geography. New minerals, materials, and metals will move to the core of economic activity and they have their own economic geography.

Taking geography and endowments together will recast international economic relationships. These could bring further tensions on top of those resulting from the change to a multi-polar world which we are already experiencing. And economic power from dominant positions in new technologies and in access to processing and mining the new resources could cause real difficulties in supply chains, security, and relationships. All these issues can be anticipated and managed if there is a shared view of the necessity of change and the opportunities it can bring. International economic policy must embody and build on an understanding of the future transformation of the world's economic geography.

The new economic geography of climate change presents both opportunities and threats. The history of industrial revolutions and changing global economic activity shows that the economic and social effects can be difficult and divisive if not carefully and collaboratively managed. While such management has traditionally been a national issue, the global nature of climate change and its response will require a broader perspective. The decline of industries such as shipbuilding and steel in the UK and USA as economies in East Asia rose from the 1960s onwards, and the impact of China's rise since the mid-1980s, led to deep economic and social consequences with severe economic and social tensions within and across nations. Whilst some major industrial changes come from technology (including the internet and digitalisation) and others come from changing relative prices, such as renewables relative to gas and coal, the rise of new producers and competitors is a major part of the story.

These stresses and divisive effects have political as well as social consequences and played a role in the Brexit vote in the UK and the election of populist governments in Europe and the USA. While the new economic geography will bring many benefits and opportunities, the global transformation will also present severe challenges in terms of rising international tensions. The policies of the second Trump administration constitute a further example of the changes and stresses which can follow as we move, in large measure, to a multi-polar world, a move driven by changing economic strengths. Policy-makers, economists, and communities must look ahead. The new economic geography will bring new international challenges. Building it in collaboration could transform our ability to live together in the future.

Notes

1   Although if carbon absorption becomes excessive, there are serious risks of acidification.

2   The CBAM levies a duty corresponding, in principle, to the carbon content of an import if a corresponding explicit or implicit tax is not paid at the origin (for certain goods).

3   Extractive industries will eventually run into rising marginal costs, notwithstanding technical progress.

4   For example, Influence Map (2022) reported that the oil and gas industry in the USA used the Ukraine war to promote long-term expansion policies in the sector. *The Guardian* reported on this lobbying effort, noting that the American Petroleum Institute (API) posted a string of tweets calling for the White House to 'ensure energy security at home and abroad' by allowing more oil and gas drilling on public lands, extending drilling in US waters and slashing regulations faced by fossil fuel firms (Milman, 2022, paragraph 2).

5   Economic diversification has long been a sound development strategy in countries rich in fossil fuels.

6   The IRA was, in large measure, rolled back with the passing of the so-called 'Big, Beautiful Bill' in Congress, signed by the President on 4 July 2025; the tax credits for renewables and other clean technologies were mostly removed. When discussing levelised cost of energy (LCOE) comparisons between intermittent renewables and natural gas, it is important to note that the comparison does not account for storage or possible grid modifications to accommodate renewables in the power system.

7   Currently over 1,000 GW per year (Wood Mackenzie, 2023). And China is, per annum, installing around 300 GW of renewable capacity domestically (S&P Global, 2024). To understand magnitudes here, note that total electricity capacity (all sources) in the UK is around 100 GW (Ember, 2024a) (see also Chapter 9).

8   The Chinese firm Envision, founded by Lei Zhang, an LSE alumnus and a donor to the LSE for its Global School of Sustainability, is involved in both the Somerset and Sunderland ventures (Campbell and Parker, 2023; Owen, 2023).

9   I was present when he made this observation in January 2025 at the World Economic Forum in Davos.

10 See endnote 9.

11 These analyses are broadly consistent with studies by the OECD (2024), IRENA (2024), and the ETC (2024).

9. International action for sustainable development: investment, finance, and collaboration

Building sustainable, resilient, and inclusive development for all countries, and thereby realising the new growth story across the world, requires international collaboration. It must be founded on the Paris Agreement if it is to deliver the global action necessary to provide a manageable climate for future generations. It has at its core investment and innovation.

The major increase in global investment necessary will require cooperation in creating the conditions for investment and its finance. It is in the self-interest of all countries that emissions are reduced across the world, that natural capital is protected, and that resilience is created. Investments in clean technologies, infrastructure, and biodiversity and the environment in one region bring benefits across the world both in reducing climate risk and in driving sustainable growth.

The goal is sustainable development; investment is at the heart of the necessary action; investment requires favourable conditions; and it requires finance. That is the logical structure of action and of our argument. Thus our calculations of finance are in terms of finance for investment; it is finance with a purpose. And the purpose of the investment is sustainable development.

Finance is of overwhelming importance in realising the investment at the urgency and scale which are now necessary, and finance is centre stage in this chapter. But other areas of collaboration are necessary too, including in creating the conditions for investment, in technology and trade, in linking action on climate and biodiversity, and in managing the overshooting of 1.5 °C, which is now inevitable.

Global cooperation can be undermined by lack of trust. Restoring the trust between developed and developing countries that has been eroded is the subject of Section 9.1; the section also explores opportunities for leadership from emerging markets and developing countries (EMDCs). The investment imperative that underpins the new growth story is set out quantitatively, by geography and sector, in Section 9.2. The analysis of necessary investment and finance is based on the work of the Independent High-Level Expert Group on Climate Finance (IHLEG), led by Amar Bhattacharya, Vera Songwe, and me for the last four COP presidencies (COPs 26–29), working closely with Éléonore Soubeyran. This research is central to the argument of this book. Scaling up investment and finance is an imperative to respond effectively to the climate and biodiversity crises. And it is also feasible. In so doing, it will deliver a new story of growth and development that is a much more attractive path than the dirty, destructive models that dominated the past. It is an opportunity.

Climate now sits at the core of many international agendas – from the UN to multilateral development banks (MDBs) to the Organisation for Economic Co-operation and Development (OECD). This shared climate threat offers a unique opportunity for both constructive action and collaboration; and tackling it together will build trust across divides. A trust that could bring greater hope in taking on other crucial challenges and divisions, including in conflict, biodiversity, health, and poverty in all its dimensions. International cooperation is required in four key areas:

1. **Mobilising finance:** Aligning investment with climate, growth and development goals, and financing that investment, discussed in Section 9.3.
2. **Technology and innovation:** Accelerating clean technology creation and diffusion, discussed in Section 9.4.
3. **Natural capital, biodiversity, land, and oceans:** Protecting ecosystems and carbon sinks, discussed in Section 9.5.
4. **Managing overshoot:** Preparing for climate impacts and potential geoengineering, discussed in Section 9.6.

These areas require cooperation to deliver resources at the necessary speed and magnitude. Climate policy can no longer be the domain of environment ministers alone – it must be embedded in the highest levels of economic and political decision-making. Achieving systemic change requires leadership from presidents, prime ministers, finance ministers, and economic policy-makers. The necessary transformation involves the whole of economy and society. And the central issues in this chapter must involve all countries. These ideas are drawn together in the final section, Section 9.7, which also offers a crucial example for international collaboration.

9.1 Future foundations: restoring trust and building new leadership

Building a global approach to climate action has involved many divisions and tensions, particularly between richer and poorer nations. At the heart of the divide is fractured trust – rooted in historical injustices, unfulfilled promises, and perceived conflicting economic interests. Without addressing these tensions, international cooperation will remain fragile. Restoring trust requires clear commitments, reliable action, and greater equity in global collaboration.

As the risks of climate change have become reality, particularly for EMDCs, and as the opportunities of climate action have begun to reveal themselves, we have seen EMDCs step up to take leadership in international climate discourse and action. We look at trust and leadership in turn.

Rebuilding trust

As we saw in Chapters 3 and 4, in recent years, a series of crises have deepened distrust in the developing world in relation to the developed. The Covid-19 pandemic, the continuing effects of the Ukraine war, and the persistent failure to deliver on climate finance commitments have reinforced scepticism about global solidarity. Developing countries, many already burdened by debt and worsening climate impacts, now question whether global commitments are of any substance. Without a willingness from wealthier nations to help resolve debt crises and foster and deliver promised financing, rebuilding trust will remain an uphill battle. Financing, as we have seen, is required for building new clean infrastructure and industries, fostering more sustainable agriculture, and protecting and investing in forests. It is necessary for building resilience into capital and activities as the climate changes.

For many developing nations, climate change is not an abstract future threat – it is a lived reality. Notwithstanding their small contribution to the origins of the problem of climate change, they are hit earliest and hardest by rising seas, storm surges, devastating droughts, overwhelming heat, and extreme weather events. Yet when disasters happen, financial and technical support all too often remain slow, insufficient, and overly complex. Despite growing international recognition of responsibility to help with 'loss and damage', many poorer nations struggle to access funding without navigating layers of bureaucracy and conditions that delay assistance. Rebuilding trust requires more than financial pledges – it demands fast, flexible, and reliable support when disasters strike.

Geopolitical tensions further complicate international climate action. The rivalry between major powers, particularly the USA and China, casts a shadow over global cooperation. Both nations have made strides in domestic climate policy – China through its rapid expansion of clean technology, the USA, until recently, through the Inflation Reduction Act (IRA)[1] – but they often remain at odds on trade, technology, security, and influence. Such tensions can spill over into climate negotiations. While rivalry and competition can drive progress – reducing costs, achieving high quality, and making rapid deployment – political hostility also fosters uncertainty and can obstruct coordinated action.

Beyond the USA–China dynamic, other global tensions shape climate diplomacy. Europe, despite past leadership on climate action, faces internal divisions and external trade disputes, particularly with China. India remains cautious, with some raising concerns that ambitious climate action could slow its economic growth, even as it increasingly recognises the potential of the green economy. Prime Minister Narendra Modi's leadership of the G20 placed cooperation and sustainable growth at the centre of discussions, yet India's difficult relationship with China and frequent opposition to trade measures in the World Trade Organization (WTO) highlight the underlying tensions that complicate international cooperation. At the same time, developing nations are growing more vocal in their criticisms of Western protectionism in green industries, as well as China's dominance in key products such as solar photovoltaics (PV) and batteries.

As we saw in Chapter 8, the transition to clean energy has triggered a race for rare earth elements and other critical resources. Such resource competition raises new geopolitical and economic concerns. China has established a dominant position in the processing and trade of many of these materials, prompting fears that it could control access to crucial supply chains. Western nations are now scrambling to diversify their sources, intensifying economic rivalries that threaten to complicate climate cooperation. Some see this jostling for control as 'quasi-colonial' positioning for dominance over future key international levers by more powerful nations, further straining trust.

These tensions regularly surface in climate negotiations, where perceived short-term national interests and strategic geopolitical agendas often overshadow long-term environmental goals. Disagreements over the responsibilities of emerging economies versus historically high-emitting nations remain contentious, often reflecting deeper geopolitical rifts. Nevertheless, the Paris Agreement showed that cooperation is possible when nations recognise their shared interests. It was a remarkable achievement in reaching a shared understanding that all must act, and demonstrating that trust, though fragile, can be built through diplomacy. Multilateralism can succeed where there is a common understanding of an immense shared threat and the possibility of shared prosperity.

At the top of the list in building trust and supporting the transition in the developing world is low-cost and readily accessible finance at the necessary scale. Showing how this can be realised is the central purpose of this chapter. Restoring trust in international climate cooperation requires more than just financial pledges – it demands timely and reliable action. Too often, commitments have been made without effective follow-through, thus fuelling frustration and disengagement. A new approach to growth and development must emphasise partnership and offer and facilitate funding that is structured to support and empower, rather than burden, developing nations.

Fractured trust and perceived conflicting interests are barriers to effective global climate action, but they are not insurmountable. Delivery on past commitments and creating mechanisms for accountability and transparency in a collaborative spirit are key, and so too is steadiness and longevity of commitment. The transition is not a zero-sum game – it holds the potential for shared prosperity if approached with genuine cooperation. Strengthening international trust requires a recognition that climate action is not just an obligation but an opportunity to reshape the global economy for a sustainable future. Governments will change and there will be setbacks. But commitment from the public, the private sector, and international institutions can bring some stability. Ultimately, the power of the shared threat and the prospects for shared prosperity are the main forces that will help the world stay the course in a collaborative way.

The future of climate leadership

Before the Paris Agreement (2015), many developing countries saw climate action as a rich-country responsibility. They feared it would hinder growth, development, and poverty reduction. As discussed, especially in Chapters 1 and 4, that perception is changing. The rising scale and costs of climate impacts, the recognition that most of the growth of future emissions will be in EMDCs, and the realisation that climate action can drive economic development and growth have transformed perceptions and attitudes. EMDCs are exercising real influence in international governance. For the first time, four consecutive G20 presidencies (2022–2025) have been led by emerging economies: Indonesia, India, Brazil, and South Africa. These presidencies have, in large measure, focused on sustainable development, with an emphasis on the development opportunities. The next G20 presidency in 2026 is scheduled to be the USA.

India's 2023 G20 presidency focused on sustainable development, climate mitigation and resilience, and biodiversity. All this was encouraged by a growing understanding that tackling climate and biodiversity challenges can be central to promoting economic growth. As part of an emphasis on inclusiveness, India stressed the importance of enhancing access to financial services for underserved populations (G20 India, 2023).[2] Brazil's 2024 G20 presidency continued this momentum, prioritising hunger, poverty reduction, and global governance reform. Brazil launched the Global Alliance Against Hunger and Poverty and called for more ambitious nationally determined contributions (NDCs) ahead of COP30, which it will host in 2025. The Global Bioeconomy Initiative, a Brazilian-led effort to integrate biodiversity into sustainable development, is likely to gain traction at COP30. South Africa's 2025 G20 presidency will further advance the priorities of the Global South.

With a second Trump presidency starting in 2025 in the USA, progress in international negotiations involving the USA may be slow. That makes 2025 a critical window for EMDCs to accelerate their leadership. If any country steps back in the international arena, it is likely that others will step forward into vacated space, making the most of new opportunities. The path forward on climate collaboration depends, in large measure, on whether EMDCs can translate their growing influence into tangible financial and policy shifts. Their leadership in investment, policy, and global institutions signals an important transformation in international climate politics. As the world moves beyond fossil fuels, the agenda is increasingly being shaped not in Washington or Brussels, but in Beijing, Jakarta, New Delhi, Brasilia, Pretoria, and beyond.

9.2 The investment imperative: what is needed where

Achieving the goals of the Paris Agreement and unlocking the opportunities of a new growth story require a strong increase in investment, with urgency and scale. Chapter 5 outlined five investment priorities for climate action: (a) clean energy transition, (b) adaptation and resilience, (c) funding for loss and damage, (d) preserving, restoring, and investing in natural capital, including sustainable agriculture, and (e) a just transition. These investments are not just about meeting climate targets – they are fundamental to economic growth and progress on the Sustainable Development Goals (SDGs), within which the Paris Agreement objectives are embedded. The figures for necessary investment are large, in part a consequence of years of delayed action. Had we started two decades earlier, the required increase in the rate of investment would have been lower. But crucially we must recognise that these are investments with high development returns, beyond their fundamental role in reducing climate risk. Hesitation now is not an option. This transition is not just possible – it is essential. There is both an investment opportunity and an investment imperative.

It is important to keep in mind throughout this chapter the basic logic of our argument in this book:

1. The goal is sustainable development as embodied in the Paris Agreement and the SDGs.
2. The driving force in achieving that goal is investment.
3. The conditions for that investment must be created.
4. Investment and its components require the right kind of finance on the right scale at the right time.

Let us now look at the extent of the investment and finance that will be required and in what categories.

Investment needs by 2030 and 2035

The IHLEG on Climate Finance report for COP29 (Bhattacharya et al., 2024) found that these climate-related investments must reach, globally, US\$6.3–6.7 trillion annually by 2030, rising to US\$7–8.1 trillion by 2035.[3] The calculations were based on the investments necessary in the five key areas to achieve the goals of the Paris Agreement (including staying well below 2 °C, enhancing resilience, and protecting natural capital) aligned with progress towards related SDGs, and in the context of a growth path consistent with International Monetary Fund (IMF) projections to 2035. This logic defines our notion of 'investment needs'. 'Climate finance' is finance associated with the five key areas. Thus the numbers on investment and finance in this chapter are *deduced* from what is necessary. They are not some kind of opening bid in a negotiation.

These investments deliver much more than the Paris Agreement narrowly interpreted in terms of temperature, resilience, and natural capital. Given that the relevant investments will directly influence progress towards the SDGs, along with delivery on Paris, it is important to take careful account of those effects in analysing necessary investments. The eight directly related SDGs are clean water/sanitation (SDG 6), affordable clean energy (SDG 7), decent work and economic growth (SDG 8), industry, innovation, and infrastructure (SDG 9), sustainable cities and communities (SDG11), responsible consumption and production (SDG 12), climate action (SDG 13), and life on land (SDG 15). It is clear that all eight are directly influenced by investments in the areas described. However, indirectly through the processes of development and growth, these investments will affect all 17 of the SDGs.

We should emphasise that for the Paris targets, we should describe a whole path of investment and emissions rather than one or two years. Concentrations influence warming and concentrations are shaped by cumulative flows of emissions. But by focusing on particular years, here 2030 and 2035, it is easier to understand magnitudes and breakdowns.

Let us now look at these investments in more detail. By 2030, investment must flow roughly as follows (Bhattacharya et al., 2024):

- US$2.7–2.8 trillion into advanced economies.
- US$1.3–1.4 trillion into China.
- US$2.3–2.5 trillion into EMDCs (other than China).

Figure 9.1: Total climate investment needs by economic regions for 2030 and 2035

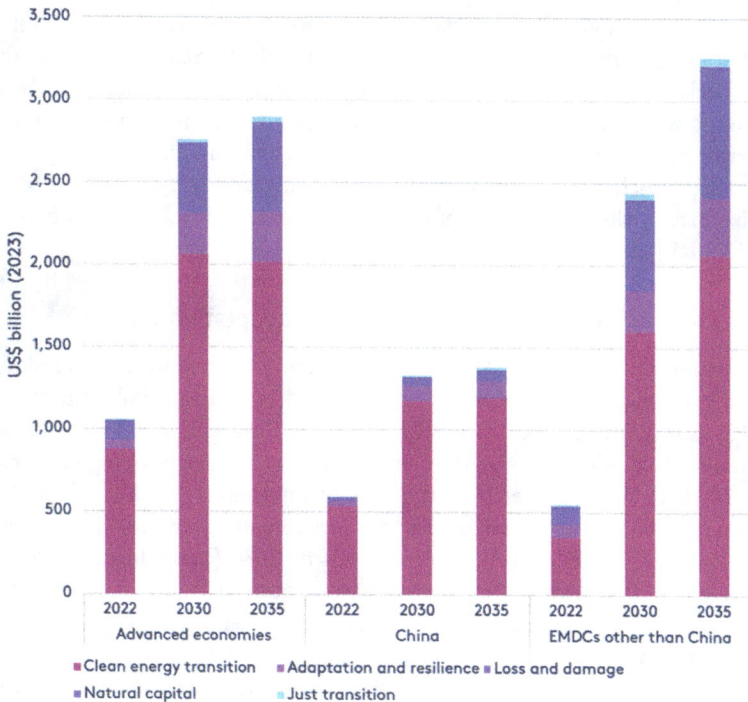

Source: Figure 1.1 in Bhattacharya et al. (2024, p. 14). Copyright 2024 The authors. Reproduced with permission.

Figure 9.1 illustrates how investment needs will shift across regions over time. By 2035, EMDCs (other than China) will require more investment than developed countries – more than double China's needs.[4] With such large investment needs, urgent frontloaded action before 2030 is critical to achieving global climate and development goals. Investment needs are particularly strong for EMDCs, which already face disproportionate climate impacts, reduced capacity, and barriers to investment in general. Yet, EMDCs (other than China) accounted for just 15% of global clean energy investment in 2023 (International Energy Agency [IEA], 2024a).[5] And, though total power sector investment in EMDCs (other than China) grew by 12% in 2023, its growth rate still lagged behind the 16% average experienced in advanced economies and China (IEA, 2024b).

From 2018 to 2021/2, global climate finance more than doubled (Climate Policy Initiative [CPI], 2024), reaching a record high of US$1.9 trillion in 2023, and likely exceeding US$2 trillion in 2024 – equivalent to 1% of global GDP (CPI, 2025). However, only 19% of global climate finance from 2018 to 2022 was in EMDCs (other than China), with developed economies (45%) and China (36%) absorbing the majority (CPI, 2024).

Private investors remain wary of EMDC markets due, in large measure, to perceived risks. This drives up borrowing costs and limits investment flows. EMDCs face mounting financial pressures as they try to scale up climate investment, as many already struggle with high debt, inflation, and unstable currencies. Climate change is worsening their economic vulnerability, with extreme weather events pushing them deeper into financial distress. Rising interest rates further limit their access to capital markets, forcing them to delay much-needed projects. Without purposive action, the financial imbalance in climate investment between developed and developing nations will remain severe.

Categories of investment needs: mitigation, adaptation, and more

Mitigation attracted over 90% of global climate finance in 2023, predominantly in energy and transport, as measured by Climate Policy Initiative (CPI) (2025). The United Nations Environment Programme (UNEP) (2023a) estimates that adaptation investment requirements in developing countries are 10 to 18 times higher than the flow of international public adaptation finance in 2021. Funding for loss and damage remains even more inadequate, covering less than 0.2% of projected needs by 2030 (Lowy Institute, 2024). Table 9.1 breaks down global annual investment needs and those of EMDCs (other than China) by 2030 in the five key areas identified – see also Chapter 5. The importance of integrating mitigation, adaptation, and development was stressed in Chapters 2 and 5, and it is not always easy or sensible to try to distinguish mitigation investments from adaptation. Nevertheless, the gross inadequacy of adaptation and loss and damage should be clear from these numbers.

Table 9.1: Key climate investment needs by 2030

Investment area	Key insights	Annual investment needs by 2030
Clean energy transition	**Clean energy investment[6] is concentrated in advanced economies and China:** 70% of global solar and wind resources lie in the Global South (excluding China) (Singh and Bond, 2024), yet they accounted for only 10% of the growth in clean energy investment from 2015 to 2024 (IEA, 2025). Africa received only 2% of global clean energy investment in 2023 (IEA, 2023b).	**Global:** US$4.8 trillion
		EMDCs (other than China): US$1.6 trillion
Adaptation and resilience	**Many EMDCs face large adaptation investment needs.** Lack of investment in this area threatens decades of development progress.	**Global:** US$0.59 trillion
		EMDCs (other than China): US$0.25 trillion
Loss and damage	**Rising climate-related disasters** are causing escalating economic and non-economic losses, particularly in EMDCs.	**Global:** US$0.63 trillion
		EMDCs (other than China): US$0.25 trillion
Natural capital and sustainable agriculture	Nature-based solutions (NbS),[7] such as reforestation and sustainable farming, **reduce emissions and strengthen biodiversity, food security, and economic resilience.**	**Global:** US$0.4 trillion
		EMDCs (other than China): US$0.3 trillion
Just transition	**The shift to low-carbon economies** must embody support for communities and workers impacted by the transition.	**Global:** US$0.07 trillion
		EMDCs (other than China): US$0.04 trillion

Note: Figures from Bhattacharya et al. (2024), unless otherwise cited.
Source: Author's elaboration.

The clean energy transition constitutes the largest category of investment needs. By 2035, EMDCs (other than China) will account for nearly half of the increase in clean energy investment needed to meet global climate goals. Currently, much of the clean energy investment of EMDCs (other than China) is concentrated in larger economies like India and Brazil. Sub-Saharan Africa receives the least investment, despite its large potential for renewable energy. It attracted only US$13 per capita in energy transition investments during 2020–2023; this is 40 times less than the global average (International Renewable Energy Agency [IRENA], 2024). These nations, experiencing rapid population and economic growth, face rising energy demand and must build sustainable energy systems from the ground up.

Adaptation remains critically underfunded, reaching US$76 billion in 2022 globally, which is still grossly insufficient and represents only 5% of total climate finance (CPI, 2024).[8] This figure then decreased to US$65 billion in 2023 – the most recent year for which data are available (CPI, 2025). Some estimates suggest that adaptation finance flows to developing countries must increase by a factor of four (Global Centre on Adaptation and CPI, 2024). Yet the Country Climate and Development Reports (CCDRs) show that many high-return investments focused on resilience struggle to secure funding (World Bank, 2023). A recent report estimated that, on average, the benefits of 'the right adaptation measures' can outweigh costs by a ratio of 10:1 (Swiss Re, 2023). Previously, the Global Commission on Adaptation (2019) found that the benefit–cost ratio on investments in improved resilience ranges from 2:1 to 10:1; in other words, the Global Commission on Adaptation argued that US$1 spent on adaptation can yield net economic benefits of US$2–10. These benefits accrue from decreased risks and losses and from the greater innovation and productivity resulting from strengthened resilience and reduced risk.

Sound adaptation is always good development (CPI, 2023b). However, whilst some adaptation investments can be profitable and attract private investment, there are also many adaptation investments that are not financially attractive to private sector investors due to difficulties in capturing returns. For example, flood control across a city or region can deliver great benefits to the public as a whole in that city or region while not necessarily directing financial returns to a private investor. Many adaptation projects imply large upfront costs with long payback timeframes, hindering private flows. In the period between 2019 and 2022, less than 3% of global adaptation efforts have been financed by the private sector (Global Centre on Adaptation and CPI, 2023). Thus, many adaptation investments will have to be made by governments and public organisations, although pressure on public resources is usually intense, especially in developing countries. These challenges of finance emphasise the importance of identifying and pursuing projects which integrate development, mitigation, and adaptation.

Loss and damage costs are also escalating rapidly and will rise sharply as climate risks escalate. In 2022, EMDC economies lost over US$109 billion from major climate events alone, excluding non-economic impacts (Richards et al., 2023). The Vulnerable Twenty (V20) – a group of 74 countries particularly exposed to climate change – reported that 55 of its members lost US$525 billion between

2000 and 2019 due to climate change's effects on temperature and precipitation patterns (V20, 2022). The establishment of the Loss and Damage Fund at COP27 was a step forward, but funding remains far below the growing needs of vulnerable nations. A combined total of US$792 million has been pledged by 19 nations to the Fund, but this represents less than 0.3% of the expected loss and damage costs by 2030 (Bhattacharya et al., 2024).

Investments in natural capital and sustainable agriculture offer great opportunities for growth, emissions reductions, and biodiversity. Although 90% of the global investment opportunity in nature conservation lies in developing countries, 80% of nature finance flows to developed economies (Bhattacharya et al., 2023). This is another example of broad inefficiencies and inadequacies in climate finance – imbalances in volume, sectoral allocation, and financial instruments. The challenge is not just about increasing funding but also about restructuring where and how it flows.

Fostering a just transition as countries shift to low-carbon economies will involve addressing the social and economic impacts of the transition, particularly dislocations of activities in regions dependent on fossil fuel industries. Relative prices will change and households will need to make new investments, for example in heating/cooling and in transport. Policies to protect poorer groups in the transition will be crucial to both tackling injustice and sustaining political support, as we have discussed in Chapter 6.

Figure 9.2 describes the requirements across the five priority areas for EMDCs (other than China). We must be clear on two points regarding these numbers. First, funding for mitigation and the energy transition must increase strongly. Second, the very weak adaptation funding is not an argument for reducing mitigation funding. If that happened, development opportunities would be lost and the challenges of adaptation would be far bigger in the future. The expansion of funding for mitigation and the energy transition should occur alongside even

Figure 9.2: Investment/spending requirements for climate and sustainable development in EMDCs other than China (US$ billion per year by 2030, increment from current in parentheses)

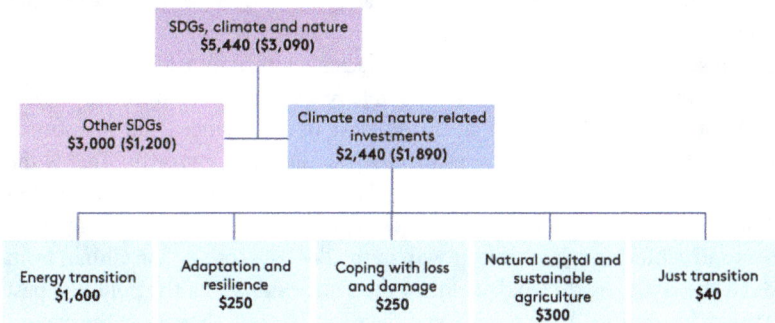

Source: Figure 1.3 in Bhattacharya et al. (2024, p. 17). Copyright 2024 The authors. Reproduced with permission.

greater proportional growth in funding for adaptation and resilience, loss and damage, and nature. We must avoid falling into the trap of setting mitigation and adaptation investment against each other. So many investments advance both mitigation and adaptation. Thus it is much better to say, for example, that at least 50% of investments should have a strong adaptation element than to say adaptation and mitigation investment should be 50:50. Second, in the design and delivery of climate action, mitigation, adaptation, and development should be integrated. To separate out the three activities is an analytical mistake. And it is a practical mistake because separating radically different budgetary and administrative measures could disrupt that integration.

Accelerating investment: the urgent need for transformation

The investment push is not just about climate action – it is about economic transformation. Many EMDCs have a chance to leapfrog outdated energy models and move directly to clean power. But seizing these opportunities requires immediate, large-scale investment and the creation of enabling environments for these investments. Accelerating that investment must start now. Delaying action will drive up future costs. The longer we wait, the steeper and more expensive the transition will become. Investments made before 2030 will be powerful determinants of whether we can achieve a smooth, manageable shift to a low-carbon global economy. We are already compressing into 20 years action which could have been taken over 40 years if we had started 20 years ago, when it was already obvious we had to act (see, e.g., Stern, 2006). Further delay or hesitation will make the transition still more difficult.

9.3 Mobilising finance: international collaboration

Climate finance[9] must expand dramatically to meet the urgent need for investment, particularly in EMDCs (other than China). The requirements for climate investment were discussed in Chapter 5 and set out in Table 9.1 and Figures 9.1 and 9.2. The EMDCs will drive future economic growth and emissions reductions, and they have growing adaptation needs as the threats from climate change grow ever more severe. Yet, their needs for climate finance remain severely underfunded. Without decisive international action, the gap between their financial needs and available resources will continue to widen.

We must emphasise again that the purpose of the finance is for investment and that the purpose of the investment is for sustainable growth. That is the logic of our analyses. The finance we calculate is not some kind of compensation or bill that is due from the rich countries. It is finance necessary for delivery on Paris and achieving sustainable growth in the EMDCs (other than China) from which the whole world will benefit. At the same time, given the polluting past of the rich countries, their wealth, and their self-interest, there is, in the view of many – including me – a moral obligation to generate this support.

Macroeconomic and fiscal policies shape how nations allocate resources, foster investment, and drive economic growth. In today's interconnected world, no economy operates in isolation. A global response to climate change demands strong international cooperation, particularly in aligning financial systems with sustainability goals. The United Nations Framework Convention on Climate Change (UNFCCC) remains the primary platform for global climate agreements. However, the UN lacks financial resources and acts more as a political vehicle for the creation of agreements than as a direct investor in climate action. Nevertheless, it plays a key strategic role in setting direction and fostering collaboration. But, the real core and drivers of international public action for climate finance are the G20, MDBs, and central banks. They can generate and shape investment flows and reduce the financial risks of climate action. In so doing their relationships with the private sector are key. Indeed, it is the partnership of the private sector with the official public flows which provides the overall total and structure of the international flows. Private sector investment and its finance are at the heart of the story of transformation, both domestic and international.

The G20, MDBs, and central banks, working with development finance institutions (DFIs), form the backbone of the public part of the global financial system. They influence the flow of private capital, public capital, debt relief mechanisms, and the policy frameworks that determine whether countries can successfully finance the transition to a low-carbon economy. They also play a crucial role in shaping the conditions for investment.

Identifying and quantifying the financing pathways

Mobilising the trillions needed for climate action requires a fundamental shift in global finance. Finance must be available at scale, accessible to developing economies, and structured to be predictable and affordable. What is needed is the right kind of finance, on the right scale, in the right place, at the right time. The EMDCs (other than China) embody both the greatest challenges in financing and the most significant opportunities for transforming emissions, development, and poverty reduction – as we have consistently argued in this book.

Creating a new framework for climate finance for EMDCs (other than China) should be top of the agenda for climate action. The financing requirements are set out in Figure 9.3. The flows of the different kinds of capital combine to form the overall total. The disaggregation into different types of finance is based on the different kinds of investment. For example, much of the finance for electricity generation would be private but most of adaptation would be public. The numbers in Figure 9.3 indicate that, to achieve the trillions needed, external sources of finance for private sector investment must grow 15-fold, MDB financing must triple, and concessional finance must increase 14-fold. These numbers are daunting, but they are feasible. Failure on any major scale would be catastrophic. Without intervention to foster investment and bring the necessary finance, developing economies will remain trapped in a cycle of underinvestment, worsening climate impacts, and mounting financial instability.

Figure 9.3: Mobilising the necessary financing for EMDCs other than China (US$ billion per year by 2030, increment from current in parentheses)

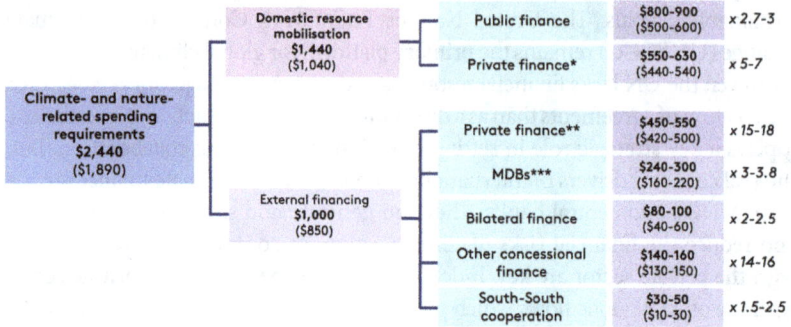

Note: *Includes household savings. **A significant proportion of this private finance would be directly and indirectly catalysed by MDBs, other DFIs, and bilateral finance. ***Includes multilateral climate funds.
Source: Bhattacharya et al. (2024, p. 5). Copyright 2024 The authors. Reproduced with permission.

Scaling up finance requires a coherent strategy that blends four key sources: domestic resource mobilisation (DRM) (public and private), international private, MDB and development finance institutions (DFIs), and concessional finance. These are in rank order of magnitude. Each plays a distinct but complementary role in managing and reducing risk and enabling the right capital to reach where it is needed. It is a mistake to look only at the aggregate numbers. How the components combine and complement each other is critical to generating the different types and scale of finance needed for the different types and scale of investment.

The breakdowns in Figure 9.3 should not be interpreted with excessive precision. It is the order of magnitude of total investments to be financed and the rough breakdowns of the components which matter. Thus, we can think of something around US$2.4 trillion for the total and 60% domestic. Of the external of around US$1 trillion, we should have around a half private, a quarter MDB and DFI, and a quarter concessional. What matters is the overall scale and how the different elements fit together and mutually support each other.

DRM remains the bedrock of climate finance, accounting in 2022 for about 70% of current flows of finance in the areas under examination here. In Figure 9.3 we have domestic financing at US$1.4 trillion annually by 2030. As a proportion of the total of US$2.4 trillion, that would be around 60%, a little lower than the current 70%. We made the assumption of 60% given the scale of the expansion and the public nature of some of the relevant investments. Achieving the expansion of DRM will require strong and purposive action. Such action will reduce perceived risk and boost investor confidence.

Key measures for public revenues include raising tax revenues, implementing carbon pricing, and phasing out fossil fuel and other toxic subsidies. On the private side the challenge is to increase the scale and availability of private saving, including by expanding domestical capital markets.[10] Strong international cooperation is essential to support capacity building in tax design and administration. The arrival of digitisation and more sophisticated IT systems (such as Aadhaar or the system of Unique Identification Numbers in India) in many countries will be an important aid to the process of increasing both public and private DRM. These new systems can expand the tax base, reduce evasion, and improve compliance. And they can provide a reliable information basis which can help underpin private markets. Strengthening tax systems, expanding digital finance, and developing local capital markets can reduce reliance on volatile foreign capital. And strong DRM can reduce perceived risk and attract more stable foreign flows, private and public.

In terms of *external private finance*, EMDCs (other than China) will require US$1 trillion in external finance annually by 2030 to meet their US$2.4 trillion total investment needs, about half of which will have to be contributed by private finance.[11] Mobilising this level of private finance requires a 15- to 18-fold increase in external private finance flows from 2022 levels, which hinges on enhancing project bankability and reducing the cost of capital. These will, in turn, be powerfully influenced by expanding the role, including financial flows, of MDBs, through MDBs forging closer relationships with the private sector, taking more risk, and becoming easier to work with – in other words, *Better, Bolder, and Bigger.*[12]

Private sector (both domestic and external) involvement will be crucial in a number of activities, particularly in energy, transport, and industrial decarbonisation. However, investors hesitate due to concerns over project risk, including around development, execution, and revenue; perceived country risk; currency volatility and overall political instability; and policy uncertainty. To overcome these barriers and concerns, governments will need clear investment roadmaps and regulatory stability in the context of a generally possible investment climate. Public–private partnerships, credit guarantees, and structured contracts – such as power purchase agreements (PPAs) and feed-in tariffs (FiTs) – can create confidence in stable revenue streams for renewable energy projects, making them more attractive to investors. Insurance mechanisms, currency hedging facilities, and blended finance structures can further reduce financial risks.

MDBs play an absolutely central role. They can support the host government in creating the conditions for investment including the strategic direction and consistent programmes which are crucial for investor confidence. Some of this will be through the development of country platforms which provide guidance on and structures for direction of travel, coordination, and the ability to deal with problems as they arise.

They also, and critically, through their financial instruments help reduce, manage, and share risk, bringing down the cost of capital and facilitating both private and public investment. And their presence itself in strategies, programmes, and projects gives confidence to potential investors on consistency and clarity of policy. We emphasised the importance of policy clarity in Chapter 6. Governments are a key source of policy risk and the presence of an MDB reduces that risk. The role of the MDBs is discussed further later in this section.

Bilateral climate finance provided through Official Development Assistance (ODA) from advanced economies has played an important role in the broader climate finance system. Though a relatively small component of overall flows, it is critical for supporting important investments, particularly in areas where private sector investment is likely to be limited. Investments that may be of real value but do not easily generate direct revenue returns will require finance with little or no debt service. ODA can also help by providing guarantees to other lenders; this can greatly reduce the price of capital that those other lenders require for their participation. As part of a combination of sources with low- or zero-debt service, ODA contributions should at least double from their 2022 level of US\$43 billion annually. Amongst priorities for use should be investment in adaptation and resilience, and tackling loss and damage. Nevertheless, for the first time in six years, in 2024, ODA from official donors (Development Assistance Committee [DAC] member countries) decreased (by around 7%) in real terms compared to 2023 (Organisation for Economic Co-operation and Development [OECD], 2025a). Further reductions are expected in 2025 relative to 2024 (OECD, 2025b).

Further sources of *low- or zero-cost finance* will be necessary to meet the scale of action required, particularly in relation to projects or programmes which struggle to create revenue flows or where guarantees or other risk sharing is needed. Given the limited prospects from ODA, innovative financing mechanisms will likely be needed. Voluntary carbon markets (VCMs) offer one potential source. VCMs could be more effective if they shifted from projects to programmatic funding. A major company, for example, could fund a fraction of a large initiative, aligning with national development strategies. Such an approach would allow these flows to be integrated into development planning and financing. And the counterfactuals which are necessary for VCMs would be more easily established. For example, on a project VCM, say for stopping deforestation, it is possible that the project simply shifts deforestation to another location. In a programmatic initiative it is easier to compare directly with a business-as-usual counterfactual.

Levies on high-emitting activities such as aviation and shipping could generate substantial revenue. The Global Solidarity Levies Task Force, launched at COP28, is exploring international taxation options, with proposals expected by COP30 in late 2025. They are examining, inter alia, possibilities for taxation of maritime and aviation activities. These are emitting activities, largely international, which are for the most part not subject to international taxation. They therefore fit well as tax bases for this global public good.

Concessional finance remains indispensable for adaptation and resilience, loss and damage, natural capital, and just transitions – areas where private capital is scarce. The Amazon Fund demonstrates how grant-based financing can support forest conservation and sustainable livelihoods. It is financed mainly through donations from countries, with Norway and Germany being the largest contributors. These funds are given based on Brazil's efforts to reduce deforestation. Similarly, blue bonds – such as the Seychelles' Blue Bond, issued with World Bank support – finance sustainable fisheries. Expanding these instruments can mobilise private and institutional investors while maintaining a focus on long-term sustainability.

Finance within the EMDCs, sometimes called South–South cooperation, could generate substantial low- or zero-cost flows. China and some fossil-fuel-rich countries are already moving in that direction. China launched the Belt and Road Initiative in 2013. In recent years it has increasingly shifted its activities toward low-carbon development. The United Arab Emirates (UAE) initiated funding for green energy in 2013, including via the Abu Dhabi Fund for Development. Following COP28 in Dubai it initiated Alterra, the world's largest private climate investment fund.

Progress is emerging in international climate finance, with initiatives such as the COP28 Global Climate Finance Framework working to coordinate efforts. However, major obstacles remain. There is only slow progress in tackling high capital costs in EMDCs and weak private capital flows. Overcoming these challenges requires coordinated global action, including reforms to MDBs, stronger bilateral commitments, and new low-cost finance sources in order to unlock the required scale of financing. For the necessary scale and transformation, international financial institutions (IFIs), MDBs' shareholders, and private sector actors must work together. Deepened collaboration and coordinated leadership within the group of MDBs will be necessary to expand their role on the scale necessary in de-risking investments, lowering borrowing costs, and attracting private capital.

Figure 9.4: Climate finance goals agreed at COPs

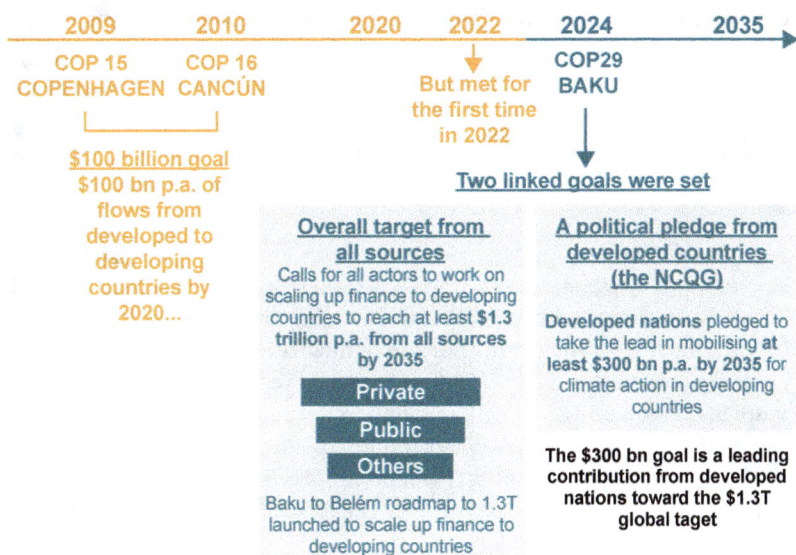

| 2009 | 2010 | 2020 | 2022 | 2024 | 2035 |

COP 15
COPENHAGEN

COP 16
CANCÚN

But met for the first time in 2022

COP29
BAKU

$100 billion goal
$100 bn p.a. of flows from developed to developing countries by 2020...

Two linked goals were set

Overall target from all sources
Calls for all actors to work on scaling up finance to developing countries to reach at least **$1.3 trillion p.a. from all sources by 2035**

Private

Public

Others

Baku to Belém roadmap to 1.3T launched to scale up finance to developing countries

A political pledge from developed countries (the NCQG)

Developed nations pledged to take the lead in mobilising at least $300 bn p.a. by 2035 for climate action in developing countries

The $300 bn goal is a leading contribution from developed nations toward the $1.3T global taget

Source: Authors' elaboration.

The pace, scale, and composition of climate finance necessary for EMDCs is fairly clear. The challenge now is building international action from a low starting point and at the speed required.

Building international action and climate finance: the COPs and the goals

COP29 in Baku (November 2024) was designated as the 'finance COP'. A key objective was the agreement of a New Collective Quantified Goal (NCQG) to replace the US$100 billion per annum commitment. This commitment originated in COP15 in Copenhagen (2009), became a commitment by developed countries at COP16 in Cancún (2010), and was reaffirmed at COP21 in Paris (2015). However, it was achieved only in 2022, two years later than promised (2020). See Figure 9.4 for a timeline on finance commitments at the COPs.

It is important to recognise that the US$100 billion per annum initiated at COP15 was a *negotiated* figure in contrast to the *deduced* figures in this chapter. Meles Zenawi (Prime Minister of Ethiopia and speaking for the African Union) and I (as an independent friend of Meles) negotiated this with Mike Froman and Hilary Clinton, representing the USA. The need that Meles and I calculated was much higher and was focused on public flows. The final negotiated language described the flows as 'public and private'.

The framework for COP29 negotiations was based on the report titled *Raising Ambition and Accelerating Delivery of Climate Finance* (Bhattacharya et al., 2024), produced by the IHLEG on Climate Finance. The outcome from Baku was mixed. The fact that agreement was reached was of real importance, since failure could have undermined the whole Paris framework. However, the agreement which was reached fell short of requirements in two major ways:

1. It focused on 2035 instead of 2030, despite the need for immediate acceleration.
2. The agreed NCQG falls short of the financing necessary to meet climate and development targets as described in this chapter.

Raising Ambition and Accelerating Delivery of Climate Finance (Bhattacharya et al., 2024) estimates that EMDCs (other than China) need around US$1.3 trillion annually in external financing by 2035.[13] However, the NCQG – intended to define the portion of finance attributable to developed countries – was set at just US$300 billion per year by 2035, as shown in Figure 9.4. Based on the COP29 calculation framework (which was not the same as that adopted for the US$100 billion calculations), this figure would have been at least US$500 billion per annum.[14] For a deeper analysis of these financing gaps, see Bhattacharya et al. (2024). Although there have been steps forward, leaps are needed.

The roles of the key actors

Amongst the international 'G' groupings, it was the G7 who first addressed climate policy in 2005 under UK Prime Minister Tony Blair. The weak interest and progress at that meeting led to the launch of *The Economics of Climate Change: The Stern Review* by PM Blair and Chancellor of the Exchequer (Finance Minister) Gordon Brown. The Chancellor, the PM, and I took the view that the slow progress and the lack of interest on the part of prime ministers and presidents at that meeting were in large part the result of the absence of a clear economic analysis of the necessity for action and what action would entail. This was in contrast with strong progress on Africa at that meeting, with its agreement to double aid for Africa between 2005 and 2010 (which was largely achieved). That progress was built on the report for the G7 titled *Our Common Interest*, which I led, and a very effective campaign, 'Make Poverty History', led by Bono and Bob Geldof.

It took much longer for the G20 to engage. However, recent G20 presidencies – including India in 2023, Brazil in 2024, and South Africa in 2025 – have placed sustainable development, biodiversity, and green finance at the forefront, as noted in Section 9.1. The G20 plays an important role in climate discussions, representing economies responsible for around 80% of global GDP and of greenhouse gas (GHG) emissions. Unlike the UNFCCC, however, the G20 excludes many of the world's poorest and most climate-vulnerable nations. India made an important step in 2023 by securing permanent African Union membership within the G20. The UNFCCC, which includes nearly all countries, is a place where the poorest countries can be heard. And, as we saw in Chapter 1, they did indeed play a strong role in the Paris Agreement. But the UNFCCC, for most countries, involves environment ministers rather than finance ministers. The UNFCCC can point to the need for resources, but it is not an effective place to mobilise resources.

At the national level, finance ministries and central banks are increasingly embedding climate considerations into their core functions and collaborating internationally in these tasks. The Coalition of Finance Ministers for Climate Action, whose work was introduced in Chapter 4, exemplifies this shift. Both the Grantham Research Institute on Climate Change and the Environment at the LSE and the World Bank have supported its creation, providing a global platform for finance ministries to exchange experiences, share best practices, and integrate climate action into fiscal policies.[15] Finance ministries are now central players in climate policy. It is these ministries that determine and influence tax structures, investment incentives, and national spending priorities. Their growing commitment reflects the recognition that climate action can drive sustainable economic growth, boosting resilience, productivity, and long-term prosperity. And it is a major source of potential instability and risk. As more countries embed climate considerations into financial strategies, momentum for systemic change is advancing. There will be setbacks for this type of collaboration, such as Donald Trump's second presidency, but the direction of travel is clear.

Central banks and financial regulators are beginning to recognise climate change as a systemic financial risk. Mark Carney as Governor of the Bank of England was an early leader with his speech in September 2015 on 'the tragedy of the horizon', in which he pointed out and analysed the links between climate change and financial stability. This helped establish the Network for Greening the Financial System (NGFS).[16] The IMF has been, under the leadership of Managing Directors Christine Lagarde and Kristalina Georgieva, incorporating climate considerations into macroeconomic assessments and lending programmes.

Climate-related financial risks – including extreme weather disruptions, stranded assets, and volatility in fossil fuel markets – pose major threats to economic stability. Many EMDCs already face crippling debt burdens, high inflation, and capital flight. Rising interest rates have further constrained their ability to invest in clean energy and adaptation, and nature projects. The IMF has a key role in helping stabilise debt-stricken economies and facilitating sustainable fiscal policies. Whilst many central banks at the national level are moving in that direction, there remains pushback from 'institutional fundamentalists' who insist that central banks and the IMF should both have narrow mandates and interpret them in narrow ways – thus to exclude, in this case, climate issues. That is very strange, in my view, when part of the rubric is financial stability and climate change can have powerful effects on financial stability (see also Chapters 7 and 10).

International accounting conventions on the treatment of risk and capital dominations and requirement can cause real problems. As we saw in Section 6.4, the Basel rules – a set of banking standards designed to underpin international banking system stability – can inadvertently discourage lending to sectors that are crucial for the energy transition. Climate-aligned investments, particularly in EMDCs, are often classified as high risk under existing regulatory frameworks, which means banks must hold more capital to support such loans. This makes these investments less attractive. Conversely, climate-related risks of carbon-intensive assets tend to be underpriced under Basel Pillar 1, creating incentives for financial institutions to continue financing existing activities that are misaligned with climate goals, or to support new investments that contribute to carbon lock-in (Kammourieh and Songwe, 2024). The result is a structural mismatch between financial regulation and climate investment needs, creating another barrier to scaling climate finance where it is most urgently needed.

Centrality of MDBs

Most importantly, it is the MDBs that are at the core of turning climate opportunities and commitments into real investments. They can provide finance at rates close to market benchmark rates; help manage, share, and reduce financial risks; and help attract private sector investment, particularly in clean energy, infrastructure, nature, and climate adaptation. Some have already taken steps toward greener portfolios. In 2020, the European Bank for Reconstruction and Development (EBRD) pledged to allocate over 50% of its total annual investment to green projects by 2025 – a goal it achieved in 2021, four years ahead of schedule, and matched in subsequent years (EBRD, 2023). The Asian Development Bank (ADB) committed US$100 billion in climate finance from 2019 to 2030 (ADB, 2021). The World Bank's Climate Change Action Plan (2021–2025) set a target of 45% climate finance across its total lending portfolio (World Bank, 2021). Notwithstanding Trump's second presidency and the fact that the USA is the largest shareholder, I hope the World Bank will continue to make efforts to demonstrate that pursuing resilient and low-emission growth is in the interest of all clients. That is indeed what this book and the work of the World Bank itself have shown. Other MDBs with zero or small US shareholding can expand with fewer concerns about the position on climate of the current US administration. For example, the Asian Infrastructure Investment Bank (AIIB), the European Investment Bank (EIB), the New Development Bank (NDB), the Islamic Development Bank (IsDB), and Development Bank of Latin America and the Caribbean (CAF) have zero US shareholding.

The financing capacity of the MDBs as a whole should at least triple by 2030,[17] and quadruple by 2035 (Bhattacharya et al., 2024) for delivery on Paris (as explained above). These are the implications of the finance numbers presented in Table 9.1 and Figure 9.2. As emphasised above, private finance can be scaled up through innovative risk-sharing mechanisms, credit enhancements, and regulatory reforms. Domestic financial systems in EMDCs will need to be strengthened to attract long-term domestic institutional capital, a crucial element of DRM. In so doing, external capital will also be attracted. There is a great deal the MDBs can do here in helping build the local investment climate and reduce country risks, thus expanding both DRM and external flows, both private and public. A range of financial instruments – including equity, guarantees, long-term flexible loans, and so on –should be part of the package the MDBs bring.

The MDBs' leverage makes them extraordinarily cost-effective – every US$1 of paid-in capital is multiplied several times through their balance sheets, generating far greater financing than direct government spending. Their financial structure relies not only on paid-in capital from member countries, but also on callable capital, which member countries commit to provide in the case that regular capital is insufficient. Historically, this capital is not called upon and serves instead as a critical guarantee against which MDBs can borrow in international markets. Combined with the MDBs' strong credit ratings, it allows the MDBs to borrow at very low rates, meaning every dollar of paid-in capital is leveraged to generate many more dollars in lending capacity.

In 2021, the G20 (under the Italian presidency) commissioned an independent review on capital adequacy frameworks for the MDBs (the G20 CAF review). In its report published in 2022, it recommended that they expand their lending – meaning greater leveraging of their capital – to sustainably increase funding for global development (see G20 Italy and G20 Indonesia, 2022) and showed how MDBs could work their balance sheet harder. Combining these insights with its own work, the G20 MDB report of 2023 (commissioned under the G20 presidency of India) recommended a tripling of MDB financing flows (also see below).

Further, in the two *Triple Agenda* reports (see Singh and Summers, 2023a, 2023b), it was argued that a tripling of MDB financing from pre-Covid levels, from around US$100 billion to US$300 billion per annum, could be achieved by more efficient use of their balance sheets (as recommended by the CAF Review, see G20 Italy and G20 Indonesia, 2022), combined with greater use of guarantees and other instruments, plus an extra US$60 billion of paid-in capital. Progress by the MDBs in the past couple of years has been strong in relation to these projections, but nevertheless the extra US$60 billion in paid-in capital is still needed to achieve the sustained tripling of annual lending (personal communication from those working on the calculations for the Triple Agenda reports). To illustrate the power of the leverage, if countries paid in US$10 billion a year over 5–10 years, that would be around US$500 million a year for a country with a 5% share, such as the UK. That would allow, with the CAF measures, a tripling of their financing, from around US$100 billion per annum to around US$300 billion per annum.

The MDBs can help improve investment conditions by providing policy-based financing that strengthens macroeconomic stability and institutional reforms. The value of the MDBs is particularly evident in their ability to de-risk private investment. A clear example is the World Bank's Scaling Solar programme in Zambia, which combined concessional loans, government-backed guarantees, and private sector participation to deliver record-low solar tariffs. Private long-term investment projects in the African energy sector face interest rates that are far higher than for comparable investments in richer economies and in China; using project surveys the IEA estimates differences in the debt cost of capital of some 5–6% for South Africa (IEA, 2023a), and credit margins reported by the World Bank would suggest that the gap can be upwards of 20% in some other African countries. Finance and support from MDBs can radically reduce the interest gap, based not least on the fact that the credit history of African infrastructure projects has been, across many countries, broadly comparable to that in richer economies (GEMs, 2025).

It is important to recognise that the MDBs will have to work in all EMDCs (other than China). Major infrastructure investments will take place in middle-income countries in the coming two decades. If they are high-carbon, the implication for poorer countries, particularly in Africa, would be devastating. At the same time the poorest countries have great need for building resilience and capital to leapfrog onto a clean development path. Thus an emphasis on overcoming world poverty implies investment in all these countries and not just in those that are currently the poorest.

In 2023, the G20 Independent Expert Group (IEG),[18] chaired by N.K. Singh and Larry Summers, called for MDBs to adopt a 'triple mandate': end extreme poverty, drive shared prosperity, and support global public goods (Singh and Summers, 2023a). As noted, it recommended a tripling of MDB financing by 2030 and continuing to expand beyond that. Meeting global development and climate goals will require an additional US$3 trillion per year in EMDCs (other than China) by 2030 – an additional US$1.8 trillion for climate investments and US$1.2 trillion for broader SDG targets like health and education. The second report of the Singh–Summers IEG (October 2023) was *Better, Bolder, and Bigger* (Singh and Summers, 2023b). The 'bigger' means expansion; the 'bolder' means the MDBs taking more risk and working more closely with the private sector; and the 'better' means better functioning of the MDBs as a group and making themselves much easier to work with.

Beyond strictly financial support, the MDBs and other international institutions can step up to provide technical assistance and capacity-building in support of climate policy in EMDCs. By leveraging the global networks within which they operate and experiences from other countries, they can assist in developing national climate action plans, improving regulatory frameworks, and integrating climate considerations into economic policies. Often, attractive projects are there in abundance but are stymied by obstacles and impediments to getting things done, for example via red tape, corruption, or difficulty with the acquisition of land and with connections to and availability of infrastructure. The MDBs can play a key role in facilitating knowledge sharing, providing data-driven insights, and supporting the adoption of innovative technologies and sustainable practices. Non-finance support can be highly complementary to ramping up financing in the most effective and targeted ways. For all these reasons the multilateral financial institutions are critical to the scale and composition of the finance which is necessary. While some international financial mechanisms show movement and promise, a far greater sense of urgency is needed.

The MDBs, as emphasised, have a particular role in relation to risk. In understanding how to help in dealing with risk, it is useful to distinguish three elements: risk reductions, risk management, and risk sharing. We can explain through the example of wildfire. Risk of fire is reduced by care with barbecues and cigarette ends. Fires are managed through fire services and infrastructure of water availability. Risks are shared via insurance. Action on all three together can unlock investment at scale by substantially reducing the costs of finance. For example, government-backed PPAs in renewable energy provide long-term revenue confidence and reduce risk, thus attracting capital and reducing its cost. Risk management can come from liquidity buffers or green credit lines, which can safeguard investments from regulatory shifts and currency fluctuations. The African Risk Capacity (ARC), a group which is a specialised agency of the African Union, shares or pools sovereign insurance premiums across countries, providing rapid disaster response financing and preventing financial distress. The MDBs can provide first-loss guarantees, where they agree to absorb the first layer of potential losses, so other investors feel safer investing. Blended finance structures can play a major role in risk-sharing. An example is the Global Energy Alliance for People and Planet (GEAPP), bringing together philanthropic capital, MDB resources, and private investment to fund clean energy projects that would otherwise struggle to secure financing.

Expanding these models and approaches in dealing with risk is critical to making climate investments both scalable and affordable. There is overlap and complementarity between instruments in relation to risk reduction, management, and sharing. The labelling is not crucial. There is a rich set of instruments available, both separately and in combination. The MDBs are central here but, as the examples show, there is much that other agents can do.

Climate finance comes predominantly in the form of debt, making up around 55% of global flows in 2023 (CPI, 2025). Many EMDCs are already burdened with unsustainable debt, leaving little fiscal space for climate action.

With rising interest rates, many face prohibitive borrowing costs, pushing them further from the capital they need for climate action. Alternatives to debt for new resources are needed; so too is action to reduce the burden of existing debt.

Multilateral initiatives like the G20's Debt Service Suspension Initiative (DSSI) and the Common Framework for Debt Treatments offer debt relief to low-income countries, helping them redirect resources towards climate action. These efforts need to be more ambitious, longer term, and better integrated into national economic policies.

Alternative financial instruments should be tailored in relation to the financial difficulties or barriers present in each context. Grants can support early-stage projects that struggle to attract commercial investment. Revenues from carbon pricing and taxation – such as levies on international shipping emissions – could provide stable funding streams while discouraging carbon-intensive activities. It is important also to reduce the many toxic subsidies which incentivise damage to the environment and encourage emissions; and in so doing there is the possibility of generating large revenues. These include many of the subsidies associated with chemical fertilisers and fossil fuels (see Damania et al., 2023; Sutton et al., 2024).

Carbon finance, which monetises avoided emissions, can unlock capital for potentially high-impact projects with limited revenue potential.[19] These are some examples of other innovative instruments being developed:

- **Debt-for-nature swaps** help nations reduce debt by committing to conservation efforts, linking financial stability with environmental goals. By refinancing costly debt with lower-interest loans tied to environmental action, countries free up funds for sustainable development.
- **Green bonds** provide a channel for governments and corporations to fund climate programmes by attracting investments from environmentally conscious investors.
- **Blended finance** combines public, private, and philanthropic funds to manage and share risks, bring in private sector capital, and lower the cost of capital.

The expansion of the MDBs should draw heavily on blended finance, combining public and private capital to de-risk investments. Guarantees – one of the most effective tools for mobilising private investment – should be expanded strongly. They represented just 4% of MDB commitments between 2016 and 2020, even though they attract five times more private capital than loans (Blended Finance Taskforce, 2023). The Sub-National Climate Finance Initiative (SCF) offers a promising model, aiming to leverage US$750 million in blended equity funding with a 20:1 private-to-public finance ratio. By blending concessional finance with commercial capital, the SCF mobilises private investment for climate-resilient infrastructure projects at the sub-national level and provides technical assistance to local governments.

Special drawing rights (SDRs) – an IMF-created reserve asset – could become a powerful tool for climate finance. SDRs are not a currency but serve as a claim on freely usable foreign exchange reserves (International Monetary Fund [IMF], 2021). Developed countries receive most SDRs, since they are allocated in proportion to the IMF quota or share of each country, but they can channel them to climate-vulnerable countries through mechanisms like the Resilience and Sustainability Trust (RST) or the Poverty Reduction and Growth Trust (PRGT) and, potentially, through MDBs (Zattler, 2024). During the pandemic, the IMF allocated US$650 billion in SDRs, of which, under the quota-based systems, US$500 billion went to developed economies. G7 and G20 members later pledged to re-channel US$100 billion to assist poorer nations. Expanding and restructuring allocations of SDRs could provide much-needed low-cost liquidity for climate action in developing countries.

Mobilising capital for resilience

Climate-vulnerable nations face devastating economic losses from extreme weather events – Dominica's losses due to Hurricane Maria in 2017 were more than 200% of its GDP and Grenada experienced losses of around 200% of its GDP due to Hurricane Ivan in 2004 (Hurley et al., 2024). Without support, they borrow at higher rates, sink deeper into debt, and become even less able to invest in resilience. Proposals such as the 2024 Bridgetown Initiative, led by Prime Minister Mia Mottley of Barbados, offer promising solutions. By embedding disaster clauses in debt contracts, governments could pause repayments when extreme weather strikes, freeing up resources for recovery. More broadly, finance should be restructured to reflect the reality that climate shocks are not one-time crises but ongoing economic threats. Contracts which are disaster-contingent, reliable, and fast disbursing are crucial to break these vicious cycles.

Private capital mobilisation can play a key role in improving financial resilience through innovative financial instruments, such as parametric insurance. While traditional insurance pays out based on the evaluation of losses after an event, parametric insurance pays a fixed amount when predefined, objective, and measurable intensity parameters are met, such as storm wind speed. The intensity threshold is designed to reflect the potential loss from the event that was insured. This enables quick payouts to be made as soon as the event intensity threshold is met, offering an alternative or complementary coverage for losses that are many times excluded from traditional insurance due to excluded risks, limited insurance capacity, or widespread damage events (Garcia Ocampo and Lopez Moreira, 2024). An example is the 2017/2018 Catastrophic Risk Insurance Programme in the Philippines, which provided rapid liquidity to 25 provinces after climate shocks, transferring typhoon and earthquake risks from the country to the international reinsurance market in local currency through a parametric insurance product with a multi-layered risk financing structure.

Parametric insurance engages the private sector in two main ways: first, as a capital provider, where private investors, frequently alongside public institutions such as the MDBs or governmental institutions, provide capital for the insurance mechanism itself; second, as a consumer, purchasing coverage to manage the risks of their own operations. In doing so, businesses operating in economic sectors and geographical regions particularly exposed to climate change can enhance their own economic viability. In the case of the Philippines, three main actors were involved in the provision of insurance under the Catastrophic Risk Insurance Programme: a state-owned insurance company, which acted as the primary insurer; the World Bank Treasury, which acted as a re-insurer; and a panel of international insurance companies, from where the World Bank drew capital through catastrophe bonds (high-yield and high-risk bonds). Through its involvement, the World Bank facilitated the programme's placement in the international insurance market, attracting global reinsurers and attaining a competitive price (McNally et al., 2024).

While climate finance struggles to go to scale, vast sums continue to flow to support fossil fuels and environmentally-damaging practices. In 2022, explicit fossil fuel subsidies totalled US$1.3 trillion (Black et al., 2023), with a further US$5.7 trillion in implicit subsidies, arising from not fully accounting for environmental costs and foregone consumption tax revenues. Thus, there was a total of US$7 trillion of subsidies for fossil fuels in 2022 (Black et al., 2023).[20] The estimated scale of environmentally-harmful (explicit and implicit) subsidies from agriculture, fisheries, and forestry was US$840 billion in 2023 (Koplow and Steenblik, 2024). These subsidies not only strain public budgets but actively undermine climate and environmental goals. Redirecting even a fraction of these funds toward clean energy, adaptation and resilience, loss and damage, nature, and a just transition could transform global climate finance.[21]

The challenge is not just mobilising capital but structuring and combining capital from different sources so that they reinforce each other. Large-scale renewable energy projects will mainly rely on private capital but in many cases will require guarantees and concessional loans backed by the MDBs to lower capital costs. Most adaptation and natural capital projects and programmes will need concessional finance, but can still attract private investment through instruments like carbon credits and sovereign climate bonds.

Although the architecture for international climate finance is taking shape, at the same time emissions continue to rise, and investment in the transition and the new growth story remains far too low. Financial institutions are moving too slowly to support the scale of investment needed for climate action. As we have seen, it is developing countries that are seeing particularly low capital flows, even though their opportunities and needs are the greatest. Thus, they are unable to invest in clean energy, adaptation, nature, and more broadly sustainable growth on anything like the scale necessary. Expanding MDB financing, restructuring debt, and unlocking private investment should be at the top of global priorities. The urgency should be clear. The longer the delay, the greater the climate risk. The longer the delay, the more likely that we cross tipping points (see Chapter 2). And the longer the delay, the greater the lost opportunities in fostering growth and reducing poverty.

9.4 Technology, industrial policy, trade, and innovation

Technological advance has been impressive. Yet, there remain important challenges in achieving low or zero carbon in a number of key activities. National and international actions around industrial policy can accelerate diffusion, bring sectoral transformation, and drive new technologies forward. And trade and openness will be critical in driving demand for new products forward and the process of diffusion.

Technology

The race to lead in key clean technologies is well underway. As highlighted in Chapters 4, 5, and 8, competition is already fierce in areas such as solar and wind power, batteries and storage, and electric vehicles and it is rapidly intensifying in emerging sectors like green hydrogen and green steel. Digital technologies and artificial intelligence (AI) have come at an opportune time in this race and will help accelerate the pace and effectiveness with which the clean energy transition occurs.

Low- or zero-carbon technologies provide lower costs of production than emitting and polluting technologies across around a third of emissions. That proportion will grow rapidly over the next decade (see Chapters 4, 5, and 8). But in some important areas, such as cement, aviation, and parts of agriculture (particularly pastoral), whilst progress is being made, technologies where the clean or low-emitting is cheaper than the dirty are still a long way off. Technology has achieved a great deal, and much more is coming through quickly, but there are substantial areas where strong research and development (R&D) and innovation are needed. These difficult sectors and activities are areas of priority for both public and private research. The stronger the policy and action across the world, the stronger will be the perceived potential demand, the more intense the R&D, and the earlier the results.

Technological progress, under the impetus of strong competition, is moving very rapidly in power – both generation and storage – and in surface transport. It is moving quickly in heating, cooling, and buildings. Together, these cover around 40% of emissions (UNEP, 2024). There is much that good public policy can do to keep that progress strong and to continue the very rapid fall in costs in these areas. Stimulating demand for the clean products can come through taxing or pricing carbon, regulating emissions and pollution, providing affordable and available capital for low-carbon investments in homes, product policies, enabling complementary infrastructure such as EV-charging stations, and so on. That is a task for all countries. Coordinating internationally on such policies, so that global demand for the relevant products and processes emerges still more strongly, could have powerful influence on the supply side of new technologies and the associated R&D. Of course, the fact that these areas are moving quickly does not imply that we should slacken support for R&D. There is real momentum on which we can build, yet many major technological challenges remain.

Given that for some areas – such as cement, aviation, and pastoral agriculture – it will prove difficult or costly to find zero-carbon options, carbon dioxide removal (CDR) will unavoidably be an important element in climate action, as will, possibly, solar radiation management (SRM). These are also priority areas for R&D – see Section 9.6.

The pressure and possibilities for change will be strongly influenced by both domestic and international market structures. Thus, industrial and trade policy are key elements in shaping future costs and supply chains. These are issues to which we now turn.

Industrial policy

In this context of rapid technological change, there is a growing recognition both of a strategic imperative to cut emissions swiftly, and of the new opportunities for growth. This is reflected in a wave of country-level green industrial policies. These include China's 14th Five-Year Plan (2021–2025) on Modern Energy System Planning and 15th Five-Year Plan, to be published in 2026 (covering 2026–2030). Other examples are the EU's Green Deal and Net Zero Industry Act (NZIA) and Clean Industrial Deal; India's Union Budget, energy access, and green hydrogen policies (Garg, 2022); and, until recently, the US IRA.[22] These all combine incentives and finance for green technology and investments with legal, regulatory, and policy support. The structure and future of these strategies, and the competitive race they embody, will likely have profound implications for geopolitics, technical change, economic and industrial geography, trade flows, and supply chains, as we saw in Chapter 8. But change will also involve structural dislocations, and how these play through and are managed will have strong implications for the ability to tackle the climate and biodiversity crises and to create the new path of growth.

The new wave of green industrial policies has the potential to accelerate the energy transition globally, unlocking many of the drivers of growth core to our new growth story. Well-designed policies can support innovation and learning-by-doing and boost investments in manufacturing and deployment of clean technology at scale, thus driving down costs and enabling global diffusion of low-emissions technologies. National industrial strategies and policies will have international impacts in terms of the scale and consistency of demand. Elements of difference and of coherence have their role to play. Differences in approaches facilitate the emergence of new possibilities. Coherence, for example in the specifications for municipal buses, can allow scale and cost reductions. The combinations of differences and coherence will drive the dynamics. Thus, national and international interactions between groups of entrepreneurs, architects, engineers, mayors, and finance ministries will all be part of a process of acceleration.

The overall impact of green industrial policies on innovation, competitiveness, and trade depends on the way policies are designed. For example, some of the subsidies included in the IRA (most now discarded in the major bill signed on 4 July 2025 by President Trump) and the EU's NZIA are conditional on content produced within the country or region. These local-content requirements (LCRs), or other instruments that include geographically discriminatory provisions, might, over time, produce domestic scale and cost reductions. That is the 'infant industry' argument. But history and theory have taught us that they could also lead to an inefficient restructuring of supply chains, arising from the diversion of resources to where the subsidy is implemented. In this way it can decrease international competition, distort trade, and increase the prices of technologies both in the short and medium term, making the energy transition temporarily more expensive – and potentially slowing down decarbonisation.

Once protection is introduced, those who benefit will likely press for it to continue. There are many examples in economic history – from agricultural subsidies in the EU under their Common Agricultural Policy since the 1960s to the US steel industry being protected by tariffs under President George W. Bush in the early 2000s. 'Temporary' protection can become long-lasting in practice. For a protectionist policy to be effective and worth pursuing, it should make the domestic production it encourages competitive on world markets. If it eventually does, there could be a global benefit of more vibrant competition.

We have seen that policies to stimulate green demand, such as Germany's pioneering renewable energy policies – particularly its feed-in tariffs in the early 2000s, which fostered domestic demand for solar PV – led to technological advancements and cost reductions across the world. In particular, it enabled Chinese manufacturers to scale up production, driving down costs at rates which were not predicted at the time (Quitzow, 2013). The judicious combination of domestic industrial policies and openness to trade will be at the heart of the cost reductions and technological progress – see the next subsection on openness.

The LCR type of incentive structure could be interpreted as trade-distortive subsidies, which are prohibited under WTO rules (see WTO, 1994, Article 4 ASCM; WTO, 1986, GATT Article III:4). They may also trigger trade tensions and protectionist responses in other countries, making international trade in green technology more fragmented and less efficient, slowing down technological diffusion and effective climate action. In May 2024, the Biden administration in the USA raised tariffs on Chinese-made EVs to 100%, more or less banning them from the US market. In June 2024, the EU also imposed tariffs on Chinese EVs, but at lower rates than the USA. Under the second Trump administration, in 2025, the USA imposed new tariffs applying to all nations, directed with particular intensity at China but also strongly at the EU. China and the EU have retaliated with tariffs of their own against goods from the USA, sparking fears of a global trade war, increase in inflation, and a potential economic recession. No doubt future negotiations on these tariffs and trade will play through during Trump's second presidency, but there is real danger in all this resulting in a collection of national policies which will, taken together across nations, reduce competition, increase costs, and dampen innovation relative to a more open world.

It is the combination of fostering the growth of demand, promoting new sources of supply, and domestic and international competition that will drive costs down. Thus, LCRs should be time limited. There are valid arguments for providing public support to domestic manufacturing of green technologies at early stages – prior to going to scale and learning-by-doing, when perceived risk may be strongest – eventually the domestic support should be phased out. In this respect it is crucial for countries to have a well-defined programme over time for subsidy or exclusion schemes, if the lock-in of inefficiencies is to be avoided. Such a programme should include criteria guiding the phase-down of subsides and exclusions. An example would be criteria guiding the reduction of the subsidy when the cost of the subsidised activity or technology reaches a certain level, after a given degree of diffusion of products into the market, or after a given date. Such a transparent and structural approach will give the guidance and clarity on future policy that can allow for future planning and risk management for investors. This is an example of what we meant in Chapter 6 by saying that, for investor confidence, policies should be 'predictably flexible'.

In the past, subsidies in some countries have been removed with little notice. An example of an abrupt subsidy removal was the Spanish government's sudden decision to eliminate solar PV subsidies in 2010, just a few years after introducing them in 2007 and after having created the expectation that they would last for decades (other renewable energy subsidies were also cut at a similar time). This decision, made after the 2008–2009 financial crisis, shook investor confidence and led to the country being internationally sued more than 50 times (International Institute for Sustainable Development [IISD], 2010; UNEP, 2023a). Such policy behaviour makes future policies less effective, as investor confidence in public policy is undermined.

This new wave of industrial policies – including China's five-year plans, the EU's green deals, India's green initiatives, and the US IRA – can also lead to subsidy races across countries, as demonstrated by the reactions of the EU to the IRA and Chinese subsidies. The risk here is that by engaging in subsidy races with other industrial nations, countries will end up needlessly driving the number of subsidies up, leading to inefficient and wasteful allocation of public finance and making the policy more expensive for the world as a whole. There is a risk that only high-income countries with stronger fiscal systems and greater access to financial markets will be able to engage in the 'green subsidies race'. This could result in reduced green investment and some countries being left behind (particularly EMDCs other than China), which could make their transition more costly and politically contentious. Slowing their transition would be damaging to all.

The strongest potential structure for producing rapid technological change lies in a world where, in key industries, there are multiple high-scale competitive producers. Of particular importance in the green transition are solar and wind, batteries, EVs, green hydrogen and steel, certain minerals, microchips in relation to AI, and AI itself. The number of high-scale producers will be limited by market size. However, maintaining a strong level of competition in these industries may require at least four or five producing countries, possibly more in some sectors. And strong within-country competition is crucial too. We have seen how powerful that was in bringing down the cost of solar panels and EVs in China.

Building these structures will require competition and collaboration across nations. It will involve the WTO, the MDBs, and international sectoral bodies. These are times for creativity and the avoidance of formulaic views, with a spirit of innovation and openness and the common good of tackling climate change and building a new growth story.

Trade and openness

As just argued, a constructive and dynamic trade structure for the world in the coming years around green technologies and investment for the transition would be one that embodies the competition that can encourage innovation and low costs; fosters the exploitation of economies of scale; allows countries, particularly poor countries, access to lowest-cost products; and offers supply chains that are robust and resilient. Enhanced international collaboration on trade should seek advances on all four of these dimensions.

Given the imperative of investment in the transition, the opportunities for growth, and the rise of tariffs and protectionism as of 2025, there is an urgent need for clear principles on trade for preserving and encouraging competition whilst allowing countries to build the capacity that can deliver that competition (WTO, 2022). This is a time to revisit, clarify, and strengthen arrangements for openness in the face of tension and disruption. This process requires a shared understanding of strategies and a recognition that whilst support mechanisms will be necessary, they should be time limited, with limitations operating in relation to transparent principles.

It means being open to inward investment from countries with specialised expertise, such as in batteries or microchips. A country seeking to build EVs cheaply and quickly should recognise that China is currently in the lead in building batteries and should consider collaboration with and openness to investment from China. There was a time, when I was in China in the late 1980s, when China's strategy was openness to inward investment and joint ventures so that they could catch up technologically. China is now ahead in many technologies.[23] Other countries can now gain knowledge from inward investment from China. And inward investments in China will now also likely involve great learning by the inward investor.

The G20, the WTO, the OECD, and the IFIs should work to establish these procompetitive principles, which could underpin openness, whilst building clean and resilient new activities and supply chains. Shared recognition of the structures that are broadly beneficial to the world on the criteria described is a crucial first step along the way. There are surely lessons from the profoundly damaging effects of tariff wars and 'beggar-thy-neighbour' policies of the 1930s. International collaboration on these issues should not be delayed until the damage from new tariff wars of 2025 appears.

Various other country networks and mechanisms can support this competitive, low-cost, high-scale, and resilient future. Sectoral clubs and climate clubs, such as the Global Arrangement on Sustainable Steel and Aluminium and the G7 Climate Club, can enable greater ambition, provide vehicles for sharing technology, and mobilise private finance channels. Such networks can also help address concerns around competitiveness and trade-related action, directly or through the WTO forums. The Coalition of Trade Ministers on Climate can play a role in boosting international cooperation on climate, trade, and sustainable development. And, similarly, the Coalition of

Figure 9.5: Carbon leakage and the EU's CBAM explained

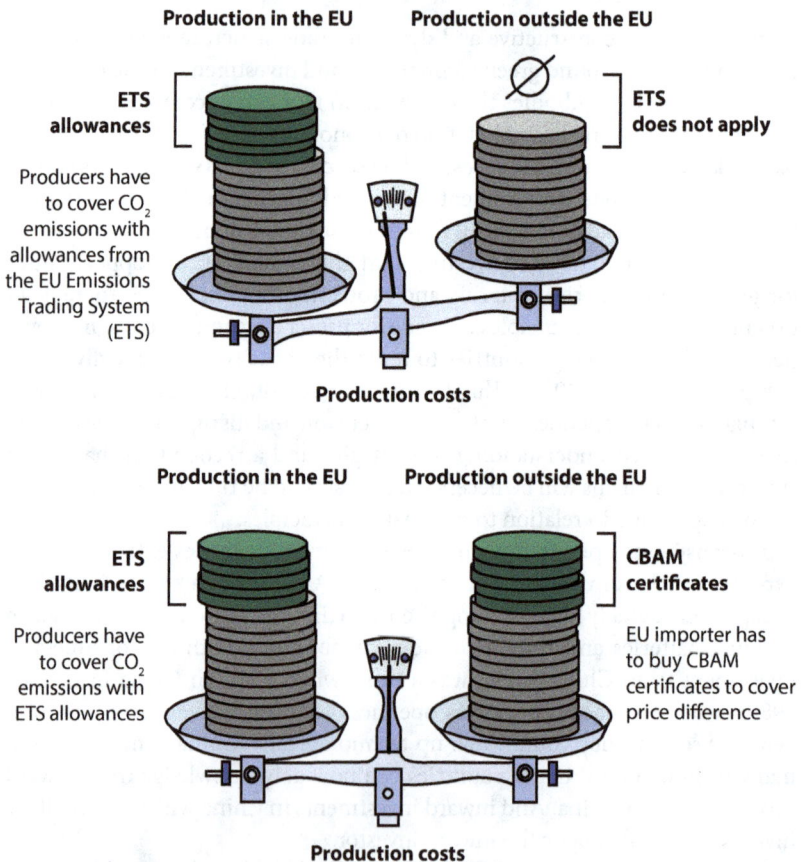

Source: Author's elaboration based on European Council (2025).

Finance Ministers for Climate Action and the NGFS, as we discussed earlier.

Together they can provide valuable structures to create coherent strategies, and work to incentivise and support the greening of activities in EMDCs.

Policies can be designed to directly cover trade-related carbon emissions. The EU's Carbon Border Adjustment Mechanism (CBAM) imposes a border carbon price on some carbon-intensive goods covered by the EU Emissions Trading System (ETS) that are imported into the region (see Figure 9.5). The goal of CBAM is to align the carbon pricing of imported goods with that of products manufactured within the EU, thereby levelling the playing field with domestic industries while supporting the EU's climate objectives and encouraging cleaner production practices in non-EU countries. CBAM is a different mechanism from an LCR and is not necessarily protectionist. It is a policy that seeks to limit market failure. But to prevent it from developing into a protectionist vehicle, clarity and simplicity are critical. Thus, it is important to keep it focused on just a few key sectors and products and base it on clear and explicit criteria. At present CBAM covers cement, electricity, fertilisers, iron and steel, and aluminium and hydrogen. The plan is for it to cover all EU ETS sectors by 2030 (European Parliament, 2023). Care will be necessary to avoid the smuggling in of protectionism as its base expands.

The WTO, together with these coalitions and clubs, could help generate progress along all four dimensions of innovation, scale, access to lowest cost, and resilient supply chains. But progress will require strong support for the WTO from its members and constructive collaboration amongst the coalitions, networks, and clubs. This could be an example where a recognition of the importance of collaborating around climate could have the benefit of boosting some much-needed collaboration around trade and the WTO. As emphasised at a number of points in this book, the need to collaborate on climate can boost collaboration elsewhere.

The year 2025 is a fractious and difficult moment for trade in the whirlwind of tariffs and Trump's second presidency. But these processes of innovation and change are crucial to tackling the climate and biodiversity crises and creating a new story of growth; they will have to cover the next few decades. Analytical and strategic work is crucial now. And if the USA steps back on trade and climate, others can step forward and build lasting action for the future.

Let us now examine some specifics of strategies and policy, together with examples that could put these principles into action to accelerate technological change, innovation, and investment.

Unlocking global innovation

Enhanced international collaboration on technology can foster innovation, bring down costs, accelerate adoption and diffusion, and promote the availability and affordability of green technologies for all countries. Many of the opportunities of the new growth story are tied to international cooperation. For example, intermittency is much more easily handled with an extensive grid which can move electricity across geographies. Sharing surplus or low-cost clean electricity among regions is mutually beneficial. The UK collaborates with neighbouring countries like Norway (mostly wind) (North Sea Link, n.d.), France (mostly nuclear) (Department of Energy Security and Net Zero [DESNZ], 2023), and soon Morocco (solar and wind) (Xlinks, n.d.). Cooperation of this kind across green development opportunities to unlock the new growth story requires a strategy embodying many or most of the aspects discussed in the following paragraphs.

Aligning technology goals. Recent research shows that the chances of realising a positive tipping cascade in clean technology that can drive costs down are stronger when actions are aligned internationally so that there are prospects of large and dynamic markets with shared standards (Systemiq, 2023). For example, if the three largest car markets (European, North American, and Chinese) were aligned to require all new car sales to be zero emissions by 2035, then it would bring forward the cost-parity date of EVs and internal combustion engines by up to five years (Lam and Mercure, 2022). Whilst the immediate prospect of US collaboration under Trump's second presidency is weak, others can move forward.

Green trade liberalisation, reforms, and diffusion. Green trade liberalisation and reforms of trade-related measures are important to enable the diffusion of clean technologies to developing countries. Other options to enhance technology adoption include the use of de-risking mechanisms such as loan guarantees in developing economies (IMF, 2021). However, this is no longer solely an issue of technology transfer from developed to developing. Technological advance is taking place geographically across the world, and across sectors from AI to agriculture. As we have seen, China is playing a particularly important role. Openness to these advantages in technology and products is crucial to the pace and efficiency of the transition. An example of international collaboration to create, reduce the cost of, and diffuse technologies is the work of the CGIAR (formerly the Consultative Group for International Agricultural Research) on agriculture. It has, for more than half a century, developed and shared new technologies and crops in agriculture. It was a key driver of the 'green revolution' in wheat in the 1960s and 1970s. And in recent years, it has focused on climate resilience and lowering emissions (see, e.g., Amahnui et al., 2025; CGIAR, 2021).

Reducing the cost of capital, especially for green technologies. Much technological advance – although not all (skilled labour matters too) – is embodied in capital. Many of the required technologies are capital intensive and many developing countries have to pay a high cost of capital. Hence bringing down the cost of capital will be important, via investment demand, in taking technology forward globally and particularly in developing countries. The role of MDBs, as we have already seen, will be critical. See Section 9.7 for an example of great potential importance.

Innovation and diffusion frameworks. Innovation and diffusion frameworks linking up firms, governments, and civil society (Aghion et al., 2021) can help foster private sector investments in clean and innovative assets to move quickly.

'Mission-oriented' programmes for innovation along the lines analysed and pro-posed by Chris Freeman – as we saw in Chapter 7 – as well as broader policies aimed at creating investment and innovation-led growth can be strong drivers of technology that could help deliver advances in environmental sustainability. In so doing, they could follow the successful examples of mission programmes that have driven innovation and diffusion forward in other contexts, including in medicine and defence (Freeman, 1995; Mazzucato, 2018; Stern and Valero, 2021) and in agriculture, as noted in relation to the CGIAR.

R&D coordination. International coordination in R&D will likely be neces-sary in some key relevant areas. Examples could be where AI could be applied to accelerate research into synthetic biology; fusion; carbon capture, utilisa-tion, and storage (CCUS); and direct air capture (DAC). A further important example for AI and data concerns resilience and adaptation, where infor-mation at the local level, and tailored to local detail, is crucial to timely and effective action. That requires high-resolution capacity in combination with strong local data. These will need international collaboration both in com-puting power and in collecting and integrating data. Some, such as Professor Tim Palmer and others, have been pressing for a CERN for climate (Palmer, 2024).[24] Breakthroughs on fundamental particle physics at the European Organization for Nuclear Research (CERN) were made possible by building hardware and a collection of talent and scientists and technologies on a scale that would not have been possible in a single country.

Collaboration networks. One example of a collaboration network is the Break-through Agenda, which brings together 45 countries and businesses to accelerate the development and deployment of clean technologies and drive down costs by 2030, so that 'clean is cheaper' (IEA et al., 2022). Collaboration can take the form of joint innovation (e.g., Mission Innovation, 2021), by converging around standards (e.g., Global Cement and Concrete Association [GCCA], 2024; Inter-national Partnership for Hydrogen and Fuel Cells [IPHE], 2023), and by creat-ing buyers' clubs or coordinated green procurement programmes for pioneering products and processes (e.g., First Movers Coalition [FMC], 2024; Industrial Deep Decarbonisation Initiative [UNIDO], 2024). Other examples are the coor-dination of procurement across cities, such as in the C40 (C40, 2024). If many cities decide they want certain kinds of electric buses or traffic management sys-tems, R&D and innovation will take place in response, economies of scale will be realised, and costs will fall (C40, 2015; Jena and Trivedi, 2023).

9.5: Aligning global climate and biodiversity action

Unlocking innovation on a global scale will help speed up the clean energy transition that is so vital both to achieving net zero and to bringing forward an era of sustainable growth. What we must not forget is the entanglement of the climate crisis with the biodiversity crisis. This interconnectedness demands that we align global action on climate with global action on biodi-versity. Although there is momentum in global agreements for biodiversity,

financial flows still fall far short of what is necessary and international cooperation remains weak.

Most of the world's vital ecosystems – tropical rainforests, oceans, and the
atmosphere – function as global commons. Their degradation in one place
affects the entire planet and a protected forest in one country means little if
deforestation shifts to another. International coordination is essential.

Among the agreements that address these topics are the UNFCCC, the
Convention on Biological Diversity (CBD), and the UN Convention to
Combat Desertification (UNCCD) – the three Conventions that emerged
from the Rio Earth Summit of 1992. Whilst the UNFCCC has had the highest
profile, all three are important. And their common origins in Rio in 1992
reflect their interwoven relationship. Trade measures like the Convention
on International Trade in Endangered Species (CITES) regulate markets for
endangered wildlife. The Antarctic Treaty (UN, 1959) has helped preserve one
of the world's last untouched wildernesses.

Despite these efforts, results have also often fallen short on biodiversity as
they have on climate. Just a year after the 2021 Glasgow Declaration, where
145 world leaders pledged to halt deforestation by 2030, deforestation rates had
already exceeded the target by over one million hectares (Goldman et al., 2023).
Similarly, the Global Biodiversity Framework (GBF) aims to conserve 30% of
the world's land and seas by 2030. It seeks to mobilise finance to fill an annual
estimated gap of US$700 billion by 2030 (The Nature Conservancy [TNC], 2020).
The two strategies in the Framework are to (a) reduce or repurpose harmful
subsidies and incentives by US$500 billion, and (b) increase biodiversity-
related funding from public and private sources by at least US$200 billion per
year – with at least US$20 billion per year by 2025 and US$30 billion per year by
2030 of these international flows being directed to developing countries (UNEP,
2025). So far (writing in Spring 2025) progress on the GBF is not promising.
Some global initiatives are summarised in Table 9.2.

That progress is inadequate in relation to the biodiversity crisis does not
mean the total absence of action, as Table 9.2 illustrates. Finance flows
into biodiversity have increased. For example, global multilateral financial
flows[25] into biodiversity-related development increased from US$1.4 bil-
lion in 2015 to US$11.3 billion in 2022, and private finance mobilised by,
or alongside, development finance increased from US$0.09 billion in 2016
to US$1.76 billion in 2022 (Biodiversity Finance Trends, 2024). However,
the vast majority is spent in developed countries, when the bulk of the
opportunity is in the developing world (Bhattacharya et al., 2023). Indeed,
EMDCs (other than China) hold around 90% of the investment opportunity
in protecting and restoring nature between 2020 and 2030. But substantial
increases in investment are required to take advantage of these opportunities. Measures such as transitioning agricultural practices towards the
regenerative, protecting and restoring ecosystems, and reducing waste pollution are all needed (Songwe et al., 2022).

Table 9.2: International action on climate and biodiversity since 1987

Initiative	Year	Description and aims
The Montreal Protocol (UNEP, 1987)	1987	Treaty on phasing out HCFCs. A longstanding example of success in international cooperation, leading to restoration of the ozone layer. Specific, voluntary, user standards, supported by voluntary funding. Advantage of availability (through R&D and innovation) of alternatives.
The Kigali Amendment (UNEP, 2016)	2016	Amendment to the Montreal Protocol to phase down hydrofluorocarbons (HFCs), which are potent GHGs.
The Glasgow Leaders' Declaration on Forests (UNCCD, 2021)	2021	Heads of 145 countries committed to halt and reverse forest loss by 2030. This was followed by a summit of the leaders of the eight Amazon Basin countries (2023), the first in 14 years, resulting in an agreement to join forces to protect the crucial rainforest from 'reaching the point of no return'.
The Kunming–Montreal Global Biodiversity Framework (GBF) (CBD, 2022)	2022	A landmark treaty adopted by nearly 190 countries at Biodiversity COP15. Aims to set in motion an economic transition that puts biodiversity on a path to recovery by 2030 and to reach the global vision of a world living in harmony with nature by 2050. Range of targets for 2030, including the headline target of '30×30', an ambition to conserve 30% of the world's land and seas for nature by 2030.
The High Seas Treaty (UN, 2023b)	2023	Adopted by UN members to protect oceans and sustainably use marine biodiversity. While the treaty does not set specific conservation targets, it can help nations designate marine protected areas, requires environmental impact assessments for potentially harmful activities, and provides for the transfer of marine technology to developing countries.
COP28 UAE Declaration on Climate, Nature, and People (UNFCCC, 2023)	2023	Signed by 18 countries in 2023. Emphasises the importance of integrating climate action and biodiversity conservation to achieve the goals of the Paris Agreement and the GBF.

Source: Author's elaboration.

As we saw in Chapter 2, the climate and biodiversity crises are deeply inter-connected. For example, deforestation accelerates climate change through the release of carbon, while rising temperatures and changing precipitation and seasons disrupt ecosystems and threaten species survival. Policies and financing must reflect this overlap. And they have common causes, particularly via the burning of fossil fuels. This overlap and interweaving points to policies and mechanisms which can help tackle both. For example, VCMs could pri-oritise carbon reduction projects that also restore biodiversity by including premiums in the carbon price for such projects; that would allow climate finance to support both goals. Agricultural climate-related regulations should integrate considerations of ecosystem health. Energy, transport, and water regulations should also embed strong environmental standards.

The means to increase funding into biodiversity are, broadly speaking, the same as those for climate action and include international financial resources, DRM, private finance, blended finance, low- or zero-cost flows, and innova-tive instruments as described above. Private-sector involvement in financing is limited by the inability to generate revenue flows in many cases, but there are important examples (see Chapter 6) where such flows can be generated. Governments nationally and internationally, and IFIs, can play a role both in de-risking and in providing incentives for the private sector to get involved.

Urgent action is necessary in biodiversity hotspots like the Amazon, Congo Basin, and Indonesian rainforests, which contain more than 50% of the world's remaining primary tropical forests (Fleck, 2022). These hotspots often span multiple countries; the Amazon spans eight, the Congo Basin six. And the effects of their degradation – such as the release of vast amounts of CO_2 and the undermining of global ecosystem resilience – are global. Action therefore needs to be coordinated at the supranational level. Conservation in these regions is often hindered by domestic political pressures and eco-nomic interests tied to deforestation (Jackson, 2014; Milmanda and Garay, 2019; Tegegne et al., 2016). Strengthening governance, increasing financial incentives for conservation, and countering the influence of vested interests are key to progress. Increasingly, satellite observation can provide very precise measurements of the state of forests and land and thus provide a database for governance and for measurement of performance. For example, Planet Labs, an Earth-imaging company, has created a Forest Carbon Monitoring tool that offers high-resolution satellite images, allowing for deforestation monitoring and carbon measurement (Anderson et al., 2024). ICEYE, a micro-satellite manufacturer, uses Synthetic Aperture Radar (SAR) technology to detect deforestation and forest degradation even through cloud cover, in all weather conditions, and at night (ICEYE, 2025).

Beyond government action, businesses and civil society play a crucial role in conservation and restoration. Companies that shape global supply chains can drive large-scale change through procurement standards and sustainability commitments. The Tropical Forest Alliance, for instance, works with businesses to eliminate deforestation-linked commodities from supply chains. Consumer demand also exerts pressure – certification schemes like the Rainforest Alliance and UTZ have shifted production practices in sectors such as cocoa, coffee, and palm oil. Yet, unintended consequences can arise. A study by Krauss and Krishnan (2016) found that when the German chocolate company Floral switched from paying premiums for organic cocoa to requiring UTZ certification[26] – it seems that the latter has more stringent social criteria but less stringent ecological criteria – most Nicaraguan farmers felt they had to adapt to the certification priorities of Floral. But many farmers disagreed with this change, with some claiming that the costs to meet the new standard were not being covered by the premium paid. Consumer backlash later forced the company to reintroduce organic certification. This highlights the need for carefully designed market mechanisms and focused economic analyses that examine and account for entire supply chains (see Dietz and Grabs, 2022).

Global movements on biodiversity and environment have proven powerful agents of change, shaping public opinion, influencing policy, and helping to enforce environmental laws:

The World Wildlife Fund (WWF) is a leader in global conservation efforts, working with governments and businesses to expand protected areas.

Greenpeace combines research, advocacy, and direct action to pressure companies and governments to adopt stronger environmental policies.

The World Resources Institute (WRI) uses data-driven approaches, such as Global Forest Watch, which provide real-time satellite monitoring of deforestation. Such tools have enabled communities, governments, and activists to detect illegal logging in protected areas within days.

ClientEarth, a legal NGO, has successfully challenged governments in court. In 2022, it won a landmark case against the UK government's inadequate net zero strategy, forcing revisions. In 2024, it mounted a second challenge alongside Friends of the Earth and the Good Law Project – and won again. The High Court found the UK government's climate strategy, the Carbon Budget Delivery Plan, unlawful, ruling that its assumption that the policies outlined would achieve the intended emissions reductions was wrong (ClientEarth, 2024).

Community-led conservation movements have also reshaped policy. Kids for Tigers, an educational outreach programme in India, evolved into one of the country's largest environmental movements, showing how grassroots activism can drive long-term change. I have had the privilege of seeing them in action in Madhya Pradesh, and the children's enthusiasm and knowledge is inspiring. In Canada, a 20-year campaign by Greenpeace and Indigenous leaders led to the protection of the Great Bear Rainforest, one of the world's last intact temperate rainforests.

Integrating nature's real value into economic decision-making is critical. Natural ecosystems provide immense benefits, but, in large measure, traditional economic analyses or models fail to account for them. A key feature of policy reforms and action should be to reflect the intrinsic and economic value of biodiversity. That transforms conservation from simply a moral imperative to a core part of sustainable development strategies. The Dasgupta Review proposes supranational institutional structures to enable the compensation of countries for conserving critical ecosystems or imposing fees for access to shared resources like fisheries and the high seas. Some progress has been made: international fishing quotas have shown partial success (UN Trade and Development [UNCTAD], 2024) and protected area agreements, like those governing the Amazon and Antarctic regions, offer some examples of cooperative conservation. However, implementation remains difficult due to national resistance to external taxation or external governance structures and the opposition of, often powerful, vested interests.

Stronger international leadership and cooperation are essential if the biodiversity crisis is to be tackled effectively. Protecting nature is not a luxury or a side issue – it is a fundamental necessity for economic stability, food security, and climate resilience. Without urgent action, we risk destabilising the very systems that sustain life. The interweaving with the climate crisis provides an opportunity for action on biodiversity since structures on climate are more advanced and measurement issues, difficult though they are, are less complex than with biodiversity.

9.6 Overshooting, negative emissions, geoengineering

Given the likelihood of overshooting temperature targets, as we are already very close to 1.5 °C, negative emissions technologies (NETs) – which remove CO_2 from the atmosphere – will be essential. Some emissions, such as those from some forms of agriculture and some industrial processes, will be difficult to eliminate entirely. In other cases, the complete elimination of emissions might be very expensive. Offsetting them through carbon dioxide removal (CDR) is critical to stabilising temperatures and eventually reversing overshoot. Solar radiation management (SRM) is also an area that deserves careful scrutiny, both because it likely carries many risks and because as we approach possible tipping points it might provide a way of buying time.

Both CDR and SRM require international cooperation. In the case of CDR it will be crucial to provide international prices to incentivise action. The whole world benefits from CDR but there has to be some kind of price to provide incentives for action. Alternatively, governments could get together to commit to working individually and collectively to remove carbon by public action. In the case of SRM, similar arguments apply. In addition, with SRM there may be a need for international agreement to control the possibility of individual actors or countries acting on their own. We discuss CDR and SRM in turn.

Scaling carbon dioxide removal

Several CDR methods are either in use or under development. These range from well-established, land-based approaches like afforestation and soil carbon sequestration to algae (natural climate solutions, or NCS), technologies such as bioenergy with carbon capture and storage (BECCS), and direct air capture (DAC). However, the large-scale deployment of CDR necessary to reduce global CO_2 lev-

Figure 9.6: Taxonomy of CDR methods

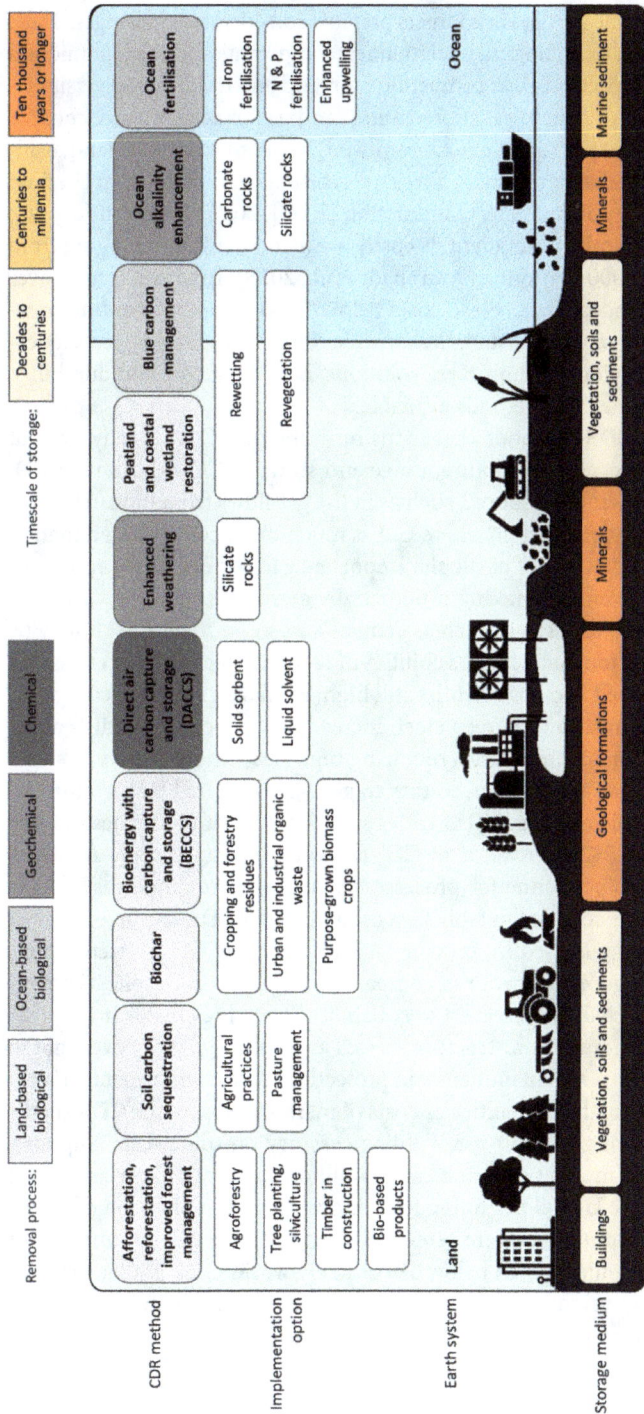

Note: Methods are categorised based on removal process (grey shades) and storage medium (for which timescales of storage are given, yellow/brown shades). Main implementation options are included for each CDR method. Note that specific land-based implementation options can be associated with several CDR methods; for example, agroforestry can support soil carbon sequestration and provide biomass for biochar or BECCS. I would add algae explicitly to the methods listed here.

Source: Box 8, Figure 1 in Babiker et al. (2022, p. 1262). Copyright 2022 IPCC. Reproduced with permission. Use of IPCC figure(s) is at the User's sole risk. Under no circumstances shall the IPCC, WMO or UNEP be liable for any loss, damage, liability or expense incurred or suffered that is claimed to have resulted from the use of any IPCC figure(s), without limitation, any fault, error, omission, interruption or delay with respect thereto. Nothing herein shall constitute or be considered to be a limitation upon or a waiver of the privileges and immunities of WMO or UNEP, which are specifically reserved.

els on sufficient scale for the Paris targets presents considerable challenges. These include high costs, technological uncertainties, and potential socioeconomic and environmental impacts, such as competition for land and risks to biodiversity.

Natural methods, such as afforestation and reforestation, are cheaper (around US$10–50 per tonne of CO_2 removed) but constrained by land availability and vulnerable to reversal – forests can be cut or burned down (Bednar et al., 2023; Energy Transitions Commission [ETC], 2022). Direct air capture (DAC), while basically permanent, is costly, with current prices ranging from US$600 to US$1,000 per tonne (Azarabadi et al., 2023). Innovation may, over two or three decades, bring these costs below US$100 per tonne, but large-scale deployment remains distant. Meanwhile, BECCS and soil-based sequestration could provide medium-term solutions, balancing cost and durability. Figure 9.6 illustrates the methods available.

Unlike other CDR methods that focus on removing CO_2 already present in the atmosphere, carbon capture, usage and storage (CCUS) captures CO_2 emissions directly at their source, such as in the gas flows from industrial processes and energy production, where CO_2 is much more concentrated than in the atmosphere. CCUS is of particular importance in sectors where emissions are difficult to eliminate, making it potentially a crucial tool for decarbonising 'hard-to-abate' industries, such as cement and some chemicals. It may be necessary in steel too, but here possibilities of creating steel without emissions are being developed and costs, whilst still high, are falling. Green hydrogen is a key element in relation to green steel. Increasingly, green electricity can be used in some chemical industries requiring high heat where a few years ago, it was thought gases would be necessary to generate required temperatures.

The cost of implementing CCUS varies significantly, with estimates ranging from US$15 to US$25 per tonne of CO_2 for some industrial applications, to US$40 to US$120 per tonne for processes involving more dilute gas streams (IEA, 2021). These costs are notably lower than those for DAC, where, in most cases, much energy is used in drawing air with very low CO_2 concentrations through filters. Figure 9.7 provides a comparison of methods. Despite its potential, CCUS faces challenges related to scalability. It requires major investment in transport and storage infrastructure, as well as strong regulatory oversight to deliver long-term CO_2 containment and protection from environmental risks.

Notwithstanding the difficulties or costs of many or most of the CDR methods, it is now clear that their use will be necessary on some scale. Suppose, for example, that in 2050 emissions are 15 billion tonnes CO_2 per annum. If it costs on average US$80 per tonne for removal (using a collection of methods) and 15 billion tonnes were removed, then the total per annum cost of US$1.2 trillion would be well below 1% of likely world GDP at that time. By that point, many might agree that this is a price worth paying. Of course, other arithmetic around quantity and price is possible.

Figure 9.7: Differences between CDR and CCUS

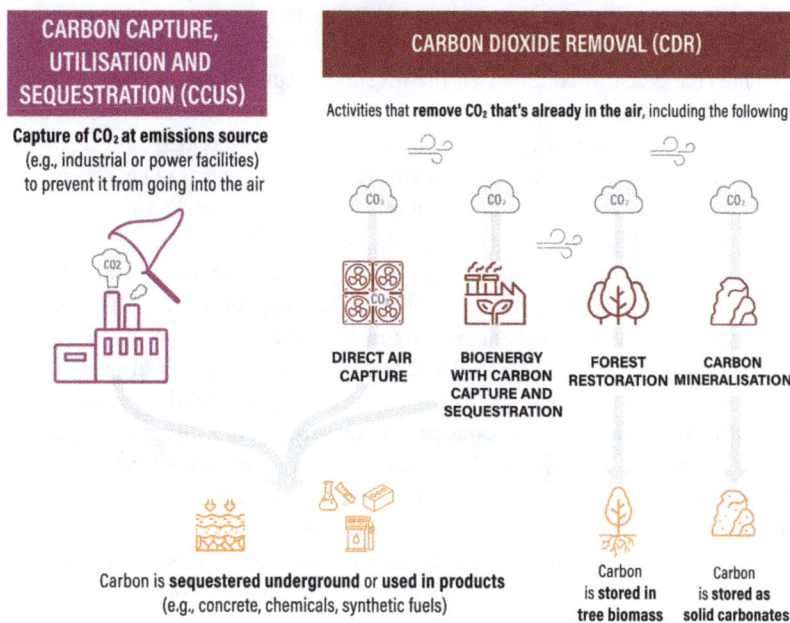

Note: For CO_2 used in products, the lifetime of the product determines the duration of CO_2 storage; some utilisation options only provide temporary storage.
Source: Lebling et al. (2023). Copyright 2023 World Resources Institute. Adapted with permission.

Geoengineering: a controversial last resort?

In the context of potential overshoot and the challenges associated with negative emissions, geoengineering has emerged as an increasingly prominent, but contentious, issue within climate policy. Unlike CDR, which seeks to reduce atmospheric CO_2, geoengineering encompasses a range of techniques designed to directly alter the Earth's climate system.

The most discussed form of geoengineering is SRM, which aims to reflect a portion of the sun's energy back into space to cool the planet. This is done through techniques like stratospheric aerosol injection, where particles are dispersed into the upper atmosphere to reflect sunlight, and marine cloud brightening, which aims to increase the reflectivity of clouds over the ocean. While SRM could theoretically lower global temperatures, it does not address the root cause of climate change – excessive GHG emissions – and carries significant environment, ethical, and political risks. Importantly, there is still very little evidence and research on the risks and impacts of SRM (UNEP, 2023b).

One major concern is that the sudden termination of SRM could trigger abrupt and extreme warming. Additionally, SRM methods could disrupt precipitation patterns, weaken monsoons, and exacerbate droughts in some regions while causing flooding in others (Haywood et al., 2023). The effects would be unevenly distributed, raising concerns over political conflicts and climate injustice. Another risk is the potential for moral hazard – by reducing the immediate impacts of climate change, SRM could lessen the urgency to cut emissions. Thus, its presence could accentuate the problem it is attempting to 'solve'. Also deeply troubling, because SRM does not reduce GHG emissions, it could divert focus and action on rising CO_2 levels, allowing environmental harms from increased emissions and concentrations, such as ocean acidification, to continue unchecked.

The costs of SRM are highly uncertain, and its potential damage could be catastrophic. While research on climate intervention techniques should continue, SRM should, in my view, remain a last resort, at least until its risks are far better understood, to be used only in combination with deep emissions cuts and carbon removal (UNEP, 2023b). There are severe governance challenges in regulating these technologies internationally. Without clear rules and oversight, nations – or even private actors – could unilaterally deploy geoengineering. The risks surely demand international cooperation to establish research, governance, and decision-making frameworks before pressure for the use of such interventions becomes overwhelming.

Avoiding dangerous overshoot requires accelerating emissions reductions today, not relying on future technological fixes. Negative emissions technologies will be necessary but should be scaled responsibly, prioritising natural and cost-effective solutions before deploying expensive and, possibly, higher-risk methods. Meanwhile, geoengineering remains a speculative and dangerous gamble, at least until we know much more. The world must act now to cut emissions, develop robust carbon removal systems, and avoid moral hazards that delay the transition to a sustainable economy. The longer we delay, the more severe will be the climate risks and the stronger the pressure to use potentially dangerous geoengineering technologies. Nevertheless, we have to recognise that we are about to pass 1.5 °C and are likely approaching tipping points. We will need to know something of how dangerous geoengineering could be. That requires research.

9.7 Concluding remarks: a global response to a global challenge

The crises of climate change and biodiversity loss demand an urgent, coordinated global response. Their scale and complexity require broad and deep international collaboration – fragmented or delayed action will amplify risks and undermine all other development goals. We have shown that international collaboration must begin with a shared recognition of urgency. All the SDGs are important and complementary, but climate and biodiversity goals are supremely time sensitive. Delay in tackling climate and biodiversity risks irreversible damage, making it harder – if not impossible – to sustain progress across other critical areas such as health, poverty reduction, and economic stability.

Throughout this chapter, we identified four priority areas where strong, coordinated international efforts are essential:

1. **Mobilising finance systems:** Mobilising investment at scale and enabling EMDCs (other than China) to access capital at affordable rates; aligning finance with its core purpose – driving investment.
2. **Technology and trade:** Shaping global markets, fostering innovation and diffusion, and securing supply chains for clean energy and critical resources.
3. **Natural capital and biodiversity:** Strengthening conservation, restoring ecosystems, and embedding nature in economic decision-making.
4. **Managing overshoot:** Developing responsible strategies for negative emissions and potential geoengineering.

Each of these areas presents technical, economic, and political complexities. Progress will require careful negotiation, the building of trust, and alignment of national interests with global action and stability. And we should recognise that coordinated action in these areas can also ease broader geopolitical tensions. Climate policy often serves as an entry point for diplomacy, even among nations with strained relationships – the USA and China have provided clear examples over time.

Among these priorities, low-cost finance is critical and offers the fastest route to real progress. Scaling up climate and development finance is not just feasible – it delivers exceptional value for money. I have outlined the investment and financial flows needed for a Paris-aligned world, demonstrating that climate action and development goals are mutually reinforcing, and that the cost of inaction far outweighs the cost of action. Indeed, we should speak in this context of investment rather than cost. These are investments with strong returns in terms of development as well as investments that reduce the immense risks of climate change.

Developing nations, however, remain deeply sceptical on finance, particularly due to past unfulfilled financial commitments. A serious, large-scale commitment to and delivery of climate finance – from both public and private, international and domestic sources – could rebuild trust and unlock cooperation across other areas. Finance should be the foundation on which international collaboration moves forward. The world must act rapidly and decisively. Climate action is no longer just about avoiding disaster – it is also about seizing a historic opportunity to build a cleaner, more resilient, and more prosperous global economy. The investment case has never been stronger. The technological momentum is accelerating. The economic and social rewards from the new path are immensely attractive. The risks of delay are stark. A new economic and industrial revolution is unfolding – countries that engage and invest will prosper; those that dither will bear the heaviest costs. The path forward is clear. *Act now, act together, and act at scale.*

Let me conclude this chapter on international collaboration on an optimistic note by offering an example of international climate action which could be implemented now with huge rewards: the bringing of solar-powered electricity to Africa at speed and scale.

China currently has the capacity to produce more than 1,000 GW of solar PV panels per year (Wood Mackenzie, 2023; Xue, 2025). Nearly 300 GW of renewable capacity per year has been installed in China in recent years (S&P Global, 2024). Half of sub-Saharan Africa, around 600 million people, lack access to electricity. The total installed electricity capacity is around 250 GW (BloombergNEF [BNEF], 2024). Thus there is several hundred GW per year of solar capacity in China that could be exported. Just 100–150 GW per year in Africa would treble its capacity within a few years and could bring the vast majority of Africa access to electricity by the end of the decade.

China can produce solar PV capacity at US$0.10 per watt or less (Wood Mackenzie, 2024). Other countries are closer to US$0.30 per watt. In general, necessary capital for gas-powered electricity is in the region of US$0.60 per watt and US$1.30 per watt (IEA, 2024a; Lazard, 2024)[27] and then it is necessary to buy the gas to run the power station. These simple numbers tell us that there is a huge opportunity to provide much of Africa with access to clean, low-cost electricity within a decade. Storage would be needed too, but the cost of batteries has also fallen very rapidly and sufficient capacity exists, particularly in China. In much of Africa, round-the-clock solar could be produced at US$0.02 to US$0.05 per kilowatt hour by 2030 (IEA, 2022). Gas-powered electricity in Africa would be more like US$0.03 to US$0.08 per kilowatt hour now, and likely rising to US$0.11 by 2030 (IEA, 2022).

Further, decentralised solar does not require the extensive grid structure associated with larger-scale fossil fuel plants. And decentralised solar capacity at the village or farm level can be planned in a way that will be consistent with future grid structures (see Box 5.1 on such decentralised solar projects in Africa in Chapter 5). Professor Jim McDonald – until recently President of the Royal Academy of Engineering – has led such work in Malawi; this type of design is referred to as 'swarming' (see UN, 2021; University of Strathclyde, 2022).

There are many who have proposed such initiatives. The United Nations Development Programme (UNDP) supports swarm grid projects in developing countries such as Vanuatu and Thailand. These decentralised grids consist of power cubes, resembling large car batteries, which are charged by a solar array. The stored energy is distributed to households or community buildings through underground cables (UN, 2021). The private sector has also engaged in swarming. SolShare, for example, a climate-tech company based in Bangladesh, has developed a peer-to-peer energy trading platform that uses smart meters to enable real-time electricity sharing. This creates a decentralised microgrid that directs excess solar power to households with energy deficits, improving access while reducing waste (Solshare, 2023). But now that the cost of PV panels has fallen, in 2025, so low, it is time to move quickly and on scale.

Finance for such a programme should come from the MDBs. China's development banks could play a strong role too. Low-cost capital is critical to keeping down cost. Procurement should be by competitive tender. Many companies would compete for contracts on such scale, although given capacity and costs it is likely that most of the successful companies in such competition would be Chinese. That should be accepted if it is indeed the outcome of competitive tendering. The examples of these large-scale operations and the size of the future markets could encourage solar PV producers in other countries to expand and lower costs. Some support for these developments, as part of industrial policy in some countries, would prepare for greater competition in future expansion. Whilst the capacity is there to supply at such low cost, and the need in Africa is so intense, grasping the opportunity is an economic and environmental 'no-brainer'.

This example represents a huge opportunity to bring zero-carbon and low-cost energy, economic inclusion, and climate-resilient and sustainable development to some of the poorest countries of the world. Without electricity there is no access to the internet and the digital world. Such access improves financial inclusion. It brings safety to women who might otherwise have to venture over long distances to get firewood. It enables children to study and small businesses to open in the evenings. It allows refrigeration and longer storage of food and medicines. Now is the moment to show that the world is capable of coming together to bring a massive surge in development opportunities to hundreds of millions. There is no better way of demonstrating the huge potential returns from international collaboration. Taking this opportunity together as a world could open the door to a new era.

Notes

[1] The fate of the IRA under President Donald Trump, who took office for the second time in early 2025, remains to be seen. However, the so-called 'Big, Beautiful Bill', signed by President Trump on 4 July 2025, seems to have reversed most of the IRA's support for green activities.

[2] Finance Minister Nirmala Sitharaman led efforts to reform the MDBs, commissioning two influential reports on how MDBs could scale up finance for sustainable development – mentioned in Chapter 4 as the reports on the Triple Agenda for MDBs. Both were led by N.K. Singh and Larry Summers, and I was a member of their Independent Expert Group – see Singh and Summers (2023a, 2023b). The findings are discussed in Section 9.3.

[3] Investment figures are in 2023 US dollars.

[4] These estimates are consistent with growth projections from institutions such as the IMF and the World Bank, and comprise investments in the energy transition consistent with net zero, adaptation and resilience, coping with loss and damage, protection of natural capital, and just transition. The estimate for the energy sector, which dominates climate investment, aligns with projections from the International Energy

Agency (IEA, 2024a) and the Energy Transition Commission (ETC, 2024). For sources, see Bhattacharya et al. (2024).

5 Note that the IEA definition of 'global clean energy investment' overlaps with, but is not identical to, that used by the CPI in calculating 'global climate finance'.

6 Clean energy investment refers to the 'ongoing capital spending on assets' (IEA, 2024c, p. 3) in clean power (renewables, grids, and storage), end use (includes energy efficiency and electrification for the main end-use sectors – transport, buildings, industry), clean fuels, and CCUS (IEA, 2024a).

7 According to UNEP (2023b, p. v), NbS can be defined as 'actions to protect, conserve, restore, sustainably use and manage natural or modified terrestrial, freshwater, coastal and marine ecosystems, which address social, economic and environmental challenges effectively and adaptively, while simultaneously providing human well-being, ecosystem services and resilience and biodiversity benefits.' I find 'solutions' an unattractive word here, since it suggests completeness and finality. Climate and development do not generally fit with that kind of definiteness. But the term is widely used and cannot always be avoided.

8 Though it is still very weak in relation to need, adaptation finance has been increasing, doubling between 2018 and 2022. The amounts in the last years, according to CPI (2024, 2025), are US$35 billion in 2018; US$42 billion in 2019; US$56 billion in 2020; US$61 billion in 2021; US$76 billion in 2022; and US$65 billion in 2023. During this period (2018–2022), the distribution of adaptation finance among country-development groupings was the following: advanced economies 7%; China 35%; EMDCs (excluding LDCs and China) 37%; LDCs 19%; and transregional 1%.

9 Climate finance refers to the allocation of funds to activities, programmes, or projects aimed at tackling climate change through adaptation or mitigation (Soubeyran and Macquarie, 2023). We have defined it for the purpose of our calculations in relation to the five investment categories presented in Table 9.1.

10 See Chapter 6 and the work of the IMF (Black et al., 2023), World Bank (Damania et al., 2023), and OECD (2024) on both public and private resource mobilisation.

11 The estimated public–private breakdown comes from looking at the nature of investments. Electricity generation may be mostly private, but natural capital is likely to be mostly public, for example.

12 That was the title of the report, chaired by N.K. Singh and Larry Summers, of the Independent Expert Group for the Indian G20 Presidency on MDBs (Singh and Summers, 2023b). I was a member of that group.

[13] This is based on the application of similar methods as used in the creation of figures for 2030 – see Figure 9.3 and associated text.

[14] A key difference was that the US$300 billion figure included all flows from MDBs, whereas previous calculations on delivery of US$100 billion (agreed at Cancún and Paris) were based on the developed countries' share in the MDB flows.

[15] Pekka Moran of the Finnish Finance Ministry played a key role in the creation of this Coalition of Finance Ministers and in its initial foundation in the 'Helsinki Principles' (Coalition of Finance Ministers for Climate Action, 2019). Nick Godfrey and Anika Heckwolf have been key to the work of GRI in supporting the Coalition, working closely with Amar Bhattacharya and myself. See also Coalition of Finance Ministers for Climate Action (2023).

[16] This network is working to integrate climate risks into monetary policy and financial regulation. In 2025 includes around 140 member institutions.

[17] See the work of Bhattacharya, Songwe, and Stern for the COP26, 27, 28, and 29 presidencies (Bhattacharya et al., 2023, 2024; Songwe et al., 2022) and that of the Singh–Summers G20 Independent Expert Group (Singh and Summers, 2023a, 2023b).

[18] Vera Songwe and I were members of that group and Amar Bhattacharya was central to its analytical work.

[19] According to CPI (2022, p. 52), carbon finance (as distinct from *climate* finance) 'is a type of results-based finance mechanism which involve contracts to trade emission reductions on carbon markets in the form of quotas or carbon credits (e.g., Certified Emission Reduction [CER]). Emission reductions are usually verified by a third-party auditor.' Results-based financing instruments give financial rewards in exchange for the achievement of pre-agreed results. Payments (in this case in the form of tradable credits) for a project occur when the project achieves the said results (Escalanate and Orrego, 2021). The credits can be sold on carbon markets to parties that are looking to offset their emissions, as such presenting monetary benefits to the emissions-reducing party.

[20] In these calculations China is the largest, on US$2.2 trillion per annum, with the USA at US$0.75 trillion per annum.

[21] See, e.g., the World Bank's work on toxic subsidies and alternative ways forward (Damania et al., 2023; Sutton et al., 2024). See also the OECD's (2024) and the IMF's (Black et al., 2023) analyses on fossil fuel subsidies.

[22] The IRA was launched under the Biden administration in 2022 (see Chapter 8).

[23] See, e.g., the work of ASPI (Leung et al., 2024).

[24] CERN is a particle physics laboratory that marked a change in how particle accelerator activities were conducted, from national-level activities to being based on international collaboration. This transition took place as it became evident that to make further advances in the field, machines beyond national capabilities were needed. The idea of a CERN for climate involves the creation of a computing facility dedicated to international collaboration on climate models of sufficient capacity to give, for example, much stronger local resolution on impacts than is currently possible.

[25] 'Multilateral financial flows' include finance from MDBs and other multilateral institutions, for example Global Environment Facility (GEF), Adaptation Fund, and UNEP.

[26] The UTZ label is a certification of sustainability for farming. It has been particularly important for coffee and cocoa.

[27] The IEA (2024a) provides estimates of technological costs for the USA, the EU, China, and India. The Lazard (2024) estimates are for the USA.

Part IV.
Galvanising action

10. Fallacies and confusions; obstacles and the risk of failure

The first three parts of this book have outlined the scale and nature of the challenge we face, the imperative for action, the new approach to growth and development that this action embodies, and the necessity for global cooperation in response to an inherently global problem. There is a path forward, but we must also recognise that taking action with the required scale and urgency will not be easy. In many ways, the required transformation is unprecedented in economic history.

In this brief concluding part of the book, we focus on how the necessary action can be galvanised. In doing so, in this chapter, we also provide a version of a summary of the arguments of the book, which is structured in a particular way: first, why some common arguments for inaction are mistaken and, second, the key obstacles to be overcome in taking action. It ends by emphasising that the size of the obstacles and the scale of the task are daunting, and reminds of the consequences of failure. Chapter 11 completes the summary by distilling the arguments on how the obstacles can be overcome, how the path can be charted and delivered, and why success is possible.

We begin, in Section 10.1, by examining, in the light of the arguments developed throughout the book, common fallacious arguments for inaction and delay. However, it is also important to recognise that even those advocating for strong action can sometimes propose misleading and misguided approaches. Confused arguments or implausible strategies weaken the case for urgent action. We explore these issues briefly in Section 10.2.

While it is crucial to dispel fallacies and confusion, we must also clearly identify the major difficulties and obstacles we face. The necessary investments and systemic changes are vast, complex, and urgent, and the processes involved are far from straightforward. Failures can occur in many ways. The most challenging problems are outlined in Section 10.3, along with suggestions, drawing from this book, on how they might be tackled. In discussing the challenges of implementation, it is helpful to have the problems, discussed at various points in the book, assembled in one place. Section 10.4 emphasises, as part of implementation, the importance of two areas which hitherto have received too little attention in public discussion: adaptation and biodiversity. The focus on these areas is strengthening, but much greater attention is needed. Concluding remarks, including concerning possible failure, are in Section 10.5.

10.1 Fallacies from advocates of weak or delayed action

There are many who try to denigrate the case for strong and urgent climate action. Sometimes that is motivated by the perception that climate action is some kind of 'woke' cause. Alternatively, some seem to suggest that steering a transition to a low-carbon economy is akin to the central planners, with their troublesome command-and-control techniques for resource allocation, à la Gosplan, returning from the 20th century to upend the market economy, but this time dressed in green. Some see change as uncomfortable and an assault on their way of life. In other cases, opposition to climate action arises as opportunism for perceived short-term political gain. Sometimes, it comes from protection of vested interests. Others claim their opposition comes from a careful assessment of priorities which does not put climate at the top of the list. Or that commitment to fighting poverty points elsewhere. No doubt there are other motivations.

The opposition has taken different forms over time. Outright science denial has, broadly speaking, declined, but still lives on, including in Donald Trump's second administration (from early 2025). There remains considerable opposition, and the arguments have to be confronted directly.

Whatever the motivation, there are essentially three main groups of arguments, which are summarised in Table 10.1: playing down the science and playing up the costs; arguing that fossil fuels provide energy security, which trumps climate; and arguing that climate action will be detrimental for vulnerable populations. Whilst the groups are clearly conceptually distinct ('too costly', 'energy security', 'poverty and vulnerability'), there is some overlap. In Table 10.1 the fallacies have been presented so as to capture common statements – similar arguments are often expressed in somewhat different ways. I examine these three groups of arguments in turn, following the structure in the table.

Downplaying the science and exaggerating the costs of action

The first group of arguments suggests that the risks described by the science have been exaggerated, and goes on to exaggerate the costs of action. These arguments appear in various forms, with examples listed in the first block of Table 10.1. Most of the fallacies in the first group have been addressed in the preceding sections of this book, so we will not provide detailed refutations here. Instead, we will briefly illustrate their flaws and refer back to the relevant parts of the text.

The notion that climate science indicates weak effects has been thoroughly discredited by successive reports by the Intergovernmental Panel on Climate Change (IPCC) and 200 years of climate science (see Chapter 2). The impacts of climate change are materialising more rapidly and intensely than anticipated. The argument that the 'costs of action' are prohibitive is increasingly untenable as technological advancements have accelerated and costs have come down (see Chapter 5). Indeed, we now understand that, rather than viewing climate action as a cost, we should recognise it as investment, innovation, and structural and systemic change, driving a new form of growth. This perspective has been central to this book. However, while the view that climate action is in conflict with economic growth is both fading and misleading, it persists. The belief that fossil fuels are essential for growth often stems from the idea that energy is essential for growth. While the latter is generally true, the former is not; energy does not need fossil fuels.

The argument that we should prioritise growth now, and only later address the environmental damage from greenhouse gas (GHG) emissions and other pollution, fails on two fronts. High-carbon growth will trigger potentially immense – and, in some cases, irreversible – damages. It risks crossing tipping points that would lead to dynamic instabilities, taking climate and biodiversity beyond our control. Thus, the 'clean up later' approach is not only unrealistic but also dangerous (see Chapters 2 and 4). For similar reasons, the notion that we can grow in an emitting and polluting manner, incur climate change, and then adapt later when we are wealthier is fundamentally flawed.

The final argument in the first block of the table suggests that rapid action is costly. This argument might appear to have some merit in the sense that had we acted sooner, the task would have been easier, because investments could have been spread over a longer period. But the logic fails when we consider that delaying action still further would incur immense costs due to the risks of damage from climate change which would be locked in by slow progress. In Chapters 1 and 2 we argued that if strong action had been taken 20–30 years ago, we could have made the transition over 50 years rather than 20–30. That delay and the missing of opportunities has made the cost of further delay extremely high. We will address some of the difficulties and obstacles we now face in Section 10.3.

Energy security

The second set of arguments centres around energy security. Critics argue that renewables are intermittent, unreliable, and difficult to integrate into grid and distribution systems. Additionally, it is often claimed that the best way to achieve energy security is through domestic fossil fuel production. While energy security is indeed a critical issue for every country, the arguments summarised in the Table are fundamentally flawed. Significant progress has been made in organising grid systems to accommodate high levels of renewables through demand-and-supply management across time and space, coupled with grid-level storage (International Energy Agency [IEA], 2023). This does require investment in grids and storage, but such investment not only facilitates the integration of renewables but also delivers substantial efficiency gains and cost reductions over time periods and geographies. Well-integrated grids, multiple sources of supply, and good energy markets can now accommodate and make the most of very high proportions of renewable energy (Energy Transitions Commission [ETC], 2021 and 2025).[1]

Table 10.1: Flawed arguments for delaying climate action, contrasted with reality

	Fallacy	Reality
1. Playing down the science and playing up the costs	• The science suggests modest effects, distant in time. • The costs of action are high. • The dirty is cheaper and going clean sacrifices growth in living standards. • Climate action diminishes competitiveness and slows growth. • Development requires energy and energy requires carbon. • We can grow first and clean up later. • Fossil fuels can drive growth; we will be richer when climate effects come through and we can pay for adaptation. • Rapid action is costly; we should postpone net zero and make changes slowly. • It is cheaper to adapt than to mitigate.	• Two hundred years of climate science supports the contrary. • There have been rapid falls in the costs of clean technologies. • Technological advancements have accelerated, making the clean cheaper than the dirty over much of emissions, with strong further progress on the way. • Climate action can be a profitable investment that drives growth. • Energy does not require carbon. • High-carbon growth is dangerous, with high costs, tipping points, and irreversibilities. • Delaying decarbonisation is more costly than acting now. • Delaying mitigation will result in many climate impacts being extremely costly or unmanageable.
2. Fossil fuels provide energy security, which trumps climate	• Renewables are unreliable, and grid and distribution systems are limited in their ability to include them. • Energy security comes from investing in fossil fuels.	• Notable progress has been made in integrating renewables into energy systems, in markets to balance supply and demand, and in storage. AI is making this still more effective. • History shows that fossil fuels do not guarantee energy security, as dependence has led to energy insecurity and instability. • Fossil fuels are processed and traded on international markets and are not 'local products'.

Fallacy	Reality
3. Climate action will be detrimental for poor and vulnerable populations	
• Climate action is an elite and upper-class issue which diverts attention and resources from challenges of cost-of-living/poverty. • Climate action in emerging markets and developing countries (EMDCs) will prevent growth, overcoming poverty, and catching up.	• Climate action can drive economic growth and improve living standards for impoverished individuals and countries through, for example, the development of green technologies and jobs, by expanding access to energy, and lowering its costs.
• Climate action will inevitably harm poor people through job losses, dislocation, and increased energy costs.	• New activities create new jobs. • Increasingly the clean is cheaper than the dirty. • Transitions can be managed through sound policy that ensures a just and equitable transition.
• Fossil fuel profits fund public action for growth and for poor people.	• Oil and gas development is betting against successful decarbonisation. • Fossil fuel activities could lead to stranded assets.
• The historical responsibility of developed countries for much of past emissions implies that EMDCs should not act until rich countries reduce their emissions and provide EMDCs with resources.	• The bulk of future growth will be in EMDCs, and necessary global emissions reductions cannot be achieved without their action. That action is in their interests. • Rich countries do indeed have an obligation to act and to support.

Source: Author's elaboration.

History shows that reliance on fossil fuels has, for most countries, led to energy insecurity. The oil crises of the 1970s and the recent Russia–Ukraine war resulted in sudden, dramatic price rises and availability issues that disrupted the global economy. Wind and solar energy are domestically available in all countries and reliance on renewables, along with some nuclear energy, can indeed provide energy security and reduce dependence on global fossil fuel markets. History and technological advances tell us that the objective of energy security is best achieved by moving away from fossil fuels.

Producing additional oil and gas in a country like the UK does not guarantee energy security. These resources are refined and traded on global markets, and a small country has little influence over world prices or its own supply (Ekins, 2024). It is implausible that, in the event of an oil or gas emergency, the UK could requisition domestic oil for domestic use. For example, UK oil from the North Sea is generally exported for refining. These are truly global markets.

Given that the world still relies heavily on fossil fuels for energy – around 80% of the total global energy supply (IRENA, 2024) – we must recognise that during the transition, energy crises resulting from disruptions in fossil fuel markets will require short-term adjustments in consumption, in production, and in alternative fossil fuel sources. For example, when Russian natural gas supplies were cut off during the Russia–Ukraine war, Germany temporarily increased its reliance on coal, a particularly damaging consequence given the high emissions associated with coal. By 'short term' in this context, we mean a period of two to three years. However, the arguments here on the importance of the transition away from fossil fuels are not narrowly about the short term but rather about the creation of a medium and longer term where renewables and nuclear energy predominate, drastically reducing vulnerability to global crises rooted in fossil fuel markets. This 'longer term' could be just two decades away in most countries (and less in some, such as the UK), as rapid progress towards renewables is already underway and accelerating. In the medium term, up to 2040, delivering on the Paris Agreement will require substantial reductions in fossil fuel production and consumption (ETC, 2023). Further, new explorations do not typically yield production results in two or three years; it is more likely to take 10 or 15 years, by which time production and consumption should be declining. For many countries, new exploration could lead to stranded assets (see Chapter 8).

The potential lock-in effect of continued fossil fuel exploration goes beyond the risks of stranded assets. It threatens to entrench high-carbon patterns of urban development and infrastructure that are less productive, less liveable, and more vulnerable to climate risks. Continued reliance on fossil fuels and existing structures can be associated with urban sprawl and the construction of inefficient buildings that increase energy consumption and reduce quality of life. This same locking in also undermines the resilience needed to adapt to climate change impacts. Investments in long-lived infrastructure – such as highways, ports, and power plants – often fail to incorporate the resilience required to withstand extreme heat, flooding, or sea-level rise. We must avoid locking ourselves into costlier, less efficient, and dangerous infrastructure and ways of living.

Many countries still depend on coal for electricity and heavy industry. Among fossil fuels, coal is the most polluting in terms of carbon emissions per unit of energy. It also causes significant harm in other ways, particularly through air pollution. In addition, as the coal market is global, similar arguments regarding energy security apply to coal as they do to oil and gas. However, the energy security argument for coal does have more substance than for oil and gas because, due to transportation costs, coal is generally used much closer to its points of extraction.

Nonetheless, the energy security arguments for coal are not fundamentally different. The particularly severe environmental impact of coal, on climate as well as air pollution, tells us that reliance on coal must be rapidly reduced. Energy security from renewables and nuclear can be stronger than that provided by coal.

While our focus in this subsection is on energy security, we should remind ourselves again that, as discussed in Chapters 4, 5, and 9, renewables already offer a cost advantage over fossil fuels in many applications, including power and transport, without requiring subsidies or carbon pricing – provided that the cost of capital is manageable. However, we should be clear that carbon pricing is, in general, good economic policy, because the emission of GHGs is a destructive externality.

Income distribution and poverty reduction

The third category of arguments from those advocating weak or delayed action revolves around income distribution and poverty reduction. One common argument is that climate and environment issues are elite concerns, diverting attention from poverty reduction efforts. This is closely related to the claim that climate action stifles economic growth, thereby preventing poorer countries from catching up. As discussed in Chapter 5, this argument is fundamentally flawed; climate action can drive economic growth.

These arguments based on distribution often imply that proponents of climate action are indifferent to improving the living standards of poor people or countries. However, these assertions are either false or misguided for the same reasons: well-designed climate action can promote both growth and poverty reduction, while the failure to take climate action hits poor people the hardest and earliest. Those advocating climate action are very often the keenest supporters of strong policy to reduce poverty; they have economic logic as well as, in my view, values on their side.

Another argument here suggests that climate action will inevitably harm poor people through job losses, dislocation, or increased energy costs. The answer to this lies in sound policy-making that delivers a just transition. It is true that a poorly managed transition could harm some disadvantaged groups.

Therefore, it is essential to invest in people and places, to create new and better opportunities. And where necessary, providing support through direct transfers, or other price or subsidy methods. A just and equitable transition requires good policy (see Section 6.5). And the gains from dealing with the many relevant market failures associated with GHGs (see Chapters 5 and 6) through prices and taxes imply that there will be resources to support those who suffer dislocations.

A final argument concerns the profits generated from fossil fuels and their potential role in funding clean growth, the transition, and broader development initiatives. For example, countries like Brazil and Mexico might argue that profits from their state-owned oil and gas companies could finance investments in clean infrastructure and industries, health, education, and a just transition. Similarly, several African and Central American countries discovering new oil and gas reserves may question why they should not utilise these profits to fund sustainable and equitable growth.

These are legitimate and important questions about who should benefit from fossil fuel profits or rents, some of which will continue during the transition to a clean economy, and about how these profits or rents should be used. However, these issues should not be confused with the mistaken notion that all nations with access to fossil fuel reserves should continue their exploitation. Such a path would be environmentally disastrous for everyone.

One key concern is that so many of those with fossil fuel reserves, or those considering the development of new discoveries, believe that exploiting these assets will be a profitable endeavour. However, if oil and gas consumption is reduced at a rate consistent with the Paris Agreement, many existing uses of, and potential developments in, these fossil fuels will become unprofitable (Mercure et al., 2021). Those facing high costs for power-plant construction and for oil and gas extraction, will likely find themselves with stranded assets. Thus, if we follow a path which is consistent with the Paris Agreement, many of those investors will be deluding themselves on their profitability. In essence, pressing ahead with oil and gas development is akin to betting against the successful management of climate change.

The important issues on the use of profits and rents from oil and gas call for international dialogue and action regarding how and where oil and gas production should be scaled back. While much of this process will be driven by market forces – where those with the lowest production costs will endure the longest as consumption and prices decline – richer countries, including those in the Gulf, could take the lead in reducing production as demand decreases. This would allow residual profits and rents to be channelled, as far as possible, to poorer nations.[2] Although one should temper expectations regarding the success of such a process, it is a discussion that must occur. In the meantime, the risk of stranded assets resulting from excessive oil and gas investment globally should be highlighted clearly and effectively.

Simultaneously, as argued throughout this book, investment in clean infrastructure and industry should be robustly supported through all available avenues, particularly in developing countries where the renewable opportunities are so strong and the cost of capital is often high. The ownership of oil and gas assets does not guarantee their competitiveness against renewables in electricity generation. We know that, in most cases, fossil-fuel-based power generation is more expensive, given a reasonable cost of capital, and this discrepancy will only grow as renewable energy costs continue to fall. And further, electric vehicles (EVs) run on renewable electricity will be both cleaner and cheaper than vehicles powered by internal combustion engines; indeed, in large measure they already are.

Coal presents a somewhat different challenge. Given its large GHG emissions, the global community must rapidly move away from coal. Moreover, the high cost of transporting coal means it is less integrated into the global market. In this context, international support and finance are the appropriate response to assist poorer countries in transitioning away from coal, managing a just transition, and reaping the cost and health benefits of cleaner technologies.

10.2 Confusion and misdirection

The previous section focused on largely flawed arguments for delaying climate action. This section examines confusion and misdirection from those who, while recognising the severe dangers of inaction, advocate for routes for action that are either wasteful, damaging, overly simplistic, or likely to provoke political backlash that could hinder more practical solutions – or, in some cases, all of these problems simultaneously.

We will consider the following arguments in turn: abandoning capitalism, abandoning growth, relying solely on carbon pricing, prioritising adaptation over mitigation, and refusing to discuss geoengineering or carbon removal. Table 10.2 sets out the confused arguments, along with a brief clarification. Following the structure in the table, we examine each one in turn.

Abandoning capitalism

One argument posits that rampant capitalism inevitably leads to behaviour which is dominated by narrow self-interest, which neglects the impacts on others and the environment, and which relentlessly pursues accumulation and growth without regard for broader consequences (Magdoff and Foster, 2011). However, a core problem with abandoning capitalism and markets is the absence of a credible, constructive, or environmentally friendly alternative. If we abandon capitalism, the question arises: what mechanism will replace the market when it comes to decision-making and determining resource allocation? Historical alternatives to capitalism, such as feudalism, centrally planned communist societies, or traditional indigenous economies, do not seem relevant or effective for managing the complexities of a modern economy, which relies heavily on technological innovation and physical capital.

Table 10.2: Confusion and misdirection in tackling climate change, with clarifications

	Confusion	Clarification
1. Capitalism must be abandoned	• Disregarding the environment is innate to capitalism and that will not change.	• Attempts to abandon markets and private entrepreneurship have failed, for understandable reasons. • The challenge is not substituting capitalism with something else but rather harnessing the power of markets and innovation.
2. Growth must be abandoned	• Growth damages the environment; therefore, it must be abandoned. • Long-run planetary boundaries imply abandoning growth now.	• This perspective confuses the idea of indefinite growth based on current patterns and methods with the need to manage a transition to a different path of development as quickly as possible. The former would be very destructive, indeed impossible. • The challenge is to decouple environmental degradation and economic activity.
3. There is a simple solution to climate challenges	• There is one market failure of over-riding importance: emissions of GHGs. • Economics tells us that this is unambiguously the most efficient policy to pursue.	• The GHG failure is indeed crucial, but there are at least five other key relevant market failures (see Chapter 6). • There is also the need to achieve systemic changes, which requires action beyond one price or tax.
	• Technology will solve everything.	• Technology opens new avenues for growth, but challenges are severe around economic, social, and political obstacles.

	Confusion	Clarification
4. Adaptation must be prioritised over mitigation	• The priority is to adapt, particularly for poor countries. • Mitigation is a costly diversion, as poor countries have lower emissions. • There is a competition between mitigation and adaptation.	• Strong adaptation efforts sometimes cannot handle the scale of climate risks even now. • Relegating mitigation will lead to stronger climate impacts, surpassing tipping points or dynamic instabilities; climate impacts that will likely be unmanageable. • Climate action can often deliver both mitigation and adaptation benefits in some programmes, as well as development.
5. Solar radiation management (SRM) is too dangerous to be discussed, and carbon dioxide removal (CDR) will discourage action	• Discussing SRM or geoengineering leads to delay and it is clearly too dangerous to even contemplate. • Similarly, discussing CDR will lead to postponement of action.	• Since delay in action is exposing humanity to tipping points and dynamic instabilities, all options must be explored: conducting well-informed analysis is key for decision-making. • We need to examine SRM to understand its risks, which are indeed likely to be severe. But that requires analysis. • Innovation can reduce the cost of CDR methods. • The role of CDR is complementary to mitigation efforts, but still essential, given the likelihood of over-shooting the Paris targets.

Source: Author's elaboration.

In other words, there is no viable practical alternative to an economy that utilises markets for the great majority of investments, decisions, and actions.

There are some who may believe that a practical alternative to market capitalism could develop with changed values, risks, and new technologies, such as AI, that together could help solve the many problems around market failure, information, power distribution, and so on. I have my doubts. But it is surely clear that such a new model of how to run an economy is most unlikely to emerge in the next two or three decades that are so critical for the climate and biodiversity crises.

Let us briefly review the historical lessons on alternatives to capitalism from an environmental perspective. First, reverting to feudalism as an economic model is not an option. It depended on land and labour being the core factors of production, and dominant landowners, with only minor roles for the accumulation of capital and technological changes.

Second, societies where powerful dictators control resources via central planning have generally performed poorly in environmental stewardship and in raising living standards. These systems typically fail not only in terms of freedom and creativity, but also in the broader dimensions of sustainable development. China, for example, experienced significant economic growth in the three decades following the disastrous policies of Mao Zedong (including around the 'Great Leap Forward' and the 'Cultural Revolution'), but much of this growth was accompanied by severe environmental degradation. Only in the last 15 or 20 years has China begun to seriously address environmental issues, and it has done that at the same time as continuing the move towards markets. Environment is now centre stage. Economically, China should now be viewed as largely a market economy, even though politically, it has a high degree of centralised power and control. But in its recent history, careful thinking about incentives and standards has come alongside concern for consequences for the environment.

The former Soviet Union is another prominent example. Central planning, as seen in the Soviet model from 1917 to 1989, was notoriously wasteful and environmentally destructive. As an illustration, it has been calculated that in 1988 the Soviet Union was polluting 1.5 times more than the USA per GNP unit (Shahgedanova and Burt, 1994).[3] This was related to the energy intensity of its economy, which was double that of Western Europe and 30% higher than in the USA (Bashmakov and Chupyatov, 1991). Energy use was particularly inefficient, since energy prices were centrally determined and outdated compared to world energy prices and domestic energy production costs. There was little in the way of economic incentives to invest in energy efficiency (Cooper and Schipper, 1991). Energy use was wasteful as well as polluting.

Third, some indigenous societies, without capitalist economies, have lived in harmony with the environment for very long periods. For example, Aboriginal societies in Australia appeared to have maintained environmental balance for over 50,000 years. In this sense they were extraordinarily successful, with activities and interactions shaped by culture, conventions, mutual understanding, and deep local and environmental knowledge. However, given the complexities of modern economies, with their strong role for the accumulation of capital and technological change, and the much higher population densities relative to natural resources, recreating such societies on a large scale seems impossible. Nevertheless, studying these societies can offer valuable insights into managing our natural capital and collaboration. There is so much to learn from them. But these lessons do not suggest that we should abandon capitalism altogether.

The challenge, then, is to govern, modify, and shape a capitalist market economy in a way that takes much greater account of the environment, resilience, and social cohesion. Different societies can do that in different ways. This involves reflecting on our goals, institutions, rights, and freedoms – and how they can be formulated and enhanced – and tackling issues of social cohesion, inclusion, participation, and inequalities. These considerations and objectives, including the reduction of poverty, the creation of shared prosperity, and the achievement of sustainability, have been central to our discussion of public policy throughout this book. Thus, the issue is not about abandoning capitalism and the market economy; it is about harnessing the power of markets and entrepreneurial spirit in much more constructive ways, guided by societal goals and needs, participation, public discussion, and careful analysis. There will be Indian, Chinese, other Asian, European, African, North American, South American, Oceanic, and many other ways of doing this, drawing on different cultures, histories, and economic structures.

Abandoning growth

Some argue that economic growth will inevitably be destructive in a world with finite natural resources. And many suggest we should stop growth now, particularly in richer countries. In its most simplistic version, this perspective confuses the concept of indefinite growth, based on current patterns and methods,[4] with the need to manage a transition over the next two or three decades (see also Section 5.3). The former would indeed be very destructive, indeed impossible. Further, those who make the argument often insist on using narrow definitions of growth in this argument, focusing solely on conventional GDP. They suggest that those who indicate the importance and possibility of growth think only in terms of GDP. Such confusions around growth are both misleading and damaging.

The policy and action agenda on climate, biodiversity, and sustainability is primarily concerned with actions over the next two or three decades. While climate action will certainly continue beyond 2050, the immediate priority is to drive the necessary changes quickly and to establish a new approach to growth and development in the coming two or three decades. As we have argued throughout this book, this new approach – through investments, innovations, resource efficiencies, structural and systemic changes, creativity, and improved health – can raise living standards, including real income levels, across the board. When pursued effectively, this strategy advances the Sustainable Development Goals (SDGs) in all their dimensions, including consumption and income. It is absolutely not about 'indefinite growth', whatever that means; it is far broader in its concepts, activities, and measurement than GDP.

Let us be clear: the way forward involves both raising living standards, including incomes, and the reduction of GHG emissions, pollution, and biodiversity loss. The challenge is to break the relationship between economic activity and environmental degradation and destruction. We have demonstrated how this can be achieved. Reaching net zero does not, I hope, imply an absurd scenario of zero population or zero output. Modest decreases in consumption and output will not take world emissions from where we are now, at close to 60 GtCO$_2$ equivalent per annum, to anywhere near net zero. The task in the next two decades is to do things very differently rather than simply to do a bit less of them. Of course, our approach described here strongly supports the more efficient use of resources, less waste, shifts in diets, and more environmentally friendly lifestyles. However, this does not necessarily mean reduced consumption. Instead, it can and should mean better consumption and improving most dimensions of well-being.

There are some, however, who draw on economic history in an attempt to show that a new clean growth story seems impossible. For example, Robert Gordon's (2016) *The Rise and Fall of American Growth* argues that the growth experienced in the 20th century was a unique historical event which is unlikely to be repeated. Gordon suggests that the most important technological and other discoveries have been made and that future discoveries will inevitably be less powerful in their ability to drive growth.

There are also those who argue that energy consumption must rise and that movement away from fossil fuels is destined to be limited (Smil, 2017, 2010). A recent book, *More and More and More*, by the historian Jean-Baptiste Fressoz (2024), continues this pessimistic perspective on energy transitions in arguing that energy and resource consumption rose through all past transitions and implying that resource consumption will inevitably rise in future transitions. These approaches suggest that economic history tells us that we will experience 'energy addition', with new energy sources added, but not energy transition in the sense of abandoning the use of earlier sources of energy.

Fressoz's analysis does carry interesting insights into past transitions, including the importance of looking at resource use across all activities and supply chains. Solar panels and wind turbines need resources and energy in their construction. But there is a fundamental difference between the future energy transition we have been describing and past transitions. The future transition has its basic source of energy in renewables not natural resources, particularly fossil fuels. Of course wind turbines and solar panels use resources, but on a much smaller scale than coal, oil, and gas.

It is true that the new route will be very different from the past, but to argue that it is impossible because it has not been done before is simply illogical.

It will be difficult, and we have discussed and will discuss obstacles, but we can show how they can be overcome. The narrow determinism of a rigid economic history argument – that is, 'not done in the past so cannot be done in the future' – is a counsel of despair. We can show how it can be done. That kind of argument suggests that all we can do is to head for destruction more slowly by cutting back on consumption. It asserts, effectively, that we are doomed.[5]

The argument that environmental concern necessitates the abandonment of growth in the coming decades is also politically untenable and likely to result in failure, particularly in the developing world. Many of these countries, understandably and firmly, put the question: 'You in the rich world became wealthy through dirty growth, you filled the atmosphere with GHGs, and now you tell us that we cannot raise our own incomes and escape poverty?' It is morally, analytically, practically, and politically wrong for those in the rich world to make an argument that even appears to suggest this. Such a suggestion would undermine the prospects for collective action. As we have emphasised, the key policy and action questions focus on how to develop a new approach to sustainable, resilient, and inclusive development that raises incomes while breaking the past destructive relationships between economic activity and the environment. We have demonstrated how this can be accomplished. It will not be easy, but failure to act urgently and on scale will inexorably undermine development and poverty reduction.

We must be clear, and I emphasise again, that this is not a story about indefinite growth. Our discussion focuses on change in the next two decades or so. As we implement the necessary changes, innovations, investments, and new structures, we will create vastly different prospects for the future. By mid-century, the questions surrounding development will likely look very different, as we trust – or hope – that new, less destructive approaches to development will be firmly in place.

All of this does not dispute the idea that the 'carrying capacity' of our planet's ecosystems is finite. Indeed, it is true that the Earth cannot support an unlimited number of people or an unbounded set of economic activities, even if those activities are less harmful than in the past. Scholars such as Partha Dasgupta (2021a) have rightly emphasised this point. Others have discussed the concept of a 'doughnut' model, with the outer rim representing ecological constraints and the inner rim reflecting the minimum necessary living standards (Raworth, 2017). Johan Rockström and others have set out important and convincing analytical work on planetary boundaries (Rockström et al., 2009).[6] These boundaries are indeed of real importance for strategies for development.

Our challenge is to change radically in the next two decades so that our future can be sustainable – and, in so doing, we must understand and recognise the boundaries. During this period, we can and must combine rising living standards, particularly in poorer regions of the world, with significant reductions in the environmental damage caused by our economic activities. In fact, we can find ways for these activities to enhance our environment. Such actions do not necessarily push us beyond, or further beyond, any overall environmental or planetary boundary. On the contrary, by drastically reducing the damage from consumption or production, they can and should move further away from these boundaries, or back towards them where they have already been crossed.

The fastest and most effective way to avoid – or reverse – crossing these boundaries is through radical change in how we conduct our activities, rather than simply halting or cutting consumption. That being said, part of the new approaches to growth would indeed involve radical reduction in waste and damaging forms of consumption. And the challenge of the 'carrying capacity' – what Dasgupta calls the 'impact inequality'[7] – will not go away. But our actions in the coming two decades can make it much less difficult to find a sustainable equilibrium in the medium term.

Simplistic or formulaic one-shot 'policy solutions'

In examining the confusions that can hinder effective climate action, we must also consider the notion that a single, simple solution can resolve the complex challenge of climate change. A prime example of such a 'silver bullet' proposition is the idea that implementing the right carbon price is all we need. As discussed in Chapters 2 and 6, there is indeed a compelling case for a carbon price, and it is a policy position that I, along with many economists, strongly support. However, we also identified five other key market failures that are directly relevant to fostering the necessary innovation and investment. Moreover, systemic changes – such as in cities, land use, or transport systems – cannot be delivered by a carbon price alone, or even by a few other market corrections. Those policy actions on market failures are crucial, but systemic change also requires a strategic approach where multiple agents get together to chart a way forward, particularly where the dynamics and expectations can be complex. Cities, for example, cannot change without such collaboration.

Therefore, it is misleading, and indeed incorrect, to claim that 'economics tells us' the most efficient way to tackle GHG emissions is solely through a carbon price, and to imply that is all we have to do. It is important not to misunderstand what is being said here: the carbon price should indeed be central to policy, but we will need much more in the policy and action packages to create the new path of development and growth that is required.

There are other, slightly broader but still overly simplistic, approaches that are sometimes proposed. One such approach is the belief that 'technology will solve everything'. While technological advancements are indeed crucial – and the progress in areas such as solar energy, wind power, batteries, and EVs has been remarkable, helping to define and open up the new avenues for growth – we have also seen that many of the most challenging problems are economic, social, or political. There are many market failures and vested interests that technology alone cannot overcome, as crucial as technology no doubt is (also see Section 10.3). As we saw in Chapter 9, it is also a dangerous mistake to think that the problems can all be fixed by geoengineering (also see below).

Prioritising adaptation over mitigation

This argument often begins by highlighting the world's failure to reduce emissions effectively. It then suggests that since this failure is likely to continue, we must now prioritise adaptation over mitigation. The argument on the importance of adaptation is surely sound. We have indeed been far too slow in cutting emissions, and adaptation and building resilience are now critical to protecting lives and livelihoods, especially in poor countries that are particularly vulnerable to the effects of the global shortfall in mitigation efforts.

The confusion lies in the idea that strong adaptation efforts can handle whatever climate challenges arise and, further, that there is some kind of horse race between adaptation and mitigation, with adaptation being the 'right horse' to back. This argument is often coupled with the correct and important observation that poor countries contribute relatively small amounts to global GHG emissions and are disproportionately vulnerable to climate change.

However, relegating mitigation to the background in favour of adaptation will lead to future impacts that would likely be unmanageable. There are real risks of tipping points and dynamic instabilities that could result in climates and environments where large parts of the world become uninhabitable. This risk applies to all countries, but is particularly severe for poor countries and communities. Any slacking of mitigation efforts could leave poorer countries with climate risks that no amount of local adjustment could deal with: in many cases, 'adaptation' would be death or migrating. We are already moving into that territory – we will be beyond 1.5 °C within a few years – where the risk of tipping points is rising strongly (see Chapter 2), but the further we go, the higher those risks. Reducing mitigation efforts is profoundly dangerous.

While poorer countries currently contribute relatively little in terms of emissions, failing to prioritise mitigation would lock them into a path of dirty growth, potentially making them the major polluters of the future. That said, there is a compelling practical and moral case for significantly increasing investment in, and support for, adaptation and resilience in poor and vulnerable countries. The mistake lies in suggesting that we can ease back our efforts on mitigation. The opposite is true: these must be accelerated, as we saw in Chapters 2 and 4. The ever-increasing need for adaptation is a reflection of our dismal underperformance on mitigation.

The narrative, however, is not merely about a false horse race between adaptation and mitigation. As discussed in Chapters 4 and 5, there are numerous examples where climate action delivers mitigation, adaptation, and development simultaneously: from mangrove restoration and the rehabilitation of degraded land to decentralised solar power and public transport systems. At the heart of the policy challenge now is the expansion and pursuit of these types of investments. But to emphasise again, it is clear, and important to recognise, that adaptation has had far too little attention and investment; this failure is emphasised in Section 10.4 below.

Refusing to discuss geoengineering or carbon removal

Some argue that solar radiation management (SRM) – which involves preventing energy from entering the atmosphere – is too dangerous a subject even to discuss. The reasons cited include the fear that such discussions might reduce efforts on mitigation (the so-called 'moral hazard'), the potential for unknown or unintended consequences that could be catastrophic, and the governance issues inherent in decisions of such global consequence potentially being made by a small group of people. All three of these concerns are valid and significant (see Chapter 9).

However, the delay in reducing emissions has brought us perilously close to tipping points and dynamic instabilities in the climate system, which could lead to irreversible and catastrophic outcomes. In this context, it would be reckless to dismiss the exploration of actions that might 'buy time'. Such research is best conducted before the pressures for activating SRM become overwhelming. The analysis might well conclude that the risks associated with SRM are so severe that it should not be seen as a viable option. Indeed, I think that conclusion is likely. But this should be a conclusion based on thorough, well-informed analysis. Gathering information on the possible consequences will not be easy, because a global experiment could itself be dangerous. However, there is potential in modelling. And studies of past volcanic eruptions which cooled the earth via dust in the atmosphere carry some information. Further, local experiments in cloud brightening may be possible and useful.[8]

None of this should detract from the core argument of this book. The most effective actions are those that foster a sustainable, resilient, and inclusive approach to growth and development. We must act with purpose and urgency. The choice and responsibility lie in our hands.

In addition to reducing emissions, CDR will also play a crucial role in reaching the Paris Agreement goals. Different CDR methods exist, as we saw in Section 9.6, which can be grouped mainly into three (ETC, 2022): first, natural climate solutions (NCS) such as reforestation, land restoration, and algae; second, engineered solutions such as direct air capture (DAC) and carbon capture and storage (CCS); and third, hybrid solutions such as biomass with carbon removal and storage (BiCRS or BECCS). CDR methods such as afforestation and reforestation are already being implemented. They should be pursued more energetically and, if done well, would carry environmental and other benefits way beyond carbon reduction. There are further options whose feasibility and risks are more uncertain, as they are still in early development, such as ocean fertilisation (ETC, 2022).

Each method has different potential co-benefits, costs, and risks, and none will replace deep emissions reductions. Though their price is expected to decrease with innovation and implementation, most engineered CDR methods are expensive – particularly, at present, DAC. The risk exists that they are not, on cost grounds, implemented at large scale. Natural CDR is less costly but faces different risks, such as land competition and the potential impermanence of nature-based solutions (ETC, 2022).

Notwithstanding the difficulties, CDR is likely to have an essential role, since it will not be possible to eliminate CO_2 emissions completely (ETC, 2022), for example from some activities associated with industrial production and agriculture (Intergovernmental Panel on Climate Change [IPCC], 2023). Virtually all IPCC mitigation scenarios compatible with the Paris Agreement assume deployments of CDR which are large compared to its current use. Therefore, it is necessary to rapidly upscale CDR development and bring down costs. The question is then which methods will be deployed, when, how, and by whom (Babiker et al., 2022).

As an illustration of scale and cost, suppose that by 2050 overall emissions, allowing for sources and sinks, had come down from 60 to 20 $GtCO_2e$. If direct air capture costs US\$200 per tonne (it is currently around US\$500 or more), then a cost of US\$4 trillion per annum would be involved. That might be around 2% of GDP then.

10.3 Obstacles, action to tackle them, and the research agenda

This book has outlined a clear path forward: the prospects of taking strong action are potentially very attractive, while the dangers of delay are immense. However, we must recognise that the scale, pace, and nature of the action required will not be easy to deliver. There are numerous and substantial obstacles and difficulties to overcome along the way. The majority of these challenges lie within the realms of economics, society, and politics. It is largely through the social sciences that we can understand these obstacles and devise strategies to overcome them.

Whilst the problems associated with economics, society, and politics are emphasised here – indeed, we argue that they are the thorniest issues – we should not overlook that there is much to do in science and technology, notwithstanding the remarkable progress that has occurred. The world will need substantial advances in key hard-to-abate sectors, including agriculture, aviation, maritime, cement, and steel. There are promising signs in each area, but there is still much to be done. And we will need much cheaper CDR.

In this section we highlight the most significant and challenging obstacles and discuss how they might be tackled. These challenges are summarised in headline form in Table 10.3. They have been identified earlier in the book, but they are gathered here for summary and overall assessment. Thus, this section serves as a summary of many of the key conclusions and messages of this book.

The obstacles can be broadly categorised into five interwoven areas: lack of investor confidence in strategic direction and problems with the investment climate; finance and macroeconomic challenges; dislocation, structural change, and the need for a just transition; changing behaviours; and politics and political economy, both national and international. There are ways to overcome each of them, separately and together. Tackling these difficulties requires both commitment and a shared societal and political understanding of what can be done, of the urgency and necessity of action, and of the potential rewards. This section will also underscore where the social sciences – particularly economics and political economy – should focus attention and research. Whilst not included as obstacles in Table 10.3, adaptation and biodiversity have received far too little attention and investment. These issues are taken up in Section 10.4.

The five areas and sets of issues of Table 10.3 constitute an urgent and challenging agenda for policy-makers and economists and social scientists alike. Those working on these issues must both act as, or support and inform, policy-makers, and think as researchers simultaneously, clearly, and swiftly. Action and research must proceed together. Delay is dangerous. But so too is misguided action. Thinking and doing must be combined and simultaneous.

Table 10.3 points towards a research agenda in terms of careful analysis of the obstacles and the construction of policies and actions to deal with them. We will assemble those research issues in the next chapter, including some comments on methodological approaches that might be possible. See in particular Table 11.1. Both Table 10.3 and the research agenda, Table 11.1, are linked to the action agenda set out at the end of Chapter 4.

The entries in Table 10.3 concern obstacles and barriers. The necessary research issues include, in addition to understanding these obstacles and barriers, policy design and incentive structures to overcome market failures and to guide investment, innovation, and research and development (R&D) directions, including around natural capital. These issues are of great importance and are included in Table 11.1.

Strategic clarity and the investment climate

The first, and crucial, task in driving the necessary change is creating the conditions that can guide and encourage investors in the required directions. These conditions must recognise and reflect the scale, nature, and urgency of the innovative investments that are at the heart of generating a new form of sustainable growth. As we have argued, this growth can lead to all of increased productivity, reduced emissions, greater resilience, and rising living standards across the board. Facilitating and fostering this investment is the most important strategic and policy objective in tackling climate change and generating the new growth story.

Whether one looks to Keynes or Hayek, or simply listens to investors, it is clear that uncertainty is a major deterrent to investment. Government-induced policy risk can kill or severely reduce investment. A difficult infrastructure environment in terms of power, transport, or water, for example, can be a major discouragement, as can a bureaucratic or regulatory environment that creates obstacles or delays. Predatory practices by corrupt officials or criminal elements (such as protection rackets, ransoms, hijacking, or highway robberies) can also act as severe deterrents to investment. Policies, infrastructure, institutions, and governance all shape investment and its perceived viability.

In this context, clarity of strategic direction is particularly important. Much of the necessary investment will be long term, particularly in infrastructure. Stability in strategy and policy, along with shared commitment across the political spectrum, can enhance investor confidence. Conversely, their absence can be a major hindrance. Legal and institutional frameworks can bolster confidence in strategic directions. The way in which courts around the world have recently been upholding the obligations of governments and firms regarding climate and environment protection is an example (Setzer and Higham, 2024). See, in particular, the clear and strong opinion on climate given by the International Court of Justice in July 2025.

Institutional structures, such as the UK's Climate Change Committee and relevant legislation, can also contribute to confidence in strategic direction. Strong, legally-binding net zero national commitments, accompanied by clear and credible transition pathways, are central to reducing uncertainty and fostering market stability (Bhattacharya et al., 2023). Political and public support for pledging and implementing measures to achieve net zero objectives can be strengthened by communicating how net zero policies are aligned with broader development actions and goals, such as improving employment opportunities, supporting domestic industries, enhancing energy security, and making cities more accessible (Averchenkova and Chan, 2023).

Understanding the obstacles to investment will be a key part of strategic policy in fostering investment in terms of actions to overcome them. That is why the careful development of the investment climate, sound policy, and credibility are so important. Information from private firms and financial institutions on their own understanding of obstacles will be key elements in creating and supporting the understanding of policy-makers. These issues are discussed in Chapter 6 and, from an international perspective, in Chapter 9. The country platform (see Chapter 9) around which key agents – public, private, civil society and international – can share ideas and coordinate can be a strong foundation.

Finance and macroeconomic challenges

Recognising and fostering the scale, nature, and urgency of the necessary investments is the starting point. The next action element, as just emphasised, is to create the conditions for that investment. However, investment requires finance.

In Chapters 5 and 9, we have examined the scale and nature of the finance required. Given the magnitude of the necessary investments, the scale of the required finance will be substantial. That starts with domestic resource mobilisation, public and private, which will, as we saw, be the biggest element in overall finance. But external finance for EMDCs (other than China) will be crucial. As we discussed in Chapter 9, external finance should flow at approximately US$1 trillion per year by 2030, compared with US$150 billion in recent years (Bhattacharya et al., 2024; Songwe et al., 2022). The three broad components of this US$1 trillion per annum in external finance are, in order of size: private

Table 10.3: Key obstacles and barriers to action

	Obstacles	Key actions
1. Lack of confidence in strategic direction and investment climate	Lack of confidence in: • government policy and institutions. • revenue flows. • getting things done, e.g., planning permissions, skilled labour.	• Conditions must be adequate to encourage investments. • Clarity of strategic direction is critical.
2. A. Finance barriers: availability, affordability, accessibility	• Availability and cost of capital for poorer countries. • Scale of supply of affordable capital. • Problems of access.	• Investment requires adequate finance sources. Major expansion necessary. • The relation between multilateral development banks (MDBs) and the private sector must change – actions are clear but political will is needed.
2. B. Macro: debt	• Sizeable increase in investment could cut consumption, cause balance of payment problems, lead to inflation. • Many governments would have to be able to borrow and satisfy markets and many face problems of already high debt.	• Sound macroeconomic management and fiscal rules are required. Action on debt of some EMDCs. • Support from development finance institutions (DFIs) is crucial.
3. Dislocation, structural change, and the need for a just transition	• Areas, cities or sectors could face major dislocation, resulting in opposition. • There may be pushback due to the perceived costs of clean relative to dirty. • There may be pushback on levying carbon prices/taxes and removing fossil fuel subsidies.	• Potential dislocations should be tackled, and benefits must be distributed. Importance of community participation in designing and implementing a just transition. Major investment or support likely necessary. • Effective policy-making requires strong collaboration among the academic sector, governments, the private sector, and communities.

	Obstacles	Key actions
4. Changing behaviours	• Many of the necessary behavioural changes may be seen as difficult, disruptive, and costly.	• Transitions require public discussion and accessible information on changes. • Communication on opportunities of new clean tech and activities is needed. Help in making change easier, especially financially and organisationally.
5. Politics and political economy: national and international	Politicians may: • given the difficulties, delay action. • decide the difficulties are in the shorter term and the pay-offs in the longer term, and back away. • argue that they do not get pressured strongly by constituents or citizens on these issues. • face campaigns from vested interests, particularly around fossil fuels. Internationally: • The developing world may view the developed world with mistrust and see a failure to meet moral obligations to both reduce emissions and provide support.	• Achieving change requires mobilising forces in favour. • Civil society and politicians can pressure for action. • Policy should manage costs associated with change, and involve public participation. Particular attention to vulnerable groups. • Leaders need to communicate a vision of a better future. Early wins are important. • International discussion and agreement is required, including around rich-country support.

Note: Adaptation and biodiversity are major issues which have received too little attention and investment. See section 10.4.

Source: Author's elaboration.

sector flows, MDBs and DFIs, and low- or zero-cost finance such as Official Development Assistance (ODA), voluntary carbon markets (VCMs), special drawing rights (SDRs), philanthropy, and new forms of taxation. Achieving this by 2030 will require a 15-fold increase in private flows, a tripling of flows from MDBs and DFIs, and a significant expansion of low- or zero-cost finance. Low- or zero-cost financing of US$200–300 billion per year will be needed by 2030, requiring a doubling of bilateral contributions from advanced economy partners and a 14- to 16-fold increase in other sources, including VCMs, use of SDRs, solidarity levies on internationally polluting activities, other international taxes, and debt swaps (Bhattacharya et al., 2024). These expansions are very large, but they are necessary to respond to the crises on the scale required and, at the same time, create the new growth story. And we saw in Chapter 9 how these expansions could be realised. They are feasible.

As argued in Chapter 5, the scale of finance now necessary arises from the scale of investment necessary. That scale of investment itself arises from two factors: the scale of the task and the great risks of delay in action. This is now a transformation which must take place in two or three decades. Had the world started in earnest on that transformation two decades ago, when the challenges were already clear, the necessary per annum increase in investment would have been much lower. Further delay into the future would make that problem still more acute. At the same time, we must emphasise that these necessary increases in investment will bring high economic and social returns, in addition to reducing the immense risks of climate change.

As we have argued, host countries must create the conditions for the investment if it is to materialise. And they will also, as already emphasised, be the most important source of finance. In most countries, domestic investment is primarily financed by domestic savings, both private and public. In EMDCs (other than China) the proportion financed internally may be smaller, particularly for poorer countries. In the analyses above, we took the figure of US$1.4 trillion per annum internal out of a total of US$2.4 trillion per annum of overall climate finance investment, or nearly 60%.

The bulk of the financing is expected to come from the private sector, but this requires substantial progress in de-risking and tackling the cost of capital. DFIs, and especially the MDBs, have a crucial role to play not only in catalysing private finance but also in supporting urgently needed public investments. Transforming how the MDBs engage with the private sector and operate as a coordinated system requires major reforms. The commitment, energy, and direction for the necessary changes and expansion of finance from these institutions must come both from their shareholders and from their leadership. But as we have shown in Chapter 9, these expansions are possible with comparatively small injections of paid-in capital. An injection of US$60 billion of paid-in capital, for example via US$6 billion, each year for 10 years, could, with other reforms of capital use,[9] release a tripling of per annum MDB finance, increasing from around US$100 billion to US$300 billion per annum.

Ultimately, this is largely a matter of political will. If this political will is to emerge, it will be built on an understanding of the necessity of delivering on the Paris Agreement and creating the new path of growth and the investment and finance required to achieve this. I hope that this book contributes to this understanding. The finance ministers and heads of state of G20 countries have a special role to play.

In my view, the necessary actions have been identified, and the detailed analysis for implementation is being developed. The challenges now lie in building the political will and tackling the practicalities of implementation for both private and public institutions, domestically and internationally. That is a key obstacle. Overcoming it will not be easy, but the consequences of failure – namely, the failure to meet the Paris temperature targets of well below 2 °C, with efforts for 1.5 °C – will be devastating, particularly for the poorest people. Dismissing these proposed changes as 'unrealistic' or 'too difficult' is, in my view, the opposite of realism. It will be tough either way, but the much tougher path is one of delay and inaction.

For many countries, macroeconomic and debt issues may undermine confidence in their ability to raise investment and maintain macroeconomic stability. A key consequence of this is a high cost of capital. Overcoming the macro challenges is a central obstacle. Sound macroeconomic management, sensible fiscal rules, and support from domestic financial institutions and development banks are crucial here. Central to sound macroeconomic policy and fiscal rules in this context is the principle that *sound investment, financed by capital at a reasonable cost and tenor, is macroeconomically responsible.* Such investment can reduce debt-to-GDP ratios over time. See, for example, the two reports of the Independent Experts Group for the India G20 Presidency, chaired by N.K. Singh and Larry Summers, of which I was a member (Singh and Summers, 2023a, 2023b). That being said, some of the poorest countries with severe debt will need some kind of debt restructuring. The International Monetary Fund (IMF) can and should play a leading role here; it will need the backing of the major G20 countries.

Dislocation, structural change, and a just transition

The scale of change required, as argued throughout this book, will inevitably lead to dislocation, with some groups potentially facing substantial losses. Examples include the potential consequences of the closure of coal mines and coal-fired power stations, the phasing out of dirty steel plants, the decline of industries producing parts for internal combustion engines, and the rising costs of fossil fuels for many. Individuals who seek to invest in new forms of heating or cooling will need to find the capital and manage the process of making improvements. Additionally, some people may resist the expansion of solar or onshore wind projects or transmission lines near their homes, which they fear will reduce the value of their properties. All of these factors contribute to the dislocation caused by such transformative changes. Managing dislocation is a central obstacle.

For many of those affected, there will be a sense – often understandable – that their rights to development or to work towards improving their standard of living have been infringed. These effects of the transition, particularly when they impact poorer communities, demand special attention.

Policies, institutions, and structures will differ depending on the context. However, some common themes emerge. For example, the investments necessary for the transition will be undertaken by various types of households, firms, cities, and governments. Generally, the cost of capital is higher for poorer individuals. Therefore, in most countries, it will be important to design systems that enable poorer people to have access to lower-cost finance, for example to invest in making their homes more energy efficient, to transition to new heating methods, to purchase EVs, or to adopt new agricultural techniques. These schemes will operate more effectively and efficiently if they are designed on a large scale and with public underwriting, since it is costly for a private financial institution to carry out scrutiny of borrowers for small-size loans.

A good starting principle is to foster and manage change by initiating public discussions on how to respond to the dislocations, in particular by creating opportunities. Those affected are likely to have valuable ideas. Creating opportunities will usually involve investing in the people and places most affected by change. In some cases, social protection and support will be necessary. But the focus in the first instance should be on creating new opportunities, which are likely, over time, to be both more fulfilling for individuals and lighter on the public purse.

In concluding this brief discussion of dislocation and a just transition, it is important to recognise that public policy cannot entirely prevent some people from losing out. The potential gains for global communities are immense, and halting progress because some will lose out is dangerous to all. The task of public policy is to find ways of extending the benefits of these immense gains to as many people as possible and to work to keep the potential losses as low as possible; for some, social protection and support will be necessary. The participation of communities in the design of public policy will be critical to making this process as effective as possible. It is important that gains are seen directly as early as possible if political support is to be built and sustained.

Some wealthier groups or countries, particularly those who benefit from economic rents associated with fossil fuels, will see their wealth diminish over time as the world moves away from fossil fuels. This outcome is inherent in the strategy and is an unavoidable part of the transition. These vested interests will resist change. Understanding that political economy, national and international, will be critical to success. But it is hard to argue that the rents that arise in one industrial or technological structure constitute indefinite rights, particularly if the activities are so damaging to others. A shared recognition of the great returns to driving change, and the dangers if we fail to change, will be vital to the creation of a political environment where opposition from vested interests can be overcome. There is more on political economy later.

Changes in behaviours

Much of the analysis of economic policy assumes that economic agents, whether consumers or firms, fully understand and pursue their preferences and best interests. Thus, such analyses focus on shaping incentive structures to influence actions and behaviours of people who make their choices in this way. For example, if public policy aims to reduce fossil fuel use, it might raise the prices of these fuels, or if it seeks to encourage the adoption of solar panels, it could offer subsidies. Standards and regulation can also be used to guide behaviour. These are sound principles of public policy, as we set out and embraced in Chapter 6. These incentive-based approaches are likely to do most of the 'heavy lifting' in shaping choices.

However, some attempts to influence behaviour go beyond such approaches. The insights from behavioural economics and institutional economics can also offer substantial further potential for public policy. There is work to be done on how to apply these insights to public policy.

The process of change may require individuals to engage with institutions and systems that have not previously played a prominent role in their lives. In this process of embarking on the new, understanding and using information is a key behavioural issue. Examples include navigating new electricity tariffs, understanding and managing EVs and their charging needs, adopting new farming methods, or adjusting to changed public transport systems. Such unfamiliarity, coupled with a lack of understanding and information, can lead to anxiety and resistance to change. Indeed, two of the outstanding pioneers of behavioural economics/psychology, Danny Kahneman and Amos Tversky (1991), showed that people often exhibit a strong aversion to perceived losses relative to where they currently stand, meaning they are more concerned about avoiding losses than they are about achieving equivalent gains. Kahneman and Tversky referred to this phenomenon as 'loss aversion'.

In this context, understanding institutional design and behaviour will be crucial in making these structures as accessible and user-friendly as possible. One example of a method to guide behaviour in this context is the 'default option'. If, for example, under a defined-contribution pension scheme, participants have to select preferred investment avenues for their pension, then the option that is specified as the default can exert a powerful influence on choices (Beshears et al., 2006). There is an extensive and valuable literature on behavioural and institutional economics (Hodgson, 2009; Morgan et al., 2010). While this is not the place for a detailed review, it is important to emphasise that this is an important area for action, innovation, and research. This is part of the research agenda set out in the next chapter (see also Chapter 6).

One useful set of lessons from behavioural economics comes from one of its pioneers, Cass Sunstein, who helped develop the concept of the 'nudge' (Thaler and Sunstein, 2008).[10] If people are cautious about adopting new behaviours, a nudge can help guide them in a direction that ultimately benefits both themselves and the community. One example is that prompting people to become organ donors while paying for their car tax added 100,000 donors to the register in a single year (Loosemore, 2014).

An example of a combination of incentive structures and behavioural change in supporting community action, from my own experience, is the recycling of Christmas trees in early January in South Harting, a village near where I live in Sussex. Local residents were invited to bring their used Christmas trees to the local school playground, where they could be shredded and turned into useful chippings. The event was highly convenient, with clear information on when and where to go, and it was also a social occasion where neighbours could exchange holiday greetings. It is an example of cost management, information, convenience, and community organisation in influencing behaviour.

Studying behaviour in different contexts is central because the effectiveness of specific interventions often varies across regions, communities, and different groups within them. There is a problem here in that much of the behavioural science research has been conducted in Western developed countries (Vlasceanu et al., 2024). In the context of developing nations, understanding behaviour relative to adaptation is of particular importance. The BASIN project at the Grantham Research Institute at LSE (where I work) aims to explore how behavioural and psychological sciences can inform decision-making for equitable and climate-resilient water security in sub-Saharan Africa (London School of Economics [LSE], n.d.). The research within the project will test the application of behavioural and psychological insights on adaptation for specific countries in sub-Saharan African, including Burkina Faso, Malawi, and Tanzania – nations that have experienced frequent flood and drought events since the mid-2000s. Among the aims of the project is identifying key levers to enhance community adaptation behaviour, considering behaviour determinants at both the individual and the organisational level. This requires identifying the factors and issues that drive individual choices, for example how risks are perceived by individuals and the access communities have to affordable and effective flood mitigation measures.

While behavioural studies focused on individual actions are valuable, they represent only part of the necessary responses. Their contribution to addressing climate change should not be interpreted as implying that the root causes of challenges are solely individual. For example, although considering individual risk perception is critical when designing policies for household flood protection, vulnerability to floods is part of broader systemic issues and will have to be tackled by public action at a district or regional level (Ingram and Gannon, 2024). That can involve, for example, building flood protection barriers or public warning schemes. Indeed, behavioural change can and must go way beyond 'nudges'. The emphasis on interventions at the individual level, such as choices around personal consumption, has at times obscured the deeper structural and institutional forces that shape behaviour, and has sometimes embodied an overestimation of what individuals can achieve on their own because of a nudge (Chater and Loewenstein, 2023).[11] For example, system-level factors such as the built infrastructure can perpetuate certain behaviours (e.g., in the case of transportation), cultural and political factors can affect climate behaviour (e.g., by influencing societal norms), and legislation can support or hinder private investment consistent with net zero (Nielsen et al., 2024).

Overall, the big effects in shifting demands and consumption and production patterns are likely to arise from prices, subsidies, regulation, standards, capital markets, improved information, and so on, as we saw in Chapter 6, or from direct public investment in infrastructure. But these actions and policies around major investments are likely to achieve better results by understanding human behaviour and psychology. Recognising this, multiple governments have established units to develop behavioural insights to be incorporated in policy-making. For example, the UK created the Behavioural Insights Team (BIT), or 'Nudge Unit', in 2010 within the Cabinet Office, under David Halpern (Afif, 2017).

Politics, political economy, and vested interests

Change is often met with resistance, including from those who foresee potential dislocations, losses, or inconveniences affecting them; those who simply dislike change; or those who buy into conspiracy theories about government control. Some may argue that the costs and disruptions of change outweigh the supposed benefits. Substantial vested interests, particularly those associated with fossil fuel industries, will resist change as they perceive threats to their economic rents.

These forces can form a formidable opposition to change. Or they can build support for arguments for a slower approach. Driving the process of change is inherently political – and, in many cases, intensely so. Achieving change requires the mobilisation of forces to push for change. Fortunately, there are many who understand the threats posed by unmanaged climate change and recognise the substantial opportunities presented by new development paths. And they do express their views politically. Yet, the momentum for change they offer, whilst significant, is not yet sufficient to generate the pace of change that is necessary.

Over the last decade or two, young people, who stand to benefit greatly from these changes, have been particularly vocal in pressing for action. In many countries, climate and biodiversity issues are taught in schools and, for the past decade, students entering universities already have some understanding of the issues and how we can respond as a world. Additionally, many from older generations recognise the moral responsibility of preserving and investing in the environmental heritage and natural capital they will pass on to the next generation.

Political leadership occurs at different levels. People come together more readily in cities, towns, or rural areas where the sense of community may be stronger. City alliances have played a crucial role in driving climate action, for example the US Climate Alliance for Cities during Trump's first presidency (2017–2021). See also the work of C40, where mayors can exchange ideas not only on how to get things done but also on how to build political support. And they can gain enthusiasm and courage in political leadership from each other.

However, there are politicians who see short-term political advantage in capitalising on the discomfort of those who find potential changes unsettling or see it as against their interests. For example, in the final years of the most recent Conservative government in the UK, from 2023 to July 2024, when they lost the election, their leadership attempted to use resistance to climate action as a 'wedge issue' to differentiate themselves from Labour, the Liberal Democrats, and other parties. There are signs that the new leadership of the Conservative Party will do the same in opposition.

Political movements such as the *gilets jaunes* in France during 2018–2019 protested strongly and effectively against increased fuel taxes. Similarly, during 2023–2024, German farmers protested against cuts in tractor diesel fuel subsidies (Connolly, 2024), which led to revisions of policy (Neubert, 2024). See also Chapter 6. Such opposition can make politicians hesitate.

Success in leading change in the face of such political challenges requires four key elements: packaging of policies, management of change, public participation, and communication. First, it involves crafting sound policy packages that combine multiple actions which work together. It is not wise or just to impose a single policy that carries negative impacts on a particular group, especially poorer groups, without offering something else to improve their situation. In the case of the *gilets jaunes*, for example, the fuel price rise was seen as another in a series of policies that disproportionately affected rural poor people (such as stricter maintenance rules on the vehicles essential to their rural livelihoods). Second, it is crucial to find ways to help people adapt to change over time and manage the associated costs. These could include phasing in policies to allow preparation for change, low-cost adjustment loans, or, in the case of transport, incentives to trade in polluting vehicles. Third, public participation in determining collaboratively how best to manage change can improve both policy design and acceptability. Fourth, leaders need courage and skill in communication, including the ability to present the change as exciting and attractive, to show that it will be inclusive in its benefits, and to offer a vision of a better future. Political support and the making of investment both require positivity. At the same time, openness to discussion and showing understanding and commitment to those affected are of great importance to the acceptability of change. In all this, timely delivery and the tangibility of benefit will be critical to political traction. Renewables do lower costs of electricity, but people have to see and experience those reduced costs in real time.

The resistance from the fossil fuel industry poses a significant obstacle (see Section 8.2). This industry is well funded and generally plays its cards effectively. This is not the place for an exhaustive account but, as highlighted in Section 7.5, in their book *Merchants of Doubt*, Naomi Oreskes and Erik Conway (2010) demonstrate clearly and strongly how the fossil fuel industry vigorously and often ruthlessly pursues its own interests through whatever methods it chooses. They detail, for example, how this industry, particularly in the USA, sought to undermine climate science by sowing doubt about its findings, even as their own internal scientists confirmed both the validity of the science and the severity of the consequences of inaction.

The fossil fuel industry, particularly in the USA, also provides substantial funding to politicians who may act in its interests (Goldberg et al., 2020; Influence Map, 2019). The period of rising oil and gas price rises and market disruption following Russia's invasion of Ukraine in the Spring of 2022 was another opportunity for the oil and gas lobby to push back against the transition, using spurious arguments related to energy security (as discussed earlier in this chapter). The political economy surrounding these issues can be intense, and research into how it operates and manifests itself is a key component for leading the transition successfully. Such research can involve threats and personal risk, and requires strength and determination as well as research skills.[12]

Success in leading change also requires standing up to those who seek to halt progress. And to do so successfully and manage to keep climate action going, it is central to work in coalitions. As Jean Chrétien (2025) argued recently, 'To fight back against a big, powerful bully, you need strength in numbers.' At the international level, the impact of setbacks on climate – such as Trump withdrawing the USA from the Paris Agreement for a second time in early 2025 – depends on the reaction of the rest of the world and their willingness to keep collaborating and acting without, in this case, the US national leadership. We currently live in a fragmented world where 'not all problems require leadership from a single dominant country', as the authors of the book *New World, New Rules* put it in a recent article (Papaconstantinou and Pisani-Ferry, 2025). For example, China and the EU can step up to lead on climate action, occupying the vacant space that Trump leaves for others to take. And we are seeing such leadership in Brazil, both in its G20 presidency of 2024 and in its presidency of COP30 in 2025.

10.4 Crucial issues that get too little attention: adaptation and biodiversity

In discussion of climate action, including in this book, adaptation and natural capital or biodiversity are not forgotten. But they receive too little attention. As a partial counterweight here, I want to give them special emphasis for further work. At the same time, it is important to stress that these issues are particularly location and context specific. Climate action involves real detail and strategies and policies that not only embody big ideas but also take account of the special circumstances of each country or place. A valuable example of detailed country work is in the Country Climate and Development Reviews (CCDRs) carried out by the World Bank working with individual countries.

Adaptation and loss and damage

Climate action encompasses more than just a successful transition and the associated investments. The climate has already shifted in dangerous directions, and further changes are inevitable. Understanding how to adapt and investing in adaptation are now major and urgent challenges. Additionally, responding to the loss and damage that will occur, even with sensible adaptation and resilience measures in place, is a key global responsibility; it is global emissions that have caused the problems.

The research agenda in this area is extensive and challenging. Three key points from our earlier discussions that are particularly important for research and action are highlighted here. First, there is a critical need for information with sufficient precision to enable local action. That action involves not only building and designing more robustly, but also the ability to act in a timely way when disasters occur or are imminent. Currently, our climate impact models and methods do not provide the local resolution with the precision necessary to guide effective decision-making and local action for many stakeholders. To take an example close to home in the UK, while we know that more and larger storm surges are expected on the River Thames, and while that has led to the strengthening of the flood defences, more detailed information is required to help households understand the specific measures they should take based on their individual circumstances. Similarly, when it comes to warnings of imminent danger, for example typhoons or cyclones in the Philippines or Bay of Bengal, the timelines and precision of information are crucial for enabling effective evasive action.

Second, restorative action after disasters have struck often requires fast-disbursing resources. More work is needed to develop financial structures that can quickly and effectively deliver resources to those affected. The valuable work around the Bridgetown Agenda is one such example that requires further exploration and action, including its emphasis on disaster clauses in loan structures which allow debt reduction or repayment postponement in the event of natural disasters.[13]

As mentioned in Section 9.3, parametric insurance can also facilitate faster disbursement, including in anticipation of threatening events. For example, insurance cover could pay out to farmers if there was no rain for a given period, if temperatures were above a given level for a given time, or if rainfall exceeded a certain level, and so on. Operating in this way can be less costly to organise (less litigation or dispute) and facilitate faster delivery.

Third, integrating development, mitigation, and adaptation should be central to the strategy for building a new approach to sustainable, resilient, and inclusive growth and development. While we have many examples of successful integration (see Section 5.2), many more are necessary. Combining the three makes investments more productive, enhances public support, and is more attractive to private finance than adaptation alone.

Demonstrating how these three elements can be successfully combined will help counter the misleading narrative of a false competition between climate action and development, or between mitigation and adaptation. Such narratives distort the issues and hinder progress. The CCDRs provide a valuable set of examples of analyses of international action at the country level.

In Chapter 9, it was argued that adaptation is badly underfinanced. That is no doubt correct. But we should at the same time recognise that the experience of climate change has brought a greater focus on disaster risk management. And there is increasing awareness of the importance of designing resilience into infrastructure and supply chains.

Natural capital and biodiversity

In recent years, natural capital and biodiversity have finally moved up the international and national agendas. However, this progress has been much too slow. Climate change, biodiversity loss, and the degradation of natural capital are inextricably linked, in terms of both the common factors driving these threats and the interactions that mutually reinforce them.

This interconnection and the common causes have been a major theme of this book – see Chapters 4 and 8. I will not reiterate the arguments here. However, I would strongly emphasise the importance of understanding these issues together and formulating policies that tackle them together. Climate change is sometimes seen as easier to grasp, partly because its primary causes – emissions of GHGs – are more straightforward to measure and manage. Ecosystems, on the other hand, are inherently more complex. But if we fail to act on and manage climate, biodiversity, and natural capital together, we will fail on all fronts.

COP30 in Brazil in 2025 will bring climate and biodiversity together still more prominently than in past COPs. One example of an innovation that could make an important contribution is the proposed Tropical Forests Forever Facility (United Nations Environment Programme Finance Initiative [UNEP FI], 2025). This provides new financial structures where countries, philanthropy, and the private sector can come together to create a fund which will generate returns over time to potentially provide a continual flow of funding to sustain and invest in forests. It could provide real scale. This is the kind of creativity the challenges require.

Table 10.4 summarises the main challenges related to adaptation and to natural capital, and outlines some key action points to begin overcoming the challenges.

Table 10.4: Key challenges and actions related to adaptation and natural capital

	Challenges	Key action
Adaptation	• Insufficient detail in information on possible climate impacts.	• Improvement of databases, climate impact models, and methods and computing power.
	• The delivery of resources quickly where disaster strikes.	• Development of financial structures to disburse needed resources fast. • The ability to organise transport of assistance/resources and people can be crucial.
	• Difficulties in attracting private capital.	• Integration of development, mitigation, and adaptation. • Improvement of insurance models, including parametric insurance (see Chapter 9).
Natural capital	• Has only recently moved up the agenda; less clarity in sets of actions than for climate. • Problems of restructuring systems and responsibilities. • Valuing natural capital and integrating into market decisions. • Difficulties in attracting private capital.	• Integration of climate change, biodiversity loss, and degradation. • Improvement of data, validation of techniques. • Strengthening of institutions and markets to provide sound incentives.

Source: Author's elaboration.

10.5 Concluding remarks: dispelling fallacies and overcoming obstacles to action

This chapter has been a partial summary of the book in a particular form: showing the faults in the arguments for inaction and identifying the difficulties that will be encountered in taking action. Both are important in galvanising and guiding action. Thus, the first purpose of this chapter has been to dispel fallacies that support arguments for delay and inaction. The second objective was to address confused arguments that advocate for the wrong types of action or where the weakness of the arguments undermines the case for action. These first two purposes are important in building the case for timely action, dealing with faulty arguments for delay, and building the political case for action. The third, and of particular importance for future research and action, was to highlight the more difficult or intractable problems that must be tackled if the world is to proceed with the necessary actions at the required pace. The fourth purpose was to emphasise the key, but often underemphasised, challenges of adaptation and biodiversity; these must be core parts of climate action.

Action, obstacles, and research

Putting these four elements together, I have attempted to weave together threads from the preceding chapters and arguments in a structured manner. Of these four purposes, the third and fourth generate both an action agenda (see Table 4.2) and a research agenda. That research agenda is drawn together in the next chapter (see Table 11.1).

The focus here has been on the problems within the realms of economics, society, and politics. These issues must be examined, understood, and tackled through serious social science research and analysis, as well as political and public action. In my view, in taking action, the most difficult and complex problems lie in these areas. However, it is crucial not to be complacent about the role of technology in relation to investment and innovation, or the role of science in addressing the risks we face and how they can be mitigated. As discussed in Chapters 4 and 5, while there has been strong progress in key technological areas, such as renewable energy and storage, much more progress is needed in areas such as reducing the costs of zero-carbon fuels for air and maritime transport, advancing various industrial processes (including steel and cement), advancing lower-carbon and more resilient agriculture, creating and managing energy grids, and similarly the technological side of cities, transport, and land systems. Particularly in the areas of adaptation and building resilience, we need significant advancements and refinements in the science. Further, we must deepen our scientific and technological understanding of SRM and work to reduce the costs of CDR.

Combating misinformation, which is growing, and dispelling the fallacies, which are so often repeated and recycled, should be a major element of the action agenda. That task is not analysed in detail here, but I hope the book and this chapter provide some of the ammunition and resources needed. The growth story will be key, but so too will be deepening the understanding of the immense risks of delay. It is indeed a crucial task and one which I hope the new Global School of Sustainability, which began at the LSE in January 2025, will take on with analytical rigour and constructive vigour.

Finally, a major issue that involves the intersection, interweaving, and mutual reinforcement of economic, social, and political action with technology is the role of AI. Much of the necessary change involves the transformation and management of complex systems, including energy, transport, cities, land, and water. AI can be of great importance in the discovery of new materials and the location of resources. Much of this work is data-intensive, fertile ground for AI. This potential has rightly been attracting increasing attention. In my view, AI and the green transition, and their mutual reinforcement, will be the core of the new growth story for the 21st century.

The risk of failure

The transition, associated investment, and systemic and structural change could and should have begun much earlier. A central message of the Stern Review (2006), was that delay is dangerous. Delay has meant that the increase in global investment in income must now be bigger than if we had started in earnest two decades ago. The world is now trying to do in two decades what could have been done over four decades had we started earlier. That means that the urgency and scale of the task are intense. Delay has become still more dangerous. Because of that urgency and scale, the economic and political obstacles are now still bigger, notwithstanding the advances of technology. Those obstacles can be overcome, but we must recognise that the risk of failure in relation to the Paris goals and in achieving a manageable climate is severe.

It is our task to show how to overcome obstacles to action. The prize – in terms of a safer planet and a new path of growth and development – is immense. Correspondingly, the risks of failure involve catastrophe in terms of lives and livelihoods on a global scale. Every decimal point of temperature increase counts; the longer we delay and the more feeble our response, the more difficult the task. And the greater the intensity of the consequences of failure.

Learning will play a vital role in both the green transition and the broader growth story it entails. In Chapter 7, this learning was emphasised in the context of government policy, but it will be equally important in business models and the technological dimensions of decarbonisation. At present, we cannot map exactly how the world will reach net zero along the full path, and particularly around mid-century; we do not yet have all the answers. What we do know is that the goal is achievable, and we understand the immediate and near-term actions required to move us forward. This is a story of structural change: long term, evolving, and at times disruptive. As we act on what we know, and what we know how to do, today, we must remain open to new insights and innovations that will help carry us forward in later phases of the transition. Throughout this process, we will continue to learn and develop more effective technologies and strategies to transform economic and social systems and build a more prosperous and safer world.

Notes

[1] Power systems can reach variable renewable energy (VRE) penetration rates of 75% to 90% if actions are taken to establish effective system operations supported by an adequate energy market (ETC, 2021). That estimated percentage may rise over time with experience and technical advance (see ETC 2025).

2 Strictly speaking, the minimum-global-cost way of carrying through the transition would be for higher cost producers for oil and gas to drop out before lower cost. In principle, profits arising in low-cost countries could be redistributed. However, politically it might be easier to adjust via production allocation rather than revenue transfers.

3 This research considered the spatial and temporal changes in emissions and ambient concentration of four pollutants: suspended particles, sulphur dioxide, nitrogen oxides, and carbon monoxide.

4 Formally, in many economic models, there are infinite horizons and long-run exponential growth in the sense that output expands like $(1+g)$ to the power t or exp (gt). That is a formal mathematical specification of 'indefinite growth'. In *The Limits to Growth* (Meadows et al., 1972), the argument was that exponential growth would eventually run into fixed constraints. Greta Thunberg spoke at the UN on 23 September 2019: 'We are in the beginning of a mass extinction and all you can talk about is money and fairy tales of eternal economic growth' (Thunberg, 2019).

5 Interestingly, the analysis of Fressoz does not go into the 21st century and thus he misses the extraordinary progress in the last quarter of a century (Tooze, 2025).

6 These interdependent boundaries include climate change, novel entities (e.g., microplastics), stratospheric depletion, atmospheric aerosol loading, ocean acidification, biogeochemical flows, freshwater change, land system change, and biosphere integrity. For more detail on each boundary, see Stockholm Resilience Centre (2023). Many of them have already been crossed.

7 The Dasgupta Review (2021b) refers to the imbalance between human demands on the ecosystem and the supply provided by nature as the 'impact inequality'. If the former exceeds the latter, we run down our natural capital. We cannot do that indefinitely.

8 If clouds are made brighter, they could reflect a small fraction of incoming energy back into space. This could be achieved by adding aerosols in locations where clouds form.

9 See the report of the G20 group on the Capital Adequacy Framework (CAF) (G20 Italy and G20 Indonesia, 2022). And Chapter 9.

10 See Halpern's book *Inside the Nudge Unit: How small changes can make a big difference* (2015).

[11] Further, within consumption-focused studies, research tends to concentrate on everyday activities rather than on actions that might be less frequent but highly impactful, such as when it comes to decisions on acquiring durable assets, such as homes, and activities like air travel. Other examples of understudied research areas include the dynamics influencing the behaviour of those who communicate about climate change, those who make investment decisions, or, for example, landowners who could use their land to restore natural habitats. Another example is why individuals might choose to act in favour of delaying or hampering climate action (Nielsen et al., 2024).

[12] Threats and harassment have occurred to a number of friends and colleagues working in this area. And to me.

[13] The Bridgetown Initiative is a proposal to reform the international financial architecture – particularly the way in which wealthier countries support developing nations in confronting the effects of climate change. It is named after the capital of Barbados, where the initiative was originated. See Bridgetown Initiative (2022, 2024).

11. Prospects for success: opportunity, urgency, and multilateralism

In the first part of our concluding section, Chapter 10, we explored the fallacies, confusions, and obstacles that impede climate action. Examining and tackling these challenges and difficulties is at the core of the arguments of this book. And we emphasised not only the magnitude and nature of the obstacles, together with how they could be overcome, but also the risk of failure in the light of the delays in action over the last two decades. However, we close on a positive note: there is a new path of development that is far more attractive and sustainable than the dirty, destructive paths of the past. The journey, the creation of this new path, will not be easy, but it is feasible, and this book has tried to show how it can be done. This journey represents the growth story of the 21st century.

Delay is dangerous. Indeed past delays have made the challenges more difficult. Further hesitation or denial in the face of necessary change is the most dangerous response of all; in this sense, it is also the most unrealistic. Continuing on a course that is leading us towards disaster is neither wise nor realistic, no matter how inconvenient the truth may be. Blundering towards catastrophe through hesitation or denial is simply reckless.

In this concluding chapter, Section 11.1 reflects on the lessons learned since *The Economics of Climate Change: The Stern Review*, 20 years after we began that work in 2005. The conclusion that inaction is far more costly than action has only grown stronger. We have also learned in these 20 years much about the way forward, particularly the dynamics and appeal of this new growth and development model. Those reflections were set out in the Introduction, and only a brief summary is given here.

Section 11.2 sets out what is involved in this new path of development, outlining the policies and actions needed to guide the transition. This transition will undoubtedly involve real challenges. It will require substantial investment, innovation, and structural and systemic change. A sound, strong, and credible strategy will be crucial, and the details of policy will matter greatly. Although there is much still to learn, we know enough to press forward with determination. The transition has already begun. International collaboration will be essential, as this is a crisis global in its origins, effects, and response. With the right strategy, policy, and international collaboration, the world can create this new growth story for the 21st century – one that is sustainable, resilient, and inclusive.

Section 11.3 emphasises the importance of multilateralism in this supremely international endeavour. That is not easy in a world that has become increasingly fractious. But an immense threat can bring people and communities together, particularly where the response has great intrinsic attraction. And collaboration around climate and biodiversity could make collaboration elsewhere less difficult.

Finally, in Section 11.4, I emphasise not only the urgency of this transition but also the optimism that it is feasible; it can indeed become the growth story of this century. Multilateralism, commitment, creativity, and determination will be critical to success. There will be many obstacles and numerous potential routes to failure. Optimism about what can be done is not necessarily optimism about what will be done. Failure remains a deeply concerning possibility. A key objective of this book is to contribute to the translation of potential success into the reality of a new way forward.

11.1 Retrospect: developments since the Stern Review

The two decades since we began work on the Stern Review (2006) have been eventful, marked by disruptive events such as the global financial crisis, Brexit in the UK, the two Trump administrations in the USA, the Covid-19 pandemic, and the war in Ukraine (see Chapter 4). This period has also taught us much about the crises in climate and biodiversity, and how we can respond to them. Section 1.1 in Chapter 1 sets out my reflections and developing views. The following is a short synopsis of those thoughts.

In summary, the central conclusion of the Stern Review – that inaction in the face of climate change is far more costly than action – has only been reinforced. This conclusion is now even stronger, for three key reasons. First, the science has become ever more alarming, in part because progress has been far too slow. Second, technological advancements have been far more promising than anticipated. Third, the political landscape has shifted, with climate change moving to the forefront of international discussions, as demonstrated by landmark agreements and, in particular, the Paris Agreement in 2015. However, the political climate remains challenging, with progress often accompanied by, or interspersed with, setbacks.

The scientific message on the urgency of action has never been clearer or stronger (see Chapters 2 and 4). As people worldwide experience the direct impacts of climate change – through more frequent and intense floods, storm surges, heatwaves, droughts, wildfires, and more – public understanding and acceptance of the science continue to grow (see Section 4.1). But, far from falling towards net zero, emissions have continued to rise. The task is now still more urgent and on a larger scale than if purposive action had begun two decades ago. When the Stern Review was first published, some accused us of being alarmist. In hindsight, we understated the risks.

Technological progress outpaced expectations from two decades ago, including those of the Stern Review. We did not foresee the dramatic reduction in the cost of electricity from solar and wind power, which has fallen by a factor of 10 or 15 or more. The clean is already cheaper than the dirty for sectors that cause around a third of emissions. The electrification of power, transport, heating, cooling, and much of industry will drive the new economy, with zero-carbon electricity at its core. The pace of technological change has been remarkable.

On the political front, the Paris Agreement at COP21 in December 2015 was a crucial milestone, a landmark agreement that signalled the world's recognition of the severity of the climate crisis and a collective readiness to act (see Chapters 2, 4, and 9). It provided a strategic direction that boosted innovation and investment in new technologies and underscored the alignment of economic, environmental, and social priorities. The political journey has not been without setbacks. The global financial crisis of 2008–2009 diverted political attention for several years. In 2017 President Trump withdrew the USA from the Paris Agreement. President Biden rejoined in 2021 and Trump has again withdrawn in 2025. The Covid-19 pandemic in 2020–2023 further monopolised political bandwidth and focus. The war in Ukraine, beginning in Spring 2022, triggered an energy crisis that led to some backlash to the transition, particularly from the fossil fuel industry. Notwithstanding the setbacks, global commitment has persisted, albeit with some exceptions.

The possibilities offered by technology and the growing realisation that the transition holds genuine prospects for growth and healthier lives have been vital in maintaining political momentum. The increasing engagement with this story was a crucial factor in achieving the Paris Agreement. Today, the case for this new growth story is even stronger, as I have tried to demonstrate throughout this book. Nevertheless, there has been, and will continue to be, pushback from vested interests, particularly within the fossil fuel industry (see Chapters 7 and 10). Further, as the transition gains momentum and real changes must be implemented in real time, managing a just transition, alongside political pressure and leadership, will be critical to success. As emphasised in Chapter 10, even though we can see how to overcome them, the obstacles are large.

Notwithstanding some stuttering and pushback, momentum is strong, with the private sector's recognition of opportunity being a key driver. Many private investors see the transition as the greatest commercial opportunity of our age.[1] The falling cost of the clean implies that it will increasingly be the technology of choice, whether or not the investor is concerned about climate change. And that observation is underscored if we take account of the increased riskiness of the dirty.

Overall, the conclusions of the Stern Review have not only passed the test of time, they have grown stronger. And we now see still more clearly the dynamics, potential, and attractiveness of the new path of growth and development.

11.2 Prospect: fostering action and an agenda for economics and the social sciences

The opportunities in the new growth story should now be evident. In saying so, it is important to emphasise once more (as discussed in Chapter 5) that we are referring to the next two or three decades, and a fundamentally new form of growth and development. This is absolutely not about indefinite growth under the old model; such a path would be reckless and ultimately unsustainable, given its destructive impact on the environment. This new growth story is indeed feasible, but the pressing question remains: can the world act on the scale and with the urgency required?

Optimism about what the world can achieve must be translated into what the world will actually do. The possibility of creating a better world at the necessary pace is both real and appealing, but that does not guarantee it will happen. Many, including myself, are deeply worried about what the world will do. It may well blunder into catastrophe. The challenge is to turn possibility into action. Here, political leadership, public pressure, private sector creativity, and sound economic analysis must come together. This book sends strong messages to the economics profession, my own field, that they have to do a better job. The profession has suffered from 'intellectual lock-in' or inertia associated with a commitment to models that might be convenient and familiar but do not fit the problem. Thus they have often perpetuated misleading arguments. They are now, finally, moving in a better direction.

The biggest obstacles to transformation to a new form of sustainable, resilient, and equitable growth and development lie in economics, politics, and society. Our subjects must embrace a new conceptual agenda on the objectives, strategies, and dynamics of development. Figure 11.1 sets out such an agenda for economics and the social sciences, drawing on the themes and lessons in this book. Economics and the social sciences have never been more important, but new thinking is necessary across a range of key concepts, perspectives, and analytical approaches. The urgency implies that we must research and act at the same time.

Figure 11.1 summarises the challenges facing economics, politics, and society in terms of analytics and action. It reflects the perspective of an academic economist who has spent the last two decades working on these issues analytically, politically, administratively, and internationally. I should also emphasise that my career before the Stern Review was dedicated, as it still is, to the analysis and implementation of growth, development, and structural change, particularly in poorer countries. As I hope this book demonstrates, I see climate action and the achievement of sustainable, resilient, and inclusive growth and development as together at the core of the overall story of growth and development and not separate subjects. Further, this framework reflects not only a lifetime of experience in academic work in the theory and practice of development, but also 10 years (1993–2003) as first Chief Economist of the European Reconstruction and Development Bank (EBRD) and then

Figure 11.1: An overarching research agenda for economics and the social sciences

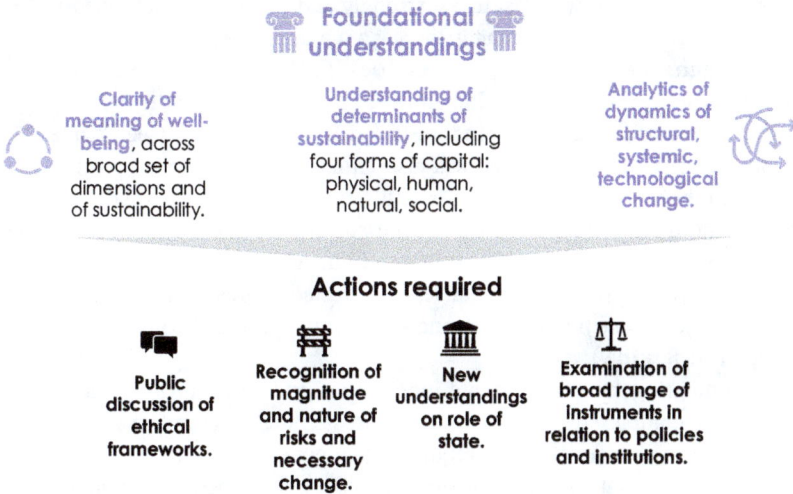

Foundational understandings

Clarity of meaning of well-being, across broad set of dimensions and of sustainability.

Understanding of determinants of sustainability, including four forms of capital: physical, human, natural, social.

Analytics of dynamics of structural, systemic, technological change.

Actions required

Public discussion of ethical frameworks.

Recognition of magnitude and nature of risks and necessary change.

New understandings on role of state.

Examination of broad range of instruments in relation to policies and institutions.

Source: Author's elaboration.

Chief Economist of the World Bank, plus three and a half years in the UK Treasury, the Finance Ministry (2003–2007), including as Head of the Government Economic Service and of the Stern Review.

Let us briefly explore the key elements presented in Figure 11.1, how they fit together, and what they imply for action. The sense of purpose that must drive major change should begin with clear objectives. These objectives should centre on well-being and sustainability (as discussed in Chapters 3 and 6). When considering development objectives, the international community has largely coalesced around the Sustainable Development Goals (SDGs) (adopted by the UN in 2015). These goals encompass a broad range of issues: income and food, health, education, environment and sustainability, equity across all these dimensions (including gender equity), and international collaboration. And they apply, as sustainability must, to all countries and not just the developing world, as was the case with the Millenium Development Goals (MDGs).[2] The adoption of the SDGs was driven by a growing recognition of the threats to sustainability posed by paths of growth which, despite their achievements in raising income, reducing poverty, and prolonging life expectancy, have caused severe environmental damage (as discussed in Chapter 2); the environment is itself an objective but also a means, and its damage threatens to undermine or reverse progress on all fronts. To reiterate, our definition of sustainability is providing future generations with opportunities at least as good as our own, assuming they adopt a similar approach towards generations that follow.

Clarity on objectives should be accompanied by an understanding of the factors and forces at work. This includes a focus on physical, human, natural, and social/cultural capital, all of which shape our opportunities. The implications for action to advance sustainability and tackle the crises of climate and biodiversity loss are clear: we must invest strongly and urgently in all four forms of capital. In terms of physical capital, there is a major priority for infrastructure in emerging markets and developing countries (EMDCs) other than China, particularly energy infrastructure. In natural capital, we must halt degradation and invest in restoration, including in forests, land use, and oceans (see Chapters 5, 6, 8, and 9). These investments will also enhance human capital, particularly in health, but we must also invest in skills, education, and knowledge for the new economy. The transition will test our social capital, which includes social cohesion, mutual trust, and the strength of our institutions, since it will bring dislocation, disruption, and challenges from vested interests – hence the importance of a just transition and a strong sense of purpose and commitment (see Chapters 6, 8, and 9).

Implementing these investments, innovations, and restructurings will involve navigating complex dynamics around fundamental structural, systemic, and technological change. This will require active government involvement and a new role for the state (see Chapters 6 and 7). Success in these areas will depend on transparency and participation across all levels of society, including public discussion on ethical frameworks, objectives, and strategies for building policy and overcoming obstacles. Public discussion, if it is to lead to commitment, will require a shared understanding of the magnitude and nature of the risks so that the urgency and necessity of action is widely recognised. It will also require confidence in the attractiveness of the new path. Without that commitment it will not be easy to overcome the many obstacles along the way.

The overarching agenda set out in Figure 11.1 is translated into more detailed research tasks, together with some pointers to possible methods, in Tables 11.1 (understanding the structure of the problems and challenges) and 11.2 (formulating policies and actions in response). Broadly speaking, Table 11.1 is on the analytics and understanding of the challenges, structures, risks, and dynamics, and Table 11.2 is on the action agenda. However, the analytics are organised to feed into policy and action, so there is inevitably some overlap.

Table 11.1 focuses first on the objectives, the risks, and the structure of investment across different forms of capital necessary to pursue the objectives as effectively as possible given country circumstances and risks. The second element of Table 11.1 examines the role of investment in structural change, taking account of macro circumstances; the third focuses on natural capital in particular, and the fourth on the challenges of adaptation.

In each case the methods will involve a number of analytical perspectives and collection of models. Indeed, the challenge is to build the understanding of the determinants and dynamics of investment and change in the context of risk, in a way that can guide action and policies to deliver the necessary investment. The complexities are such that we will need a range of analytical perspectives, including case studies and scenarios, dynamic narratives, and more formal modelling. No single model can do the work of building understanding here.

And, in particular, forcing the questions into or onto the standard work horses of exogenous growth theory and computable general equilibrium models, and expecting them to offer a useful overarching story, would be – and has been – seriously misleading. There will be important lessons from economic history. And from cost–benefit analyses (CBAs) provided, they are set in the context of systemic change and not narrowly incremental in a largely static background (which is often the standard approach to CBA). See comments (and notes) on Table 11.1 on scope of research methods.

The agenda set out in Tables 11.1 and 11.2 is very different from that addressed by integrated assessment models (IAMs), which have been very prominent in much of economic research. They are largely general equilibrium models, often with exogenous indefinite growth and little structural change. They are largely silent on the issues that are at the heart of the challenge (see Chapters 3 and 4).

Table 11.2 concerns policies and actions. The first set of these focuses on foundations in terms of objectives, an understanding of the structures (as in Table 11.1), and the principles and theories of policy, particularly taking into account risk, market failure, and dynamics, all of which are crucial in this context. Policies and actions for transforming systems are of the essence. It will be important to bring policy and action insights from behavioural and institutional economics. The second set embodies a broader view of politics and narratives in forming commitment to action, taking into account how political economy can disrupt and oppose. The sense of direction and purpose of change which are intrinsic to the tasks involved requires a re-examination of the role of the state in providing both that sense of direction and keeping the necessary pace (Chapter 7). Combining the sense of direction with individual rights, choices, and freedoms will be a vital element in and of itself as well as crucial to the success of the transition.

The climate and biodiversity crises are global problems; international action is thus at the heart of the story, explored particularly in Part III of this book and summarised in Table 11.2. Analysing and understanding mutual obligations and roles in a constructive and collaborative way will be key elements of the foundations of action. Private finance and multilateral financial institutions will be critical, particularly in combination, and there will be much analytical work necessary to underpin effective action, especially around both the scale and the management of risk. Here, in particular, we must combine acting at pace with creating the sound analytical foundations necessary for effective criteria. This is all about thinking, acting, and learning at the same time.

As the final part of Table 11.2 we have the political, integrative skills, and 'gardener's craft' of putting multiple pieces of analysis and different perspectives together into decision-making. There is no formulaic model for the aggregation of all the perspectives, analyses, and pressures into decisions. But it is important to bring all these together in a structured way. Wise judgement of what are key elements and ingredients to success is itself a key ingredient. Some messages and perspectives should come through clearly and strongly, but generally no one particular model should dominate. There is no substitute here for wise and determined leadership, and sound policy and political judgement.

Table 11.1: Research agenda: the structure of the challenges and opportunities

I. Investment, risks, sustainable development, transformation

Tasks/issues	Methods
• Articulating objectives.	• Examining basic philosophies, public discussion, and international agreements, including the Paris Agreement and SDGs.
• Understanding strategies to tackle the climate/biodiversity crises and create a new path of growth and development (including necessary timing, scale, and geography of investment and change). Centrality of investment: it is both imperative and an opportunity.	• Linking objectives and strategies to capital targets, and paths and capital to investment programmes.
• Understanding the nature and magnitude of risks, and how to manage and react.	• Understanding foundations and nature of policy with extreme risks.
• Identifying necessary investments and understanding the obstacles to investment and structural change.	• Constructing industry studies.
• Analysing the role of infrastructure in change and enabling change.	• Building scenarios.
• Understanding how to build the necessary skills for transition.	• Carrying out cost analyses and modelling.
• Understanding the nature of and how to act on the issues around transition and health.	• Understanding complementarities and networks.
• Integrating analyses of climate and biodiversity.	• Understanding sources of possible delay and associated risks in transition.
• Integrating development, mitigation, and adaptation.	• Building creative alternatives to standard computable general equilibrium modelling and standard exogenous growth models.
• Investigating strategies to tackle specific economic, social, and political barriers to investment and change: involvement and role of private sector, investment climate, and country platforms.	• Understanding disequilibrium dynamics.
• Researching ethical and practical issues around intertemporal and intratemporal distribution across dimensions of development.	• Drawing lessons from economic history.

Source: Author's elaboration.

II. Structures, systems, macro, finance	
Tasks/issues	**Methods**
• Understanding key necessary elements of structural change (sectors), technologies and systemic change (including cities, land, energy, transport, and water), and how to foster structural and systems change. • Understanding interactions between climate and biodiversity in relation to key systems. • Investigating ways to increase investment for climate action and development while addressing macroeconomic and debt challenges; financial regulation. • How to expand domestic savings/domestic resource mobilisation: public finances, financial markets, and toxic subsidies.	• Analysis of dynamic forces shaping structural change. • Role of strategies, plans, and incentives. • Sectors and complementarities. • Income distribution, taxation, and savings. • Macro-balances and dynamics. • Linking public finances and dynamics of investment, returns, and savings. • Debt dynamics, including flows and assets. • Role of domestic financial markets. • Understanding effects of national and international financial regulations, and how to design them. • Funding of capital markets. • New opportunities in public finance. • Role of AI.

III. Natural capital and biodiversity	
Tasks/issues	**Methods**
• Deepening understanding of interweaving of challenges in climate, biodiversity, and development and key points and nature of policy interaction. • Creating and building information, valuation, and markets. • Understanding strategies to unlock private investment in natural capital initiatives that simultaneously address climate change, biodiversity loss, and ecosystem degradation.	• Understanding and integrating overall role of biome in sustainable development. • Understanding services from natural capital. • Data, observation, and measurement. • Principles and practice of valuation. • Modelling interactions: climate, biodiversity, and economy. • Dynamics and triggers for tipping points. • Design and analytics of market structures, regulation, and finance.

IV. Adaptation and dislocation

Tasks/issues	Methods
• Understanding and developing investments and strategies for resilience and interweaving with development. • Developing strategies to unlock private investment in initiatives that simultaneously drive development, mitigation, and adaptation. • Developing strategies, policies, institutions and other structures to manage dislocations, manage real and perceived costs, and foster a just transition. • Improving adaptation data availability and accuracy and improving climate impact models. • Identifying financial structures for the rapid deployment of resources where disaster strikes.	• Analysis of how to anticipate dislocations. • Disequilibrium and adjustment. • Sequencing of actions. • Role and methods of participation. • Analytics of combining opportunity and protection.

Notes:
• All elements should be guided by evidence and data and by appropriate analysis of the data. The use of data in this context poses challenges because of the centrality of endogenous structural and systemic change: statistical and econometric methods often assume fixed system structures. Changing structures is of the essence.
• The 'Methods' column does not offer specific research methods (that would be a lengthy process) but points to broad issues and approaches. Major thematic approaches for research methods must include: (a) structural/systemic change; (b) pace and processes of change, as time is of the essence; (c) evidence bases, recognising that the future will be structurally different from past; (d) disequilibrium, dislocation, and potential instabilities; (e) recognition that in many cases the potential scale of risks is outside the usual scope of economic analysis; and (f) importance of carrying an overall integrated picture of transition and change whilst maintaining focus on detailed issues.

Source: Author's elaboration.

A robust agenda for action is set out in Figure 11.2. It should be founded in strong analysis but we must act, teach, and learn at the same time given the urgency and difficulty of the tasks. In particular, the economics profession has a leading role to play in understanding the magnitude and nature of risk, analysing the dynamics of fundamental change, and identifying the broad range of instruments – policy, institutional, behavioural – that can foster change and underpin the action agenda.

If this work is done well by all five groups of politicians, public servants, the private sector, communities, and academics, then it is possible to translate optimism on what we can do into optimism about what we will do. The advances in technology have been remarkable and continue to create immense opportunities. That progress can and must continue. It can build on great momentum in research and development and innovation in universities, in research establishments, and, especially, in the private sector. Scientists have been, and will continue to be, crucial in understanding the drivers and impacts of climate change, guiding the pace and methods for mitigation, and helping the world adapt to its effects. Politicians, public servants, the private sector, and communities can and must do their work. The academic challenges now lie primarily in the realms of economics, politics, and society, where the main obstacles to achieving the new growth story will be found. With all five groups playing their part professionally, in society and across countries, it will be possible to create an effective global response to what is fundamentally a global problem.

Figure 11.2: Agenda for action to be pursued by politicians, public servants, the private sector, communities, and academics

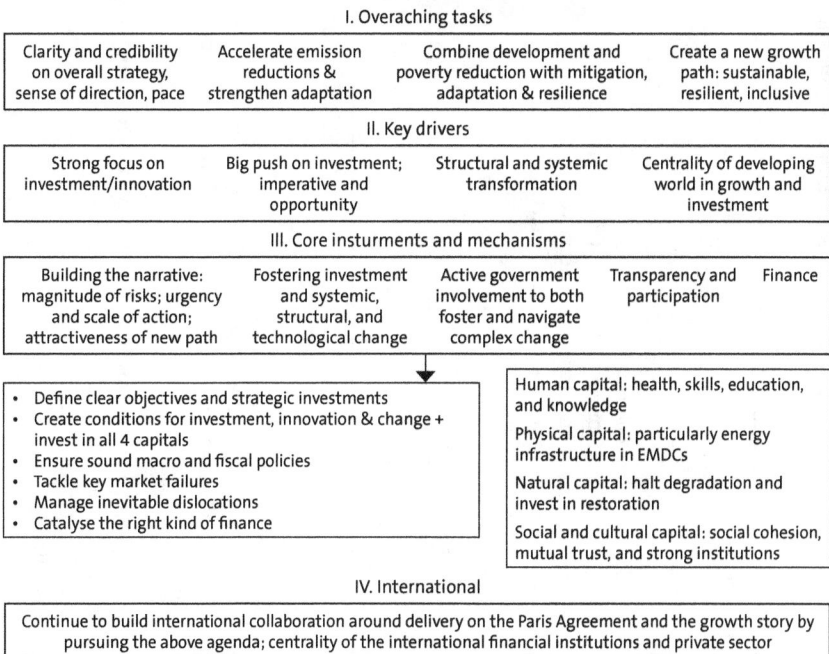

I. Overaching tasks			
Clarity and credibility on overall strategy, sense of direction, pace	Accelerate emission reductions & strengthen adaptation	Combine development and poverty reduction with mitigation, adaptation & resilience	Create a new growth path: sustainable, resilient, inclusive

II. Key drivers			
Strong focus on investment/innovation	Big push on investment; imperative and opportunity	Structural and systemic transformation	Centrality of developing world in growth and investment

III. Core insturments and mechanisms				
Building the narrative: magnitude of risks; urgency and scale of action; attractiveness of new path	Fostering investment and systemic, structural, and technological change	Active government involvement to both foster and navigate complex change	Transparency and participation	Finance

- Define clear objectives and strategic investments
- Create conditions for investment, innovation & change + invest in all 4 capitals
- Ensure sound macro and fiscal policies
- Tackle key market failures
- Manage inevitable dislocations
- Catalyse the right kind of finance

Human capital: health, skills, education, and knowledge

Physical capital: particularly energy infrastructure in EMDCs

Natural capital: halt degradation and invest in restoration

Social and cultural capital: social cohesion, mutual trust, and strong institutions

IV. International

Continue to build international collaboration around delivery on the Paris Agreement and the growth story by pursuing the above agenda; centrality of the international financial institutions and private sector

Source: Author's elaboration.

Table 11.2: Research agenda: rising to the challenges and taking the opportunities

I. Policies, institutions, law, behaviour	
Tasks/issues	Methods
• Values, objectives, ethics, freedoms, risk. • Fostering and enabling private sector markets, innovation, industrial strategy; importance of credibility of policy and institutions. • Tackling market failures (see Chapter 6): carbon prices and markets, research and development (R&D), capital markets, etc. • Transforming systems: design and incentives. • Identifying and overcoming sectoral obstacles. • Aligning law with environment and transition. • Aligning regulation with pace and direction of change. • Linking and integrating micro, structural/systemic, and macro (see Part I, Section II above); linking and integrating with nature/biodiversity (Section III above); linking and integrating with adaptation/dislocation (Section IV above).	• Moral philosophy. • Theories of choice, risk, and justice. • Intertemporal valuations and discounting. • Public economics as if time matters (Stern, 2018). • Interactions of market failure and dynamics. • Building policy credibility, role of institutions. • Designing, managing, and optimising systems. • Designing institutions. • Psychology/behavioural economics. • Combining market and institutional design. • Overall growth and macro strategy shapes micro/structural; micro/structural shapes overall growth and macro. • Role of AI.
II. Political economy, narratives, role of the state	
Tasks/issues	Methods
• Understanding political and social obstacles to change. • Building and sharing the narrative of the new growth story. • Understanding how to manage, shape, and leverage politics and political economy dimensions, including political support and leadership, to enable necessary change. • Identifying pathways to influence behaviour, including drawing on behavioural and institutional economics. • Examining and clarifying choice, freedom, and responsibilities in the transition. • Facilitating citizens' assemblies and participation. • Deepening understanding of changing role of the state.	• Identification of gainers and losers, and path dynamics. • Mechanisms of enabling and obstructing. • Public policy and adaptive/endogenous preferences. • Behavioural/institutional economics. • Methods of political economy.

III. International action: geography, ethics, finance

Task/issues	Methods
• Understanding mutual dependence around emissions, biodiversity. • Analysing and fostering the new economic geography of clean endowments and particular minerals. • Building the resilience of supply chains: innovation, collaboration, and competition. • Opportunities and challenges across nations in different circumstances. • Mutual obligations: history, income, and wealth; ethics in relation to emissions and finance. • Finance: scale, purpose, matching, and public–private. • Transforming actions of international financial institutions (IFIs), particularly finance: scale and work with the private sector; international regulation. • Role of UN and related institutions.	• Principles of location. • Dynamics of international market structures and innovation. • Analysis of supply chains. • Moral philosophy underpinning international obligations. • Finance to manage risk, match needs, and build collaboration.

IV. Putting it all together

Task/issues	Methods
• Integrating micro, macro/investment, structural dynamics, system change, international. • Bringing modelling, scenarios, and risk analyses together. • Cross-cutting: digitisation, data, and artificial intelligence (AI). • Role of finance ministries, governments, and international institutions in overall strategies and decisions. • Using judgement from political experience; understanding local, national, and international environment and economic history.	• Understanding systems and functioning of governments. • Analysing the political economy of interest groups. • Placing institutions at centre stage. • Incentive structures in organisations. • Combining systems. • Integrating private investment and public action. • Different possibilities for role of the state. • Political philosophy. • Structured wisdom of experience.

Source: Author's elaboration.

As emphasised in Chapter 10, the dissemination of misinformation and the repetition of fallacies present major obstacles to action. Overcoming those obstacles, understanding them, and responding to them is a key challenge for both research and action. This will be a major subject for the new Global School of Sustainability at the LSE.

11.3 Multilateralism

This is a decisive decade for the world. We must invest, innovate, and pursue change in all countries. However, the challenges we face are quintessentially global, with global causes and global impacts. Collaboration across nations will be crucial (see Chapters 3, 4, and 9). Multilateralism has never been more important. At the same time, we must recognise that recent years have seen a rise in international tensions and a weakening of key international institutions such as the UN and the World Trade Organization (WTO). And President Donald Trump's second administration (2025–) is likely to be less internationalist and climate- or environment-oriented than that of his predecessor.

However, we must also recognise that climate change and/or biodiversity is often the first issue where nations can see a common interest and the value of collaboration. And these are issues that would usually stretch across the term of a government or administration in all countries. Progress in multilateralism of special significance was made at COP16 in 2010, COP21 in 2015, and COP26 in 2022. COP16 rescued multilateralism on climate from potential collapse after Copenhagen COP15 and established the principle of financial obligations of rich countries. COP21 in Paris was where the basic overall framework and targets for climate change were set. COP26 brought in the private sector in a major way.

During a particularly tense period between the USA and China, both nations came together with a joint declaration at COP26 in late 2021. When Presidents Xi Jinping and Joe Biden shook hands at Indonesia's G20 summit in autumn 2022, climate was the first topic mentioned as an area for cooperation. Moreover, successful collaboration on climate could set a positive tone for better collaboration in other areas. Collaboration on that front between the USA and China may decrease during the second Trump presidency, but it is likely to increase between China, India, Europe, and across and within the Global South. As I write (Spring 2025), we are already seeing movement here.[3]

The key elements around which collaboration can be built are as follows. First, there must be a shared understanding that the climate and biodiversity crises are global in nature and all must be involved. Achieving net zero emissions globally requires nearly all countries to reach net zero. Further, it is the cumulative emissions across countries and over time that determine total concentrations – and thus climate change – hence the importance of working together internationally for both the short and the long period. This understanding of the centrality of the involvement of all nations is now much stronger than it was at the time of the United Nations Framework Convention on Climate Change (UNFCCC), adopted at the Rio Earth Summit in 1992, and the Copenhagen Summit of 2009 (COP15). It was only after COP15 that the interpretation of common but differentiated responsibility (CBDR) began to evolve, with many developing countries beginning to move away from the notion that they should do very little until rich countries had done much more and had provided substantial finance.

Second, international action will progress more quickly if there is a shared recognition that climate action generates a new and attractive growth story. Understanding of this growth story is becoming more widespread. Contributing to its development has been central to my work over the last decade. I hope this book builds further on that effort. However, it is far too early to declare victory in this argument. While we are making progress, there is still much work to do, and there are many with vested interests and misguided analyses who would prefer to stay on old paths and reject this argument. Increasingly, I trust, we will see examples of success, where countries and regions achieve strong growth alongside significant emissions reductions, with green investment as a key driver of that growth.

Third, promoting and achieving investment on the necessary pace and scale and particularly in EMDCs – especially its financing – is the core factor in success. CBDR surely does not mean that poorer countries should do very little on climate action. However, given the disparities in past emissions, the vulnerability of poor countries, the wealth of the rich countries, and the challenges of financing the large increase in investments necessary (particularly in poor countries), it surely does mean that rich countries should step up strongly on climate finance. Further, it is in their self-interest to do so, because they too will suffer from the severe effects of climate change. They should recognise that climate change will increasingly be shaped by the nature of investment and growth choices in poorer countries, given the likelihood of their faster growth and the fact that their infrastructure will be shaped in the next two decades. Moreover, strong growth in the developing world is essential in the fight against poverty, but if that growth is carbon intensive it will be deeply damaging to all. And we have shown (Chapter 9 in this volume; Bhattacharya et al., 2023; Bhattacharya et al., 2024; Singh and Summers, 2023a, 2023b; Songwe et al., 2022) that working more strongly through the IFIs and intensifying their collaboration with the private sector is by far the most rapid and cost-effective way of mobilising the finance required on the necessary scale.

Multilateralism will also be crucial in areas such as trade and the conditions for investment, as explored in Chapter 8. Trade and competition bring down costs. The world needs more players in green industries and technologies; some early support for the development of new players will be necessary. However, this should occur in ways that avoid entrenched protectionism and do not discriminate against developing countries. Now is the time to utilise strong global production capacity in areas such as solar photovoltaic (PV), particularly in China, to drive a rapid increase in renewable power investment in the poor countries of the world. It is there that well over half of renewable power opportunities are concentrated, but these countries receive only a small fraction (2 or 3%) of the investment. All countries should support international institutions, including the UN, the WTO, and the IFIs, in realising this rapid increase in renewables. It is of huge importance to their development and our shared future. And we argued in Chapter 9, a major expansion of solar in Africa using the large capacity for production of low-cost solar panels in China is a global opportunity of immense potential significance. The MDBs could play a major role both in low-cost finance and in the politics and practicalities of action. It could both power Africa and represent a tremendous example of the returns to collaboration.

This is a crucial moment for world leadership, one that requires looking beyond current tensions and difficulties, including those between China and the USA. Climate change and biodiversity loss are the most urgent areas for global action and the most promising for collaboration. There are understandable reasons for mistrust on the side of developing countries regarding the intentions and commitment of rich countries – arising from issues such as the inequitable distribution of Covid-19 vaccines, the slow delivery of climate finance (see Chapter 9), and the effects of the Ukraine war on the costs of food, fertilisers, and energy in the developing world.

For all these reasons, new leadership in enhancing multilateralism should come from the richer nations. However, the EMDCs must be in the vanguard of both policy formation and action. These countries are where growth will be strongest, but they are also most vulnerable. They are rightly becoming more influential within international institutions, and they can push the richer countries to act more decisively while playing a stronger role themselves. We are beginning to see this through the recent G20 leaderships of Indonesia (2022), India (2023), and Brazil (2024). We hope this will continue with South Africa's G20 presidency in 2025. Smaller countries can also be influential. For example, small island states were very effective in keeping the 1.5 °C target on the table at Paris COP21. Recently, with the leadership of Barbados, these states have been advocating for reforms in climate finance to provide better protection against extreme weather events. Building multilateralism requires constant attention from all involved. That should be shaped by clear and strong arguments for its necessity in tackling these global crises, in achieving the SDGs, and in fostering sustainable, resilient, and inclusive development.

11.4 Concluding remarks: 'Yes, we can'; success is possible

Throughout the book I have emphasised the necessity, magnitude, urgency, and difficulties of the task we face in tackling the climate and biodiversity crises. The task will not be easy – far from it. There will be setbacks and powerful forces in opposition. But let us finish on some reasons for optimism. This is not to downplay the difficulties ahead but to recognise that there are five powerful forces currently at work that can accelerate action. These five forces are world savings, technological change, international agreements, the

Figure 11.3: Five forces that present a special opportunity to deliver climate action at scale

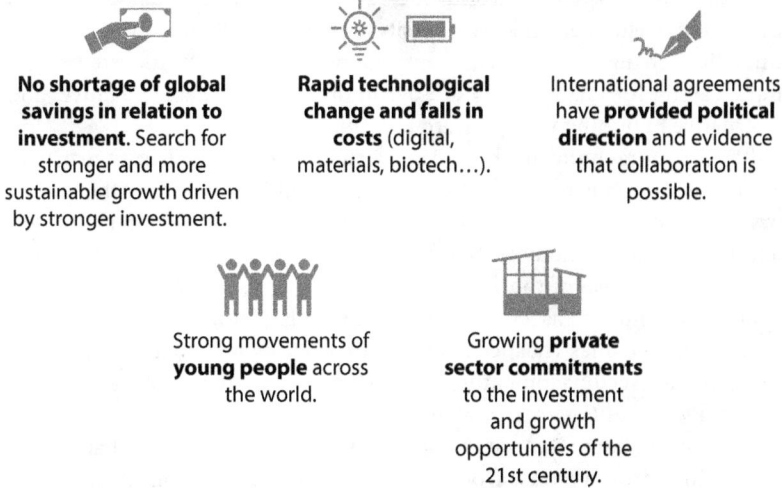

No shortage of global savings in relation to investment. Search for stronger and more sustainable growth driven by stronger investment.

Rapid technological change and falls in costs (digital, materials, biotech…).

International agreements have **provided political direction** and evidence that collaboration is possible.

Strong movements of **young people** across the world.

Growing **private sector commitments** to the investment and growth opportunites of the 21st century.

Source: Author's elaboration.

mobilisation of young people, and the private sector (see Figure 11.3). They can help generate the momentum needed to tackle these crises and realise a new story of growth and development.

These forces have all been discussed throughout this book, so we will summarise them only briefly here. First, since the global financial crisis of 2008–2009, and arguably even before, the world saw an excess of planned savings in relation to planned investment which continued at least up to the start of the Covid-19 pandemic. Given high savings rates in Asia and stronger growth rates there than elsewhere, it may reappear and present an opportunity for a global expansion in investment. Public policy and international cooperation will be necessary to translate this potential into concrete action on a country-by-country basis (see Chapter 9 and Section 10.3), particularly as many countries face macroeconomic and debt constraints. If it does not reappear, then public policy will need to foster an increase in world savings to accompany the strong increase in investment which is now necessary.

Second, technological change and innovation have led to rapid falls in the costs of sustainable technologies, a trend that continues and is expanding across most sectors of the economy (see Chapters 2, 4, and 5). The clean is already cheaper than the dirty across one-third of emissions and that fraction will likely rise rapidly over the coming decade (Systemiq, 2020).

Third, whilst coordinated international action will always have its difficulties, the recognition of the intensity of the challenges of climate and sustainability have fostered shared international objectives and commitment, as evidenced by the Paris Agreement and the adoption of the SDGs in 2015. IFIs are increasingly prioritising sustainability in their work.

Fourth, young people worldwide have recognised that action on sustainability, climate, and biodiversity is vital for their future and are demanding change. Moreover, many young people have moved beyond simply calling for change; they are becoming knowledgeable, creative, and constructive in finding ways forward and solutions.

Fifth, the momentum in the private sector is powerful. Private firms are increasingly recognising the investment opportunities in creating the growth story of the future. They are investing and innovating at pace and scale, understanding that the technologies of the 21st century will be more profitable and less risky than those of the 19th and 20th centuries. However, success will require momentum to accelerate, particularly in EMDCs (other than China). As the clean becomes cheaper than the dirty across an expanding range of sectors, private sector momentum will continue to grow.

These five forces together provide the world with a unique opportunity to turn optimism about what we can achieve into real action, into what we will achieve. To borrow a phrase from Barack Obama: 'Yes, we can.' The path will be difficult, but success is possible and failure would be catastrophic, indeed existential for many. The prize of a safer planet and a new and much better path of growth and development is immense.

Notes

[1] See Chapter 1 for references.

[2] The MDGs consisted of goals corresponding to eight areas of action with specific targets and a date of 2015 for achieving those targets. They were put together following the Millenium Declaration of September 2000.

[3] Europe's reactions to Trump's second presidency have included reaffirming commitment to the recommendations of the Draghi Report (Draghi, 2024). China's 15th Five-Year Plan (2026–2030) is looking increasingly likely to emphasise sustainable development. Many of my interactions at the World Economic Forum (WEF) in Davos, January 2025, pointed the same way.

Reference list

Acknowledgements

Stern, N. (2006). *The Economics of Climate Change: The Stern Review.* HM Treasury. https://webarchive.nationalarchives.gov.uk/ukgwa/20100407172811/ https:/www.hm-treasury.gov.uk/stern_review_report.htm [Published in book form by Cambridge University Press in January 2007.]

Stern, N. (2015). *Why Are We Waiting?: The Logic, Urgency, and Promise of Tackling Climate Change* (1st ed.). The MIT Press.

Stern, N. (2024). *A Growth Story for the 21st Century: building sustainable, resilient, and equitable development – Lionel Robbins lectures 2024: Slides and videos.* Grantham Research Institute on Climate Change and the Environment, London School of Economics and Political Science. https://www.lse.ac.uk/granthaminstitute/publication/lionel_robbins2024/

Introduction

African Union Commission (AUC), and Organisation for Economic Co-operation and Development (OECD). (2024). *Africa's Development Dynamics 2024: Skills, Jobs and Productivity.* OECD Publishing. https://doi.org/10.1787/df06c7a4-en

Ambasz. D., Sanchez Tapia, I., and Kwauk, C. (2024). Skilling 'youth on the move' to help power the green economy. *World Bank Blogs.* https://perma.cc/3BED-S3B8

Brown, O. (2008). *Migration and climate change.* International Organization for Migration. https://perma.cc/N2L5-A7WP

Burgen, S. (2024, December 18). 'If 1.5m Germans have them there must be something in it': how balcony solar is taking off. *The Guardian.* https://perma.cc/67PD-3WST

Carney, M. (2020). *The Road to Glasgow.* The Bank of England. https://perma.cc/5EBG-GN9E

Carrington, D. (2023, September 30). 'We're not doomed yet': climate scientists Michael Mann on our last chance to save human civilisation. *The Guardian.* https://perma.cc/CAN6-ZHVS

Chandler, D.L. (2023). Riddle solved: Why was Roman concrete so durable? *MIT News Office.* https://perma.cc/M3A5-7LRJ

Clement, V. W. C., Rigaud, K. K., de Sherbinin, A., Jones, B. R., Adamo, S., Schewe, J., Sadiq, N., and Shabahat, E. S. (2021). *Groundswell: Acting on internal climate migration (Part II)*. World Bank Group. http://documents .worldbank.org/curated/en/540941631203608570

Climate Action Tracker. (2023, December). *Climate Action Tracker*. C40 Knowledge Hub. https://www.c40knowledgehub.org/s/article/Climate -Action-Tracker?language=en_US

Copernicus. (2025). *The 2024 Annual Climate Summary: Global climate highlights 2024*. https://perma.cc/VS6J-YGC4

Copernicus Climate Change Service (C3S) and World Meteorological Organization (WMO). (2025). *European State of the Climate 2024*. https://doi.org/10.24381/14j9-s541

Dávila, J. and Daste, D. (2013). 15. Aerial cable-cars in Medellin, Colombia: social inclusion and reduced emissions. In M. Swilling, B. Robinson, S. Marvin, and M. Hodson (Eds.), *City-Level Decoupling: Urban Resource Flows and the Governance of Infrastructure Transitions* (pp. 45–48). UN Environment, GRID-Arendal. https://perma.cc/TTT8-C38K

Ellis, E.C. (2024, March 5). The Anthropocene is not an epoch – but the age of humans is most definitely underway. *The Conversation*. https://perma .cc/8MA6-4WUP

European Commission. (2023). *Delivering the European Green Deal: On the path to a climate-neutral Europe by 2050*. https://perma.cc/Q8Q7-36FG

GrEEn (2020). *Boosting Green Employment and Enterprise Opportunities in Ghana*. Partnership for Sustainable Development. https://perma.cc /W5FT-GPWL

Institute for Economics and Peace. (2020). *Over one billion people at threat of being displaced by 2050 due to environmental change, conflict and civil unrest*. https://perma.cc/NQ9T-RCY3

Intergovernmental Panel on Climate Change. (2021). *Summary for poli-cymakers*. In V. Masson-Delmotte, P. Zhai, A. Pirani, S. L. Connors, C. Péan, S. Berger, N. Caud, Y. Chen, L. Goldfarb, M. I. Gomis, M. Huang, K. Leitzell, E. Lonnoy, J. B. R. Matthews, T. K. Maycock, T. Waterfield, O. Yelekçi, R. Yu, and B. Zhou (Eds.), *Climate Change 2021: The physical science basis. Contribution of Working Group I to the Sixth Assessment Report of the Intergovernmental Panel on Climate Change*. Cambridge University Press. https://doi.org/10.1017/9781009157896.001

Intergovernmental Panel on Climate Change. (2023). Summary for policy-makers. In H. Lee and J. Romero (Eds.), *Climate Change 2023: Synthesis report: Contribution of Working Groups I, II and III to the Sixth Assess-*

ment Report of the Intergovernmental Panel on Climate Change (pp. 1–34). Cambridge University Press. https://doi.org/10.59327/IPCC/AR6 -9789291691647.001

International Energy Agency (IEA). (2025, April 14). *Kenya's energy sector is making strides toward universal electricity access, clean cooking solutions and renewable energy development.* https://perma.cc/KG5A-YBUW

International Labour Organization (ILO). (2024). *Global employment trends for youth 2024.* https://www.ilo.org/publications/major-publications /global-employment-trends-youth-2024

Jacobson, M. Z. (2025). *60 Countries/Territories Whose Electricity Generation in 2023 was 50–100% Wind-Water-Solar (WWS) (Including 12 With 98.4–100% WWS Generation) and 11 U.S. States That Produced the Equivalent of 53.2-118% of the Electricity They Consumed With WWS in 2024.* https://perma.cc/3A67-TX7A

Kouamé, A. T. (2024). *Gearing up for India's Rapid Urban Transformation.* World Bank Group. https://perma.cc/62JF-9KUN

Miller, K. G., Wright, J. D., Browning, J. V., Kulpecz, A., Kominz, M., Naish, T. R., Cramer, B. S., Rosenthal, Y., Peltier, W. R., and Sosdian, S. (2012). High tide of the warm Pliocene: Implications of global sea level for Antarctic deglaciation. *Geology, 40*(5), 407–410. https://doi.org/10.1130/G32869.1

Ministry of Energy of Chile. (2022). *H2V Hidrogeno Verde: Un Proyecto Pais.* https://energia.gob.cl/sites/default/files/guia_hidrogeno_abril.pdf

Ministry of Steel, Government of India. (2024). *Year-End Review 2024: Ministry of Steel.* https://perma.cc/2YWY-46ZP

Mirzabaev, A., Wu, J., Evans, J., García-Oliva, F., Hussein, I. A. G., Iqbal, M. H., Kimutai, J., Knowles, T., Meza, F., Nedjraoui, D., Tena, F., Türkeş, M., Vázquez, R. J., and Weltz, M. (2019). Desertification. In P. R. Shukla, J. Skea, E. Calvo Buendia, V. Masson-Delmotte, H.-O. Pörtner, D. C. Roberts, P. Zhai, R. Slade, S. Connors, R. van Diemen, M. Ferrat, E. Haughey, S. Luz, S. Neogi, M. Pathak, J. Petzold, J. Portugal Pereira, P. Vyas, E. Huntley, K. Kissick, M. Belkacemi, and J. Malley (Eds.), *Climate change and land: An IPCC special report on climate change, desertification, land degradation, sustainable land management, food security, and greenhouse gas fluxes in terrestrial ecosystems.* Cambridge University Press. https:// doi.org/10.1017/9781009157988.005

Modi, N. (2021, November 1). *National statement by Prime Minister Shri Narendra Modi at COP26 Summit in Glasgow.* Ministry of External Affairs, Government of India. https://www.mea.gov.in/Speeches -Statements.htm?dtl/34466/National+Statement+by+Prime+Minister+ Shri+Narendra+Modi+at+COP26+Summit+in+Glasgow

Radford, C. and Field, A. (2023). Green hydrogen in Africa: A continent of possibilities? *White and Case.* https://perma.cc/LVN9-XCE8

REN21. (2024). *Renewables 2024 Global Status Report Collection.* https://www.ren21.net/gsr-2024/

Ruto, W. (2022). We are at a crossroads in history: Africa can and must be a leader in clean energy. *The Guardian.* https://perma.cc/Y6V6-96U3

Segal, M. (2022). *BlackRock hires McKinsey sustainability leader to head new climate transition opportunities unit.* ESG Today. https://perma.cc/C6FM-BBMD

Seymour, L., Maragh, J., Sabatini, P., di Tommaso, M., Weaver, J.C., and Masic, A. (2023). Hot mixing: Mechanistic insights into the durability of ancient Roman concrete. *Science Advances, 9*(1). https://doi.org/10.1126/sciadv.add1602

Songwe, V., Stern, N., and Bhattacharya, A. (2022). *Finance for climate action: Scaling up investment for climate and development.* Grantham Research Institute on Climate Change and the Environment, London School of Economics and Political Science. https://perma.cc/V454-EKF5

Stern, N. (2006). *The Economics of Climate Change: The Stern Review.* HM Treasury. https://webarchive.nationalarchives.gov.uk/ukgwa/20100407172811/https:/www.hm-treasury.gov.uk/stern_review_report.htm

Stern, N. (2007). *The economics of climate change: The Stern Review.* Cambridge University Press.

Stern, N. (2015). *Why are we waiting?: The logic, urgency, and promise of tackling climate change* (1st ed.). The MIT Press.

Stern, N. (2024). *A Growth Story for the 21st Century: building sustainable, resilient, and equitable development – Lionel Robbins lectures 2024: Slides and videos.* Grantham Research Institute on Climate Change and the Environment, London School of Economics and Political Science. https://www.lse.ac.uk/granthaminstitute/publication/lionel_robbins2024/

Stern, N. and Stiglitz, J. (2023). Climate change and growth. *Industrial and Corporate Change, 32*(2), 277–303. https://doi.org/10.1093/icc/dtad008

Summers, L. H., and Singh, N. K. (2023, July). The multilateral development banks the world needs. *Project Syndicate.* https://perma.cc/F5UL-P94G

Systemiq. (2024). *Material Improvements: Building a Better World with Lower-Carbon Materials.* https://perma.cc/MXD3-QQJF

Thurber, M. (2024). *Assessing Energy Technology Leapfrogs.* Energy for Growth Hub. https://perma.cc/QC57-L47X

United Nations Framework Convention on Climate Change. (2015). *Paris Agreement*. United Nations. https://unfccc.int/sites/default/files/english _paris_agreement.pdf

World Economic Forum. (2025). *Unleashing the Full Potential of Industrial Clusters: Infrastructure Solutions for Clean Energies* (White Paper). https://perma.cc/CQ65-YFRY

World Meteorological Organization (WMO) (2025a). *State of the Global Climate 2024*. https://wmo.int/sites/default/files/2025-03/WMO-1368 -2024_en.pdf

World Meteorological Organization. (2025b). *State of the Climate in Asia 2024*. https://library.wmo.int/records/item/69575-state-of-the-climate- in-asia-2024

Zhang, L., Zheng, M., Zhao, D., and Feng, Y. (2024). A review of novel self-healing concrete technologies. *Journal of Building Engineering, 89*. https://doi.org/10.1016/j.jobe.2024.109331

PART I

Chapter 1

African Union Commission (AUC), and Organisation for Economic Co-operation and Development (OECD). (2024). *Africa's Development Dynamics 2024: Skills, Jobs and Productivity*. OECD Publishing. https://doi.org/10.1787/df06c7a4-en

Aghion, P. and Howitt, P. (1992). A model of growth through creative destruction. *Econometrica, 60*(2), 323–351. https://doi.org/10.2307/2951599

Arrow, K. J. (1962). The economic implications of learning by doing. *The Review of Economic Studies, 29*(3), 155–173. https://doi.org/10.2307/2295952

Betts, P. (2025). *The Climate Diplomat: A Personal History of the COP Conferences*. Profile Books.

Bodansky, D. (2011). A tale of two architectures: The once and future U.N. Climate Change Regime. *SSRN*. http://dx.doi.org/10.2139/ssrn.1773865

Bodansky, D. (2016). The Paris Climate Change Agreement: A new hope? *American Journal of International Law, 110*(2), 288–319. https://doi.org/10.5305/amerjintelaw.110.2.0288

Bodansky, D., Brunnée, J., and Rajamani, L. (2017). *International climate change law*. Oxford University Press.

Brundtland Commission. (1987). *Our Common future*. Oxford University Press. https://perma.cc/3EUZ-B6FE

Capstick, S., Whitmarsh, L., Poortinga, W., Pidgeon, N., and Upham P. (2015). International trends in public perceptions of climate change over the past quarter century. *Wires Climate Change, Royal Geographical Society, 6*(1), 35–61. https://doi.org/10.1002/wcc.321

Chancel, L., Piketty, T., Saez, E., and Zucman, G. (Eds.). (2022). *World inequality report 2022*. World Inequality Lab. https://perma.cc/26RA-QHXF

Chen, D., Rojas, M., Samset, B. H., Cobb, K., Diongue Niang, A., Edwards, P., Emori, S., Faria, S. H., Hawkins, E., Hope, P., Huybrechts, P., Meinshausen, M., Mustafa, S. K., Plattner, G.-K., and Tréguier, A.-M. (2021). Framing, context, and methods. In V. Masson-Delmotte, P. Zhai, A. Pirani, S. L. Connors, C. Péan, S. Berger, N. Caud, Y. Chen, L. Goldfarb, M. I. Gomis, M. Huang, K. Leitzell, E. Lonnoy, J. B. R. Matthews, T. K. Maycock, T. Waterfield, O. Yelekçi, R. Yu, and B. Zhou (Eds.). (2021). *Climate Change 2021: The physical science basis. Contribution of Working*

Group I to the Sixth Assessment Report of the Intergovernmental Panel on Climate Change (pp. 147–286). Cambridge University Press. https://doi.org/10.1017/9781009157896.003

Erman, A., de Vries Robbé, S.A., Thies, S.F., Kabir, K., and Maruo, M. (2021). *Gender Dimensions of Disaster Risk and Resilience*. International Bank for Reconstruction and Development, and The World Bank. http://documents.worldbank.org/curated/en/926731614372544454

Freedom House. (2023). *Freedom in the World 2023: Marking 50 years in the struggle for democracy*. https://perma.cc/8VUN-C57Z

Harrod, R.F. (1939). An essay in dynamic theory. *The Economic Journal, 49*(193), 14–33. https://doi.org/10.2307/2225181

Herre, B. (2022). *The world has recently become less democratic*. Our World in Data. https://perma.cc/B3F9-R6CB

Hirschman, A. (1958). *The Strategy of Economic Development*. Yale University Press.

Intergovernmental Panel on Climate Change (IPCC). (1990). *Climate Change: The IPCC scientific assessment*. [J.T. Houghton, G.K. Jenkins, J.J. Ephraums (Eds.)]. Cambridge University Press. https://perma.cc/539E-AJ7E

Intergovernmental Panel on Climate Change. (1996). *Climate Change 1995: The science of climate change (Contribution of Working Group I to the Second Assessment Report of the Intergovernmental Panel on Climate Change)*. [J.T. Houghton, L.G. Meira Filho, B.A. Callander, N. Harris, A. Kattenberg, and K. Maskell (Eds.)]. Cambridge University Press. https://perma.cc/F2WK-XP3U

Intergovernmental Panel on Climate Change. (2001). *Climate Change 2001: Impacts, adaptation, and vulnerability (Contribution of Working Group II to the Third Assessment Report of the Intergovernmental Panel on Climate Change)*. [J.J. McCarthy, O.F. Canziani, N.L., Leary, D.J. Dokken, and K.S. White (Eds.)]. Cambridge University Press. https://perma.cc/52MJ-PKG9

Intergovernmental Panel on Climate Change. (2018). *Global Warming of 1.5°C: An IPCC special report on the impacts of global warming of 1.5°C above pre-industrial levels and related global greenhouse gas emission pathways, in the context of strengthening the global response to the threat of climate change, sustainable development, and efforts to eradicate poverty*. [V., Masson-Delmotte, P. Zhai, H.-O. Pörtner, D. Roberts, J. Skea, P.R. Shukla, A. Pirani, W. Moufouma-Okia, C. Péan, R. Pidcock, S. Connors, J.B.R. Matthews, Y. Chen, X. Zhou, M.I. Gomis, E. Lonnoy,

T. Maycock, M. Tignor, and T. Waterfield (Eds.)]. Cambridge University Press. https://doi.org/10.1017/9781009157940

International Labour Organization. (2024). *Asia and the Pacific Brief: Global employment trends for youth 2024.* https://www.ilo.org/publications /employment-trends-youth-asia-and-pacific

Kaltenborn, M., Krajewski, M., and Kuhn, H. (Eds.). (2020). *Sustainable Development Goals and Human Rights.* Springer Nature. https://doi.org /10.1007/978-3-030-30469-0

Kuznets, S. (1934). *National Income, 1929–1932.* US Bureau of Foreign and Domestic Commerce, Seventy-Third Congress. Government Printing Office. https://perma.cc/D5JL-QJLV

Lewis, W. A. (1954). Economic development with unlimited supplies of labour. *The Manchester School, 22*(2), 139–191. https://doi.org/10.1111 /j.1467-9957.1954.tb00021.x

Lewis, W. A. (1955). *The theory of economic growth.* Routledge.

Meadows, D. H., Meadows, D. L., Randers, J., and Behrens III, W. W. (1972). *The limits to growth: A report for the Club of Rome's project on the predicament of mankind.* Universe Books.

New Climate Economy. (2014). *Better growth, better climate: The New Climate Economy Report.* World Resources Institute. https://perma.cc/5EFZ -8AZF

New Climate Economy. (2015). *Seizing the global opportunity: Partnerships for better growth and a better climate.* World Resources Institute. https:// perma.cc/2FAV-CDHF

New Climate Economy. (2016). *The sustainable infrastructure imperative: Financing for better growth and development.* World Resources Institute. https://perma.cc/LNR3-7L23

New Climate Economy. (2018). *Unlocking the inclusive growth story of the 21st century: Accelerating climate action in urgent times.* World Resources Institute. https://perma.cc/NH8T-PQW3

New Climate Economy. (2020). *A new economy for a new era: Elements for building a more efficient and resilient economy in Brazil.* World Resources Institute. https://perma.cc/QHS9-C3C9

Nurkse, R. (1953). *Problems of capital formation in underdeveloped countries.* Oxford University Press.

O'Donnell, J. (2024, 8 May). Google DeepMind's new AlphaFold can model a much larger slice of biological life. *MIT Technology Review*. https://perma.cc/9488-ZC5R

Organisation for Economic Co-operation and Development (OECD). (2023). *Artificial Intelligence in Science: Challenges, opportunities and the future of research*. OECD Publishing. https://doi.org/10.1787/a8d820bd-en

Our World in Data. (2023). *Gender gap in primary, secondary, and tertiary education, world.* https://ourworldindata.org/grapher/gender-gap-education-levels

Romer, P. M. (1986). Increasing returns and long-run growth. *Journal of Political Economy, 94*(5), 1002–1037. https://doi.org/10.1086/261420

Roser, M. (2019). *Which countries achieved economic growth? And why does it matter?* Our World in Data. https://perma.cc/8FEY-RUNB

Royal Swedish Academy of Sciences. (2024, 2 January). *Press release.* Nobel Prize Outreach AB. https://perma.cc/3GVK-FDHT

Schumpeter, J. A. (1942). *The theory of competitive price.* Harvard University Press.

Sen, A. (1999). *Development as freedom.* Oxford University Press.

Solow, R. (1956). A contribution to the theory of economic growth. *The Quarterly Journal of Economics, 70*(1), 65–94. https://doi.org/10.2307/1884513

Stern, N. (2006). *The economics of climate change: the Stern review.* HM Treasury. https://webarchive.nationalarchives.gov.uk/ukgwa/20100407172811/https:/www.hm-treasury.gov.uk/stern_review_report.htm

Stern, N. (2007). *The economics of climate change: the Stern review.* Cambridge University Press.

Stern, N., and Romani, M. (2023). *The global growth story of the 21st century: Driven by investment and innovation in green technologies and artificial intelligence.* Grantham Research Institute on Climate Change and the Environment, London School of Economics and Political Science, and Systemiq. https://perma.cc/5BA7-6Y4Q

Stern, N., and Romani, M. (2025). *What is AI's role in the climate transition and how can it drive growth?* World Economic Forum. https://www.weforum.org/stories/2025/01/artificial-intelligence-climate-transition-drive-growth/

Stern, T. (2024). *Landing the Paris climate agreement: how it happened, why it matters, and what comes next.* MIT Press.

Systemiq. (2020). *The Paris Effect: How the Climate Agreement is reshaping the global economy.* https://perma.cc/A953-7VJZ

United Nations (UN). (2022a). *Life expectancy at birth – alls, period, estimates* [Dataset]. World Population Prospects 2022. Processed by Our World in Data. https://ourworldindata.org/grapher/life-expectancy-at -birth-including-the-un-projections

United Nations Climate Change. (2015a, 12 December). *Paris Agreement is adopted by consensus at COP21* [Photograph]. Flickr. https://www.flickr .com/photos/unfccc/23692333176/

United Nations Climate Change. (2015b, 12 December). *Item M25/0210 – COP21 COP Closing plenary 20151212 1930-2230 Floor* [Video]. Digital Archive. https://archive.unfccc.int/cop21-cop-closing-plenary-20151212- 1930-2230-floor

United Nations Convention to Combat Desertification (UNCCD). (2024). *Global issues: Decolonization.* United Nations. https://perma.cc/M4EL -6GCW

United Nations Department of Economic and Social Affairs (UNDESA). (2015). *Youth population trends and sustainable development.* United Nations. https://perma.cc/RZ2C-LZYU

United Nations Department of Economic and Social Affairs. (2024). *World population prospects 2024.* United Nations. https://perma.cc/EH56 -UFXN

United Nations Development Programme (UNDP). (1990). *Human Development Report 1990.* United Nations. https://perma.cc/3FR9-Q6TU

United Nations Economic Commission for Africa (UNECA). (2024, 9 January). *As Africa's population crosses 1.5 billion, the demographic window is opening; Getting the dividend requires more time and stronger effort.* https://perma.cc/HH42-QWGH

United Nations Environment Programme (UNEP). (2023). *Emissions gap report 2023: Broken record – Temperatures hit new highs, yet world fails to cut emissions (again).* https://doi.org/10.59117/20.500.11822/43922

United Nations Framework Convention on Climate Change (UNFCCC). (1992). *United Nations Framework Convention on Climate Change.* United Nations. https://unfccc.int/resource/docs/convkp/conveng.pdf

United Nations Framework Convention on Climate Change. (2015). *Paris Agreement.* United Nations. https://unfccc.int/sites/default/files/english _paris_agreement.pdf

United Nations General Assembly. (2015). *Transforming our world: The 2030 agenda for sustainable development* (A/RES/70/1). https://perma.cc /GVQ8-KZWS

United Nations Women (UN Women). (2021). *Beyond vulnerability to gender equality and women's empowerment and leadership in disaster risk reduction: Critical actions for the United Nations system.* UN Women, United Nations Population Fund, and United Nations Office for Disaster Risk Reduction. https://perma.cc/TM5X-DX5E

United Nations. (2023a). *Sustainable Development Goals.* https://www.un.org /sustainabledevelopment

United Nations. (2023b). *The State of Food Security and Nutrition in the World.* Food and Agriculture Organization, International Fund for Agricultural Development, United Nations Children's Fund, World Food Programme, World Health Organization. https://doi.org/10.4060/cc3017en

United Nations. (2023c). *Sustainable Development Goals Report Special Edition: Towards a Rescue Plan for People and Planet.* https://perma.cc/24BU -TLEJ

United Nations. (2024a). *Life expectancy at birth – Alls, period, estimates* [Dataset]. World Population Prospects 2024. Processed by Our World in Data. https://ourworldindata.org/grapher/life-expectancy-at-birth-in-cluding-the-un-projections

United Nations. (2024b). *Population – Sex: all – Age: all – Variant: estimates* [Dataset]. World Population Prospects 2024. Processed by Our World in Data. https://ourworldindata.org/grapher/population-with-un -projections

Vlasceanu, M., Doell, K. C., Bak-Coleman, J. B., Todorova, B., Berke-bile-Weinberg, M. M., Grayson, S. J., Patel, Y., Goldwert, D., Pei, Y., Chakroff, A., Pronizius, E., van den Broek, K.L., Vlasceanu, D., Constantino, S., Morais, M.J., Schumann, P., Rathje, S., Fang, K., Aglioti, S.M., and Van Bavel, J.J. (2024). Addressing climate change with behavioral science: A global intervention tournament in 63 countries. *Science Advances, 10*(6). https://doi.org/10.1126/sciadv.adj5778

World Bank. (2022). *Correcting Course: Poverty and shared prosperity.* https://hdl.handle.net/10986/37739

World Bank. (2024a). *Lower secondary completion rate, total (% of relevant age group)* [Data]. https://perma.cc/G64C-9BL2

World Bank. (2024b, October 15). *Poverty overview.* https://perma.cc/2ETC -SA3Y

World Bank. (2024c). *Digital progress and trends report 2023.* https://doi.org /10.1596/978-1-4648-2049-6

World Bank Poverty and Inequality Platform. (2023). *Total of population living in extreme poverty by world region* [Dataset]. World Bank Poverty and Inequality Platform. https://ourworldindata.org/grapher/total -population-living-in-extreme-poverty-by-world-region

World Economic Forum (WEF). (2024). *The Global Risks Report 2024.* https://www3.weforum.org/docs/WEF_The_Global_Risks_Report_2024 .pdf

Yonzani, N., Friedman, J., Hill, R., Mitchell Jolliffe, D., Lakner, C., and Gerszon Mahler, D. (2022). *Estimates of global poverty from WWII to the fall of the Berlin Wall.* World Bank Blogs. https://perma.cc/58VS-YS98

Zenghelis, D., Serin, E., Stern, N. H., Valero, A., Van Reenen, J., and Ward, B. (2024). *Boosting growth and productivity in the United Kingdom through investments in the sustainable economy.* Grantham Research Institute on Climate Change and the Environment, London School of Economics and Political Science. https://perma.cc/X5B5-LPRX

Chapter 2

Agence France-Presse. (2024, November 16). More than 650,000 people flee as Super Typhoon Man-yi hits Philippines. *The Guardian*. https://perma.cc/R2AM-YRZN

Aivalioti, S. (2015). *Electricity sector adaptation to heat waves* (Columbia Public Law Research Paper No. 14-439). Columbia Law School, Sabin Center for Climate Change Law. https://ssrn.com/abstract=2563037

Aschwanden, A., Fahnestock, M. A., Truffer, M., Brinkerhoff, D. J., Hock, R., Khroulev, C., Mottram, R., and Khan, S. A. (2019). Contribution of the Greenland Ice Sheet to sea level over the next millennium. *Science Advances*, 5(6). https://doi.org/10.1126/sciadv.aav9396

Bayelsa State Oil and Environment Commission. (2023). *The human and environmental cost of Big Oil in Bayelsa, Nigeria*. https://perma.cc/HT8J-V98Y

Banerjee, N. (2015). Exxon's Oil Industry Peers Knew About Climate Dangers in the 1970s, Too. *Inside Climate News*. https://perma.cc/UZ3R-LJXJ

Birkmann, J., Liwenga, E., Pandey, R., Boyd, E., Djalante, R., Gemenne, F., Leal Filho, W., Pinho, P. F., Stringer, L., and Wrathall, D. (2022). Poverty, livelihoods, and sustainable development. In H.-O. Pörtner, D. C. Roberts, M. Tignor, E. S. Poloczanska, K. Mintenbeck, A. Alegría, M. Craig, S. Langsdorf, S. Löschke, V. Möller, A. Okem, and B. Rama (Eds.), *Climate change 2022: Impacts, adaptation, and vulnerability: Contribution of Working Group II to the Sixth Assessment Report of the Intergovernmental Panel on Climate Change* (pp. 1171–1274). Cambridge University Press. https://doi.org/10.1017/9781009325844.010

Brown, O. (2008). *Migration and climate change*. International Organization for Migration. https://perma.cc/N2L5-A7WP

Burkett, V. R., Wilcox, D. A., Stottlemyer, R., Barrow, W., Fagre, D., Baron, J., Price, J., Nielsen, J. L., Allen, C. D., Peterson, D. L., Ruggerone, G., and Doyle, T. (2005). Nonlinear dynamics in ecosystem response to climatic change: Case studies and policy implications. *Ecological Complexity*, 2(4), 357–394. https://doi.org/10.1016/j.ecocom.2005.04.010

Canadell, J. G., Monteiro, P. M. S., Costa, M. H., Cotrim da Cunha, L., Cox, P. M., Eliseev, A. V., Henson, S., Ishii, M., Jaccard, S., Koven, C., Lohila, A., Patra, P. K., Piao, S., Rogelj, J., Syampungani, S., Zaehle, S., and Zickfeld, K. (2021). Global carbon and other biogeochemical cycles and feedbacks. In V. Masson-Delmotte, P. Zhai, A. Pirani, S. L. Connors, C. Péan, S. Berger, N. Caud, Y. Chen, L. Goldfarb, M. I. Gomis, M. Huang, K. Leitzell, E. Lonnoy, J. B. R. Matthews, T. K. Maycock, T. Waterfield, O. Yelekçi, R. Yu, and B. Zhou (Eds.), *Climate change 2021: The physical science basis* (pp. 673–816). Cambridge University Press. https://doi.org/10.1017/9781009157896.007

Caretta, M. A., Mukherji, A., Arfanuzzaman, M., Betts, R. A., Gelfan, A., Hirabayashi, Y., Lissner, T. K., Liu, J., Lopez Gunn, E., Morgan, R., Mwanga, S., and Supratid, S. (2022). Water. In H.-O. Pörtner, D. C. Roberts, M. Tignor, E. S. Poloczanska, K. Mintenbeck, A. Alegría, M. Craig, S. Langsdorf, S. Löschke, V. Möller, A. Okem, and B. Rama (Eds.), *Climate Change 2022: Impacts, adaptation and vulnerability. Contribution of Working Group II to the Sixth Assessment Report of the Intergovernmental Panel on Climate Change* (pp. 551–712). Cambridge University Press. https://doi.org/10.1017/9781009325844.006

Carleton, T. A., and Hsiang, S. M. (2016). Social and economic impacts of climate. *Science, 353*(6304). https://doi.org/10.1126/science.aad9837

Castano Isaza, J., Lee, S. M., Dani, S. S., Beck, M.W., Narayan, S., Losada, I.J., Espejo Hermosa, A., Ortega, S.T., Herrero, S.A., Mandal, A., Smith, R.A., Edwards, T., Kinlocke, R., Mitchell, S., Webber, M., Trench, C. Francis, P., and Spence, A. (2019). *Forces of Nature: Assessment and Economic Valuation of Coastal Protection Services Provided by Mangroves in Jamaica.* World Bank Group. http://documents.worldbank.org/curated/en/357921613108097096

Chersich, M.F., Pham, M.D., Areal, A., Haghighi, M.M., Manyuchi, A., Swift, C. P., Wernecke, B., Robinson, M., Hetem, R., Boeckmann, M., and Hajat, S. (2020). Associations between high temperatures in pregnancy and risk of preterm birth, low birth weight, and stillbirths: systematic review and meta-analysis. *The BMJ, 371.* https://doi.org/10.1136/bmj.m3811

Childs, M. L., Lyberger, K., Harris, M., Burke, M., and Mordecai, E. A. (2025). *Climate warming is expanding dengue burden in the Americas and Asia.* https://doi.org/10.1101/2024.01.08.24301015

Cissé, G., McLeman, R., Adams, H., Aldunce, P., Bowen, K., Campbell-Lendrum, D., Clayton, S., Ebi, K. L., Hess, J., Huang, C., Liu, Q., McGregor, G., Semenza, J., and Tirado, M. C. (2022). Health, wellbeing, and the changing structure of communities. In H.-O. Pörtner, D. C. Roberts, M. Tignor, E. S. Poloczanska, K. Mintenbeck, A. Alegría, M. Craig, S. Langsdorf, S. Löschke, V. Möller, A. Okem, and B. Rama (Eds.), *Climate Change 2022: Impacts, adaptation and vulnerability. Contribution of Working Group II to the Sixth Assessment Report of the Intergovernmental Panel on Climate Change* (pp. 1041–1170). Cambridge University Press. https://doi.org/10.1017/9781009325844.009

Clement, V. W. C., Rigaud, K. K., de Sherbinin, A., Jones, B. R., Adamo, S., Schewe, J., Sadiq, N., and Shabahat, E. S. (2021). *Groundswell: Acting on internal climate migration (Part II).* World Bank Group. http://documents.worldbank.org/curated/en/540941631203608570

Climate Action Tracker. (2023, December). *Climate Action Tracker*. C40 Knowledge Hub. https://www.c40knowledgehub.org/s/article/Climate -Action-Tracker?language=en_US

Climate Change Tracker. (2024). *Current remaining carbon budget and trajectory*. https://perma.cc/W8BX-B7UK

Climate Overshoot Commission. (2023). *Reducing the risks of climate overshoot*. https://perma.cc/NVL7-R5ZH

Cooley, S. R., Klinsky, S., Morrow, D., and Satterfield, T. (2023). Sociotechnical considerations about ocean carbon dioxide removal. *Annual Review of Marine Science, 15*, 1–21. https://doi.org/10.1146/annurev-marine -032122-113850

Dasgupta, P. (2021). *The economics of biodiversity: The Dasgupta review*. HM Treasury. https://perma.cc/8UQV-KMF3

Eckstein, D., Künzel, V., and Schäfer, L. (2021). *Global Climate Risk Index 2021: Who Suffers Most from Extreme Weather Events? Weather-Related Loss Events I 2019 and 2000–2019* (Briefing Paper). GermanWatch, and MunichRe. https://perma.cc/3KJP-AVCN

European Environment Agency (EEA). (2024). *Urban adaptation in Europe: what works? Implementing climate action in European Cities* (EEA Report 14/2023). https://www.eea.europa.eu/en/analysis/publications/urban -adaptation-in-europe-what-works

Evans, S. and Viisainen, V. (2023). *Revealed: How colonial rule radically shifts historical responsibility for climate change*. Carbon Brief. https://perma.cc /5ED4-YTJC

Evison, W., Low, L.P., and O'Brien, D. (2023). *Managing nature risks: From understanding to action*. PwC Strategy and Business. https://perma.cc /E9QN-W6N5

Fowle, A. (2024, 2 September). Weather tracker: Extreme heat hits Brazil, fuelling risk of wildfires. *The Guardian*. https://perma.cc/2HAD-7EAD

Forster, P. M., Smith, C., Walsh, T., Lamb, W. F., Lamboll, R., Cassou, C., Hauser, M., Hausfather, Z., Lee, J.-Y., Palmer, M. D., von Schuckmann, K., Slangen, A. B. A., Szopa, S., Trewin, B., Yun, J., Gillett, N. P., Jenkins, S., Matthews, H. D., Raghavan, K., ... Zhai, P. (2025). Indicators of Global Climate Change 2024: annual update of key indicators of the state of the climate system and human influence. *Earth System Science Data, 17*(6), 2641–2680. https://doi.org/10.5194/essd-17-2641-2025

Gallo, E., Quijal-Zamorano, M., Méndez Turrubiates, R.F., Tonne, C., Basagaña, X, Achebak, H., and Ballester, J. (2024). Heat-related mortality in Europe during 2023 and the role of adaptation in protecting health. *Nature Medicine, 30*, 3101–3105. https://doi.org/10.1038/s41591-024-03186-1

Gautier, A. (2023). *How and when did the Greenland Ice Sheet form?* National Snow and Ice Data Center. https://nsidc.org/learn/ask-scientist/how-when-greenland-ice-formed?utm

Gleick, P. H. and Cooley, H. (2021). Fresh water scarcity. *Annual Review of Environment and Resources, 46*(1), 319–348. https://doi.org/10.1146/annurev-environ-012220-101319

Global Commission on Adaptation. (2019). *Adapt now: A global call for leadership on climate resilience.* Global Commission on Adaptation, and World Resources Institute. https://perma.cc/D5VX-CJPD

Hammill, A. and McGray, H. (2018). *Is it adaptation or development?* International Institute for Sustainable Development. https://perma.cc/49FS-49TN

Heal, G. and Park, J. (2015). *Goldilocks economies? Temperatures stress and the direct impacts of climate change* (Working Paper 21119). National Bureau of Economic Research. http://www.nber.org/papers/w21119

Hendrix, C.S., Koubi, V., Selby, J., Siddiqi, A., and von Uexkull, N. (2023). Climate change and conflict. *Nature Reviews Earth and Environment, 4*, 144–148. https://doi.org/10.1038/s43017-022-00382-w

Henrico, I., and Doboš, B. (2024). Shifting sands: the geopolitical impact of climate change on Africa's resource conflicts. *South African Geographical Journal*, 1–27. https://doi.org/10.1080/03736245.2024.2441116

Hoegh-Guldberg, O., Jacob, D., Taylor, M., Guillén Bolaños, T., Bindi, M., Brown, S., Camilloni, I.A., Diedhiou, A., Djalante, R., Ebi, K., Engelbrecht, F., Guioy, J., Hijoka, Y., Mehrotra, S., Hope, C.W., Payne, A.J., Pörtner, H.O., Seneviratne, S.I., Thomas, A., and Zhou, G. (2019). The human imperative of stabilizing global climate change at 1.5 °C. *Science, 365*(6459). https://www.doi.org/10.1126/science.aaw6974

Institute for Economics & Peace. (2020). *Over one billion people at threat of being displaced by 2050 due to environmental change, conflict and civil unrest.* https://perma.cc/NQ9T-RCY3

Intergovernmental Panel on Climate Change (IPCC). (1992). *Climate Change: The 1990 and 1992 IPCC Assessments.* WMO, and UNEP. https://perma.cc/69ED-F2FN

Intergovernmental Panel on Climate Change. (2001). *Climate Change 2001: Impacts, adaptation, and vulnerability (Contribution of Working Group II to the Third Assessment Report of the Intergovernmental Panel on Climate Change).* [JJ. McCarthy, O.F. Canziani, N.A. Leary, D.J. Dokken, and K.S. White (Eds.)]. Cambridge University Press. https://perma.cc/52MJ -PKG9

Intergovernmental Panel on Climate Change. (2018). Summary for policy-makers. In V. Masson-Delmotte, P. Zhai, H.-O. Pörtner, D. Roberts, J. Skea, P. R. Shukla, A. Pirani, W. Moufouma-Okia, C. Péan, R. Pidcock, S. Connors, J. B. R. Matthews, Y. Chen, X. Zhou, M. I. Gomis, E. Lonnoy, T. Maycock, M. Tignor, and T. Waterfield (Eds.), *Global warming of 1.5°C: An IPCC special report on the impacts of global warming of 1.5°C above pre-industrial levels and related global greenhouse gas emission pathways, in the context of strengthening the global response to the threat of climate change, sustainable development, and efforts to eradicate poverty* (pp. 3–24). Cambridge University Press. https://doi.org/10.1017/9781009157940.001

Intergovernmental Panel on Climate Change. (2019). Summary for policy-makers. In H.O. Pörtner, D. C. Roberts, V. Masson-Delmotte, P. Zhai, M. Tignor, E. Poloczanska, K. Mintenbeck, A. Alegría, M. Nicolai, A. Okem, J. Petzold, B. Rama, and N. M. Weyer (Eds.). *IPCC special report on the ocean and cryosphere in a changing climate* (pp. 3–35). Cambridge University Press. https://doi.org/10.1017/9781009157964.001

Intergovernmental Panel on Climate Change. (2021). *Climate Change 2021: The physical science basis. Contribution of Working Group I to the Sixth Assessment Report of the Intergovernmental Panel on Climate Change.* [V. Masson-Del-motte, P. Zhai, A. Pirani, S.L. Connors, C. Péan, S. Berger, N. Caud, Y. Chen, L. Goldfarb, M.I. Gomis, M. Huang, K. Leitzell, E. Lonnoy, J.B.R. Matthews, T.K. Maycock, T. Waterfield, O. Yelekçi, R. Yu, and B. Zhou (Eds.)]. Cam-bridge University Press. https://doi.org/10.1017/9781009157896

Intergovernmental Panel on Climate Change. (2022a). *Climate Change 2022: Impacts, adaptation and vulnerability. Contribution of Working Group II to the Sixth Assessment Report of the Intergovernmental Panel on Climate Change.* [H.-O. Pörtner, D.C. Roberts, M. Tignor, E.S. Poloczanska, K. Mintenbeck, A. Alegría, M. Craig, S. Langsdorf, S. Löschke, V. Möller, A. Okem, and B. Rama (Eds.)]. Cambridge University Press. https://doi.org /10.1017/9781009325844

Intergovernmental Panel on Climate Change. (2022b). Annex II: Glossary. In V. Möller, R. van Diemen, J. B. R. Matthews, C. Méndez, S. Semenov, J. S. Fuglestvedt, and A. Reisinger (Eds.), *Climate Change 2022: Impacts, adaptation and vulnerability. Contribution of Working Group II to the Sixth Assessment Report of the Intergovernmental Panel on Climate Change* (pp. 2897–2930). Cambridge University Press. https://doi.org/10 .1017/9781009325844.029

Intergovernmental Panel on Climate Change. (2023). Summary for policymakers. In H. Lee and J. Romero (Eds.). *Climate Change 2023: Synthesis report: Contribution of Working Groups I, II and III to the Sixth Assessment Report of the Intergovernmental Panel on Climate Change* (pp. 1–34). Cambridge University Press. https://doi.org/10.59327/IPCC/AR6-9789291691647.001

Intergovernmental Science-Policy Platform on Biodiversity and Ecosystem Services (IPBES). (2019). *Global assessment report on biodiversity and ecosystem services.* [E. Brondizio, S. Diaz, J. Settele, and H. T. Ngo (Eds.)]. https://doi.org/10.5281/zenodo.3831673

International Cryosphere Climate Initiative (2021). *State of the Cryosphere 2021. A Needed Decade of Urgent Action.* https://drive.google.com/file /d/1DlMG64Gs2yErkI9zS1Aiw6Yj_OpdlP4x/view

International Dalit Solidarity Network (IDSN). (2013). *Equality in Aid: Addressing Caste Discrimination in Humanitarian Response.* https:// reliefweb.int/report/world/equality-aid-addressing-caste-discrimination -humanitarian-response

International Energy Agency. (2022). *Global Methane Tracker 2022.* https:// perma.cc/X7KN-YASJ

International Labour Organization (ILO). (2024). *Heat at work: Implications for safety and health.* https://www.ilo.org/sites/default/files/2024-07/ILO _OSH_Heatstress-R16.pdf

Jang, W. S., Neff, J. C., Doro, Y. I. L., and Herrick, J. E. (2020). The Hidden Costs of Land Degradation in US Maize Agriculture. *Earth's Future,* 9(2). https://doi.org/10.1029/2020EF001641

Kang, S., and Eltahir, E. A. B. (2018). North China Plain threatened by deadly heatwaves due to climate change and irrigation. *Nature Communications, 9.* https://doi.org/10.1038/s41467-018-05252-y

Kemp, L., Xu, C., Depledge, J., Ebi, K. L., Gibbins, G., Kohler, T. A., Rockström, J., Scheffer, M., Schellnhuber, H.J., Steffen, W., and Lenton, T. M. (2022). Climate Endgame: Exploring catastrophic climate change scenarios. *Proceedings of the National Academy of Sciences, 119*(34). https://doi .org/10.1073/pnas.2108146119

Lenton, D. I., Armstrong McKay, S., Loriani, J. F., Abrams, S. J., Lade, J. F., Donges, M., Milkoreit, T., Powell, S. R., Smith, C., Zimm, J. E., Buxton, E., Bailey, L., Laybourn, A., Ghadiali, J. G., and Dyke, J. G. (Eds.). (2023). *The Global Tipping Points Report 2023.* University of Exeter. https:// report-2023.global-tipping-points.org/

Lenton, T. M., Rockström, J., Gaffney, O., Rahmstorf, S., Richardson, K., Steffen, W., and Schellnhuber, H. J. (2019). Climate tipping points—Too risky to bet against. *Nature, 575*(7784), 592–595. https://doi.org/10.1038 /d41586-019-03595-0

Lenton, T.M., Xu, C., Abrams, J.F., Ghadiali, A., Loriani, S., Sakschewski, B., Zimm, C., Ebi, K., Dunn, R., Svenning, J., and Scheffer, M. (2023). Quantifying the human cost of global warming. *Nature Sustainability, 6,* 1237–1247. https://doi.org/10.1038/s41893-023-01132-6

Madakumbura, G., Thackeray, C., Hall, A., Williams, P., Norris, J., and Sukhdeo, R. (2025, 13 January 13). *Climate change a factor in unprecedented LA fires.* University of California Los Angeles. https://perma.cc/V5VD-BCKT

Marcos, M., Amores, A., Agulles, M., Robson, J., and Feng, X. (2025). Global warming drives a threefold increase in persistence and 1 °C rise in intensity of marine heatwaves, *Proceedings of the National Academy of Sciences of the United States of America, 122*(16). https://doi.org/10.1073/pnas.2413505122

Martin, P. E., and Barker, E. F. (1932). The infrared absorption spectrum of carbon dioxide. *Physical Review, 41*(3), 291–303. https://doi.org/10.1103/PhysRev.41.291

Massachusetts Institute of Technology (MIT). (2024a). *Radiative Forcing.* MIT Climate Portal. https://climate.mit.edu/explainers/radiative-forcing

Massachusetts Institute of Technology. (2024b, 25 September). *Is methane release from the Arctic unstoppable?* MIT Climate Portal. https://climate.mit.edu/ask-mit/methane-release-arctic-unstoppable

Massachusetts Institute of Technology. (2024c, 7 November). *What would happen if the Atlantic Meridional Overturning Circulation (AMOC) collapses? How likely is it?* MIT Climate Portal. https://climate.mit.edu/ask-mit/what-would-happen-if-atlantic-meridional-overturning-circulation-amoc-collapses-how-likely

Mehryar, S. (2022, June 16). *What is the difference between climate change adaptation and resilience?* Grantham Research Institute on Climate Change and the Environment, London School of Economics and Political Sciences. https://perma.cc/6LXT-HV94

Mordecai, E. A., Cohen, J. M., Evans, M. V., Gudapati, P., Johnson, L. R., Lippi, C. A., Miazgowicz, K., Murdock, C. C., Rohr, J. R., Ryan, S. J., Savage, V., and Shocket, M. S. (2022). Detecting the impact of temperature on transmission of Zika, dengue, and chikungunya using mechanistic models. *PLOS Neglected Tropical Diseases, 11*(4). https://doi.org/10.1371/journal.pntd.0005568

Moseman, A. (2021, 10 November). *How are gases in the atmosphere analyzed and measured?* MIT Climate Portal. https://climate.mit.edu/ask-mit/how-are-gases-atmosphere-analyzed-and-measured

Moseman, A. (2024, 9 January). *What are the best- and worst-case scenarios for sea level rise?* MIT Climate Portal. https://climate.mit.edu/ask-mit/what-are-best-and-worst-case-scenarios-sea-level-rise

National Aeronautics and Space Administration (NASA). (2024). *Carbon dioxide*. NASA Global Climate Change. https://perma.cc/8YXF-SLNC

The Nature Conservancy (TNC). (2024). *The Importance of Mangroves.* https://perma.cc/JR9T-K7NF

Oreskes, N. and Conway, E. M. (2010). *Merchants of Doubt: How a handful of scientists obscured the truth on issues from tobacco smoke to global warming.* Bloomsbury Press.

Our World in Data. (2024). *CO₂ and Greenhouse Gas Emissions Data Explorer.* https://ourworldindata.org/explorers

Oxfam. (2014). *We no longer share the land: Agricultural change, land, and violence in Darfur* (Oxfam Briefing Paper 184). https://oxfamilibrary .openrepository.com

Park, R. J., Behrer, A. P., and Goodman, J. (2021). Learning is inhibited by heat exposure, both internationally and within the United States. *Nature Human Behaviour, 5*(1), 19–27. https://doi.org/10.1038/s41562-020-00959-9

Parsons, L.A., Masuda, Y.J., Kroeger, T., Shindell, D., Wolff, N.H., and Spector, J.T. (2022). Global labor loss due to humid heat exposure underestimated for outdoor workers. *Environmental Research Letters, 7*(1). https:// doi.org/10.1088/1748-9326/ac3dae

Pörtner, D. C. Roberts, M. Tignor, E. S. Poloczanska, K. Mintenbeck, A. Alegría, M. Craig, S. Langsdorf, S. Löschke, V. Möller, A. Okem, and B. Rama (Eds.), *Climate Change 2022: Impacts, adaptation, and vulnerability: Contribution of Working Group II to the Sixth Assessment Report of the Intergovernmental Panel on Climate Change* (pp. 1171–1274). Cambridge University Press. https://doi.org/10.1017/9781009325844.010

Rahiem, M. D., Rahim, H., and Ersing, R. (2021). Why did so many women die in the 2004 Aceh tsunami? Child survivor accounts of the disaster. *International Journal of Disaster Risk Reduction, 55.* https://doi.org/10 .1016/j.ijdrr.2021.102069

Rahmstorf, S. (2024). Is the Atlantic overturning circulation approaching a tipping point? *Oceanography, 37*(3), 16–29. https://doi.org/10.5670 /oceanog.2024.501

Reagan, R. (1984, 6 November). *Remarks at a reelection celebration in Los Angeles, California* [Speech transcript]. Ronald Reagan Presidential Library and Museum. https://perma.cc/2YF7-K6T3

Rentschler, J., Salhab, M., and Jafino, B. A. (2022). Flood exposure and poverty in 188 countries. *Nature Communications, 13*(1). https://doi.org/10 .1038/s41467-022-30727-4

Reuters. (2023, 19 April). Panama Canal lowers maximum depth limit on ships due to drought. *Reuters.* https://www.reuters.com/business /environment/panama-canal-lowers-maximum-depth-limit-ships-due -drought-2023-04-19/

Ritchie, P., Rosado, M., and Roser, M. (2024). *Greenhouse gas emissions.* Our World In Data. https://ourworldindata.org/greenhouse-gas-emissions

Science Media Centre (SMC). (2024). *Expert reaction to State of the Climate 2023.* Science Media Centre. https://perma.cc/PAU4-PMPZ

Seppänen, O., Fisk, W.J., and Lei, Q.H. (2006). Effect of temperature on task performance in office environment. *Ernest Orland Lawrence Berkeley National Laboratory.* https://perma.cc/4EAC-B7FU

Sharifi, A., Simangan, D. and Kaneko, S. (2021). Three decades of research on climate change and peace: a bibliometrics analysis. *Sustainable Science, 16,* 1079–1095. https://doi.org/10.1007/s11625-020-00853-3

Shine, K. P. and Perry, G. E. (2023). Radiative forcing due to carbon dioxide decomposed into its component vibrational bands. *Quarterly Journal of the Royal Meteorological Society, 149*(754), 1856–1866. https://doi.org/10 .1002/qj.4485

Swingedouw, D., Bily, A., Esquerdo, C., Borchert, L. F., Sgubin, G., Mignot, J., and Menary, M. (2021). On the risk of abrupt changes in the North Atlantic subpolar gyre in CMIP6 models. *Annals of the New York Academy of Sciences, 1504*(1), 187–201. https://doi.org/10.1111/nyas.14659

Tellman, B., Sullivan, J.A., Kuhn, C. et al. (2021) Satellite imaging reveals increased proportion of population exposed to floods, *Nature, 596,* 80–86. https://doi.org/10.1038/s41586-021-03695-w

United Nations Development Programme (UNDP). (2021). *Sunflowers making women's lives brighter on the coasts.* https://www.undp.org/bangladesh /stories/sunflowers-making-womens-lives-brighter-coasts

United Nations Development Programme. (2024). *Regional Human Development Report. Making Our Future: New directions for human development in Asia and the Pacific.* https://www.undp.org/asia-pacific/publications /making-our-future-new-directions-human-development-asia-and-pacific

United Nations Environment Programme (UNEP). (2023a). *Emission Gap Report: Broken record—Temperatures hit new highs, yet the world fails to cut emissions (again).* https://doi.org/10.59117/20.500.11822/43922

United Nations Environment Programme. (2023b). *Adaptation Gap Report 2023: Underfinanced. Underprepared. Inadequate investment and planning on climate adaptation leaves world exposed.* https://doi.org/10.59117 /20.500.11822/43796

United Nations Environment Programme. (2024a). *As shortages mount, countries hunt for novel sources of water.* https://perma.cc/5QSS-WZAZ

United Nations Environment Programme (2024b). *Emissions Gap Report 2024: No more hot air ... please! With a massive gap between rhetoric and reality, countries draft new climate commitments.* Nairobi. https://doi.org/10.59117/20.500. 11822/46404.

United Nations Framework Convention on Climate Change. (2015). *Paris Agreement.* United Nations. https://unfccc.int/sites/default/files/english_paris_agreement.pdf

United Nations Framework Convention on Climate Change. (2020). *Data for adaptation at different spatial and temporal scales.* https://unfccc.int/sites/default/files/resource/AC%20adaptation%20data%20full.pdf

United Nations Trade and Development (UNCTAD). (2021). *Review of Maritime Transport 2021.* United Nations. https://unctad.org/system/files/official-document/rmt2024_en.pdf

van Westen, R. M., Kliphuis, M., and Dijkstra, H. A. (2024). Physics-based early warning signal shows that AMOC is on tipping course. *Science Advances, 10*(6). https://doi.org/10.1126/sciadv.adk1189

Woetzel, L., Krishnan, M., Pinner, D., Samandari, H., Engel, H., Kampel, C., and von der Leyen, J. (2020, June). *Reduced dividends on natural capital?* McKinsey Global Institute. https://perma.cc/UB3Z-RMFT

Wong, T., Campos, R., Mackres, E., Staedicke, S., and Doust, M. (2024). *What Would Cities Look Like With 3 Degrees C of Warming vs. 1.5? Far More Hazardous and Vastly Unequal.* World Resources Institute. https://perma.cc/5BR5-2XRV

World Bank. (2018). *Riding the Wave: An East Asian Miracle for the 21st Century* (World Bank East Asia and Pacific Regional Report). https://doi.org/10.1596/978-1-4648-1145-6

World Bank. (2021). *The Economics of Large-scale Mangrove Conservation and Restoration in Indonesia.* https://perma.cc/9QJ9-9MWF

World Bank. (2023). *Planting Mangrove Forests Is Paying Off in Indonesia.* https://perma.cc/WXA4-BGBG

World Bank. (2024). *The Cost of Inaction: Quantifying the Impact of Climate Change on Health in Low- and Middle-Income Countries.* http://documents.worldbank.org/curated/en/099111324172540265

World Economic Forum (WEF). (2020). *Nature Risk Rising: Why the Crisis Engulfing Nature Matters for Business and the Economy* (New Nature Economy series). https://perma.cc/53D6-EVD3

World Health Organization (WHO). (2024, 30 May). *Dengue – Global situation.* https://perma.cc/XYP8-7HNV

World Meteorological Organization (WMO). (2021, 31 August). *Weather-related disasters increase over past 50 years, causing more damage but fewer deaths.* https://perma.cc/SJ6A-Q45Y

World Meteorological Organization. (2023, 22 May). *Economic costs of weather-related disasters soar, but early warnings save lives.* https://perma.cc/U34U-RSKB

World Meteorological Organization. (2025). *State of the Climate in Asia 2024.* https://library.wmo.int/records/item/69575-state-of-the-climate-in-asia-2024

World Weather Attribution (WWA). (2024). *Hot, dry and windy conditions that drove devastating Pantanal wildfires 40% more intense due to climate change.* https://perma.cc/X2EW-S9UU

World Weather Attribution. (2025a, 28 January). *Climate change increased the likelihood of wildfire disaster in highly exposed Los Angeles area.* https://perma.cc/V5Z8-S2MH

World Weather Attribution. (2025b, 27 March). *Consecutive extreme heat and flooding events in Argentina highlight the risk of managing increasingly frequent and intense hazards in a warming climate.* https://perma.cc/9N3T-SVW2

World Wildlife Fund (WWF). (2024). *Living Planet Report 2024: A system in peril.* https://perma.cc/PR5G-8CSC

Xinhua. (2025, 7 March). Argentina face new heat wave as hundreds of thousands lose power. *The Global Times.* https://www.globaltimes.cn/page/202503/1329674.shtml

Xu, C., Kohler, T.A., Lenton, T., and Scheffer, Marten. (2020). Future of the human climate niche. *Proceedings of the National Academy of Sciences, 117*(21), 11350–11355. https://doi.org/10.1073/pnas.1910114117

Zhang, B., Wang, S., and Slater, L. (2024). Anthropogenic climate change doubled the frequency of compound drought and heatwaves in low-income regions. *Communications Earth and Environment, 5*(715). https://doi.org/10.1038/s43247-024-01894-7

Zhou, N. (2019). Former Australian fire chiefs say Coalition ignored their advice because of climate change politics. *The Guardian.* https://www.theguardian.com/australia-news/2019/nov/14/former-australian-fire-chiefs-say-coalition-doesnt-like-talking-about-climate-change

Zuo, J., Pullen, S., Palmer, J., Bennetts, H., Chileshe, C., and Ma, T. (2015). Impacts of heat waves and corresponding measures: A review. *Journal of Cleaner Production, 92,* 1–12. https://doi.org/10.1016/j.jclepro.2014.12.078

Chapter 3

Besley, T. and Persson, T. (2023). The political economics of green transitions. *The Quarterly Journal of Economics, 138*(3), 1863–1906. https://doi.org/10.1093/qje/qjad006

Betts, P. (2025). *The Climate Diplomat: A Personal History of the COP Conferences.* Profile Books.

Bhattacharya, A., Songwe, V., Soubeyran, E., and Stern, N. (2023). *A climate finance framework that is fit for purpose: Decisive action to deliver on the Paris Agreement — Summary.* Grantham Research Institute on Climate Change and the Environment, London School of Economics and Political Science. https://perma.cc/ZML3-CQ8Q

Bhattacharya, A., Songwe, V., Soubeyran, E., and Stern, N. (2024). *Raising ambition and accelerating delivery of climate finance.* Grantham Research Institute on Climate Change and the Environment, London School of Economics and Political Science. https://perma.cc/STN9-LW8T

Carrington, D. (2023, 30 September). 'We're not doomed yet': climate scientists Michael Mann on our last chance to save human civilisation. *The Guardian.* https://perma.cc/CAN6-ZHVS

Carson, R. (1962). *Silent Spring.* Houghton Miffin Harcourt Company.

Chancel, L., Piketty, T., Saez, E., and Zucman, G. (Eds.). (2022). *World inequality report 2022.* World Inequality Lab. https://perma.cc/26RA-QHXF

Chichilnisky, G., Hammond, P. J., and Stern, N. (2020). Fundamental utilitarianism and intergenerational equity with extinction discounting. *Social Choice and Welfare, 54*(2/3), 397–427. https://doi.org/10.1007/s00355-019-01236-z

Climate Assembly UK. (2020). *The path to net zero.* House of Commons. https://perma.cc/Y3ET-BEMK

Convention Citoyenne pour le Climat. (2021). *Les Propositions de la Convention Citoyenne pour le Climat.* https://perma.cc/DU72-GSSF

Dasgupta, P. (2021). *The economics of biodiversity: The Dasgupta review.* HM Treasury. https://perma.cc/8UQV-KMF3

Dasgupta, P. and Heal, G. (1979). *Economic theory and exhaustible resources.* Cambridge University Press.

Evans, S. (2021). *Analysis: Which countries are historically responsible for climate change?* Carbon Brief. https://perma.cc/JBU5-A4WJ

Evans, S., and Viisainen, V. (2024). *Analysis: China's emissions have now caused more global warming than EU.* Carbon Brief. https://www.carbonbrief.org/analysis-chinas-emissions-have-now-caused-more-global-warming-than-eu/

Farrell, D.M., Suiter, J., and Harris, C. (2019). Systematizing' constitutional deliberation: The 2016–18 citizens' assembly in Ireland. *Irish Political Studies, 34*(1), 113–123. https://doi.org/10.1080/07907184.2018.1534832

Fengler, W., Gill, I., and Kharas, H. (2023). *Making emissions count in country classifications*. The Brookings Institution. https://perma.cc/Z6WR-6GVL

Fesmire, S. (2020). Pragmatist ethics and climate change. In D. Miller and B. Eggleston (Eds.), *Moral Theory and Climate Change: Ethical perspectives on a warming planet*. Routledge.

Franta, B. (2018a, 1 January). On its 100th birthday in 1959, Edward Teller warned the oil industry about global warming. *The Guardian*. https://perma.cc/PA95-HKRL

Franta, B. (2018b). Early oil industry knowledge of CO2 and global warming. *Nature Climate Change 8*, 1024–1025. https://doi.org/10.1038/s41558-018-0349-9

Global Carbon Project. (2014). *Global Carbon Budget*. https://www.globalcarbonproject.org/carbonbudget/archive/2014/GCP_budget_2014_lowres_v1.02.pdf

Hallegatte, S., Godinho, C., Rentschler, J., Avner, P., Dorband, I. I., Knudsen, C., Lemke, J., and Mealy, P. (2023). *Within reach: Navigating the political economy of decarbonization*. World Bank. http://hdl.handle.net/10986/40601

Intergovernmental Panel on Climate Change. (2014). *Climate Change 2014: Synthesis report. Contribution of Working Groups I, II, and III to the Fifth Assessment Report of the Intergovernmental Panel on Climate Change.* [Core Writing Team, R.K. Pachauri, L.A. Meyer (Eds.)]. IPCC. https://perma.cc/9QA7-BJRJ

Intergovernmental Panel on Climate Change. (2018). Summary for policymakers. In V. Masson-Delmotte, P. Zhai, H.-O. Pörtner, D. Roberts, J. Skea, P. R. Shukla, A. Pirani, W. Moufouma-Okia, C. Péan, R. Pidcock, S. Connors, J. B. R. Matthews, Y. Chen, X. Zhou, M. I. Gomis, E. Lonnoy, T. Maycock, M. Tignor, and T. Waterfield (Eds.), *Global warming of 1.5°C: An IPCC special report on the impacts of global warming of 1.5°C above pre-industrial levels and related global greenhouse gas emission pathways, in the context of strengthening the global response to the threat of climate change, sustainable development, and efforts to eradicate poverty* (pp. 3–24). Cambridge University Press. https://doi.org/10.1017/9781009157940.001

International Energy Agency (IEA), Crippa, M., Guizzardi, D., Pagani, F., Banja, M., Muntean, M., Schaaf, E., Monforti-Ferrario, F., Becker, W., Quadrelli, R., Risquez Martin, A., Taghavi-Moharamli, P., Köykkä, J., Grassi, G., Rossi, S., Melo, J., Oom, D., Branco, A., San-Miguel, J., and Pekar, F. (2024). *GHG emissions of all world countries* (JRC138862). Publications Office of the European Union. https://doi.org/10.2760/4002897

International Energy Agency (IEA). (2022). *Global Methane Tracker 2024*. https://perma.cc/K3RG-KVHD

Jamieson, D. (2007). When Utilitarians Should Be Virtue Theorists. *Utilitas, 19*(2), 160–183. https://doi.org/10.1017/S0953820807002452

Jones, M. W., Peters, G. P., Gasser, T., Andrew, R.M., Schwingshackl, C., Gütschow, J., Houghton, r. A., Friedlingstein, P., Pongratz, J., and Le Quéré, C. (2024). – with major processing by Our World in Data. "Share of global greenhouse gas emissions" [dataset]. Jones et al., "National contributions to climate change 2024.2" [original data]. https://zenodo.org/records/14054503

Kant, I. (1993) [1785]. *Groundwork of the metaphysics of morals* (J.W. Ellington, Trans.; 3rd ed.). Hackett. (Original work published 1785).

Layard, R. (2005). *Happiness: lessons from a new science*. Penguin UK.

Light, A. (2002). Contemporary environmental ethics: From metaethics to public philosophy. *Metaphilosophy, 33*(4), 426–449. https://doi.org/10.1111/1467-9973.00238

Liu, H., Evans, S., Zhang, Z., Song, W., and You, X. (2023). *The Carbon Brief Profile: China*. Carbon Brief. https://perma.cc/7GGD-2GTB

MacFarlane, R. (2025). *Is a river alive?*. Penguin Books.

Marshall, A. (1890). *Principles of economics*. Macmillan and Company.

Miller, K. G., Wright, J. D., Browning, J. V., Kulpecz, A., Kominz, M., Naish, T. R., Cramer, B. S., Rosenthal, Y., Peltier, W. R., and Sosdian, S. (2012). High tide of the warm Pliocene: Implications of global sea level for Antarctic deglaciation. *Geology, 40*(5), 407–410. https://doi.org/10.1130/G32869.1

Mirrlees, J. A. (1963). *Optimum accumulation under uncertainty* (Doctoral dissertation). University of Cambridge.

Nordhaus, W. D. (1991). To slow or not to slow: The economics of the greenhouse effect. *The Economic Journal, 101*(407), 920–937. https://doi.org/10.2307/2233864

Nordhaus, W. D. (2018, December 8). *Climate change: The ultimate challenge for economics*. Nobel Prize. https://perma.cc/M586-HEUV

Nordhaus, W.D. (1992). An Optimal Transition Path for Controlling Greenhouse Gases. *Science, 258*, 1315–1319. https://doi.org/10.1126/science.258.5086.1315

Oreskes, N. and Conway, E. M. (2010). *Merchants of Doubt: How a handful of scientists obscured the truth on issues from tobacco smoke to global warming*. Bloomsbury Press.

Organisation for Economic Co-operation and Development (OECD). (2024). *Climate finance provided and mobilised by developed countries in 2013–2022: Climate finance and the USD 100 billion goal*. OECD Publishing. https://doi.org/10.1787/19150727-en

Ostrom, E. (1990). *Governing the Commons: The evolution of institutions for collective action*. Cambridge University Press.

Ostrom, E. (2000). Collective action and the evolution of social norms. *Journal of Economic Perspectives, 14*(3), 137–158. https://doi.org/10.1257/jep.14.3.137

Palmer, C. (2014). Contested frameworks in environmental ethics. In R. Rozzi, S. Pickett, C. Palmer, J. Armesto, and J. B. Callicott (Eds.), *Linking ecology and ethics for a changing world: Values, philosophy and action* (pp. 191–206). Springer.

Pathak, M., Slade, R., Pichs-Madruga, R., Ürge-Vorsatz, D., Shukla, R., and Skea, J. (2022). *Climate Change 2022 Mitigation of Climate Change: Technical summary*. Intergovernmental Panel on Climate Change. https://doi.org/10.1017/9781009157926.002

Pigou, A.C. (1920). *The Economics of welfare*. Macmillan and Co.

Pope Francis. (2014). God forgives: Nature does not. *L'Osservatore Romano; Weekly Edition in English*. https://perma.cc/XL9J-CKBF

Pope Francis. (2015). *Laudato Si': On care for our common home*. Vatican City: Vatican Press. https://perma.cc/A443-2K5R

Rawls, J. (1971). *A theory of justice*. Harvard University Press.

Robbins, L. (1932). *An essay on the nature and significance of economic science*. Macmillan and Company.

Sands, P. (2025, April 26). Why our planet (and not just its people) should have legal rights. *The Financial Times*. https://www.ft.com/content/9fb7e995-eba8-43ae-ab95-f85cfda0d126

Solow, R. (1991). *Sustainability: An economist's perspective*. The Eighteenth J. Seward Johnson Lecture, Marine Policy Center, Woods Hole Oceanographic Institution.

Stern, N. (2014a). Ethics, equity, and the economics of climate change paper 1: Science and philosophy. *Economics and Philosophy, 30*(3), 397–444. https://doi.org/10.1017/S0266267114000297

Stern, N. (2014b). Ethics, equity, and the economics of climate change paper 2: Economics and politics. *Economics and Philosophy, 30*(3), 445–501. https://doi.org/10.1017/S0266267114000303

Stern, N. (2015). *Why are we waiting? The logic, urgency, and promise of tackling climate change* (1st ed.). The MIT Press.

Stern, N., and Xie, C. (2021). *The 14th Five-Year Plan: peaking China's greenhouse gas emissions and paving the way to carbon neutrality.* Grantham Research Institute on Climate Change and the Environment, London School of Economics and Political Science. https://perma.cc/S55G-TP6D

Stern, N., Stiglitz, J. E., and Taylor, C. (2022). The economics of immense risk, urgent action and radical change: towards new approaches to the economics of climate change. *Journal of Economic Methodology, 29*(3), 181–216. https://doi.org/10.1080/1350178X.2022.2040740

Stevens, H. (2023). The United States has caused the most global warming. When will China pass it? *Washington Post.* https://www.washingtonpost.com/climate-environment/interactive/2023/global-warming-carbon-emissions-china-us/

Stiglitz, J. E. (1982). The rate of discount for benefit—Cost analysis and the Theory of the Second Best. In R.C. Lind, K.J. Arrow, G.R. Corey, P. Dasgupta, A. Sen, T. Staufer, J.E. Stigiltz, and J.A. Stockfisch, (Eds). *Discounting for Time and Risk in Energy Policy* (pp. 151–204). RFF Press.

United Nations Environment Programme (UNEP). (2023). *Emission Gap Report: Broken record—Temperatures hit new highs, yet the world fails to cut emissions (again).* https://doi.org/10.59117/20.500.11822/43922

United Nations Environment Programme, and Climate and Clean Air Coalition. (2021). *Global Methane Assessment: Benefits and Costs of Mitigating Methane Emissions.* https://perma.cc/6SAD-GSPM

United Nations Framework Convention on Climate Change (UNFCCC). (1992). *United Nations Framework Convention on Climate Change.* United Nations. https://unfccc.int/resource/docs/convkp/conveng.pdf

Verbruggen, A. (2022). The geopolitics of trillion US$ oil and gas rents. *International Journal of Sustainable Energy Planning and Management, 36,* 3–10. https://doi.org/10.54337/ijsepm.7395

The Wall Street Journal. (2019, 16 January). Economists' Statement on Carbon Dividends. *The Wall Street Journal.* https://www.wsj.com/articles/economists-statement-on-carbon-dividends-11547682910

Weitzman, M. L. (2009). On modeling and interpreting the economics of catastrophic climate change. *The Review of Economics and Statistics, 91*(1), 1–19. https://doi.org/10.1162/rest.91.1.1

White, E. (2024, 17 January). China's emissions peak in sight as solar and electric cars boom. *The Financial Times.* https://www.ft.com/content/352c9205-c8d7-46b0-a162-582cb36f241b

Worker, J. and Palmer, N. (2021). *A guide to assessing the political economy of domestic climate change governance* (WRI Working Paper). World Resources Institute. https://doi.org/10.46830/wriwp.18.00047

You, X. (2024, 19 January). What does peak emissions mean for China — and the world? *Nature.* https://doi.org/10.1038/d41586-024-02877-6

Chapter 4

Andre, P., Boneva, T., Chopra, F., and Falk, A. (2024). Globally representative evidence on the actual and perceived support for climate action. *Nature Climate Change, 14*, 253–259. https://doi.org/10.1038/s41558-024-01925-3

Bain & Company. (2023). *Consumers say their environmental concerns are increasing due to extreme weather; study shows they're willing to change behavior, pay 12% more for sustainable products.* https://www.bain.com/about/media-center/press-releases/2023/consumers-say-their-environmental-concerns-are-increasing-due-to-extreme-weather-study-shows-theyre-willing-to-change-behavior-pay-12-more-for-sustainable-products/

Bhattacharya, A., Songwe, V., Soubeyran, E., and Stern, N. (2024). *Raising ambition and accelerating delivery of climate finance.* Grantham Research Institute on Climate Change and the Environment, London School of Economics and Political Science. https://perma.cc/H6D9-MNSX

Bian, L., Dikau, S., Miller, H., Pierfederici, R., Stern, N., and Ward, B. (2024). *China's role in accelerating the global energy transition through green supply chains and trade.* Grantham Research Institute on Climate Change and the Environment, London School of Economics and Political Science. https://perma.cc/7EK5-UZ93

Black, R., Cullen, K., Fay, B., Hale, T., Lang, J., Mahmood, S., and Smith, S.M. (2021). *Taking Stock: A global assessment of net zero targets.* Energy and Climate Intelligence Unit and Oxford Net Zero. https://perma.cc/8FBJ-8LDE

BloombergNEF. (2023). *A Power Grid Long Enough to Reach the Sun Is Key to the Climate Fight.* https://perma.cc/KF83-GLFX

BloombergNEF. (2024a). *Clean Electricity Breaks New Records; Renewables on Track for Another Strong Year. BloombergNEF.* https://about.bnef.com/insights/clean-energy/clean-electricity-breaks-new-records-renewables-on-track-for-another-strong-year-bloombergnef/

BloombergNEF. (2024b, 10 December) *Lithium-Ion Battery Pack Prices See Largest Drop Since 2017, Falling to $115 per Kilowatt-Hour.* https://perma.cc/452P-2JPT

Bond, K., Butler-Sloss, S., and Walter, D. (2024). *The cleantech revolution: It's exponential, disruptive and now.* Rocky Mountain Institute. https://perma.cc/52K8-WHRK

Butler-Sloss, S., Bond, K., and Walter, D. (2024). *The race to the top in six charts and not too many numbers.* The Electrotech Revolution. https://renewablerevolution.substack.com/p/the-race-to-the-top-in-six-charts

Carney, M. (2020). *The Road to Glasgow* [Speech transcript]. The Bank of England. https://perma.cc/5EBG-GN9E

Center for Countering Digital Hate (CCDH). (2024). *The new climate denial.* https://perma.cc/5PCL-BJGW

Centre for Research on Energy and Clean Air (CREA), and Global Energy Monitor (GEM). (2025). *When coal won't step aside: The challenge of scaling clean energy in China.* https://perma.cc/53CU-H4QT

The Church of England. (2018). *Pension Funds Challenge Major European Emitters On Climate Lobbying.* https://perma.cc/E857-JDMG

Climate Action Tracker. (2024, 17 September). *China.* https://perma.cc /HCJ8-ZT22

Climate Policy Initiative (CPI). (2021). *Framework for Sustainable Finance Integrity: A tool for guiding action across the financial system.* https:// perma.cc/7QAN-V7T2

Coalition of Finance Ministers for Climate Action. (2024). *Strengthening the Role of Ministries of Finance in Driving Climate Action: A Framework and Guide for Ministers and Ministries of Finance.* https://www .financeministersforclimate.org/sites/cape/files/inline-files/Strengthening %20the%20role%20of%20Ministries%20of%20Finance%20in%20driving %20action%20FULL%20REPORT.pdf

Cuming, V. (2024). *Emerging markets energy investment outlook 2024.* BloombergNEF, and Glasgow Financial Alliance for Net Zero. https:// perma.cc/854X-FVY7

Dayal, S. (2025, 26 March). India-China issues expected but can be addressed without conflict, says India foreign minister. *Reuters.* https:// www.reuters.com/world/india-china-issues-expected-can-be-addressed -without-conflict-says-india-foreign-2025-03-26/

Democrats Senate. (2022). *Summary of the energy security and climate change investments in the Inflation Reduction Act of 2022.* https://perma .cc/WEX8-STUK

Dennis, J. (2025, 16 January). GFANZ is in freefall – so what happens next? *BusinessGreen.* https://www.businessgreen.com/opinion/4396371/gfanz -freefall-happens

Department for Energy Security and Net Zero (DESNZ). (2023). *Electricity Generation Costs 2023.* https://perma.cc/BM5K-98ST

Depledge, J. (2005). Against the grain: The United States and the global climate change regime. *Global Change, Peace and Security, 17*(1), 11–27. https://doi.org/10.1080/0951274052000319337

Dhakal, S., Minx, J. C., Toth, F. L., Abdel-Aziz, A., Figueroa Meza, M. J., Hubacek, K., Jonckheere, I. G. C., Kim, Y.-G., Nemet, G. F., Pachauri, S., Tan, X. C., and Wiedmann, T. (2022). Emissions trends and drivers. In P. R. Shukla, J. Skea, R. Slade, A. Al Khourdajie, R. van Diemen, D. McCollum, M. Pathak, S. Some, P. Vyas, R. Fradera, M. Belkacemi, A. Hasija, G. Lisboa, S. Luz, and J. Malley (Eds.), *Climate Change 2022: Mitigation of climate change: Contribution of Working Group III to the Sixth Assessment Report of the Intergovernmental Panel on Climate Change* (pp. 215–294). Cambridge University Press. https://doi.org/10.1017/9781009157926.004215

Dominici, F. (2025). How AI could help reduce climate change and air pollution. *Forbes.* https://www.forbes.com/sites/francescadominici/2025/01/20/how-ai-could-help-reduce-climate-change-and-air-pollution/

Draghi, M. (2024a). *The future of European competitiveness: A competitiveness strategy for Europe.* European Commission. https://perma.cc/7UK6-A9ZV

Draghi, M. (2024b). *The future of European competitiveness: In-depth analysis and recommendations.* European Commission. https://perma.cc/4SG4-JAVD

Ember. (2024a). *Electricity data explorer.* Retrieved January 10, 2025, from https://ember-energy.org/data/electricity-data-explorer/

Ember. (2024b). *India.* https://perma.cc/E4QG-24RH

Energy Transitions Commission (ETC). (2023). *Financing the Transition: How to Make the Money Flow for a Net-Zero Economy.* Systemiq. https://perma.cc/E3VC-L2CC

Evans, S. and Viisainen, V. (2024, 6 March). *Analysis: Trump election win could add 4bn tonnes to US emissions by 2030.* Carbon Brief. https://perma.cc/5HLH-MLQF

Grasso, M. and Giugni, M. (Eds.). (2022). *The Routledge handbook of environmental movements.* Routledge.

Graves, P. and Wright, B. (2018). *Solar Power in Texas.* Texas Government. https://perma.cc/7AFK-ZM72

Gu, D., Andreev, K., and Dupre, M. E. (2021). Major trends in population growth around the world. *China CDC Weekly, 3*(28), 604–613. https://doi.org/10.46234/ccdcw2021.160

Guterres, A. (2023, 1 December). *Secretary-General's remarks at opening of World Climate Action Summit [as delivered]* [Speech transcript]. United Nations. https://perma.cc/D2U9-H6C3

Harvey, F. (2025, 11 April). Shipping companies to pay for carbon dioxide produced by vessels. *The Guardian.* https://perma.cc/KKW4-A5K3

Henry, C., Rockström, J., and Stern, N. (Eds.). (2020). *Standing up for a sustainable world: voices of change.* Edward Elgar.

Hickman, L. (2018). *Exclusive: BBC issues internal guidance on how to report climate change.* Carbon Brief. https://perma.cc/EVF9-2Z4Y

Institute for Energy Economics and Financial Analysis (IEEFA). (2024). *Update on India's electricity capacity, generation and investment: POW-ERup 1Q 2024.* https://perma.cc/LA5U-K842

Intergovernmental Panel on Climate Change (IPCC). (2022). *Mitigation pathways compatible with long-term goals.* In P.R. Shukla, J. Skea, R. Slade, A. Al Khourdajie, R. van Diemen, D. McCollum, M. Pathak, S. Some, P. Vyas, R. Fradera, M. Belkacemi, A. Hasija, G. Lisboa, S. Luz, and J. Malley, (Eds.). (2022). *Climate Change 2022: Mitigation of Climate Change. Contribution of Working Group III to the Sixth Assessment Report of the Intergovernmental Panel on Climate Change* (pp. 295–408). Cambridge University Press. https://doi.org/10.1017/9781009157926.005

Intergovernmental Panel on Climate Change. (2018). *Global warming of 1.5°C: An IPCC special report on the impacts of global warming of 1.5°C above pre-industrial levels and related global greenhouse gas emission pathways, in the context of strengthening the global response to the threat of climate change, sustainable development, and efforts to eradicate poverty.* [V., Masson-Delmotte, P. Zhai, H.-O. Pörtner, D. Roberts, J. Skea, P.R. Shukla, A. Pirani, W. Moufouma-Okia, C. Péan, R. Pidcock, S. Connors, J.B.R. Matthews, Y. Chen, X. Zhou, M.I. Gomis, E. Lonnoy, T. Maycock, M. Tignor, and T. Waterfield (Eds.)]. Cambridge University Press. https://doi.org/10.1017/9781009157940

International Energy Agency (IEA). (2020). *Evolution of solar PV module cost by data source, 1970–2020.* https://perma.cc/T3K4-UL4X

International Energy Agency. (2021a). *Financing Clean Energy Transitions in Emerging and Developing Economies.* IEA, World Bank, and World Economic Forum. https://perma.cc/7EB5-4R3L

International Energy Agency. (2021b). *Net zero by 2050.* https://perma.cc/6RBJ-YUUA

International Energy Agency. (2022). *Africa Energy Outlook 2022.* https://perma.cc/95Z3-N7KV

International Energy Agency (2023a). *Financing Clean Energy in Africa.* https://www.iea.org/reports/financing-clean-energy-in-africa

International Energy Agency. (2023b). *World Energy Investment 2023*. https://perma.cc/6V65-YB7K

International Energy Agency. (2024a). *Global EV Outlook 2024: Moving towards increased affordability*. https://perma.cc/9SWS-NU9Z

International Energy Agency. (2024b). *Meeting Power System Flexibility Needs in China by 2030: A market-based policy toolkit for the 15th Five-Year Plan*. https://perma.cc/999G-AYG2

International Energy Agency. (2024c). *Renewables 2023*. https://perma.cc /C9L4-FCSB

International Energy Agency. (2024d). *Brazil*. https://www.iea.org/countries /brazil

International Energy Agency. (2024e). *Chile*. https://www.iea.org/countries /chile

International Energy Agency. (2025a). *World Energy Investment 2025*. https://www.iea.org/reports/world-energy-investment-2025

International Energy Agency. (2025b). *Growth in global energy demand surged in 2024 to almost twice its recent average*. https://www.iea.org/ news/growth-in-global-energy-demand-surged-in-2024-to-almost-twice-its-recent-average

International Energy Agency. (2025c). *Energy and AI*. https://perma.cc /WE2Q-BGYN

International Energy Agency. (2025d, 5 March) *The battery industry has entered a new phase*. https://perma.cc/442Z-ADW3

International Maritime Organization (IMO). (2025, April 11). *IMO approves net-zero regulations for global shipping*. https://perma.cc/NPW4-GN35

International Monetary Fund (IMF). (2024, October). *World Economic Outlook Database, October 2024*. https://www.imf.org/en/Publications/WEO /weo-database/2024/October

International Monetary Fund. (2025). *World Economic Outlook: A Critical Juncture amid Policy Shifts*. https://www.imf.org/en/Publications/WEO/ Issues/2025/04/22/world-economic-outlook-april-2025

International Renewable Energy Agency (IRENA). (2023a). *Renewable Power Generation Costs in 2022*. https://perma.cc/883V-ZFMN

International Renewable Energy Agency. (2023b) *Renewable energy and jobs: Annual review 2023*. https://perma.cc/645L-RHAX

Ipsos. (2023). *Earth Day 2023: Public opinion on climate change*. https:// perma.cc/LL8Q-HEP3

Ives, M. C., Righetti, L., Schiele, J., De Meyer, K., Hubble-Rose, L., Teng, F., Kruitwagen, L., Tillmann-Morris, L., Wang, T., Way, R., and Hepburn, C. (2021). *A new perspective on decarbonising the global energy system* (Report No. 21-04). Smith School of Enterprise and the Environment, University of Oxford. https://perma.cc/HT7A-NT6A

Jones, L., and Hameiri, S. (2020). *Debunking the Myth of 'Debt-trap Diplomacy' How Recipient Countries Shape China's Belt and Road Initiative* (Research Paper). Chatham House. https://perma.cc/GN5U-Z78Y

Juliana v. United States, 217 F. Supp. 3d 1224 (D. Or. 2016).

Kugelman, M. (2025, March 24). India–China relations: Modi's hope for a thaw amid uncertain geopolitics. *BBC News*. https://perma.cc/42RS -GSNY

Landais, C., Jean, S., Philippon, T., Saussay, A., Schnitzer, M., Grimm, V., Malmendier, U., Truger, A., and Werding, M. (2023). *The Inflation Reduction Act: How should the EU react?* Conseil d'analyse économique, and German Council of Economic Experts. https://perma.cc/ETK9-FQLE

Leiserowitz, A., Maibach, E., Rosenthal, S., Kotcher, J., Goddard, E., Carman, J., Ballew, M., Verner, M., Marlon, J., Lee, S., Myers, T., Goldberg, M., Badullovich, N., & Thier, K. (2023). *Climate Change in the American Mind: Beliefs & Attitudes*, Yale University and George Mason University.

Lliuya v. RWE AG, Case No. 2 O 285/15 (LG Essen Dec. 15, 2016), appeal filed, Case No. I-5 U 15/17 (OLG Hamm).

Maia, S. and Demôro, L. (2022). *Power Transition Trends 2022: Coal power spikes, but progress on renewables brings hope.* BloombergNEF. https://perma.cc/C3VU-EM82

McSweeney, R. and Tandon, A. (2024). *Attribution studies: How climate change is influencing extreme weather events.* Carbon Brief. https://perma .cc/F2H4-43K2

Ministério de Minas e Energia. (2024). *Fontes renováveis responderam por 93,1% da geração de energia elétrica em 2023.* Governo do Brazil. https:// perma.cc/F47V-5NE9

Ministry of Energy, Government of Chile. (2022). *Transición Energética de Chile.* https://perma.cc/CW5J-QMD2

Mohammed, A. (2022, 7 November). *Africa is being devastated by a climate crisis it didn't cause. COP27 must help.* UN Sustainable Development Group. https://unsdg.un.org/latest/blog/africa-being-devastated-climate -crisis-it-didnt-cause-cop27-must-help

Myllyvirta, L. (2024) *Analysis: Clean energy was the top driver of China's economic growth in 2023.* Carbon Brief. https://perma.cc/UN6T-VEQK

Net Zero Tracker. (2023). *Net Zero Stocktake 2023.* NewClimate Institute, Oxford Net Zero, Energy and Climate Intelligence Unit and Data-Driven EnviroLab. https://perma.cc/GX39-R27X

Net Zero Tracker. (2024). *Net Zero Stocktake 2024.* https://perma.cc/E6UL-JHUA

Network for Greening the Financial System (NGFS). (2025, 11 March). *Membership.* https://perma.cc/QJ4X-8V73

Neubauer et al. v. Germany, BvR 2656/18 (Federal Constitutional Court [Germany], 29 Apr. 2021).

Nilsen, E. and Rigdon, R. (2024). The biggest winners of Biden's green climate policies. *CNN Climate.* https://edition.cnn.com/2024/06/16/climate/clean-energy-investment-republicans/index.html

Nisbett, N. and Spaiser, V. (2023). Moral power of youth activists – Transforming international climate politics? *Global Environmental Change, 82.* https://doi.org/10.1016/j.gloenvcha.2023.102717

Obama, B. (2017). The irreversible momentum of clean energy. *Science, 355*(6321), 126–129. https://doi.org/10.1126/science.aam6284

Öko-Institut. (2013). *Reform of the German Renewable Energy Sources Act (EEG) – Oeko-Institut's suggestions.* https://perma.cc/DC9R-AMWU

Perez-Goropze, J., Cardenas, C., and Tesfay, N. (2024, 16 October). *Emerging Markets: A Decisive Decade.* S&P Global. https://www.spglobal.com/en/research-insights/special-reports/look-forward/emerging-markets-a-decisive-decade

Pope Francis. (2014). God forgives: Nature does not. *L'Osservatore Romano; Weekly Edition in English.* https://perma.cc/XL9J-CKBF

Pope Francis. (2015). *Laudato Si': On care for our common home.* Vatican City: Vatican Press. https://perma.cc/A443-2K5R

Poushter, J., Fagan, M., and Gubbala, S. (2022). *Climate change remains top global threat across 19-country survey.* Pew Research Center. https://perma.cc/4JUJ-5Z35

PwC. (2023). *June 2023 Global Consumer Insights Pulse Survey.* https://perma.cc/PF64-8NHE

Quitzow, R. (2015). Dynamics of a policy-driven market: The co-evolution of technological innovation systems for solar photovoltaics in China and Germany. *Environmental Innovation and Societal Transitions, 17,* 126–148.

REN21. (2025). *Unpacking an African renewables-based economy.* https://www.ren21.net/wp-content/uploads/2019/05/REN21_STRAT_I_BRIEF_2025_HD-2.pdf

Rhodium Group, and Massachusetts Institute of Technology Center for Energy and Environmental Policy Research (MIT CEEPR). (2024). *Clean investment monitor: Tallying the two-year impact of the Inflation Reduction Act.* https://perma.cc/2TL5-CND8

Rystad Energy. (2024). *Brazil's well services market poised for significant growth.* https://www.rystadenergy.com/insights/brazil-s-well-services-market-poised-for-significant-growth

S&P Global (2022). *Global light duty EV sales to rise to 26.8 mil by 2030: Platts Analytics.* https://www.spglobal.com/commodity-insights/en/news-research/latest-

S&P Global. (2024, 19 November). *Sustainability insights: Rising curtailment in China power producers will push past the pain.* https://www.spglobal.com/ratings/en/research/articles/241119-sustainability-insights-rising-curtailment-in-china-power-producers-will-push-past-the-pain-13327811.

Samaranayake, N. (2021, 2 March). Chinese Belt and Road investment isn't all bad—or good. *Foreign Policy.* https://foreignpolicy.com/2021/03/02/sri-lanka-china-bri-investment-debt-trap/

SEforALL. (2025). *Kenya. Green Manufacturing Policy & Investment Guide.* https://www.seforall.org/system/files/2025-04/SEforALL-Kenya_Green_Manufacturing_Policy_and_Investment_Guide-2025-FV.pdf

Sen, A. (1999). *Development as freedom.* Oxford University Press.

Sen, A. (2009). *The Idea of justice.* Harvard University Press.

Setzer, J. and Benjamin, L. (2020). Climate litigation in the Global South: Constraints and innovations. *Transnational Environmental Law, 9*(1), 77–101. https://doi.org/10.1017/S2047102519000268

Setzer, J. and Higham, C. (2024). *Global trends in climate change litigation: 2024 snapshot.* Grantham Research Institute on Climate Change and the Environment, London School of Economics and Political Science. https://perma.cc/VH2F-4A7Q

Singh, N. K. and Summers, L. (2023a). *Strengthening multilateral development banks: The triple agenda MANDATES I FINANCE I MECHANISMS* (Report of the Independent Experts Group, Vol. 1). G20 India. https://perma.cc/6BCU-PG3P

Singh, N. K. and Summers, L. (2023b). *Strengthening multilateral development banks: The triple agenda: Better, Bolder and Bigger MDBs* (Report of the Independent Experts Group, Vol. 2). G20 India https://perma.cc/QY4T-23KA

Singh, V. and Bond, K. (2024). *Powering up the global south.* Rocky Mountain Institute. https://perma.cc/9HGP-6W77

Solar Energy Industries Association (SEIA). (2024). *Solar and Storage Industry Pushes Policy Agenda for Trump Administration, New Congress to Strengthen American Energy Leadership.* https://perma.cc/K9FP-LNJ8

Song, W., Viisainen, V, and Patel, A. (2024). *QandA: The global 'trade war' over China's booming EV industry.* Carbon Brief. https://perma.cc/QJH5-28X2

Songwe, V., Stern, N., and Bhattacharya, A. (2022). *Finance for climate action: Scaling up investment for climate and development.* Grantham Research Institute on Climate Change and the Environment, London School of Economics and Political Science. https://perma.cc/V454-EKF5

Stern, N., Romani, M., Pierfederici, R., Braun, M., Barraclough, D., Lingeswaran, S., Weirich-Benet, E., and Niemann, N. (2025). Green and intelligent: The role of AI in the climate transition. *npj Climate Action*, 4(56). https://doi.org/10.1038/s44168-025-00252-3

Stern, T. (2024). *Landing the Paris Climate Agreement: how it happened, why it matters, and what comes next.* MIT Press.

Stockholm Environment Institute (SEI), Climate Analytics, E3G, International Institute for Sustainable Development (IISD), and United Nations Environment Programme (UNEP). (2023). *The Production Gap: Phasing down or phasing up? Top fossil fuel producers plan even more extraction despite climate promises.* https://doi.org/10.51414/sei2023.050

Swanson, A. and Rappeport, A. (2024, 6 June). U.S. adds tariffs to shield struggling solar industry. *The New York Times.* https://www.nytimes.com/2024/06/06/business/economy/tariffs-solar-industry-china.html

Systemiq. (2020). *The Paris Effect: How the Climate Agreement is reshaping the global economy.* https://perma.cc/A953-7VJZ

Systemiq. (2023). *The Breakthrough Effect: How To Trigger A Cascade Of Tipping Points To Accelerate The Net Zero Transition.* https://perma.cc/WM6Y-894N

Systemiq. (2025). *Accelerating the Breakthrough of Climate Technologies: Driving exponential growth in climate technologies with positive tipping points (forthcoming – working title)*

Thunberg, G. (2019, September 23). *Speech presented at the United Nations Climate Action Summit* [Speech transcript], New York, NY. NPR. https://perma.cc/U8B6-K6DU

Thunberg, G. (2022). *The Climate Book: The facts and the future.* Viking.

Trancik, J. E., Jean, J., Kavlak, G., Klemun, M. M., Edwards, M. R., McNerney, J., Miotti, M., Brown, P.R., Mueller, J.M., and Needell, Z. A. (2015). *Technology improvement and emissions reductions as mutually reinforcing efforts: Observations from the global development of solar and wind energy.* MIT. http://hdl.handle.net/1721.1/102237

Trout, K., Muttitt, G., Lafleur, D., Van de Graaf, T., Mendelevitch, R., Mei, L., and Meinshausen, M. (2022). Existing fossil fuel extraction would warm the world beyond 1.5°C. *Environmental Research Letters, 17*(6). https://doi.org/10.1088/1748-9326/ac6228

Tyson, A., Funk, C., and Kennedy, B. (2023). *What the data says about Americans' views of climate change.* Pew Research Center. https://perma.cc/2EAE-Z2D6

United Nations Development Programme (UNDP), and University of Oxford. (2021). *The Peoples' Climate Vote.* United Nations Development Programme. https://www.undp.org/sites/g/files/zskgke326/files/publications/UNDP-Oxford-Peoples-Climate-Vote-Results.pdf

United Nations Environment Programme (UNEP). (2023). *Global Climate Litigation Report: 2023 Status Review.* https://doi.org/10.59117/20.500.11822/43008

United Nations Framework Convention on Climate Change (UNFCCC). (1997). *Kyoto Protocol to the United Nations Framework Convention on Climate Change.* https://unfccc.int/documents/2409

United Nations Framework Convention on Climate Change. (2024). *Artificial Intelligence for Climate Action in Developing Countries: Opportunities, Challenges and Risks* (Information Note). https://unfccc.int/ttclear/misc_/StaticFiles/gnwoerk_static/AI4climateaction/ea0f2596d93640349b9b6 5f4a7c7dd24/b47ef0e99cb24e57aa9ea69f0f5d6a71.pdf

Urgenda Foundation v. State of the Netherlands, ECLI:NL:HR:2019:2007 (Hoge Raad [Supreme Court of the Netherlands] 20 Dec., 2019).

van Oldenborgh, G. J., Krikken, F., Lewis, S., Leach, N. J., Lehner, F., Saunders, K. R., van Weele, M., Haustein, K., Li, S., Wallom, D., Sparrow, S., Arrighi, J., Singh, R. K., van Aalst, M. K., Philip, S. Y., Vautard, R., and Otto, F. E. L. (2021). Attribution of the Australian bushfire risk to anthropogenic climate change. *Natural Hazards and Earth System Sciences, 21,* 941–960. https://doi.org/10.5194/nhess-21-941-2021

van Twillert, N., and Halleck Vega, S. (2023). Risk or opportunity? The Belt and Road Initiative and the role of debt in the China-Central Asia-West Asia Economic Corridor. *Eurasian Geography and Economics, 64*(3), 365–377. https://doi.org/10.1080/15387216.2021.2012816

Vasdev, A. (2023, December 18). *2H 2023 LCOE update: An uneven recovery.* BloombergNEF. https://perma.cc/Y4H3-9X2K

Vinuesa, R., Azizpour, H., Leite, I., Balaam, M., Dignum, V., Domisch, S., Felländer, A., Langhans, S.D., Tegmark, M., and Nerini, F. F. (2020). The role of artificial intelligence in achieving the Sustainable Development Goals. *Nature Communications, 11,* Article 233. https://doi.org/10.1038/s41467-019-14108-y

Vlasceanu, M., Doell, K. C., Bak-Coleman, J. B., Todorova, B., Berke-bile-Weinberg, M. M., Grayson, S. J., Patel, Y., Goldwert, D., Pei, Y., Chakroff, A., Pronizius, E., van den Broek, K.L., Vlasceanu, D., Constantino, S., Morais, M.J., Schumann, P., Rathje, S., Fang, K., Aglioti, S.M., and Van Bavel, J.J. (2024). Addressing climate change with behavioral science: A global intervention tournament in 63 countries. *Science Advances, 10*(6). https://doi.org/10.1126/sciadv.adj5778

World Bank. (2019). *Belt and road economics: Opportunities and risks of transport corridors.* https://hdl.handle.net/10986/31878

World Bank. (2025). *Global Economic Prospects: Falling graduation prospects: low-income countries in the 21st century.* https://doi.org/10.1596/978-1-4648-2147-9

World Economic Forum (WEF). (2025). *Davos 2025: Special Address by Ursula von der Leyen, President of the European Commission* [Speech transcript]. https://www.weforum.org/stories/2025/01/davos-2025-special-address-by-ursela-von-der-leyen-president-of-the-european-commission/

World Resources Institute (WRI). (2015). *New climate economy's message of economic growth and climate action takes hold.* https://perma.cc/ULP9-MP3W

World Weather Attribution (WWA). (2024). *When Risks Become Reality: Extreme Weather in 2024.* https://perma.cc/BN7Q-T42J

You, X. (2024, 19 January). What does peak emissions mean for China — and the world? *Nature.* https://doi.org/10.1038/d41586-024-02877-6

Zhang, H.-B., Dai, H.-C., Lai, H.-X., and Wang, W.-T. (2017). U.S. withdrawal from the Paris Agreement: Reasons, impacts, and China's response. *Advances in Climate Change Research, 8*(4), 234–241. https://doi.org/10.1016/j.accre.2017.10.001

PART II

Chapter 5

Abduljabbar, R., Dia, H., Liyanage, S., and Bagloee, S. A. (2019). Applications of artificial intelligence in transport: An overview. *Sustainability*, *11*(1), 189. https://doi.org/10.3390/su11010189

Abel, G. J., Brottrager, M., Crespo Cuaresma, J., and Muttarak, R. (2019). Climate, conflict, and forced migration. *Global Environmental Change*, 54, 239–249. https://doi.org/10.1016/j.gloenvcha.2018.12.003

AI Innovation for Decarbonisation's Virtual Centre of Excellence (ADViCE). (2023). *AI for decarbonisation: Assessing the UK landscape for artificial intelligence and its use in decarbonisation.* https://perma.cc/UW33-C7HU

Air Quality Life Index (AQLI). (2023). *Pakistan Fact Sheet.* https://perma.cc/ZLY8-JKGE

Amazon. (2024, April 16). How Amazon is using AI to deliver customer orders with less packaging. *Amazon News.* https://perma.cc/FDL8-BZFS

Anadon, L. D., Jones, A., Peñasco, C., Sharpe, S., Grubb, M., Aggarwal, S., Barbosa Filho, N. H., Bose, R., Cabello, A., Chaudhury, S., Drummond, P., Farmer, D., Foulds, C., Freddo, D., Hepburn, C., Kapur, V., Kejun, J., Lam, A., Mercure, J.-F., Freitas, L. H. M., Royston, S., Salas, P., Viñuales, J., and Zhu, S. (2022). *Ten principles for policymaking in the energy transition: Lessons from experience.* Economics of Energy Innovation and System Transition (EEIST). https://eeist.co.uk/eeist-reports/ten-principles-for-policy-making-in-the-energy-transition/

Awe, Y. A., Larsen, B. K., and Sanchez-Triana, E. (2022). *The global health cost of pm2.5 air pollution: a case for action beyond 2021.* World Bank Group. https://doi.org/10.1596/978-1-4648-1816-5

Baker, W., Acha, S., Jennings, N., Markides, C., and Shah, N. (2022). *Decarbonisation of buildings: Insights from across Europe.* The Grantham Institute – Climate Change and the Environment, Imperial College London. http://hdl.handle.net/10044/1/100954

Balasubramanian, P., Ibanez, M., Khan, S., and Sahoo, S. (2024). Does women's economic empowerment promote human development in low- and middle-income countries? A meta-analysis. *World Development, 178.* https://doi.org/10.1016/j.worlddev.2024.106588

Bangalore, M., Bonzanigo, L., Fay, M., Hallegatte, S., Narloch, U. G., Rozenberg, J., and Vogt-Schilb, A. C. (2014). *Climate Change and Poverty: An Analytical Framework.* World Bank Group. http://documents.worldbank.org/curated/en/275231468331203291

Bangalore, M., Hallegatte, S., Vogt-Schilb, A., and Rozenberg, J. (2017). *Unbreakable: Building the resilience of the poor in the face of natural disasters*. World Bank. http://hdl.handle.net/10986/25335

Barbier, E. B. (2014). Climate change mitigation policies and poverty. *Wiley Interdisciplinary Reviews: Climate Change, 5*(4), 483–491. https://doi.org/10.1002/wcc.281

Beynon, J. and Wickstead, E. (2024). *Climate and development in three charts: An update.* Center for Global Development. https://www.cgdev.org/blog/climate-and-development-three-charts-update

Bhattacharya, A., Dooley, M., Kharas, H., Taylor, C., and Stern, N. (2022). *Financing a big investment push in emerging markets and developing economies for sustainable, resilient and inclusive recovery and growth.* Grantham Research Institute on Climate Change and the Environment, London School of Economics and Political Science, and Brookings. https://perma.cc/W9R9-VP8N

Bhattacharya, A., Songwe, V., Soubeyran, E., and Stern, N. (2023). *A climate finance framework that is fit for purpose: Decisive action to deliver on the Paris Agreement — Summary.* Grantham Research Institute on Climate Change and the Environment, London School of Economics and Political Science. https://perma.cc/ZML3-CQ8Q

Bhattacharya, A., Songwe, V., Soubeyran, E., and Stern, N. (2024). *Raising ambition and accelerating delivery of climate finance.* Grantham Research Institute on Climate Change and the Environment, London School of Economics and Political Science. https://perma.cc/H6D9-MNSX

Bhattacharya, A., Kharas, H., Rivard, C., and Soubeyran, E. (2025). *From Aid-Driven to Investment-Driven Models of Sustainable Development,* (Working Paper 194). Brookings. https://perma.cc/6XK7-MRYM

Birkmann, J., Liwenga, E., Pandey, R., Boyd, E., Djalante, R., Gemenne, F., Leal Filho, W., Pinho, P. F., Stringer, L., and Wrathall, D. (2022). Poverty, livelihoods, and sustainable development. In H.-O. Pörtner, D. C. Roberts, M. Tignor, E. S. Poloczanska, K. Mintenbeck, A. Alegría, M. Craig, S. Langsdorf, S. Löschke, V. Möller, A. Okem, and B. Rama (Eds.), *Climate change 2022: Impacts, adaptation and vulnerability. Contribution of Working Group II to the Sixth Assessment Report of the Intergovernmental Panel on Climate Change* (pp. 1171–1274). Cambridge University Press. https://doi.org/10.1017/9781009325844.010

BlackRock. (2024, July 29). *How AI is transforming investing.* https://www.blackrock.com/us/individual/insights/ai-investing

BloombergNEF. (2024). *Mobilizing Capital in and to Emerging Markets.* https://assets.bbhub.io/professional/sites/24/Mobilizing-Capital-in-and-to-EMDEs.pdf

Boserup, E. (1970). *Women's role in economic development.* London: Allen and Unwin.

Bourguignon, F. (2004). *The poverty-growth-inequality triangle* (Working Paper No. 125). Indian Council for Research on International Economic Relations (ICRIER). https://perma.cc/F26X-QB7A

Budolfson, M., Dennig, F., Errickson, F., Feindt, S., Ferranna, M., Fleurbaey, M., Klenert, D., Kornek, U., Kuruc, K., Méjean, A., Peng, W., Scovronick, N., Spears, D., Wagner, F., and Zuber, S. (2021). Climate action with revenue recycling has benefits for poverty, inequality and well-being. *Nature Climate Change, 11*(12), 1111–1116. https://doi.org/10.1038/s41558-021-01217-0

Cao, Y., Chen, J., Liu, B., Turner, A., and Zhu, C. (2021). *China zero-carbon electricity growth in the 2020s: A vital step toward carbon neutrality.* Rocky Mountain Institute (RMI), & Energy Transitions Commission (ETC). https://perma.cc/FQ3H-BJC3

Chancel, L., Piketty, T., Saez, E., and Zucman, G. (Eds.). (2022). *World inequality report 2022.* World Inequality Lab. https://perma.cc/26RA-QHXF

Chrisafis, A. (2016, February 4). French law forbids food waste by supermarkets. *The Guardian.* https://perma.cc/3JV9-XZMP

Climate Policy Initiative (CPI). (2024). *Global Landscape of Climate Finance 2024.* https://perma.cc/FJ4W-3MEV

Climate Policy Initiative. (2025). *Global Landscape of Climate Finance 2025.* https://www.climatepolicyinitiative.org/publication/global-landscape-of-climate-finance-2025/

COP28 UAE. (2023). *Global Renewables and Energy Efficiency Pledge.* https://perma.cc/T2PR-59WT

Damania, R., Balseca, E., de Fontaubert, C., Gill, J., Kim, K., Rentschler, J., Russ, J., and Zaveri, E. (2023). *Detox Development: Repurposing Environmentally Harmful Subsidies.* World Bank Group. https://hdl.handle.net/10986/39423

De Baerdemaeker, J., Hemming, S., Polder, G., Chauhan, A., Petropoulou, A., Rovira-Más, F., Moshou, D., Wyseure, G., Norton, T., Nicolai, B., Hennig-Possenti, F., and Hostens, I. (2023). *Artificial intelligence in the agri-food sector: Applications, risks, and impacts.* Panel for the Future of Science and Technology, EPRS, European Parliamentary Research Service. https://www.europarl.europa.eu/stoa/en/document/EPRS_STU(2023)734711

Dollar, D. and Kraay, A. (2002). Growth is Good for the Poor. *Journal of Economic Growth, 7*, 195–225. https://doi.org/10.1023/A:1020139631000

Duflo, E. (2012). Women empowerment and economic development. *Journal of Economic Literature, 50*(4), 1051–1079. https://doi.org/10.1257/jel.50.4.1051

Easterly, W. (1999). Life during growth. *Journal of Economic Growth, 4*, 239–276. https://doi.org/10.1023/A:1009882702130

The Economist. (2022, October 20) Will India become a green superpower? *The Economist.* https://www.economist.com/briefing/2022/10/20/will-india-become-a-green-superpower

Ember (2025). *Electricity Data Explorer.* https://ember-energy.org/data/electricity-data-explorer/?fuel=total&tab=main&chart=trend&metric=pct_share

Energy Transitions Commission (ETC). (2018). *Mission possible: Reaching net-zero carbon emissions from harder-to-abate sectors.* Systemiq. https://perma.cc/9YN7-V68S

Energy Transitions Commission. (2023a). COP28: *A high-level assessment of mitigation proposals.* https://www.energy-transitions.org/bitesize/cop28-assessment-mitigation-proposals/#:~:text=Proposals%20at%20COP28%20focus%20on,%26%202%20targets%20for%202050

Energy Transitions Commission. (2023b). *Financing the transition: how to make the money flow for a net-zero economy.* Systemiq. https://perma.cc/E3VC-L2CC

Energy Transitions Commission. (2025a). *Power Systems Transformation: Delivering Competitive, Resilient Electricity in High-Renewable Systems.* https://www.energy-transitions.org/publications/power-systems-transformation/#download-form

Energy Transitions Commission. (2025b). *Achieving zero-carbon buildings: electric, efficient and flexible.* https://www.energy-transitions.org/wp-content/uploads/2025/01/ETC_Buildings-Decarbonisation-Report_DIGITALFINAL.pdf

Eyre, N. (2021). From using heat to using work: reconceptualising the zero carbon energy transition. *Energy efficiency, 14*(77). https://doi.org/10.1007/s12053-021-09982-9

Food and Land Use Coalition (FOLU). (2019). *Growing Better: tend ten critical transitions to transform food and land use.* https://perma.cc/6YQF-2HSD

Gallagher, K.S. (2021). The coming carbon tsunami. *Foreign Affairs. https://www.foreignaffairs.com/articles/world/2021-12-14/coming-carbon-tsunami*

Global Centre on Adaptation and Climate Policy Initiative (CPI). (2023). *State and trends in climate adaptation finance 2023.* https://perma.cc/8LJ9 -VSU6

Global Commission on Adaptation (GCA). (2019). *Adapt now: A global call for leadership on climate resilience.* World Resources Institute. https:// perma.cc/GRY8-C3AP

Hallegatte, S., Vogt-Schilb, A., Rozenberg, J., Bangalore, M., and Beaudet, C. (2020). From poverty to disaster and back: a review of the literature. *Economics of Disasters and Climate Change, 4,* 223–247. https://doi.org /10.1007/s41885-020-00060-5

Hickel, J. (2017). *The Divide: A brief guide to global inequality and its solutions.* Penguin Books.

Hidalgo, D. and Gutiérrez, L. (2013). BRT and BHLS around the world: Explosive growth, large positive impacts and many issues outstanding. *Research in Transportation Economics, 39*(1), 8–13. https://doi.org/10 .1016/j.retrec.2012.05.018

Himanshu, Lanjouw, P., and Stern, N. (2018). *How lives change: Palanpur, India, and development economics.* Oxford University Press. https://doi .org/10.1093/oso/9780198806509.001.0001

Hoffmann, J., Bauer, P., Sandu, I., Wedi, N., Geenen, T., and Thiemert, D. (2023). Destination Earth – A digital twin in support of climate services. *Climate Services, 30*(3). https://doi.org/10.1016/j.cliser.2023.100394

IceNet. (2024). About IceNet. Retrieved April 23, 2025, from https://icenet .ai/

Institute for Health Metrics and Evaluation (IHME). (2021). *Global burden of disease: Results tool.* http://ghdx.healthdata.org/gbd-results-tool

Intergovernmental Oceanographic Commission (IOC-UNESCO). (2022). *State of the ocean report, pilot edition* (IOC Technical Series, No. 173). https://unesdoc.unesco.org/ark:/48223/pf0000381921

Intergovernmental Panel on Climate Change (IPCC). (2022a). Annex I: Glossary. In R. van Diemen, J. B. R. Matthews, V. Möller, J. S. Fuglestvedt, V. Masson-Delmotte, C. Méndez, A. Reisinger, and S. Semenov (Eds.), *Climate change 2022: Mitigation of climate change (Contribution of Working Group III to the Sixth Assessment Report of the Intergovernmental Panel on Climate Change)* (pp. 1–100). Cambridge University Press. https://doi.org/10.1017/9781009157926.020

Intergovernmental Panel on Climate Change. (2022b). Summary for poli-cymakers. In P. R. Shukla, J. Skea, A. Reisinger, R. Slade, R. Fradera, M. Pathak, A. Al Khourdajie, M. Belkacemi, R. van Diemen, A. Hasija, G. Lisboa, S. Luz, J. Malley, D. McCollum, S. Some, and P. Vyas (Eds.), *Climate change 2022: Mitigation of climate change* (Contribution of Working Group III to the Sixth Assessment Report of the Intergovernmental Panel on Climate Change). Cambridge University Press. https://doi.org/10.1017/9781009157926.001

Intergovernmental Science-Policy Platform on Biodiversity and Ecosystem Services (IPBES). (2019). *Global assessment report of the Intergovernmental Science-Policy Platform on Biodiversity and Ecosystem Services.* [E. Brondizio, S. Diaz, J. Settele, and H.T. Ngo (Eds.)]. https://doi.org/10.5281/zenodo.3831673

International Energy Agency (IEA). (2020). *Evolution of solar PV module cost by data source,* 1970-2020. https://perma.cc/T3K4-UL4X

International Energy Agency. (2022a). *Africa Energy Outlook.* https://www.iea.org/reports/africa-energy-outlook-2022

International Energy Agency. (2022b). *Solar PV global supply chains.* https://perma.cc/6SQA-Y6F4

International Energy Agency. (2023a). *World energy investment 2023.*https://perma.cc/EUF5-MJKT

International Energy Agency. (2023b). *Net zero roadmap: A global pathway to keep the 1.5°C goal in reach.* https://perma.cc/4UKS-DT87

International Energy Agency. (2023c). *A global target to double efficiency progress is essential to keep net zero on the table.* tps://www.iea.org/commentaries/a-global-target-to-double-efficiency-progress-is-essential-to-keep-net-zero-on-the-table

International Energy Agency. (2023d). *Energy Efficiency 2023.* https://www.iea.org/reports/energy-efficiency-2023

International Energy Agency. (2023e). *Financing clean energy in Africa.* https://www.iea.org/reports/financing-clean-energy-in-africa

International Energy Agency. (2023f). *Why AI and energy are the new power couple.* https://perma.cc/8MTL-NBJE

International Energy Agency. (2024a). *India case study.* https://www.iea.org/reports/india-case-study

International Energy Agency. (2024b). *Brasil's G20 presidency Roadmap to Increase Investment in Clean Energy in Developing Countries – an initiative by the G20 Brazil Presidency.* https://perma.cc/XT2F-QFHP

International Energy Agency. (2024c). *SDG7: Data and Projections.* https://www.iea.org/reports/sdg7-data-and-projections

International Energy Agency. (2024d). *World Energy Outlook 2024.* https://perma.cc/349B-8SLQ

International Energy Agency, International Renewable Energy Agency (IRENA), United Nations Statistics Division (UNSD), World Bank, & World Health Organisation (WHO). (2024). *Tracking SDG 7: The energy progress report 2024.* World Bank.

International Energy Agency, International Renewable Energy Agency (IRENA), United Nations Statistics Division (UNSD), World Bank, and World Health Organisation (WHO). (2023). *Tracking SDG 7: The energy progress report.* World Bank. https://perma.cc/8UMU-M9NT

International Renewable Energy Agency. (IRENA). (2018). *Global Energy Transformation: A Roadmap to 2050.* https://www.irena.org/-/media/Files/IRENA/Agency/Publication/2018/Apr/IRENA_Report_GET_2018.pdf

International Renewable Energy Agency. (2020). *Pay-As-You-Go Models: Innovation Landscape Brief.* https://perma.cc/VWT8-7C4G

International Renewable Energy Agency. (2023a). *Towards a circular steel industry.* https://perma.cc/9369-TVUB

International Renewable Energy Agency. (2023b). *World Energy Transitions Outlook 2023: 1.5°C Pathway.* https://perma.cc/75X2-THXW

International Renewable Energy Agency. and Climate Policy Initiative. (2023). *Global landscape of renewable energy finance 2023.* https://perma.cc/J6DW-MG2S

Jackson, T. (2021). *Post growth: Life after capitalism.* Polity Press.

Jain, H., Dhupper, R., Shrivastava, A., Kumar, D., and Kumari, M. (2023). AI-enabled strategies for climate change adaptation: protecting communities, infrastructure, and businesses from the impacts of climate change. *Computational Urban Science, 3*(1). https://doi.org/10.1007/s43762-023-00100-2

Just Transition Finance Lab. (2024a). *Case study: Ayana – building a skills base for India's clean energy transition.* Grantham Research Institute on Climate Change and the Environment, London School of Economics and Political Science. https://justtransitionfinance.org/wp-content/uploads/2024/05/Just-Transition-Finance-Lab_case-study_Ayana.pdf

Just Transition Finance Lab. (2024b). *Case study: SSE – working with investors to chart a just transition.* Grantham Research Institute on Climate Change and the Environment, London School of Economics and Political Science. https://perma.cc/2UZK-X7MM

Kelley, C. P., Mohtadi, S., Cane, M. A., Seager, R., and Kushnir, Y. (2015). Climate change in the Fertile Crescent and implications of the recent Syrian drought. *Proceedings of the National Academy of Sciences, 112*(11), 3241–6. https://doi.org/10.1073/pnas.1421533112

Kelly, F. (2018). *Associations of long-term average concentrations of nitrogen dioxide with mortality.* Public Health England. https://perma.cc/G3S9 -JQ74

Kose, M. A. and Ohnsorge, F. (Eds.). (2024). *Falling long-term growth prospects: Trends, expectations, and policies.* World Bank Group. https://doi .org/10.1596/978-1-4648-2000-7

Kreibiehl, S., Jung, T. Y., Battiston, S., Carvajal, P. E., Clapp, C., Dasgupta, D., Dube, N., Jachnik, R., Morita, K., Samargandi, N., and Williams, M. (2022). Investment and finance. In P. R. Shukla, J. Skea, R. Slade, A. Al Khourdajie, R. van Diemen, D. McCollum, M. Pathak, S. Some, P. Vyas, R. Fradera, M. Belkacemi, A. Hasija, G. Lisboa, S. Luz, and J. Malley (Eds.), *Climate change 2022: Mitigation of climate change. Contribution of Working Group III to the Sixth Assessment Report of the Intergovernmental Panel on Climate Change (Chapter 17).* Cambridge University Press. https://doi.org/10.1017/9781009157926.017

Lagos Metropolitan Transport Authority (LAMATA). (2022). *Sustainable Transport in Lagos.* https://www.lamata-ng.com/#

Lankes, H. P., Macquarie, R. Soubeyran, E., and Stern, N. (2023). The Relationship between Climate Action and Poverty Reduction. *The World Bank Research Observer, 39*(1), 1–46, https://doi.org/10.1093/wbro /lkad011

Lazard. (2024, June). *LCOE: Levelized Cost of Energy.* https://perma.cc/J6YJ-VSS5

Li, Y., Zhang, Y., Pan, A., Han, M., and Veglianti, E. (2022). Carbon emission reduction effects of industrial robot applications: Heterogeneity characteristics and influencing mechanisms. *Technology in Society, 70.* https://doi.org/10.1016/j.techsoc.2022.102034

Lovins, A. (2018). How big is the energy efficiency resource? *Environmental Research Letters, 13*(9). https://doi.org/10.1088/1748-9326/aad965

Lovins, A. (2021). Decarbonizing our toughest sectors—profitably. *MIT Sloan Management Review.* https://rmi.org/insight/decarbonizing-our -toughest-sectors-profitably/

Lwasa, S., Seto, K. C., Bai, X., Blanco, H., Gurney, K. R., Kılkış, Ş., Lucon, O., Murakami, J., Pan, J., Sharifi, A., and Yamagata, Y. (2022). Urban systems and other settlements. In P. R. Shukla, J. Skea, R. Slade, A. Al Khourdajie, R. van Diemen, D. McCollum, M. Pathak, S. Some, P. Vyas, R. Fradera,

M. Belkacemi, A. Hasija, G. Lisboa, S. Luz, and J. Malley (Eds.), *Climate change 2022: Mitigation of climate change. Contribution of Working Group III to the Sixth Assessment Report of the Intergovernmental Panel on Climate Change* (pp. 864–952). Cambridge University Press. https://doi.org /10.1017/9781009157926.010

Massachusetts Institute of Technology (MIT). (2023). *Four ways AI is making the power grid faster and more resilient.* https://perma.cc/G5F5-6HNJ

Matias, Y. (2024). *How we are using AI for reliable flood forecasting at a global scale.* Google. https://perma.cc/9FLJ-8P4B

Mehra, R. (1997). Women, empowerment, and economic development. *The Annals of the American Academy of Political and Social Science, 554*(1), 136–149. https://doi.org/10.1177/0002716297554001009

Mueller, N., Rojas-Rueda, D., Khreis, H., Cirach, M., Andrés, D., Ballester, J., Bartoll, X., Daher, C., Deluca, A., Echave, C., Milà, C., Márquez, S., Palou, J., Pérez, K., Tonne, C., Stevenson, M., Rueda, S., and Nieuwenhuijsen, M. (2020). Changing the urban design of cities for health: The superblock model. *Environment International, 134.* https://doi.org/10 .1016/j.envint.2019.105132

Nabuurs, G.-J., Mrabet, R., Abu Hatab, A., Bustamante, M., Clark, H., Havlík, P., House, J., Mbow, C., Ninan, K. N., Popp, A., Roe, S., Sohngen, B., and Towprayoon, S. (2022). Agriculture, forestry and other land uses (AFOLU). In P. R. Shukla, J. Skea, R. Slade, A. Al Khourdajie, R. van Diemen, D. McCollum, M. Pathak, S. Some, P. Vyas, R. Fradera, M. Belkacemi, A. Hasija, G. Lisboa, S. Luz, and J. Malley (Eds.), *Climate change 2022: Mitigation of climate change: Contribution of Working Group III to the Sixth Assessment Report of the Intergovernmental Panel on Climate Change* (pp. 317–489). Cambridge University Press. https://doi.org /10.1017/9781009157926.009

Nature Food. (2024). Food loss and waste. *Nature Food, 5*(1), 1–2. https://doi .org/10.1038/s43016-024-01041-7

New Climate Economy. (2018). *Unlocking the inclusive growth story of the 21st century: Accelerating climate action in urgent times.* World Resources Institute. https://perma.cc/X3ER-2PZX

Noble, I. R., Huq, S., Anokhin, Y. A., Carmin, J., Goudou, D., Lansigan, F. P., Osman-Elasha, B., and Villamizar, A. (2014). Adaptation needs and options. In C. B. Field, V. R. Barros, D. J. Dokken, K. J. Mach, M. D. Mastrandrea, T. E. Bilir, M. Chatterjee, K. L. Ebi, Y. O. Estrada, R. C. Genova, B. Girma, E. S. Kissel, A. N. Levy, S. MacCracken, P. R. Mastrandrea, and L. L. White (Eds.), *Climate Change 2014: Impacts, Adaptation, and Vulnerability. Part A: Global and Sectoral Aspects. Contribution of Working Group II to the Fifth Assessment Report of the Intergovernmental Panel on*

Climate Change (pp. 833–868). Cambridge University Press. https://doi .org/10.1017/9781009157926.010

OpenAI. (n.d.). *Digital Green: Digital Green uses OpenAI to increase farmer income*. OpenAI. https://openai.com/index/digital-green/

Organisation for Economic Co-operation and Development (OECD). (2021). *Towards a more resource-efficient and circular economy: The role of the G20*. OECD Publishing. https://knowledge4policy.ec.europa .eu/publication/g20-report-towards-more-resource-efficient-circular -economy-role-g20_en

Organisation for Economic Co-operation and Development. (2023). *Scaling up adaptation finance in developing countries: Challenges and opportunities for international providers*. Green Finance and Investment. OECD Publishing. https://doi.org/10.1787/b0878862-en

Pauliuk, S., Heeren, N., Berrill, P., Fishman, T., Nistad, A., Tu, Q., Wolfram, P., and Hertwich, E. G. (2021). Global scenarios of resource and emission savings from material efficiency in residential buildings and cars. *Nature Communications, 12*(1). https://doi.org/10.1038/s41467-021-25300-4

Peel, M. (2023, November 29). Google DeepMind researchers use AI tool to find 2mn new materials. *The Financial Times*. https://www.ft.com /content/f841e9e0-c9c6-49ab-b91c-6d7bea2a3940

Probst, B., Westermann, L., Anadón, L. D., and Kontoleon, A. (2021). Leveraging private investment to expand renewable power generation: Evidence on financial additionality and productivity gains from Uganda. *World Development, 140*. https://doi.org/10.1016/j.worlddev.2020.105347

Randall, T. (2024). Long-range EVs now cost less than the average new car in the us. *Bloomberg*. https://www.bloomberg.com/news/articles/2024-06- 07/long-range-evs-now-cost-less-than-the-average-us-new-car

Ritchie, H. (2024). Solar panel prices have fallen by around 20% every time global capacity doubled. *Our World in Data*. https://ourworldindata.org/ data-insights/solar-panel-prices-have-fallen-by-around-20-every-time- global-capacity-doubled

Royal College of Paediatrics and Child Health (RCPCH). (2023). *Child health inequalities and climate change in the UK*. https://perma.cc/3VS7- 9DW6

Schwarzer, H., Panhuys, C. V., and Diekmann, K. (2016). *Protecting people and the environment: Lessons learnt from Brazil's Bolsa Verde, China, Costa Rica, Ecuador, Mexico, South Africa, and 56 other experiences* (No. 995164592002676). International Labour Organisation. https://www .ilo.org/sites/default/files/wcmsp5/groups/public/%40ed_emp/%40gjp /documents/publication/wcms_516936.pdf

Scottish and Southern Energy (SSE). (2024). *Your path to a low carbon career*. https://perma.cc/2UA2-H7PU

ShareAction. (2021). *Slow Reactions: Chemical companies must transform in a low-carbon world*. https://shareaction.org/reports/slow-reactions -chemical-companies-must-transform-in-a-low-carbon-world

Smil, V. (2004). World history and energy. *Encyclopedia of Energy, 6*, 549–561. https://perma.cc/GC9F-YM2B

Songwe, V., Stern, N., and Bhattacharya, A. (2022). *Finance for climate action: Scaling up investment for climate and development*. Grantham Research Institute on Climate Change and the Environment, London School of Economics and Political Science. https://perma.cc/V454-EKF5

State of Green. (2024). *District energy: The backbone of a flexible, resilient and efficient energy system* (White Paper). https://perma.cc/A4J5-8T93

Stern et al. (2025) on AI, forthcoming

Stern, N. (2022). A time for action on climate change and a time for change in economics. *The Economic Journal, 132*(644), 1259–1289. https://doi .org/10.1093/ej/ueac005

Stern, N. and Romani, M. (2023). *The global growth story of the 21st century: Driven by investment and innovation in green technologies and artificial intelligence*. Grantham Research Institute on Climate Change and the Environment, London School of Economics and Political Science, and Systemiq. https://perma.cc/5BA7-6Y4Q

Stern, N. and Stiglitz, J. (2023). Climate change and growth. *Industrial and Corporate Change, 32*(2), 277–303. https://doi.org/10.1093/icc/dtad008

Stern, N., Tanaka, J., Bhattacharya, A., Lankes, H. P., Pierfederici, R., Rydge, J., Taylor, C., & Ward, B. (2021). *G7 leadership for sustainable, resilient, and inclusive economic recovery and growth: An independent report requested by the UK Prime Minister for the G7*. Grantham Research Institute on Climate Change and the Environment, London School of Economics and Political Science. https://perma.cc/FUE7-HC3A

Suarez, I. (2020). *Strategies that Achieve Climate Mitigation and Adaptation Simultaneously*. World Resources Institute. https://perma.cc/3J8G-XZTS

Sutton, W. R., Lotsch, A., and Prasann, A. (2024). *Recipe for a Liveable Planet: Achieving Net Zero Emissions in the Agrifood System*. World Bank Group. https://hdl.handle.net/10986/41468

Swanson, A., and Rappeport, A. (2024, June 6). U.S. adds tariffs to shield struggling solar industry. *The New York Times*. https://www.nytimes. com/2024/06/06/business/economy/tariffs-solar-industry-china.html

Systemiq. (2023a). *The Breakthrough Effect: How To Trigger A Cascade Of Tipping Points To Accelerate The Net Zero Transition.* https://perma.cc/WM6Y-894N

Systemiq. (2023b). *Financing nature: a transformation action agenda.* https://www.systemiq.earth/financing-nature/

Systemiq. (2025). *Accelerating the Breakthrough of Climate Technologies: driving exponential growth in climate technologies with positive tipping points.*

Task Force on Climate-related Financial Disclosures (TCFD). (2017). *Recommendations of the Task Force on Climate-related Financial Disclosures: Final Report.* https://perma.cc/NX6J-AYFT

The Royal College of Paediatrics and Child Health (RCPCH). (2023). *Child health inequalities and climate change in the UK.* https://perma.cc/3VS7-9DW6

Thorpe, D. (2021). *The world's most successful model for sustainable urban development?* Smart Cities Dive. https://perma.cc/HJ2H-VM43

Thunberg, G. (2019, September 23). *Speech presented at the United Nations Climate Action Summit* [Speech transcript]. NPR. https://perma.cc/U8B6-K6DU

United Nations (UN). (2020). *A/75/181/Rev.1: The 'just transition' in the economic recovery: Eradicating poverty within planetary boundaries – Interim report of the Special Rapporteur on extreme poverty and human rights.* Retrieved from https://www.ohchr.org/en/documents/thematic-reports/a75181rev1-just-transition-economic-recovery-eradicating-poverty-within

United Nations Environment Programme (UNEP). (2019). *Global Trends in Renewable Energy Investment 2019.* https://www.unep.org/resources/report/global-trends-renewable-energy-investment-2019

United Nations Environment Programme. (2022). *State of Finance for Nature 2022.* https://www.unep.org/resources/state-finance-nature-2022

United Nations Environment Programme. (2023a). *Emissions Gap Report: Nations must go further than current Paris pledges or face global warming of 2.5-2.9°C.* https://unepccc.org/emissions-gap-report-nations-must-go-further-than-current-paris-pledges-or-face-global-warming-of-2-5-2-9c/

United Nations Environment Programme. (2023b). *State of Finance for Nature 2023: The Big Nature Turnaround – Repurposing $7 trillion to combat nature loss.* https://www.unep.org/resources/state-finance-nature-2023

United Nations Environment Programme. (2024a). *Food Waste Index Report 2024. Think Eat Save Tracking Progress to Halve Global Food Waste.* https://wedocs.unep.org/20.500.11822/45230

United Nations Environment Programme. (2024b). *UNEP Food Waste Index Report 2024 Key Messages.* https://wedocs.unep.org/bitstream/handle/20 .500.11822/45275/Food-Waste-Index-2024-key-messages.pdf?sequence= 8&isAllowed=y

United Nations Framework Convention on Climate Change (UNFCCC). (2018). *2018 Biennial Assessment and Overview of Climate Finance Flows Technical Report.* https://unfccc.int/sites/default/files/resource/2018%20 BA%20Technical%20Report%20Final.pdf

United Nations Framework Convention on Climate Change. (2023). *Artificial intelligence for climate action in developing countries: opportunities, challenges and risks.* https://unfccc.int/ttclear/misc_/StaticFiles/gnwoerk _static/AI4climateaction/ea0f2596d93640349b9b65f4a7c7dd24/b47ef0e9 9cb24e57aa9ea69f0f5d6a71.pdf

United Nations Women (UNW). (2018). *Turning promises into action: Gender equality in the 2030 Agenda for Sustainable Development.* UN. https:// www.unwomen.org/en/digital-library/publications/2018/2/gender -equality-in-the-2030-agenda-for-sustainable-development-2018

van den Bergh, J. C. J. M. (2011). Environment versus growth: A criticism of 'degrowth' and a plea for 'a-growth'. *Ecological Economics, 70*(5), 881–890. https://doi.org/10.1016/j.ecolecon.2010.09.035

van Hoek, R. and Lacity, M. (2023, November 21). How Global Companies Use AI to Prevent Supply Chain Disruptions. *Harvard Business Review.* https://perma.cc/A489-35LA

Vohra, K., Vodonos, A., Schwartz, J., Marais, E. A., Sulprizio, M. P., and Mickley, L. J. (2021). Global mortality from outdoor fine particle pollution generated by fossil fuel combustion: Results from GEOS-Chem. *Environmental Research, 195.* https://doi.org/10.1016/j.envres.2021 .110754

Wakeford, J. (2018). *When mobile meets modular: pay-as-you-go solar energy in rural Africa.* LSE Blogs. https://perma.cc/EF6B-B48B

Whitehead, F. (2014, August 20). Lessons from Denmark: how district heating could improve energy security. *The Guardian.* https://perma.cc/XV9T -NBXM

Woodcock, J., Edwards, P., Tonne, C., Armstrong, B. G., Ashiru, O., Banister, D., Beevers, S., Chalabi, Z., Chowdhury, Z., Cohen, A., Franco, O. H., Haines, A., Hickman, R., Lindsay, G., Mittal, I., Mohan, D., Tiwari, G., Woodward, A., & Roberts, I. (2009). Public health benefits of strategies to reduce greenhouse-gas emissions: Urban land transport. *The Lancet*, 374(9705), 1930–1943. https://doi.org/10.1016/S0140-6736(09)61714-1

World Bank. (2019). *State and Trends of Carbon Pricing 2019*. https://hdl .handle.net/10986/31755

World Bank. (2022). *Country Climate and Development Report: Pakistan*. https://hdl.handle.net/10986/38277

World Bank. (2023a). *Country Climate and Development Report: Democratic Republic of Congo*. https://openknowledge.worldbank.org/handle/10986 /40599

World Bank. (2023b). *Country Climate and Development Report: Brazil*. https://openknowledge.worldbank.org/handle/10986/39782

World Bank. (2024). *Poverty, Prosperity, and Planet Report 2024: Pathways Out of the Polycrisis*. https://doi.org/10.1596/978-1-4648-2123-3

World Economic Forum (WEF). (2024). *Net-Zero Industry Tracker 2024 Edition*. https://perma.cc/WA3P-QDJH

Wu, H., Atamanov, A., Bundervoet, T., and Paci, P. (2024). Is economic growth less welfare enhancing in Africa? Evidence from the last forty years. *World Development 184*. https://doi.org/10.1016/j.worlddev.2024 .106759

Zero Waste Europe (ZWE). (2020). *France's law for fighting food waste: Food Waste Prevention Legislation*. https://perma.cc/MD45-6JXH

Chapter 6

Acemoglu, D., Akcigit, U., and Celik, M. A. (2014). *Young, restless and creative: Openness to disruption and creative innovations* (Working Paper No. 19894). National Bureau of Economic Research. https://www.nber.org/papers/w19894

Aghion, P., Akcigit, U., and Howitt, P. (2014). What do we learn from Schumpeterian growth theory? In P. Aghion and S. N. Durlauf (Eds.), *Handbook of Economic Growth* (Vol. 2, pp. 515–563). Elsevier.

Aghion, P., Antonin, C., and Bunel, S. (2021). *The Power of Creative Destruction: Economic upheaval and the wealth of nations.* Harvard University Press.

Aghion, P. and Howitt, P. (1992). A model of growth through creative destruction. *Econometrica, 60*(2), 323–351. https://doi.org/10.2307/2951599

Aghion, P. and Howitt, P. (2023). The creative destruction approach to growth economics. *European Review, 31*(4), 312–325. https://doi.org/10.1017/S1062798723000212

Akerlof, G. A. (1970). The market for lemons: Quality uncertainty and the market mechanism. *Quarterly Journal of Economics, 84*(3), 488–500. https://doi.org/10.2307/1879431

Arrow, K. (1963). Uncertainty and the welfare economics of medical care. *The American Economic Review, 53*(5), 941–973.

Association of American Railroads (AAR). (2024). *Chronology of America's freight railroads.* https://perma.cc/5JG9-KBV6

Bhattacharya, A., Ivanyna, M., Oman, W., and Stern, N. (2021). *Climate action to unlock the inclusive growth story of the 21st century.* International Monetary Fund. https://www.imf.org/en/Publications/WP/Issues/2021/05/26/Climate-Action-to-Unlock-the-Inclusive-Growth-Story-of-the-21st-Century-50219

Bhattacharya, A., Songwe, V., Soubeyran, E., and Stern, N. (2023). *A climate finance framework that is fit for purpose: Decisive action to deliver on the Paris Agreement — Summary.* Grantham Research Institute on Climate Change and the Environment, London School of Economics and Political Science. https://perma.cc/ZML3-CQ8Q

Bhattacharya, A., Songwe, V., Soubeyran, E., and Stern, N. (2024). *Raising ambition and accelerating delivery of climate finance.* Grantham Research Institute on Climate Change and the Environment, London School of Economics and Political Science. https://perma.cc/H6D9-MNSX

Black, S., Liu, A. A., Parry, I. W. H., and Vernon-Lin, N. (2023). *IMF fossil fuel subsidies data: 2023 update* (Working paper). International Monetary Fund. https://www.imf.org/en/Publications/WP/Issues/2023/08/22/IMF-Fossil-Fuel-Subsidies-Data-2023-Update-537281

Blended Finance Taskforce. (2023). *Better Guarantees, Better Finance: Mobilising capital for climate through fit-for-purpose guarantees.* https://perma.cc/EWU5-MJW9

Bloomberg. (2024). *Green bonds reached new heights in 2023.* https://www.bloomberg.com/professional/insights/markets/green-bonds-reached-new-heights-in-2023/

Bolton, O., Despres, M., Pereira Da Silva, L. A., Samama, F., and Svartzman, R. (2020). *The green swan: Central banking and financial stability in the age of climate change.* Bank for International Settlements. https://perma.cc/K5KT-35CL

Boudreaux, D. J. and Meiners, R. (2019). Externality: Origins and classifications. *Natural Resources Journal, 59*(1). https://digitalrepository.unm.edu/nrj/vol59/iss1/3

British Broadcasting Corporation (BBC). (2016, April 21). US solar firm SunEdison files for bankruptcy protection. *BBC News.* https://perma.cc/URX5-YBJ3

C40 Knowledge Hub. (2024). *How cities can use procurement to create and shape markets.* https://www.c40knowledgehub.org/s/article/How-cities-can-use-procurement-to-create-and-shape-markets?language=en_US

Carney, M. (2015, September 29). Breaking the tragedy of the horizon – climate change and financial stability. Speech presented at Lloyd's of London [Speech transcript]. *Bank of England.* https://perma.cc/B3UM-WW4N

Christie, J. (2017, March 14). Solar company Sungevity files for bankruptcy to press for speedy sale. *Reuters.* https://www.reuters.com/article/legal/solar-company-sungevity-files-for-bankruptcy-to-press-for-speedy-sale-idUSL2N1GR1LJ/

Climate Change Committee (CCC). (2020). *CCC insights briefing 3: The UK's net zero target.* https://perma.cc/64DF-HURN

Climate Action Tracker. (2021). *Warming projections global update.* https://perma.cc/4VHX-JFFK

Climate Policy Initiative (CPI). (2022). *Landscape of climate finance in Africa.* https://perma.cc/T8Q5-DNZN

Convergence Blended Finance. (2024). *The state of blended finance 2024: Convergence report.* https://perma.cc/Y83C-C8E4

Convergence Blended Finance. (n.d.). *Blended Finance.* https://www .convergence.finance/blended-finance

Crotty, J. (2009). Structural causes of the global financial crisis: a critical assessment of the 'new financial architecture'. *Cambridge Journal of Economics, 33*(4), 563–580. https://doi.org/10.1093/cje/bep023

Damania, R., Balseca, E., de Fontaubert, C., Gill, J., Kim, K., Rentschler, J., Russ, J., and Zaveri, E. (2023). *Detox Development: Repurposing Environmentally Harmful Subsidies.* World Bank Group. https://hdl.handle.net /10986/39423

The Danish Government's Climate Partnerships. (2022). *Climate Partnership Playbook: How to engage the private sector in climate action plans.* State of Green. https://issuu.com/stateofgreen/docs/19476-kp-playbook_120922 _skaermversion_

Denton, F., Halsnæs, K., Akimoto, K., Burch, S., Diaz Morejon, C., Farias, F., Jupesta, J., Shareef, A., Schweizer-Ries, P., Teng, F., and Zusman, E. (2022). Accelerating the transition in the context of sustainable development. In P.R. Shukla, J. Skea, R. Slade, A. Al Khourdajie, R. van Diemen, D. McCollum, M. Pathak, S. Some, P. Vyas, R. Fradera, M. Belkacemi, A. Hasija, G. Lisboa, S. Luz, and J. Malley, (Eds.), *Climate Change 2022: Mitigation of Climate Change. Contribution of Working Group III to the Sixth Assessment Report of the Intergovernmental Panel on Climate Change* (pp. 2816–2915). Cambridge University Press. https://doi.org/10.1017 /9781009157926.019

Department of Finance Philippines. (2023). *DOF reconvenes Green Force, discusses next steps on Sustainable Finance Roadmap.* https://perma.cc /DN8T-RCQB

Dikau, S., Robins, N., Smoleńska, A., van't Klooster, J., and Volz, U. (2025). Prudential net zero transition plans: The potential of a new regulatory instrument. *Journal of Banking Regulation, 26,* 85–99. https://doi.org/10 .1057/s41261-024-00247-w

Dikau, S. and Volz, U. (2021). Central bank mandates, sustainability objectives and the promotion of green finance. *Ecological economics, 184,* 107022. https://doi.org/10.1016/j.ecolecon.2021.107022

Elliott, C., Schumer, C., Gasper, R., Ross, K., and Singh, N. (2024). *Costa Rica's Pioneering Net-zero Implementation Plan Attracts Investment, Withstands Political Changes.* World Resources Institute. https://perma.cc /LGG7-KZXG

Energy Transitions Commission (ETC). (2023). *Financing the transition: How to make the money flow for a net-zero economy.* Systemiq. https:// perma.cc/6BTC-CMBS

Energy Transitions Commission. (2024). *Building grids faster: The backbone of the energy transition.* Systemiq. https://perma.cc/8B8A-LPAR

Espagne, E., Oman, W., Mercure, J.F., Svartzman, R., Volz, U., Pollitt, H., Semieniuk, G., and Campiglio, E. (2023). *Cross-Border Risks of a Global Economy in Mid-Transition* (IMF Working Paper No. 2023/184). International Monetary Fund. https://www.imf.org/en/Publications/WP/Issues/2023/09/08/Cross-Border-Risks-of-a-Global-Economy-in-Mid-Transition-538950

European Bank for Reconstruction and Development (EBRD). (1994). *Transition report: Economic transition in eastern Europe and the former Soviet Union.* https://www.ebrd.com/content/dam/ebrd_dxp/assets/pdfs/office-of-the-chief-economist/transition-report-archive/transition-report-1994/Transition-Report-1994-English.pdf

European Bank for Reconstruction and Development. (1995). *Transition report: Investment and enterprise development.* https://www.ebrd.com/content/dam/ebrd_dxp/assets/pdfs/officc-of-the-chief-economist/transition-report-archive/transition-report-1995/Transition-Report-1995-Investment-and-Enterprise-Development-Part-One-English.pdf

European Bank for Reconstruction and Development. (1996). *Transition report: Infrastructure and savings. https://www.ebrd.com/content/dam/ebrd_dxp/assets/pdfs/office-of-the-chief-economist/transition-report-archive/transition-report-1996/Transition-Report-1996-Infrastructure-and-Savings-Part-One-English.pdf*

European Bank for Reconstruction and Development. (1997). *Transition report: Enterprise performance and growth.* Retrieved from https://www.ebrd.com/home/news-and-events/publications/economics/transition-reports.html

European Bank for Reconstruction and Development. (1998). *Transition report: Financial sector in transition.* Retrieved from https://www.ebrd.com/home/news-and-events/publications/economics/transition-reports.html

European Bank for Reconstruction and Development. (1999). *Transition report: Ten years of transition.* Retrieved from https://www.ebrd.com/home/news-and-events/publications/economics/transition-reports.html

European Bank for Reconstruction and Development. (2023). *Transition Report 2023–24: Transitions big and small.* https://www.ebrd.com/home/news-and-events/publications/economics/transition-reports/transition-report-2023-24.html

European Commission. (2016). *Green bonds: New study shows extraordinary growth and signals potential in financing Europe's climate and environ-*

ment goals. https://ec.europa.eu/commission/presscorner/detail/en/ip_16_4217

Freeman, C. and Louçã, F. (2001). *As Time Goes By: from the industrial revolutions to the information revolution.* Oxford University Press.

Grubert, E. and Hastings-Simon, S. (2022). Designing the mid-transition: a review of medium-term challenges for coordinated decarbonization in the United States. *Wiley Interdisciplinary Reviews: Climate Change, 13*(3). https://doi.org/10.1002/wcc.768

Hasanbeigi, A. and Bhadbhade, N. (2023). *Green public procurement of steel in India, Japan, and South Korea.* Global Efficiency Intelligence. https://perma.cc/NM5N-AESX

Hauber, G. (2023, December 22). *Financing the JETP: Making sense of the packages.* Institute for Energy Economics and Financial Analysis. https://perma.cc/AKK4-37WZ

Hansard (1968, June 21). *Ministry of Social Security Site, Longbenton.* Vol 766. Hansard. Retrieved from https://hansard.parliament.uk/commons/1968-06-21/debates/5d6b450d-a3bc-4348-bfd5-57db0db044ad/MinistryOfSocialSecuritySiteLongbenton

Institutional Investors Group on Climate Change (IIGCC). (2024). *Making NDCs investable – the investor perspective.* https://perma.cc/G83M-GEEX

International Energy Agency (IEA). (2024). *Reducing the Cost of Capital.* https://perma.cc/P2NX-N24R

International Finance Corporation (IFC). (2024). *Reassessing Risk in Emerging Market Lending: Insights from GEMs Consortium Statistics.* World Bank Group. https://perma.cc/K6AE-TRQN

Janeway, B. (2021). Productive bubbles. *NOEMA Magazine.* https://perma.cc /26AR-338M

Johnson, C. (1982). *MITI and the Japanese miracle: The growth of industrial policy, 1925–1975.* Stanford University Press.

Johnson, C. (1999). The developmental state: Odyssey of a concept. In M. Woo-Cummings (Ed.), *The developmental state* (pp. 32–60). Cornell University Press.

Kammourieh, S. and Songwe, V. (2024). *Next generation prudential regulation for global financial system resilience.* https://perma.cc/XR77-6495

Kharas, H. and Rivard, C. (2024). *Unpacking developing country debt problems: Selected reforms to the international financial architecture.* Centre for Sustainable Development at Brookings. https://www.brookings.edu

/wp-content/uploads/2024/04/Unpacking-developing-country-debt
-service-problems.pdf

Krugman, P. (2009). *The return of depression economics and the crisis of 2008.* W. W. Norton and Company.

Lenshie, N. E., Okengwu, K., Ogbonna, C. N., and Ezeibe, C. (2021). Desertification, migration, and herder-farmer conflicts in Nigeria: Rethinking the ungoverned spaces thesis. *Small Wars & Insurgencies, 32*(8), 1221–1251. https://doi.org/10.1080/09592318.2020.1811602

Louçã, F. (2020). Chris Freeman: Forging the evolution of evolutionary economics. *Industrial and Corporate Change, 29*(4), 1037–1046. https://doi.org/10.1093/icc/dtaa017

Lynn, L. H. (1994). MITI's successes and failures in controlling Japan's technology imports. *Hitotsubashi Journal of Commerce and Management, 29*(1), 15–33.

Managi, S., Bhattacharya, A., and Bhattacharya, T. (2023). *Inclusive wealth index: A comprehensive measure of LiFE towards 'net zero'.* G20 India. https://t20ind.org/research/inclusive-wealth-index/

Marshall, A. (1890). *Principles of Economics.* Macmillan and Company.

Meade, J. E. (1952). External Economies and Diseconomies in a Competitive Situation. *The Economic Journal, 62*(245), 54–67. https://doi.org/10.2307/2227173

Ministry of International Cooperation, Arab Republic of Egypt. (2022). *Sharm El-Sheikh guidebook for just financing.* https://guidebookforjustfinancing.com/wp-content/uploads/2022/11/Sharm-El-Sheikh-Guidebook-for-Just-Financing.pdf

NatureFinance and World Bioeconomy Forum. (2023). *Financing a Sustainable Global Bioeconomy.* https://www.climatepolicyinitiative.org/wp-content/uploads/2024/08/FinancingASustainableGlobalBioeconomy.pdf

New Climate Economy. (2015). *Seizing the global opportunity: Partnerships for better growth and a better climate.* World Resources Institute. https://perma.cc/2FAV-CDHF

Octopus Energy. (2024, December 4). *How zonal pricing could make bills cheaper.* https://octopus.energy/blog/locational-zonal-pricing-explained/

Octopus Energy Generation. (2023). *End the gridlock: Quick wins to connect more renewables.* https://perma.cc/8QL3-UFBA

Organisation for Economic Co-operation and Development (OECD). (n.d.). *Public procurement.* https://www.oecd.org/en/topics/public-procurement

.html#:~:text=Public%20procurement%20expenditure%20as%20a ,countries%20between%202019%20and%202021.

Perez, C. (2002). *Technological revolutions and financial capital: the dynamics of bubbles and golden ages.* Edward Elgar.

Perez, C. (2010). Technological revolutions and techno-economic paradigms. *Cambridge Journal of Economics, 34*(1), 185–202. https://doi.org /10.1093/cje/bep051

Pigou, A. C. (1920). *The economics of welfare.* Macmillan and Co.

Pisani-Ferry, J. (2021). *Climate policy is macroeconomic policy, and the implications will be significant* (PB21-20). Peterson Institute for International Economics. https://perma.cc/Z6FT-XJAE

Reserve Bank of India (RBI). (2024). *Monetary policy report.* https://m.rbi .org.in/Scripts/PublicationsView.aspx?id=22435

Rocky Mountain Institute (RMI). (2024). *X-Change: The race to the top.* https://perma.cc/785E-4XQ8

Sachs, J. D., Lafortune, G., and Fuller, G. (2024). *The SDGs and the UN Summit of the Future: Sustainable Development Report 2024.* SDSN, Dublin University Press. https://doi.org/10.25546/108572

Sawaqed, L. and Griffin, C. (2023). *Planning for success: strategies of investment promotion agencies technical note.* World Bank. http://documents .worldbank.org/curated/en/099005002152373092

Schumpeter, J. A. (1942). *Capitalism, socialism, and democracy.* Harper & Brothers.

Schwab, K. (2016). *The fourth industrial revolution.* Portfolio Penguin.

Segerstrom, P., Anant, T., and Dinopoulos, E. (1990). A Schumpeterian model of the product cycle. *American Economic Review, 88*(5), 1077– 1092.

Shiller, R. J. (2008). *The subprime solution: how today's global crisis happened, and what to do about it.* Princeton University Press.

Singh, N. K. and Summers, L. (2023a). *Strengthening multilateral development banks: The triple agenda MANDATES I FINANCE I MECHANISMS* (Report of the Independent Experts Group, Vol. 1). G20 India. https:// perma.cc/6BCU-PG3P

Singh, N. K. and Summers, L. (2023b). *Strengthening multilateral development banks: The triple agenda: Better, Bolder and Bigger MDBs* (Report of the Independent Experts Group, Vol. 2). G20 India https://perma.cc /QY4T-23KA

Smith, C., Gandel, S., and Franklin, J. (2024). Federal Reserve halves proposed capital requirement rise for largest US banks. *The Financial Times*. https://www.ft.com/content/86fd9a80-bf46-4711-ab33-e4dcbef5eeb4

Songwe, V., Stern, N., and Bhattacharya, A. (2022). *Finance for climate action: Scaling up investment for climate and development*. Grantham Research Institute on Climate Change and the Environment, London School of Economics and Political Science. https://perma.cc/V454-EKF5

Spence, M. (1973). Job market signaling the quarterly journal of economics, 87 (3). *MIT Press, August, 355*, 374.

Stern, N. (2007). *The economics of climate change: The Stern Review*. Cambridge University Press.

Stern, N., Tanaka, J., Bhattacharya, A., Lankes, H. P., Pierfederici, R., Rydge, J., Taylor, C., and Ward, B. (2021). *G7 leadership for sustainable, resilient, and inclusive economic recovery and growth: An independent report requested by the UK Prime Minister for the G7*. Grantham Research Institute on Climate Change and the Environment, London School of Economics and Political Science. https://perma.cc/FUE7-HC3A

Stiglitz, J. (1975). The theory of 'screening', education, and the distribution of income. *The American Economic Review, 65*(3), 283–300.

Stiglitz J. (2010). *Freefall: America, Free Markets, and the Sinking of the World Economy*. W.W. Norton & Company.

Stiglitz, J. and Stern, N. (2017). *Report of the high-level commission on carbon prices*. World Bank. http://eprints.lse.ac.uk/id/eprint/87990

Task Force on Climate-related Financial Disclosures (TCFD). (2017). *Recommendations of the task force on climate-related financial disclosures: final report*. https://perma.cc/NX6J-AYFT

The Wall Street Journal. (2019). Economists' statement on carbon dividends. *The Wall Street Journal*. https://www.wsj.com/articles/economists -statement-on-carbon-dividends-11547682910

The Washington Post. (2011). Solyndra Scandal | Full Coverage of Failed Solar Startup. *The Washington Post*. https://www.washingtonpost.com /politics/specialreports/solyndra-scandal/

Weitzman, M.L. (1974). Prices *vs.* Quantities. *The Review of Economic Studies, 41*(4), 477–491

World Bank. (2004). *World Development Report 2005: A Better Investment Climate for Everyone*. http://hdl.handle.net/10986/5987

World Bank. (2021). *The changing wealth of nations 2021: managing assets for the future*. https://hdl.handle.net/10986/36400

World Resources Institute (WRI). (2021). *Spain's National Strategy to Transition Coal-Dependent Communities*. https://perma.cc/89CU-Z9P3

Zenghelis, D., Serin, E., Stern, N. H., Valero, A., Van Reenen, J., and Ward, B. (2024). *Boosting growth and productivity in the United Kingdom through investments in the sustainable economy*. Grantham Research Institute on Climate Change and the Environment, London School of Economics and Political Science. https://perma.cc/X5B5-LPRX

Chapter 7

Besley, T. and Persson, T. (2023). The political economics of green transitions. *The Quarterly Journal of Economics, 138*(3), 1863–1906. https://doi.org/10.1093/qje/qjad006

Besley, T., Marshall, J., and Persson, T. (2023). Well-being and state effectiveness. In *World Happiness Report 2023* (11th ed., Chapter 3). Sustainable Development Solutions Network. https://worldhappiness.report/ed/2023/well-being-and-state-effectiveness/

Confederation of Indian Industry. (2023, September 23). 'The World is on Fire': A session with Prof. Larry Summers [Video]. YouTube. https://www.youtube.com/watch?v=82gUow7Fysw

Drèze, J. and Stern, N. (1987). The theory of cost-benefit analysis. In A.J. Auerbach and M. Feldstein (Eds.), *Handbook of Public Economics* (Vol. 2, pp. 909–989). Elsevier.

Eskander, S.M.S.U. and Fankhauser, S. (2020). Reduction in greenhouse gas emissions from national climate legislation. *Nature Climate Change, 10*, 750–756. https://doi.org/10.1038/s41558-020-0831-z

Freeman, C. (1995). The 'National System of Innovation' in historical perspective. *Cambridge Journal of Economics, 19*(1), 5–24. https://doi.org/10.1093/oxfordjournals.cje.a035309

Mazzucato, M. (2021). *Mission Economy: A moonshot guide to changing capitalism*. Penguin UK.

Meade, J. E. (1973). *The theory of economic externalities: The control of environmental pollution and similar social costs* (Vol. 2). Martinus Nijhoff Publishers.

Oliver, J. (2024, June 11). Right to Buy scheme must be scrapped to ease UK social housing crisis, says JLL. *The Financial Times*. https://www.ft.com/content/e2381716-4c7a-4d64-9beb-31dd8b5c5255?

Oreskes, N. and Conway, E. M. (2010). *Merchants of doubt: How a handful of scientists obscured the truth on issues from tobacco smoke to global warming*. Bloomsbury Press.

Our World in Data. (2024). *Income share of the richest 1% (before tax), 1980 to 2022*. https://ourworldindata.org/grapher/income-share-top-1-before-tax-wid-extrapolations?time=1980...latest&country=USA~GBR~OWID_WRL

Perez, C. (2002). *Technological revolutions and financial capital: the dynamics of bubbles and golden ages*. Edward Elgar.

Perez, C. (2010). Technological revolutions and techno-economic paradigms. *Cambridge Journal of Economics, 34*(1), 185–202. https://doi.org/10.1093/cje/bep051

Portes, J. (2022, August 16). I worked on the privatisation of England's water in 1989. It was a failed regime. *The Guardian.* https://perma.cc/WE64-P8Z6

Setzer, J. and Higham, C. (2023). *Global trends in climate change litigation: 2023 snapshot.* Grantham Research Institute on Climate Change and the Environment, London School of Economics and Political Science https://perma.cc/P245-GF9C

Shelter. (2024). *Loss of social housing.* https://perma.cc/2UYL-UWPL

Stern, N., Romani, M., Pierfederici, R., Braun, M., Barraclough, D., Lingeswaran, S, Weirich-Benet, E., and Niemann, N. (2025). Green and intelligent: the role of AI in the climate transition. *npj Climate Action,* 4(56). https://doi.org/10.1038/s44168-025-00252-3

Systemiq. (2020). *The Paris Effect: How the Climate Agreement is reshaping the global economy.* https://perma.cc/A953-7VJZ

Tebbit, N. (1981). Speech at Conservative Party Conference [Speech transcript]. *Oxford Reference.* https://perma.cc/W6L7-54T7

Thatcher, M. (1987, September 23). Interview for Woman's Own. Interview by D. Keay. *Margaret Thatcher Foundation.* https://perma.cc/2A8A-3HM7

PART III

Chapter 8

Acemoglu, D., Johnson, S., and Robinson, J.A. (2001). The colonial origins of comparative development: An empirical investigation. *American Economic Review, 91*(5), 1369–1401. https://doi.org/10.1257/aer.91.5.1369

Ahmed, A. (2024, June 2). Why so many of the world's best chief executives are Indian. *The Times.* https://www.thetimes.com/business-money /companies/article/world-best-chief-executives-indians-comment -t55vqvpsd

Ashford, O., Baines, J., Barbanell, M., and Wang, K. (2025, April 23). *What we know about deep-sea mining – and what we don't.* World Resources Institute. https://perma.cc/TUV9-4Y2T

Barthakur, R. (2023, October 1). Valuing nature for a nature-based economy. *Asom Barta.* https://perma.cc/DN54-CFJ5

Bhattacharya, A., Songwe, V., Soubeyran, E., and Stern, N. (2024). *Raising ambition and accelerating delivery of climate finance.* Grantham Research Institute on Climate Change and the Environment, London School of Economics and Political Science. https://perma.cc/H6D9-MNSX

Bian, L., Dikau, S., Miller, H., Pierfederici, R., Stern, N., and Ward, B. (2024). *China's role in accelerating the global energy transition through green supply chains and trade.* Grantham Research Institute on Climate Change and the Environment, London School of Economics and Political Science. https://perma.cc/7EK5-UZ93

Bickenbach, F., Dohse, D., Langhammer, R., and Liu, W. (2024). *Foul play? On the scale and scope of industrial subsidies in China* (N. 173). Institute for the World Economy. https://perma.cc/X78M-MHHV

BloombergNEF. (2022). *Climatescope 2022: Power sector assessment of 136 global markets* .https://2022.global-climatescope.org/about/

Bond, K. and Butler-Sloss, S. (2023). *The renewable revolution: It's exponential, global, and this decade.* Rocky Mountain Institute. https://perma.cc /U2XK-NAMM

Boston Consulting Group (BCG). (2023). *Impact of IRA, IIJA, CHIPS, and Energy Act of 2020 on Clean Technologies: Cross-technology Summary.* https://perma.cc/6UV4-7TH4

Bromley, H. (2024). *Biodiversity Finance Factbook.* BloombergNEF. https:// assets.bbhub.io/professional/sites/24/Biodiversity-Finance-Factbook _COP16.pdf

Budaragina, M, West, D., Villena, A. F., and Phillips, A. (2024). *Powering AI in the Global South*. Tony Blair Institute for Global Change. https://perma.cc/YVA9-5MCE

Campbell, P. and Parker, G. (2023, August 4). Chinese-owned battery group involved in Tata UK gigafactory. *The Financial Times*. https://www.ft.com/content/ce2bb0fb-a84d-4ff2-b29f-75538be45a3b

Cao, Y., Chen, J., Liu, B., Turner, A., and Zhu, C. (2021). *China zero-carbon electricity growth in the 2020s: A vital step toward carbon neutrality*. Rocky Mountain Institute (RMI), and Energy Transitions Commission (ETC). https://perma.cc/FQ3H-BJC3

Climate Change Tracker. (2024). *Current remaining carbon budget and trajectory*. https://perma.cc/W8BX-B7UK

ClimateScope. (2022). Results. [data]. https://2022.global-climatescope.org/results/

CTVC. (2022, August 1). *Inflation Reduction Act's climate tech impact #111*. Newsletter. https://perma.cc/F37F-5PG9

Department of Environment and Climate Change Bhutan. (2023). *Bhutan's Long-Term Low Greenhouse Gas Emission and Climate Resilient Development Strategy (LTS)*. https://unfccc.int/sites/default/files/resource/LTS%20Report_final%20print_copy.pdf

Dhakal, S., Minx, J. C., Toth, F. L., Abdel-Aziz, A., Figueroa Meza, M. J., Hubacek, K., Jonckheere, I. G. C., Kim, Y.-G., Nemet, G. F., Pachauri, S., Tan, X. C., and Wiedmann, T. (2022). Emissions trends and drivers. In P. R. Shukla, J. Skea, R. Slade, A. Al Khourdajie, R. van Diemen, D. McCollum, M. Pathak, S. Some, P. Vyas, R. Fradera, M. Belkacemi, A. Hasija, G. Lisboa, S. Luz, and J. Malley (Eds.), *Climate change 2022: Mitigation of climate change: Contribution of Working Group III to the Sixth Assessment Report of the Intergovernmental Panel on Climate Change* (pp. 215–294). Cambridge University Press. https://doi.org/10.1017/9781009157926.004

DiPippo, G., Mazzocco, I., and Kennedy, S. (2022). *Red Ink: Estimating Chinese industrial policy spending in comparative perspective*. Centre for Strategic and International Studies. https://perma.cc/V5VF-KKVV

Ember. (2023). *Beyond Tripling: India needs $101bn additional financing for the net-zero pathway*. Ember. https://ember-energy.org/latest-insights/beyond-tripling-india/

Ember. (2024a). *Electricity data explorer*. Retrieved January 10, 2025, from https://ember-energy.org/data-explorer/

Ember. (2024b). Share of electricity generated by wind power – Ember and Energy Institute [Dataset]. In Our World in Data. Ember, 'Yearly

Electricity Data'; Energy Institute, "Statistical Review of World Energy" [Original data]. https://ourworldindata.org/grapher/share-electricity -wind

Energy Transitions Commission (ETC). (2023). *Material and resource requirements for the energy transition.* Systemiq. https://perma.cc/3HG9 -XYQM

Energy Transitions Commission. (2024) *Credible Contributions: Bolder Plans for Higher Climate Ambition in the Next Round of NDCs.* Version 1. Insights Briefing. Systemiq. https://www.energy-transitions.org /publications/credible-contributions-bolder-plans-for-ndcs

Freeman, C. and Perez, C. (1988). Structural crises of adjustment, business cycles and investment behaviour. In Dosi, et al. *Technical Change and Economic Theory* (pp. 38–66). Pinter Publishers.

Garg, V. (2022, April 9). View: Private sector driving renewable energy wave in India. *Economic Times of India.* https://perma.cc/K395-4HK3

Gilbert, M., Lang, N., Mavropoulos, G., McAdoo, M., and Konomi, T. (2024). *Jobs, National Security, and the Future of Trade.* Boston Consulting Group. https://www.bcg.com/publications/2024/jobs-national -security-and-future-of-trade

Hoicka, C. E., Graziano, M., Willard-Stepan, M, and Zhao, Y. (2025). Insights to accelerate place-based at scale renewable energy landscapes: An analytical framework to typify the emergence of renewable energy clusters along the energy value chain. *Applied Energy, 377*(Part C). https://doi.org/10.1016/j.apenergy.2024.124559

Hove, A. (2024). *Clean energy innovation in China: Fact and fiction, and implications for the future.* The Oxford Institute for Energy Studies. https://perma.cc/S8U3-79JL

Influence Map. (2022). *The US oil/gas industry and the war in Ukraine.* https://perma.cc/P3M8-UC3W

Intergovernmental Panel on Climate Change (IPCC). (2022). *Climate change 2022: Impacts, adaptation, and vulnerability. Contribution of Working Group II to the Sixth Assessment Report of the Intergovernmental Panel on Climate Change.* Cambridge University Press. https://doi.org/10.1017 /9781009325844

International Energy Agency (IEA). (2019, February 4). *Have the prices from competitive auctions become the "new normal" prices for renewables? Analysis from Renewables 2018.* https://www.iea.org/articles/have-the -prices-from-competitive-auctions-become-the-new-normal-prices-for -renewables

International Energy Agency, International Renewable Energy Agency, United Nations Statistics Division, World Bank, and World Health Organisation. (2023). *Tracking SDG 7: The energy progress report.* World Bank Group. https://perma.cc/8UMU-M9NT

International Energy Agency, International Renewable Energy Agency (IRENA), United Nations Statistics Division (UNSD), World Bank, and World Health Organisation (WHO). (2024). *Tracking SDG 7: The energy progress report 2024.* World Bank Group. https://perma.cc/D8V3-YATK

International Energy Agency. (2020). *Evolution of solar PV module cost by data source, 1970-2020.* https://perma.cc/T3K4-UL4X

International Energy Agency. (2021). *The role of critical minerals in clean energy transitions.* https://perma.cc/2G4N-7RTG

International Energy Agency. (2022a). *Africa Energy Outlook 2022.* https://perma.cc/95Z3-N7KV

International Energy Agency. (2022b). *Global Supply Chains of EV Batteries.* https://perma.cc/4T2Y-FZZB

International Energy Agency. (2024a). *Global Critical Minerals Outlook 2024.* https://perma.cc/57SH-68VR

International Energy Agency. (2024b). *Energy Technology Perspectives 2024.* https://perma.cc/EV86-T2CV

International Energy Agency. (2024c). *Clean Energy Investment for Development in Africa.* https://perma.cc/HQM2-U6LE

International Energy Agency. (2024d). *World Energy Outlook 2024.* https://perma.cc/349B-8SLQ

International Energy Agency. (2025a). *Energy and AI.* https://perma.cc/WE2Q-BGYN

International Energy Agency. (2025b). *Uruguay.* https://www.iea.org/countries/uruguay/electricity

International Renewable Energy Agency (IRENA). (2024). *A just and inclusive energy transition in emerging markets and developing economies: Energy planning, financing, sustainable fuels and social dimensions.* https://perma.cc/PN9J-EJLF

International Renewable Energy Agency. (2019). *A New World: The Geopolitics of Energy Transformation.* https://perma.cc/2DVM-QW3J

Kouamé, A. T. (2024). *Gearing up for India's Rapid Urban Transformation.* World Bank Group. https://perma.cc/5G9P-9LEB

Leruth L., Mazarei, A., Régibeau, P., and Renneboog, L. (2022). *Green energy depends on critical minerals. who controls the supply chains?* (Working Paper 22-12). Peterson Institute for International Economics. https://perma.cc/KB3L-V6Y9

Miller, H., Dikau, S., Svartzman, R, and Dees, S. (2023). *The stumbling block in 'the race of our lives': transition-critical materials, financial risks and the ngfs climate scenarios* (Working Paper No. 393). Grantham Research Institute on Climate Change and the Environment, London School of Economics and Political Science. https://perma.cc/JGS3-V44E

Milman, O. (2022). US fossil fuel industry leaps on Russia's invasion of Ukraine to argue for more drilling. *The Guardian*. https://perma.cc/Z9HX-68Z9

Mostefaoui, M., Ciais, P., McGrath, M. J., Peylin, P., Patra, P. K., and Ernst, Y. (2024). Greenhouse gas emissions and their trends over the last 3 decades across Africa. *Earth System Science Data, 16,* 245–275, https://doi.org/10.5194/essd-16-245-2024

Muiruri, P. (2024, January 25). 'Our contribution to a cleaner world': How Kenya found an extraordinary power source beneath its feet. *The Guardian*. https://perma.cc/G755-4JST

Ojea, E., Lester, S. E., and Salgueiro-Otero, D. (2020). Adaptation of fishing communities to climate-driven shifts in target species. *One Earth, 2*(6), 544-556. https://doi.org/10.1016/j.oneear.2020.05.012

Oreskes, N. and Conway, E. M. (2010). *Merchants of doubt: How a handful of scientists obscured the truth on issues from tobacco smoke to global warming.* Bloomsbury Publishing.

Organisation for Economic Co-operation and Development. (2024). *Development Co-operation Report 2024: tackling poverty and inequalities through the green transition.* OECD Publishing. https://doi.org/10.1787/357b63f7-en.

Ortiz-Ospina, E., Beltekian, D., and Roser, M. (2024). *Trade and Globalization*. Our World in Data. https://ourworldindata.org/trade-and-globalization

Otojanov, R., Fouquet, R., and Granville, B. (2022). Factor prices and induced technical change in the industrial revolution. *The Economic History Review, 76*(2), pp.599–623. https://doi.org/10.1111/ehr.13194

Our World in Data. (2024). Solar panel prices have fallen by around 20% every time global capacity doubled. https://ourworldindata.org/data-insights/solar-panel-prices-have-fallen-by-around-20-every-time-global-capacity-doubled

Owen, G. (2024, January 9). *The future of the UK auto industry.* Policy Exchange. https://perma.cc/2JBK-HP2J

Pastukhova, M. and Walker, B. (2024). *An orderly and equitable global transition away from fossil fuels: An action framework to navigate economic, financial and geopolitical volatility.* E3G. https://perma.cc/82NX-XN42

Pörtner, H.-O., Roberts, D. C., Tignor, M., Poloczanska, E. S., Mintenbeck, K., Alegría, A., Craig, M., Langsdorf, S., Löschke, S., Möller, V., Okem, A., and Rama, B. (Eds.). (2022). *Climate change 2022: Impacts, adaptation, and vulnerability. Contribution of Working Group II to the Sixth Assessment Report of the Intergovernmental Panel on Climate Change.* Cambridge University Press. https://doi.org/10.1017/9781009325844

Puyo, D. M., Panton, A. J., Sridhar, T., Stuermer, M., Ungerer, C., and Zhang, A.T. (2024). *Key Challenges Faced by Fossil Fuel Exporters during the Energy Transition* (IMF Staff Climate Note No 2024/001). International Monetary Fund. https://www.imf.org/en/Publications/staff-climate-notes /Issues/2024/03/26/Key-Challenges-Faced-by-Fossil-Fuel-Exporters -during-the-Energy-Transition-546066

Rannard, G. (2023, July 31). Europe weather: How heatwaves could for ever change summer holidays abroad. *BBC News.* https://perma.cc/S8UM-BX5J

Ritchie, H. (2021). *The price of batteries has declined by 97% in the last three decades.* Our World in Data. https://ourworldindata.org/battery-price -decline

Ritchie, H. and Rosado, P. (2024). *Most of the world's cobalt is mined in the Democratic Republic of Congo, but refined in China.* Our World in Data. https://ourworldindata.org/data-insights/most-of-the-worlds-cobalt -is-mined-in-the-democratic-republic-of-congo-but-refined-in-china #:~:text=Almost%20three-quarters%20of%20the,cobalt%20is%20made %20in%20China

Rooper, H. (2024). *Emissions growth in the developing world.* Climate Leadership Council. https://perma.cc/KY6P-K9WQ

Roser, M. (2020). *Why did renewables become so cheap so fast?* Our World in Data. https://ourworldindata.org/cheap-renewables-growth

S&P Global. (2024, November 19). *Sustainability insights: Rising curtailment in China power producers will push past the pain.* https://www.spglobal .com/ratings/en/research/articles/241119-sustainability-insights-rising -curtailment-in-china-power-producers-will-push-past-the-pain -13327811

Seong, J., White, O., Birshan, M., Woetzel, L., Lamanna, C., Condon, J., and Devesa, T. (2024). *Geopolitics and the geometry of global trade.* McKinsey. https://perma.cc/59VA-YWMC

Shi, H., Chertow, M., and Song, Y. (2010). Developing country experience with eco-industrial parks: a case study of the Tianjin Economic-Technological Development Area in China. *Journal of Cleaner Production, 18*(3), 191–199. https://doi.org/10.1016/j.jclepro.2009.10.002

Singh, V. and Bond, K. (2024). *Powering up the global south.* RMI. https://perma.cc/9HGP-6W77

Solargis. (2024). *Solar resource map, photovoltaic power potential.* Global Solar Atlas 2.0, World Bank Group. https://globalsolaratlas.info

Stanley, A. (2024). *A demographic transformation in Africa has the potential to alter the world order.* International Monetary Fund. https://www.imf.org/en/Publications/fandd/issues/2023/09/PT-african-century

Stern, N. and Romani, M. (2023). *The global growth story of the 21st century: Driven by investment and innovation in green technologies and artificial intelligence.* Grantham Research Institute on Climate Change and the Environment, London School of Economics and Political Science, and Systemiq. https://perma.cc/5BA7-6Y4Q

Swanson, A. and Rappeport, A. (2024, June 6). U.S. adds tariffs to shield struggling solar industry. *The New York Times.* https://www.nytimes.com/2024/06/06/business/economy/tariffs-solar-industry-china.html

Traide. (2023). *Challenges encountered by hydro-power dams in Ethiopia and associated business opportunities TRAIDE Ethiopia.* TRAIDE Ethiopia, and Kingdom of the Netherlands. https://perma.cc/564V-C9J9

United Nations Environment Programme (UNEP). (2020). *Emissions Gap Report 2020.* https://www.unep.org/emissions-gap-report-2020

United Nations Industrial Development Organisation (UNIDO). (2016). *Global assessment of eco-industrial parks in developing and emerging countries.* United Nations. https://perma.cc/4SQL-EXFT

United Nations Industrial Development Organisation. (2020). *Experiences and best practices of industrial park development in the People's Republic of China.* United Nations. https://perma.cc/983V-XN8E

United Nations Industrial Development Organization. (2022). *Comparative research report on the localized performance indicator systems of the international guidelines for industrial parks in China.* United Nations. https://perma.cc/762R-UPY8

Vulnerable Twenty Group (V20). (2022). *Climate vulnerable economies loss report: economic losses attributable to climate change in V20 economies over the last two decades (2000–2019).* https://perma.cc/2N8R-QYHR

Walter, D., Butler-Sloss, S., and Kingsmill, B. (2024). *The rise of batteries in six charts and not too many numbers*. The Rocky Mountain Institute. https://rmi.org/the-rise-of-batteries-in-six-charts-and-not-too-many-numbers/

Wood Mackenzie. (2023, November 07). *China dominance on global solar supply chain*. https://perma.cc/X4E5-LGMV

World Bank. (2024). *Infrastructure*. https://perma.cc/MN8F-6BMB

World Economic Forum (WEF). (2024). *Transitioning Industrial Clusters: Annual Report*. https://perma.cc/BQ4B-Y6U5

World Economic Forum. (2025a). *Unleashing the Full Potential of Industrial Clusters: Infrastructure Solutions for Clean Energies*, White Paper. https://perma.cc/8U39-TZYP

World Economic Forum. (2025b). *Transitioning Industrial Clusters*. https://initiatives.weforum.org/transitioning-industrial-clusters/home

World Meteorological Organisation (WMO). (2025). *State of the global climate 2024* (WMO-NO. 1368). https://perma.cc/J4S7-TQPR

World Trade Organization (WTO). (2023). *World Trade Report 2023*. https://perma.cc/23SC-829E

Xu, C., Kohler, T. A., Lenton, T., and Scheffer, M. (2020). Future of the human climate niche. *Proceedings of the National Academy of Sciences, 117*(21), 11350–11355. https://doi.org/10.1073/pnas.1910114117

Chapter 9

Aghion, P., Cherif, R., and Hasanov, F. (2021). *Competition, innovation, and inclusive growth* (IMF Working Paper No. 2021/080). International Monetary Fund. https://www.imf.org/en/Publications/WP/Issues/2021/03/19/Competition-Innovation-and-Inclusive-Growth-50269

Amahnui, G. A., Sylvester, J. M., Vanegas Cubillos, M., and Castro Nunez, A. (2025). *A Six-step Approach for scaling low-emission food systems: Evidence and guidelines.* International Centre for Tropical Agriculture (CIAT). https://hdl.handle.net/10568/174077

Anderson, C., Marvin, D., and Joseph, M. (2024, September 27). *Forest Carbon Monitoring: A Dove's-eye View of Global Forest Change.* Planet Pulse. https://perma.cc/9NRG-W7W7

Asian Development Bank. (2021). *ADB Raises 2019–2030 Climate Finance Ambition to $100 Billion.* https://www.adb.org/news/adb-raises-2019-2030-climate-finance-ambition-100-billion

Asian Development Bank (ADB). (2024). *Climate Finance in 2023.* https://www.adb.org/news/infographics/climate-finance-2023

Azarabadi, H., Baker, T., Dewar, A., Lesser, R., Mistry, K., Owolabi, B., Phillips, K., Pieper, C., Sudmeijer, B., and Webb, D. (2023). *Shifting the Direct Air Capture Paradigm.* Boston Consulting Group. https://www.bcg.com/publications/2023/solving-direct-air-carbon-capture-challenge

Babiker, M., Berndes, G., Blok, K., Cohen, B., Cowie, A., Geden, O., Ginzburg, V., Leip, A., Smith, P., Sugiyama, M., and Yamba, F. (2022). Cross-sectoral perspectives. In P. R. Shukla, J. Skea, R. Slade, A. Al Khourdajie, R. van Diemen, D. McCollum, M. Pathak, S. Some, P. Vyas, R. Fradera, M. Belkacemi, A. Hasija, G. Lisboa, S. Luz, and J. Malley (Eds.), *Climate change 2022: Mitigation of climate change. Contribution of Working Group III to the Sixth Assessment Report of the Intergovernmental Panel on Climate Change* (pp. 1245–1354). Cambridge University Press. https://doi.org/10.1017/9781009157926.014

Bednar, J., Höglund. R., Möllersten, K., Obersteiner, M. (2023). *The role of carbon dioxide removal in contributing to the long-term goal of the Paris Agreement.* IVL Swedish Environmental Research Institute. https://www.diva-portal.org/smash/get/diva2:1825937/FULLTEXT01.pdf

Bhattacharya, A., Songwe, V., Soubeyran, E., and Stern, N. (2023). *A climate finance framework: Decisive action to deliver on the Paris Agreement.* Grantham Research Institute on Climate Change and the Environment, London School of Economics and Political Science. https://perma.cc/ZML3-CQ8Q

Bhattacharya, A., Songwe, V., Soubeyran, E., and Stern, N. (2024). *Raising ambition and accelerating delivery of climate finance*. Grantham Research Institute on Climate Change and the Environment, London School of Economics and Political Science. https://perma.cc/H6D9-MNSX

Biodiversity Finance Trends. (2024). *Biodiversity Finance Trends 2024*. https://perma.cc/7ACQ-45D7

Black, S., Liu, A. A., Parry, I. W. H., and Vernon-Lin, N. (2023). *IMF fossil fuel subsidies data: 2023 update* (Working paper). International Monetary Fund. https://www.imf.org/en/Publications/WP/Issues/2023/08/22/IMF -Fossil-Fuel-Subsidies-Data-2023-Update-537281

Blended Finance Taskforce. (2023). *Better Guarantees, Better Finance: Mobilising capital for climate through fit-for-purpose guarantees*. https://www. systemiq.earth/wp-content/uploads/2023/06/Blended-Finance-Task- force-2023-Better-Guarantees-Better-Finance-1.pdf

BloombergNEF (BNEF). (2024). *Africa Power Transition Factbook 2024: Record clean energy investment boosts progress towards 2030 goals*. https:// perma.cc/K6M9-DVM2

The Bridgetown Initiative. (2024). *Bridgetown Initiative on the Reform of the International Development and Climate Finance Architecture (Version 3.0)*. https://www.bridgetown-initiative.org/bridgetown-initiative-3-0/

C40. (2015). Network Spotlight: 5 reasons why C40 is built on networks. Retrieved from https://www.c40.org/news/network-spotlight-5-reasons -why-c40-is-built-on-networks/

C40. (2024). *2023 Annual Report*. https://perma.cc/DL9P-2BH9

CGIAR. (2021). *CGIAR 2030 Research and Innovation Strategy: Transforming food, land, and water systems in a climate crisis*. https://hdl.handle.net /10568/110918

ClientEarth. (2024, May 3). *Landmark High Court judgment finds government's climate plan 'unlawful' - again*. https://www.clientearth.org /latest/press-office/press-releases/landmark-high-court-judgment-finds -governments-climate-plan-unlawful-again/

Climate Policy Initiative (CPI). (2022). *Climate finance innovation for Africa*. https://www.climatepolicyinitiative.org/publication/climate-finance -innovation-for-africa/

Climate Policy Initiative. (2023a). *Global landscape of climate finance 2023. Methodology*. https://perma.cc/L2NJ-A838

Climate Policy Initiative. (2023b). *Global landscape of climate finance 2023*. https://perma.cc/Z7KP-NZDN

Climate Policy Initiative. (2024). *Global landscape of climate finance 2024: Insights for COP29.* https://perma.cc/FJ4W-3MEV

Climate Policy Initiative. (2025). *Global landscape of climate finance 2025.* https://perma.cc/6EDK-2R4T

Coalition of Finance Ministers for Climate Action. (2019). *Helsinki Principles.* https://www.financeministersforclimate.org/sites/cape/files/inline-files/FM%20Coalition%20-%20Principles%20final.pdf

Coalition of Finance Ministers for Climate Action. (2023). *Strengthening the role of ministries of finance in driving climate action: a framework and guide for ministers and ministries of finance.* https://perma.cc/9D8K-N6WS

Convention on Biological Diversity. (2022). *Kunming–Montreal global biodiversity framework.* United Nations. https://www.cbd.int/gbf

Damania, R., Balseca, E., de Fontaubert, C., Gill, J., Kim, K., Rentschler, J., Russ, J., and Zaveri, E. (2023). *Detox Development: Repurposing Environmentally Harmful Subsidies.* World Bank Group. https://hdl.handle.net/10986/39423

Department of Energy Security and Net Zero (DESNZ). (2023, March 10). *New UK-France partnership to bring 'more energy security and independence'.* Press Release. https://perma.cc/CWY2-7QC7

Dietz, T. and Grabs, J. (2022). Additionality and implementation gaps in voluntary sustainability standards. *New Political Economy, 27*(2), 203--224. https://doi.org/10.1080/13563467.2021.1881473

Energy Transitions Commission (ETC). (2022). *Mind the Gap: How Carbon Dioxide Removals Must Complement Deep Decarbonisation to Keep 1.5°C Alive.* Systemiq. https://perma.cc/G4JN-YNLX

Energy Transitions Commission. (2024). *NDCs, NCQ, and Financing the Transition: Unlocking Flows for a Net-Zero Future.* Systemiq. https://perma.cc/7B2N-3XZ3

Escalanate, D. and Orrego, C. (2021). *Results-based financing: Innovative financing solutions for a climate-friendly economic recovery.* Climate Policy Initiative. https://perma.cc/3K2U-96JA

European Bank for Reconstruction and Development (EBRD). (2023). *Sustainability Report 2023.* https://perma.cc/8XV9-ASTZ

European Council. (2025, May 28). *Fit for 55: how does the EU intend to address the emissions outside of the EU?* https://www.consilium.europa.eu/en/infographics/fit-for-55-cbam-carbon-border-adjustment-mechanism/

European Parliament. (2023). *Carbon border adjustment mechanism*. PE 754.626. https://www.europarl.europa.eu/RegData/etudes/ATAG/2023/754626/EPRS_ATA(2023)754626_EN.pdf

First Movers Coalition (FMC). (2024). *First Movers Coalition Impact Brief*. World Economic Forum. https://perma.cc/HP2B-NZ55

Fleck, A. (2022). *This chart signals the need for a rainforest OPEC*. World Economic Forum. https://www.weforum.org/stories/2022/11/chart-need-rainforest-opec-zero-deforestation-forest/

Freeman, C. (1995). The 'national system of innovation' in historical perspective. *Cambridge Journal of Economics, 19*(1), 5–24. https://doi.org/10.1093/oxfordjournals.cje.a035309

G20 India Development Working Group. (2023). *2023 G20 New Delhi Update*. G20 Presidencies. http://g20.in/content/dam/gtwenty/gtwenty_new/document/G20-2023-New-Delhi-Update.pdf

G20 Italy and G20 Indonesia. (2022). *Boosting MDBs' investing capacity: An Independent Review of Multilateral Development Banks' Capital Adequacy Frameworks*. G20 Presidencies. https://perma.cc/AJ89-8KPB

Garcia Ocampo, D. and Lopez Moreira, C. (2024*). Uncertain waters: can parametric insurance help bridge NatCat protection gaps?* (FSI Insights on policy implementation no. 62). Financial Stability Institute. https://perma.cc/PA5W-6HTH

Garg, V. (2022, April 9). View: Private sector driving renewable energy wave in India. *Economic Times of India*. https://perma.cc/EQP8-REXM

Global Cement and Concrete Association (GCCA). (2024). *Cement Industry Net Zero Progress Report 2024/25*. https://perma.cc/FWZ2-QFK3

Global Center on Adaptation, and Climate Policy Initiative (CPI). (2023). *State and trends in climate adaptation finance 2023*. https://perma.cc/8LJ9-VSU6

Global Center on Adaptation, and Climate Policy Initiative. (2024). *State and trends in climate adaptation finance 2024*. https://perma.cc/Z7HF-YW8W

Global Commission on Adaptation. (2019). *Adapt now: A global call for leadership on climate resilience*. World Resources Institute. https://perma.cc/ZSM3-2B3P

Global Emerging Markets Credit Risk (GEMs). (2025). Global *Emerging Markets Credit Risk Database: Consortium of MDBs and DFIs*. European Investment Bank. https://www.gemsriskdatabase.org/

Goldman, E., Reytar, K., Carter, S., Gibbs, D., Johnson, P., Levin, D., Sims, M., and Weisse, M. (2023) *Deforestation and Restoration Targets Tracker. Global Forest Review.* World Resources Institute. https://perma.cc/ET3R-KCKK

Haywood, J. M., Jones, A., Jones, A. C., Halloran, P., and Rasch, P. J. (2023). Climate intervention using marine cloud brightening (MCB) compared with stratospheric aerosol injection (SAI) in the UKESM1 climate model. *Atmospheric Chemistry and Physics,* 23(24), 15305-15324. https://doi.org/10.5194/acp-23-15305-2023

Hurley, G., Panwar, V., Wilkinson, E., Lindsay, C., Bishop, M., and Mami, E. (2024). *Breaking the cycle of debt in Small Island Developing States* (ODI Working Paper). ODI Global. https://perma.cc/T2MH-G7S9

ICEYE. (2025). Deforestation monitoring. https://perma.cc/N69J-FCXC

International Energy Agency (IEA), International Renewable Energy Agency (IRENA), and United Nations Climate Change High-Level Champions. (2022). *The Breakthrough Agenda Report 2022.* International Energy Agency. https://perma.cc/9ZCH-AXFE

International Energy Agency. (2021). *Is carbon capture too expensive?* https://perma.cc/KA2D-T659

International Energy Agency. (2022). *Africa Energy Outlook 2022.* https://perma.cc/95Z3-N7KV

International Energy Agency. (2023a). *Cost of Capital Observatory.* https://www.iea.org/reports/cost-of-capital-observatory

International Energy Agency. (2023b). *Financing Clean Energy in Africa.* https://www.iea.org/reports/financing-clean-energy-in-africa

International Energy Agency. (2024a). *World energy outlook 2024.* https://perma.cc/349B-8SLQ

International Energy Agency. (2024b). *World energy investment 2024.* https://perma.cc/33L8-EQCT

International Energy Agency. (2024c). *World energy investment 2024: methodology annex.* https://perma.cc/T74X-3JA5

International Energy Agency. (2025). *World energy investment 2025.* https://www.iea.org/reports/world-energy-investment-2025

International Institute for Sustainable Development (IISD). (2010, September 12). *Fiscal deficit forces Spain to slash renewable energy subsidies.* https://www.iisd.org/gsi/subsidy-watch-blog/fiscal-deficit-forces-spain-slash-renewable-energy-subsidies

International Monetary Fund (IMF). (2021). *Reaching net zero emissions.* https://www.imf.org/external/np/g20/pdf/2021/062221.pdf

International Partnership for Hydrogen and Fuel Cells (IPHE). (2023). *IPHE terms of reference.* https://perma.cc/R4HE-8BEU

International Renewable Energy Agency (IRENA). (2024). *The energy transition in Africa: Opportunities for international collaboration with a focus on the G7.* https://perma.cc/TW6G-HFSR

Jackson, R. (2014). *Controlling deforestation in the Brazilian Amazon: Alta Floresta Works Toward Sustainability, 2008–2013.* Princeton University. https://perma.cc/D2FG-STUN

Jena, L. P. and Trivedi, S. (2023*). Accelerating India's electric bus adoption: Fuelled by a strategic financing facility.* Institute for Energy Economics and Financial Analysis. https://ieefa.org/resources/accelerating-indias -electric-bus-adoption-fuelled-strategic-financing-facility?utm

Kammourieh, S. and Songwe, V. (2024). *Next generation prudential regulation for global financial system resilience.* https://perma.cc/XR77-6495

Koplow, D. and Steenblik, R. (2024). *Protecting nature by reforming environmentally harmful subsidies: An update.* Earth Track. https://perma.cc /542Z-UGJD

Krauss, J. and Krishnan, A. (2016). *Global decisions and local realities: Priorities and producers' upgrading opportunities in agricultural global production networks* (UNFSS Discussion Paper No. 7). SSRN. https://ssrn.com /abstract=3236108

Lam, A. and Mercure, J-F. (2022*). Evidence for a Global Electric Vehicle Tipping Point, Economics of Energy Innovation and System Transition* (Working Paper Series Number 2022/01). University of Exeter Global Systems Institute. https://ore.exeter.ac.uk/repository/bitstream/handle /10871/129774/Lam%20et%20al_Evidence%20for%20a%20global%20EV %20TP.pdf?sequence=1&isAllowed=y

Lazard. (2024, June). *LCOE: Levelized Cost of Energy.* https://perma.cc/J6YJ -VSS5

Lebling K., Gangotra, A., Hausker, K., and Byrum, Z. (2023). *7 things to know about carbon capture, utilization and sequestration.* World Resources Institute. https://perma.cc/MP45-K2HG

Leung, J. W., Robin, S., and Cave, D. (2024). *ASPI's two decade Critical Technology Tracker: The rewards of long-term research investment.* Australian Strategic Policy Institute. https://perma.cc/QE54-V2HU

Lowy Institute. (2024). *A climate loss and damage fund that works* (Policy brief). https://unfccc.int/sites/default/files/resource/PILL -HAMMERSLEY-Climate-loss-and-damage-fund.pdf

Mazzucato, M. (2018). Mission-oriented innovation policies: Challenges and opportunities. *Industrial and Corporate Change, 27*(5), 803–815. https://doi.org/10.1093/icc/dty034

McNally, R., Acharya, S., Marker, P., and Rahardiani, D. (2024). *Assessment of Opportunities to Mobilise Private Finance into Climate Adaptation in Southeast Asia.* Oxford Policy Management, and UKaid. https://perma.cc/53LK-JBB6

Milmanda, B. F. and Garay, C. (2019). Subnational variation in forest protection in the Argentine Chaco. *World Development, 118,* 79-90. https://doi.org/10.1016/j.worlddev.2019.02.002

Mission Innovation. (2021). *Joint member statement on the launch of Mission Innovation 2.0.* https://perma.cc/B9BA-S9VK

North Sea Link. (n.d.). *Why connect Norway and the UK?* North Sea Link. https://perma.cc/HZW6-L8HV

Organisation for Economic Co-operation and Development (OECD). (2024). *OECD inventory of support measures for fossil fuels 2024: Policy trends up to 2023.* OECD Publishing. https://doi.org/10.1787/a2f063fe-en

Organisation for Economic Co-operation and Development. (2025a). *Official development assistance (ODA).* https://www.oecd.org/en/topics/official-development-assistance-oda.html

Organisation for Economic Co-operation and Development. (2025b). *Preliminary official development assistance levels in 2024: Detailed Summary Note* (DCD(2025)6). https://one.oecd.org/document/DCD(2025)6/en/pdf

Palmer, T. (2024). *IMO Prize Lecture 2024: Ensemble weather and climate prediction – From origins to AI.* World Metereological Organisation. https://perma.cc/6XDP-L5U4

Quitzow, R. (2013). *The co-evolution of policy, market and industry in the solar energy sector: a dynamic analysis of technological innovation systems for solar photovoltaics in Germany and China* (FFU-Report 06-2013). Forschungszentrum für Umweltpolitik Freie, Universität Berlin. https://refubium.fu-berlin.de/bitstream/handle/fub188/19878/Quitzow_2013_Co-evolution_China_and_Germany_FFU_Report_06-2013.pdf

Richards, J. A., Schalatek, L., Achampong, L., and White, H. (2023). *The Loss and Damage Finance Landscape: A discussion paper for the Loss and Damage community on the questions to be resolved in 2023 for ambitious progress on the Loss and Damage Fund.* Heinrich Böll Stiftung. https://us.boell.org/sites/default/files/2023-05/the_loss_and_damage_finance_landscape_hbf_ldc_15052023.pdf

S&P Global. (2024, November 19). *Sustainability insights: Rising curtailment in China power producers will push past the pain.* https://www.spglobal.com/ratings/en/research/articles/241119-sustainability-insights-rising-curtailment-in-china-power-producers-will-push-past-the-pain-13327811

Singh, N. K. and Summers, L. (2023a). *Strengthening multilateral development banks: The triple agenda MANDATES I FINANCE I MECHANISMS* (Report of the Independent Experts Group, Vol. 1). G20 India. https://perma.cc/6BCU-PG3P

Singh, N. K. and Summers, L. (2023b). *Strengthening multilateral development banks: The triple agenda BETTER I BIGGER I BOLDER* (Report of the Independent Experts Group, Vol. 2). G20 India https://perma.cc/QY4T-23KA

Singh, V. and Bond, K. (2024). *Powering up the global south.* Rocky Mountain Institute. https://perma.cc/9HGP-6W77

SolShare. (2023). *SolShare 2023 Annual Report.* SolShare. https://perma.cc/D568-N2B3

Songwe, V., Stern, N., and Bhattacharya, A. (2022). *Finance for climate action: Scaling up investment for climate and development.* Grantham Research Institute on Climate Change and the Environment, London School of Economics and Political Science. https://perma.cc/V454-EKF5

Soubeyran, E. and Macquarie, R. (2023). *What is climate finance?* Grantham Research Institute on Climate Change and the Environment, London School of Economics and Political Science. https://perma.cc/98KL-LK63

Stern, N. (2006). *The economics of climate change: The Stern Review.* HM Treasury. https://webarchive.nationalarchives.gov.uk/ukgwa/20100407172811/https:/www.hm-treasury.gov.uk/stern_review_report.htm

Stern, N. and Valero, A. (2021). Innovation, growth and the transition to net-zero emissions. *Research Policy, 50*(9), 104293. https://doi.org/10.1016/j.respol.2021.104293

Sutton, W. R., Lotsch, A., and Prasann, A. (2024). *Recipe for a Livable Planet: Achieving Net Zero Emissions in the Agrifood System.* World Bank Group. https://hdl.handle.net/10986/41468.

Swiss Re. (2023). *We need to talk about climate adaptation.* https://www.swissre.com/institute/research/topics-and-risk-dialogues/climate-and-natural-catastrophe-risk/climate-adaptation.html

Systemiq. (2023). *The breakthrough effect: how to trigger a cascade of tipping points to accelerate the net zero transition.* https://perma.cc/WM6Y-894N

Tegegne, Y. T., Lindner, M., Fobissie, K., and Kanninen, M. (2016). Evolution of drivers of deforestation and forest degradation in the Congo Basin forests: Exploring possible policy options to address forest loss. *Land Use Policy, 51*, 312–324. https://doi.org/10.1016/j.landusepol.2015.11.024

The Nature Conservancy (TNC). (2020). *Financing Nature: Closing the Global Biodiversity Financing Gap.* https://perma.cc/T694-LAXM

United Nations (UN). (1959). *Antarctic Treaty.* https://perma.cc/3GRZ-63TF

United Nations Convention to Combat Desertification (UNCCD). (2021). *Glasgow Leaders' Declaration on Forests and Land Use.* United Nations. https://perma.cc/C7EC-5986

United Nations Environment Programme (UNEP). (1987). *Montreal Protocol on Substances that Deplete the Ozone Layer.* https://perma.cc/R22V-USNL

United Nations Environment Programme. (2016). *Kigali Amendment to the Montreal Protocol on Substances that Deplete the Ozone Layer.* https://perma.cc/8Y9A-3YAK

United Nations Environment Programme. (2023a). *Adaptation Gap Report 2023: Underfinanced. Underprepared. Inadequate investment and planning on climate adaptation leaves world exposed.* https://doi.org/10.59117/20.500.11822/43796

United Nations Environment Programme. (2023b). *One Atmosphere: An independent expert review on solar radiation modification research and deployment.* https://wedocs.unep.org/handle/20.500.11822/41903

United Nations Environment Programme. (2024). *Global Status Report for Buildings and Construction: Beyond foundations: Mainstreaming sustainable solutions to cut emissions from the buildings sector.* https://doi.org/10.59117/20.500.11822/45095

United Nations Environment Programme. (2025, March 3). *Governments Adopt First Global Strategy to Finance Biodiversity: Implications for financial institutions.* United Nations. https://perma.cc/5D3B-FK8K

United Nations Framework Convention on Climate Change (UNFCCC). (2023). *COP28 UAE Declaration on Climate, Nature and People.* https://perma.cc/6PZH-C75J

United Nations Industrial Development Organisation (UNIDO) (2024). *Industrial Deep Decarbonisation: An Initiative of the Clean Energy Ministerial.* https://perma.cc/R33L-5K7P

United Nations Trade and Development (UNCTAD) (2024). *International Fisheries Access Agreements: Challenges and opportunities to optimize development impacts.* United Nations. https://unctad.org/system/files/official-document/aldc2024d2_en.pdf

United Nations. (2021, May 22). *Renewable tech brings power swarming through the world's poorest villages.* https://perma.cc/G3YU-38C4

United Nations. (2023a). *Investment dispute settlement: Spain as respondent.* https://perma.cc/2KQU-CXU7

United Nations. (2023b). *High Seas Treaty: Agreement under the United Nations Convention on the Law of the Sea on the Conservation and Sustainable Use of Marine Biological Diversity of Areas Beyond National Jurisdiction.* https://perma.cc/Z296-3JAV

University of Strathclyde. (2022). *Malawi's first solar microgrid fuels fivefold increase in school pupils.* https://perma.cc/29QQ-6LKA

Vulnerable Twenty Group (V20). (2022). *Climate vulnerable economies loss report: economic losses attributable to climate change in V20 economies over the last two decades (2000-2019).* https://perma.cc/2N8R-QYHR

Wood Mackenzie. (2023, November 07). *China dominance on global solar supply chain.* https://perma.cc/X4E5-LGMV

World Bank. (2021). *Climate Change Action Plan 2021-2025.* https://hdl.handle.net/10986/35799

World Bank. (2023). *The Development, Climate and Nature Crisis: Solutions to End Poverty on a Livable Planet.* https://openknowledge.worldbank.org/handle/10986/40652

World Trade Organization (WTO). (1986). *General Agreement on Tariffs and Trade 1994.* https://perma.cc/3K5V-F89M

World Trade Organization. (1994). *Agreement on subsidies and countervailing measures.* https://perma.cc/FBZ6-7G4C

World Trade Organization. (2022). *World Trade Report 2022.* https://perma.cc/U5UE-J2VP

Xlinks. (n.d.). *What is the Morocco – UK Power Project?.* https://perma.cc/V5ZJ-6J85

Xue, Y. (2025, January 6). China to erase excess solar-panel capacity by 2027, UBS forecasts. *South China Morning Post.* https://perma.cc/WE2S-FW7K

Zattler, J. (2024). *Getting special drawing rights right: Opportunities for re-channelling SDRs to vulnerable countries* (Policy Brief No. 9/2024). German Institute of Development and Sustainability (IDOS). https://perma.cc/ZKD7-7SPR

PART IV

Chapter 10

Afif, Z. (2017). 'Nudge units' – where they came from and what they can do. *World Bank Blogs.* https://perma.cc/T8BV-24MY

Averchenkova, A. and Chan, T. (2023). *Governance pathways to credible implementation of net zero targets.* Grantham Research Institute on Climate Change and the Environment, London School of Economics and Political Science. https://perma.cc/B9WQ-FT34

Babiker, M., Berndes, G., Blok, K., Cohen, B., Cowie, A., Geden, O., Ginzburg, V., Leip, A., Smith, P., Sugiyama, M., and Yamba, F. (2022). Cross-sectoral perspectives. In P. R. Shukla, J. Skea, R. Slade, A. Al Khourdajie, R. van Diemen, D. McCollum, M. Pathak, S. Some, P. Vyas, R. Fradera, M. Belkacemi, A. Hasija, G. Lisboa, S. Luz, and J. Malley, (Eds.), *Climate change 2022: Mitigation of climate change. Contribution of Working Group III to the Sixth Assessment Report of the Intergovernmental Panel on Climate Change* (pp. 1245–1354). Cambridge University Press. https://doi.org/10.1017/9781009157926.014

Bashmakov, I. A. and Chupyatov, V. P. (1991). *Energy conservation: The main factor for reducing greenhouse gas emissions in the former Soviet Union.* US Environmental Protection Agency. https://www.osti.gov/servlets/purl/5533118

Beshears, J., Choi, J. J., Laibson, D., and Madrian, B. C. (2006). *The importance of default options for retirement savings outcomes: evidence from the united states* (Working Paper 12009). National Bureau of Economic Research. https://doi.org/10.3386/w12009

Bhattacharya, A., Songwe, V., Soubeyran, E., and Stern, N. (2023). *A climate finance framework that is fit for purpose: Decisive action to deliver on the Paris Agreement — Summary.* Grantham Research Institute on Climate Change and the Environment, London School of Economics and Political Science. https://perma.cc/ZML3-CQ8Q

Bhattacharya, A., Songwe, V., Soubeyran, E., and Stern, N. (2024). *Raising ambition and accelerating delivery of climate finance.* Grantham Research Institute on Climate Change and the Environment, London School of Economics and Political Science. https://perma.cc/H6D9-MNSX

The Bridgetown Initiative (2022). *Urgent and Decisive Action Required for an Unprecedented Combination of Crises. The 2022 Bridgetown Initiative for the Reform of the Global Financial Architecture.* https://perma.cc/Q3AY-FRU3

The Bridgetown Initiative. (2024). *bridgetown initiative on the reform of the international development and climate finance architecture (Version 3.0).* https://www.bridgetown-initiative.org/bridgetown-initiative-3-0/

Chater, N. and Loewenstein, G. (2023). The i-frame and the s-frame: How focusing on individual-level solutions has led behavioral public policy astray. *Behavioral and Brain Sciences, 46,* 1–26. https://doi.org/10.1017/S0140525X22002023

Chrétien, J. (2025, January 11). Jean Chrétien: Canadians will never give up the best country in the world to join the U.S. *The Globe and Mail.* https://perma.cc/G4WD-4AWC

Connolly, K. (2024, January 15). Thousands of tractors block Berlin as farmers protest over fuel subsidy cuts. *The Guardian.* https://perma.cc/GH6H-4GX9

Cooper, R. Caron and Schipper, L. (1991). The Soviet energy conservation dilemma. *Energy Policy, 19*(4), 344–363. https://doi.org/10.1016/0301-4215(91)90058-V

Dasgupta, S. P. (2021a). *The economics of biodiversity: The Dasgupta review.* HM Treasury. https://perma.cc/VXZ6-WLS3

Dasgupta, S. P. (2021b). *The economics of biodiversity: The Dasgupta review. Headline messages.* HM Treasury. https://perma.cc/BEC4-BQBU

Ekins, P. (2024). *What does more North Sea oil and gas mean for UK energy supply and net zero?* Grantham Research Institute on Climate Change and the Environment, London School of Economics and Political Science. https://perma.cc/2W6N-P6SL

Energy Transitions Commission (ETC). (2021). *Making Clean Electrification Possible: 30 Years to Electrify the Global Economy.* Systemiq. https://perma.cc/NM2N-BCNR

Energy Transitions Commission. (2022). *Mind the gap: how carbon dioxide removals must complement deep decarbonisation to keep 1.5°C alive.* Systemiq. https://perma.cc/G4JN-YNLX

Energy Transitions Commission. (2023). *Fossil Fuels in Transition: Committing to the phase-down of all fossil fuels.* Systemiq. https://perma.cc/S89J-7CSS

Energy Transition Commission. (2025). *Power Systems Transformation: Delivering Competitive, Resilient Electricity in High-Renewable Systems.* https://www.energy-transitions.org/publications/power-systems-transformation/#download-form

Fressoz, J. B. (2024). *More and more and more.* Allen Lane.

G20 Italy and G20 Indonesia. (2022). *Boosting MDBs' investing capacity: An Independent Review of Multilateral Development Banks' Capital Adequacy Frameworks.* G20 Presidencies. https://perma.cc/AJ89-8KPB

Goldberg, M. H., Marlon, J. R., Wang, X., van der Linden, S., and Leiserowitz, A. (2020). Oil and gas companies invest in legislators that vote against the environment. *Proceedings of the National Academy of Sciences, 117*(10), 5111–5112. https://doi.org/10.1073/pnas.1922175117

Gordon, R. J. (2016). *The rise and fall of American growth: The U.S. standard of living since the Civil War.* Princeton University Press.

Halpern, D. (2015). *Inside the nudge unit: How small changes can make a big difference.* Random House.

Hodgson, G. M. (2009). Institutional economics into the twenty-first century. *Studi e Note di Economia, 14*(1), 3–26.

Influence Map. (2019). *Big Oil's real agenda on climate change: How the oil majors have spent $1bn since Paris on narrative capture and lobbying on climate.* Influence Map. https://perma.cc/F6L2-PG4S

Ingram, W. and Gannon, K. (2024). *Climate justice and behaviour change: Examining the role of the individual in climate adaptation and water security.* Grantham Research Institute on Climate Change and the Environment, London School of Economics and Political Science. https://perma.cc/8G7T-A7W4

Intergovernmental Panel on Climate Change (IPCC). (2023). *Climate Change 2023: Synthesis Report. Contribution of Working Groups I, II and III to the Sixth Assessment Report of the Intergovernmental Panel on Climate Change.* [Core Writing Team, H. Lee and J. Romero (eds.)]. Cambridge University Press. https://doi.org/10.59327/IPCC/AR6-9789291691647

International Energy Agency (IEA). (2023). *Electricity grids and secure energy transitions.* https://perma.cc/3HGT-VMX7

International Energy Agency. (2024). *Renewables 2024: Analysis and forecast to 2030.* https://perma.cc/4Z6P-Q8WL

International Renewable Energy Agency (IRENA). (2024). *World Energy Transitions Outlook 2024: 1.5°C Pathway.* https://www.irena.org/Publications/2024/Nov/World-Energy-Transitions-Outlook-2024

Kahneman, D. and Tversky, A. (1991). Loss aversion in riskless choice: A reference-dependent model. *The Quarterly Journal of Economics, 106*(4), 1039–1061. https://doi.org/10.2307/2937956

London School of Economics (LSE). (n.d.). *BASIN – Behavioural adaptation for water security and inclusion.* Grantham Research Institute on Climate Change and the Environment, London School of Economics and Political Science. https://perma.cc/6U32-6KDM

Loosemore, T. (2014). *One link on GOV.UK – 350,000 more organ donors.* GOV.UK Blog. https://perma.cc/P4KD-7J2K

Magdoff, F. and Foster, J. B. (2011). *What every environmentalist needs to know about capitalism: A citizen's guide to capitalism and the environment.* NYU Press.

Meadows, D. H., Meadows, D. L., Randers, J., and Behrens, W. W. III. (1972). *The limits to growth: A report for the Club of Rome's project on the predicament of mankind.* Universe Books.

Mercure, J. F., Salas, P., Vercoulen, P., Semieniuk, G., Lam, A., Pollitt, H., Holden, P. B., Vakilifard, N., Chewpreecha, U., Edwards, N. R., and Viñuales, J. E. (2021). Reframing incentives for climate policy action. *Nature Energy, 6*(12), 1133–1143. https://doi.org/10.1038/s41560-021-00934-2

Morgan, G., Campbell, J., Crouch, C., Pedersen, O. K., and Whitley, R. (Eds.). (2010). *The Oxford Handbook Of Comparative Institutional Analysis.* Oxford University Press.

Neubert, K. (2024). German farmers' association slams government relief package. *Euractiv.* https://perma.cc/F4UJ-L3K2

Nielsen, K. S., Cologna, V., Bauer, J. B., Berger, S., Brick, C., Dietz, T., Hahnel, U. J. J., Henn, L., Lange, F., Stern, P. C., and Wolske, K. S. (2024). Realizing the full potential of behavioural science for climate change mitigation. *Nature Climate Change, 14,* 322–330. https://doi.org/10.1038/s41558-024-01951-1

Oreskes, N. and Conway, E. M. (2010). *Merchants of doubt: How a handful of scientists obscured the truth on issues from tobacco smoke to global warming.* Bloomsbury Press.

Papaconstantinou, G. and Pisani-Ferry, J. (2025, January 7). Global Cooperation in the Age of Trump. *Project Syndicate.* https://perma.cc/RU5B-WV8H

Rockström, J., Steffen, W., Noone, K., Persson, Å., Chapin, F. S., Lambin, E. F., Lenton, T. M., Scheffer, M., Folke, C., Schellnhuber, H. J., Nykvist, B., de Wit, C. A., Hughes, T., van der Leeuw, S., Rodhe, H., Sörlin, S., Snyder, P. K., Costanza, R., Svedin, U., ... Foley, J. A. (2009). A safe operating space for humanity. *Nature, 461*(7263), 472–475. https://doi.org/10.1038/461472a

Setzer, J. and Higham, C. (2024). *Global trends in climate change litigation: 2024 snapshot.* Grantham Research Institute on Climate Change and the Environment, London School of Economics and Political Science. https://perma.cc/VH2F-4A7Q

Shahgedanova, M. and Burt, T. P. (1994). New data on air pollution in the former Soviet Union. *Global Environmental Change, 4*(3), 201–227. https://doi.org/10.1016/0959-3780(94)90003-5

Singh, N. K. and Summers, L. (2023a). *Strengthening multilateral development banks: The triple agenda: mandates, finance, mechanisms* (Report of the Independent Experts Group, Vol. 1). G20 India. https://perma.cc/6BCU-PG3P

Singh, N. K. and Summers, L. (2023b). *Strengthening multilateral development banks: The Triple Agenda: Better, Bolder and Bigger MDBs* (Report of the Independent Experts Group, Vol. 2). G20 India. https://perma.cc/QY4T-23KA

Smil, V. (2010). *Energy transitions: history, requirements, prospects.* Praeger.

Smil, V. (2017). *Energy and civilization: a history.* MIT Press.

Songwe, V., Stern, N., and Bhattacharya, A. (2022). *Finance for climate action: Scaling up investment for climate and development.* Grantham Research Institute on Climate Change and the Environment, London School of Economics and Political Science. https://perma.cc/V454-EKF5

Stern, N. (2006). *The economics of climate change: the stern review.* HM Treasury. https://webarchive.nationalarchives.gov.uk/ukgwa/20100407172811/https:/www.hm-treasury.gov.uk/stern_review_report.htm

Stockholm Resilience Centre. (2023). *Planetary Boundaries.* https://perma.cc/RH4U-522R

Thaler, R. H. and Sunstein, C. R. (2008). *Nudge: Improving decisions about health, wealth, and happiness.* Yale University Press.

Thunberg, G. (2019, September 23). *Speech presented at the United Nations Climate Action Summit*, New York, NY. https://perma.cc/U8B6-K6DU

Tooze, A. (2025, January 23). Trouble Transitioning. *London Review of Books, 47*(1). https://perma.cc/KV7V-AAA9

United Nations Environment Programme Finance Initiative (UNEP FI). (2025). *Trends and innovations in nature finance: what to look out for in 2025.* https://perma.cc/JHU9-JVU3

Vlasceanu, M., Doell, K. C., Bak-Coleman, J. B., Todorova, B., Berkebile-Weinberg, M. M., Grayson, S. J., Patel, Y., Goldwert, D., Pei, Y., Chakroff, A., Pronizius, E., van den Broek, K. L., Vlasceanu, D., Constantino, S., Morais, M. J., Schumann, P., Rathje, S., Fang, K., Aglioti, S. M., … Van Bavel, J. J. (2024). Addressing climate change with behavioral science: A global intervention tournament in 63 countries. *Science Advances, 10*(6). https://doi.org/10.1126/sciadv.adj5778

Chapter 11

Bhattacharya, A. and Stern, N. (2023). *Towards a sustainable, resilient and prosperous future for India: Investment, innovation and collaboration in a changing world*. Grantham Research Institute on Climate Change and the Environment, London School of Economics and Political Science. https://perma.cc/3B7W-TC7Z

Bhattacharya, A., Songwe, V., Soubeyran, E., and Stern, N. (2024). *Raising ambition and accelerating delivery of climate finance*. Grantham Research Institute on Climate Change and the Environment, London School of Economics and Political Science. https://perma.cc/H6D9-MNSX

Draghi, M. (2024). *The future of European competitiveness: A competitiveness strategy for Europe*. European Commission. https://perma.cc/7UK6-A9ZV

Singh, N. K. and Summers, L. (2023a). *Strengthening multilateral development banks: The triple agenda*: mandates, finance, mechanisms (Report of the Independent Experts Group, Vol. 1). G20 India. https://perma.cc/6BCU-PG3P

Singh, N. K. and Summers, L. (2023b). *Strengthening multilateral development banks: The triple agenda: Better, Bolder and Bigger MDBs* (Report of the Independent Experts Group, Vol. 2). G20 India https://perma.cc/QY4T-23KA

Songwe, V., Stern, N., and Bhattacharya, A. (2022). *Finance for climate action: Scaling up investment for climate and development*. Grantham Research Institute on Climate Change and the Environment, London School of Economics and Political Science. https://perma.cc/V454-EKF5

Stern, N. (2018). Public economics as if time matters: Climate change and the dynamics of policy. *Journal of Public Economics, 162*, pp. 4–17. https://doi.org/10.1016/j.jpubeco.2018.03.006

Systemiq. (2020). *The Paris Effect: How the Climate Agreement is reshaping the global economy*. https://perma.cc/A953-7VJZ

Index

[Page numbers in *italics* denote figures; those in **bold** denote tables; 'n' indicates a note]

Trump, Donald 117, 137–139, 148–149, 230,
    300, 397
  and international trade 286
  and IRA 361 n1
  projected effect of election on US GHG
    emissions *138*
trust, rebuilding 317–318
Tubiana, Laurence 32
Tyndall, John 47

U

Ukraine *see* war in Ukraine
ul Haq, Mahbub 23
UN Convention to Combat Desertification
    (UNCCD) 350
unbalanced growth 37
uncertainty, as deterrent to investment 386
UNDP *see* United Nations Development
    Programme
UNFCCC *see* United Nations Framework
    Convention on Climate Change
United Kingdom (UK)
  car manufacturers in 302
  commitment to net zero emissions
    227–228
  grid system 231
  and market fundamentalism 265–266
  during Second World War 267–268
United Nations Development Programme
    (UNDP) 360
United Nations Framework Convention on
    Climate Change (UNFCCC) 18, 90, 134,
    269–270, 282, 327, 333, 350
  creation of 23, 31, 136
  international agreements 227
United Nations (UN) 134
United States (USA) 136–139, 147–149
  Inflation Reduction Act (IRA) 137, 156
    n21, 300, 361 n1, 363 n22
  and LCOE *299*, 314 n6
universality 25
urban density *174*
urban design 173, 219
urban transport systems 190; *see also* bus
    rapid transit (BRT) systems
urbanisation 5, 39, 219
UTZ certification 352–353, 364 n26

V

V20 *see* Vulnerable Twenty
values 99
  intertemporal 97, 110 n16
  and norms 272
  shadow 97, 110 n18
vested interests 104–105, 178, 395–397
Vlasceanu, M. 42 n1, 114

voluntary carbon markets (VCMs) 239, 330,
    352
von der Leyen, Ursula 136, 148
Vulnerable Twenty (V20) 312, 324–325

W

war in Ukraine 83, 230, 289, 314 n4, 397
waste 171; *see also* food loss and waste
    (FLW)
weather forecasting 191–192
Weitzman, Martin 95, 237
welfare economics 238
Winter, Sidney 259 n5
women, rights and empowerment of
    199–200
World Bank, Climate Change Action Plan
    (2021–2025) 334–335
world inequality, in income 27
world leadership 420
World Resources Institute (WRI) 353
World Trade Organization (WTO) 347
World Wildlife Fund (WWF) 353

X

Xi Jinping 32, 137, 141
Xie Zhenhua 33, 141–142

Y

youth activism 118–119
Youth Negotiators Academy 118

Z

Zenawi, Meles 90, 117, 332
zero-carbon methods, discovery of new
    239–240
Zhang Lei 118

www.ingramcontent.com/pod-product-compliance
Lightning Source LLC
Chambersburg PA
CBHW052115230326
41598CB00079B/3702